Climate Change 1994

RADIATIVE FORCING OF CLIMATE CHANGE
and
AN EVALUATION OF THE IPCC IS92 EMISSION SCENARIOS

Climate Change 1994

Radiative Forcing of Climate Change

and

An Evaluation of the IPCC IS92 Emission Scenarios

Edited by J.T. Houghton, L.G. Meira Filho, J. Bruce, Hoesung Lee, B.A. Callander, E. Haites, N. Harris and K. Maskell.

Reports of Working Groups I and III of the Intergovernmental Panel on Climate Change, forming part of the IPCC Special Report to the first session of the Conference of the Parties to the UN Framework Convention on Climate Change

Published for the Intergovernmental Panel on Climate Change

Published by the Press Syndicate of the University of Cambridge
The Pitt Building, Trumpington Street, Cambridge CB2 1RP
40 West 20th Street, New York, NY 10011–4211, USA
10 Stamford Road, Oakleigh, Melbourne 3166, Australia

© Intergovernmental Panel on Climate Change 1995

First published 1995

Printed in Great Britain at the University Press, Cambridge

British Library cataloguing in publication data available

Library of Congress cataloguing in publication data available

ISBN 0 521 55055 6   hardback
ISBN 0 521 55962 6   paperback

Contents

Foreword vii

Part I 1

Preface to WGI Report 3
Dedication 5
Summary for Policymakers: Radiative Forcing of Climate Change 7
1 CO_2 and the Carbon Cycle 35
2 Other Trace Gases and Atmospheric Chemistry 73
3 Aerosols 127
4 Radiative Forcing 163
5 Trace Gas Radiative Forcing Indices 205

Part II 233

Preface to WGIII Report 235
Summary for Policymakers: An Evaluation of the IPCC IS92 Emission Scenarios 237
6 An Evaluation of the IPCC IS92 Emission Scenarios 247

Appendix 1 Organisation of IPCC 305
Appendix 2 List of Major IPCC Reports 307
Appendix 3 Contributors to IPCC WGI Report 309
Appendix 4 Reviewers of IPCC Working Group I Report 315
Appendix 5 Reviewers of IPCC Working Group III Report 321
Appendix 6 Acronyms 327
Appendix 7 Units 331
Appendix 8 Some Chemical Symbols Used in this Report 333
INDEX 337

Foreword

The Intergovernmental Panel on Climate Change (IPCC) was jointly established by our two organizations in 1988, under the chairmanship of Prof. Bert Bolin, in order to: (i) assess available scientific information on climate change, (ii) assess the environmental and socio-economic impacts of climate change, and (iii) formulate appropriate response strategies.

The IPCC First Assessment Report was completed in August 1990. The Report became a standard work of reference, widely used by policymakers, scientists and other experts, and represented a remarkable co-ordinated effort by hundreds of specialists from all over the world.

Following the completion of the IPCC First Assessment Report and the holding of the Second World Climate Conference (Geneva, October-November, 1990), intergovernmental negotiations began on the elaboration of a UN Framework Convention on Climate Change (UNFCCC). Appreciating that the treaty negotiators would need the most up-to-date information on climate change, the Panel in 1991 undertook to review critically the key conclusions of its 1990 Report in the light of new data and analyses. The Panel published its 1992 update a few months ahead of the UN Conference on Environment and Development (Rio de Janeiro, June, 1992), where the Convention was signed by most of the world's nations.

In 1991-93, after some reorganization and with the endorsement of UNEP and WMO, the Panel committed itself to produce a Second Assessment Report in 1995, covering the same range of topics as in 1990 as well as the new subject area of technical issues related to the economic aspects of climate change. The Panel was aware that the Conference of the Parties to the UNFCCC (CoP) was likely to meet before the 1995 report was complete. It therefore undertook to produce for the first session of the CoP, a Special Report covering selected key topics of particular interest to the UNFCCC. The Special Report consists of:

- Radiative Forcing of Climate Change, with a Summary for Policymakers
- An Evaluation of the IPCC IS92 Emission Scenarios, also with a Summary for Policymakers
- IPCC Technical Guidelines for Assessing Climate Change Impacts and Adaptations
- IPCC Phase I Guidelines for National Greenhouse Gas Inventories

The present volume, *Climate Change 1994*, contains the first two items on radiative forcing of climate change and an evaluation of emission scenarios. The two Guidelines are stand-alone documents and are being published as such.

As usual in the IPCC, success in producing this report has depended upon the enthusiasm and co-operation of busy scientists and technical experts world-wide. We are exceedingly pleased to note here the very special efforts undertaken by the IPCC in ensuring the participation of experts from the developing and transitional economy countries in its activities, in particular in the writing, reviewing and revising of its reports. This has been a worthwhile and timely capacity-building exercise. The experts have given of their time very generously, and governments have supported them, in the enormous intellectual and physical effort required, often going substantially beyond reasonable demands of duty. Without such conscientious and professional involvement the IPCC would be greatly impoverished. We express to all these experts our grateful and sincere appreciation for their commitment.

We take this opportunity to express our gratitude, for nurturing another IPCC report through to a successful completion, to: Prof. Bolin, the Chairman of the IPCC, for his able leadership; the Co-Chairmen of the three IPCC Working Groups, Sir John Houghton and Drs L.G. Meira Filho, R.T. Watson, M.C. Zinyowera, J.P. Bruce and Hoesung Lee; the Technical Support Units of the Working Groups; and to the IPCC Secretariat in Geneva under the leadership of Dr N. Sundararaman, the Secretary of the IPCC.

G.O.P. Obasi

Secretary-General
World Meteorological Organization

Elizabeth Dowdeswell

Executive Director
United Nations Environment Programme

Climate Change 1994

Part I

Radiative Forcing of Climate Change

Prepared by Working Group I

IPCC reports are formally described as "approved" or "accepted". An "approved" report has been subject to detailed, line-by-line discussion and agreement in a plenary session of the relevant IPCC Working Group. For practical reasons only short documents can be formally approved, and larger documents are "accepted" by the Working Group, signifying its view that a report presents a comprehensive, objective and balanced view of the subject matter. In this Part, the Summary for Policymakers has been approved, and Chapters 1 to 5 have been accepted by Working Group I.

Preface to WGI Report

This report is the third produced by the Scientific Assessment Working Group of IPCC. The first comprehensive report on Climate Change (1990) concluded that continued accumulation of anthropogenic greenhouse gases in the atmosphere was likely to lead to measurable climate change. The 1990 report also introduced the concept of the Global Warming Potential (GWP) which allows the cumulative warming effect of different gases to be compared. Values for the GWPs of a range of greenhouse gases were published, the values including both the direct component due to the gas itself, and the indirect component arising from the breakdown products of greenhouse gases.

The IPCC Supplementary Report (1992) confirmed the essential conclusions of the 1990 assessment concerning our understanding of climate and the factors affecting it. It reported progress in quantifying two factors other than anthropogenic greenhouse gases which influence radiative forcing: the depletion of ozone in the stratosphere (by CFCs), and the effect of aerosols produced primarily by industrial emissions but also by biomass burning and other processes. Further research in atmospheric chemistry was revealing a more complicated picture than was first thought, and the updated values of GWP in the 1992 report quoted only the direct component of the GWP, and not the indirect.

A second comprehensive assessment, spanning all working groups, will be completed in late 1995. It has been recognised, however, that the first Conference of the Parties (scheduled for March 1995) of the United Nations Framework Convention on Climate Change (UNFCCC) would require, at an earlier date, scientific and technical advice on several key issues. This 1994 report has been prepared to help meet this need and covers two main topics.

The first topic concerns the relative importance (determined here using the concept of radiative forcing) of anthropogenic increases in atmospheric concentrations of different greenhouse gases and aerosols. The latest information is presented about the sources and sinks of greenhouse gases and aerosols, and values of GWP are updated.

The second topic, concerning the stabilisation of greenhouse gas concentrations in the atmosphere, arises in the context of Article 2 of the UNFCCC:

"The ultimate objective of this Convention and any related legal instruments that the Conference of the Parties may adopt is to achieve, in accordance with the relevant provisions of the Convention, stabilization of greenhouse gas concentrations in the atmosphere at a level that would prevent dangerous anthropogenic interference with the climate system. Such a level should be achieved within a time-frame sufficient to allow ecosystems to adapt naturally to climate change, to ensure that food production is not threatened and to enable economic development to proceed in a sustainable manner."

The 1995 report from all three IPCC Working Groups will address issues raised by Article 2 more comprehensively, including the likely impacts of different levels and time-scales of stabilisation. The present report presents a preliminary investigation into levels of greenhouse gas emissions that might lead to stable atmospheric concentrations.

This report was compiled between February 1993 and September 1994 by 25 Lead Authors from 11 countries; for their enthusiasm, commitment and sheer hard work we express our grateful thanks. Over 120 contributing authors from 15 countries submitted draft text and information to the Lead Authors and over 230 reviewers from 31 countries submitted valuable suggestions for improvement during the two-stage review process, and to them also we express our sincere appreciation.

The task of keeping the whole process together and on schedule fell to the IPCC Secretariat in Geneva - Narasimhan Sundararaman (IPCC Secretary), Sam Tewungwa, Rudie Bourgeois, Cecilia Tanikie, Chantal Ettori - and to the Working Group I Technical Support Unit in Bracknell - Bruce Callander, Neil Harris, Kathy Maskell, Fay Mills, Arie Kattenberg and, in particular recognition of her careful and thorough work in preparing the text of the report for final publication, Judy Lakeman.

For their endurance, diligence, and persistent good-humour we are very grateful. Lastly we acknowledge with appreciation the work of the Graphics Section of the UK Meteorological Office, who prepared the final diagrams for this publication.

Bert Bolin
IPCC chairman

John Houghton
Co-chair (UK) IPCC WGI

L. Gylvan Meira Filho
Co-chair (Brazil) IPCC WGI

Dedication

ULRICH SIEGENTHALER: 1941 to 1994

The present IPCC 1994 report is dedicated to our long-time friend Uli Siegenthaler. Uli has made major contributions to the scientific community and to the Intergovernmental Panel on Climate Change, which has benefited from his sound scientific background, his profound understanding of climate processes, and his long-term experience in the field of Earth System Science. As a lead author, he was actively involved and interested in the IPCC debate, and his efforts helped shape this and past IPCC assessments into excellent reference books.

Uli Siegenthaler was born in 1941 in the Bernese Oberland, Switzerland, and started his scientific career studying physics at the Eidgenössische Technische Hochschule in Zürich. After receiving his master's degree, he joined the group of Prof. H. Oeschger at the Physics Institute in Bern, where he completed his thesis studies on the application of stable isotopes to water cycle studies. In Bern Uli married Ilse, and together they raised a family of two children.

His research was characterised by quality and by a readiness to cross interdisciplinary boundaries. His style was quiet and modest. At the Physics Institute of the University of Bern he was the leader of the carbon cycle modelling group and was deeply involved in the development of various carbon cycle models, well known by experts as the box-diffusion model, the outcrop-diffusion model, and the HILDA-model. He also helped shape the carbon cycle studies at Princeton University. It is a reflection of the quality of Uli's research that the Bern-carbon cycle model is used in the present and previous IPCC assessments as a reference for scenario calculations.

Certainly, his models of the oceanic uptake of anthropogenic CO_2 have set the standards for work in this area. His many review articles demonstrate his in-depth knowledge of the global carbon cycle and the problems related to the anthropogenic perturbation in particular. As well as the future evolution of the carbon cycle he was also interested in past natural variations of the atmospheric CO_2 levels, and Uli's models of ice-age CO_2 concentrations have been used by many other researchers.

Uli Siegenthaler was not only a brilliant scientist, but also an excellent teacher. He shared his broad knowledge and his scientific interest with his students teaching many different courses on Earth System Science as well as Introductory Physics and Atomic Physics. His sound theoretical background and his clear logic made every one of his lectures a special event. He was an excellent advisor: his friendly and quiet character combined with his scientific excellence created a pleasant and stimulating environment for his many students over the years, and we all could count on his thorough and honest, yet gentle, critique of our work.

Uli Siegenthaler's death in July 1994 was a great loss to us and the scientific community. For all of us, Uli was a good friend. We will sorely miss his kind and gentlemanly manner as well as his scientific creativity.

This tribute was prepared by Fortunat Joos, Thomas Stocker, Hans Oeschger and colleagues, Climate and Environmental Physics, Physics Institute, Sidlerstr. 5, CH-3012 Bern, in response to the unanimous wish of all the Lead Authors that this report should be dedicated to the memory of Uli Siegenthaler.

Summary for Policymakers:

Radiative Forcing of Climate Change

A Report of Working Group I of the Intergovernmental Panel on Climate Change

CONTENTS

Executive Summary	11
1 What is Radiative Forcing?	15
2 Carbon Dioxide (CO_2)	16
2.1 How Has the Atmospheric Concentration of CO_2 Changed in the Past?	16
2.2 Sources and Sinks of CO_2 — Our Current Knowledge of the Carbon Budget	17
2.3 CO_2 Concentrations in the Future	19
2.3.1 For a given CO_2 emission scenario, how might CO_2 concentrations change in the future?	20
2.3.2 For a given CO_2 concentration profile leading to stabilisation, what anthropogenic emissions are implied?	21
2.4 Climate Feedbacks Associated with the Carbon Cycle	24
3 Methane (CH_4)	25
4 Nitrous Oxide (N_2O)	27
5 Halocarbons	28
6 Ozone (O_3)	29
6.1 Stratospheric Ozone	29
6.2 Tropospheric Ozone	30
6.3 The Importance of NO_x	30
7 The Effect of Tropospheric Aerosols	30
8 What Else Influences Radiative Forcing?	32
8.1 Solar Variability	32
8.2 Volcanic Activity	32
9 Global Warming Potential (GWP) — a Tool for Policymakers	32

EXECUTIVE SUMMARY

Introduction

In its first Scientific Assessment of Climate Change in 1990 the IPCC concluded that the increase of greenhouse gas concentrations due to human activities would result in a warming of the Earth's surface. "Radiative forcing" is the name given to the effect which these gases have in altering the energy balance of the Earth–atmosphere system and, using this concept, the 1990 report introduced a tool for policymakers, the Global Warming Potential, which allowed the relative warming effect of different gases to be compared. Other factors, natural and human, also cause radiative forcing. The 1990 report not only examined these factors but also reviewed a wide range of information on how climate has behaved in the past and how it might change in the future as a result of human influence.

The 1992 IPCC Supplementary report reviewed the key conclusions of the 1990 report and affirmed the basic understanding of climate change contained in the 1990 report. It did, however, provide more detail on two sources of *negative* radiative forcing — depletion of ozone in the stratosphere, and aerosols derived from anthropogenic emissions.

The scope of the present report covers only those factors which cause radiative forcing of climate change, and includes updated values of Global Warming Potentials. The full range of topics related to climate, including the *response* of climate to radiative forcing, will be covered in the second IPCC Scientific Assessment, scheduled for publication in 1995.

Major new results since IPCC 1992

These new findings add to the detail of our knowledge but do not substantially change the essential results concerning radiative forcing of climate which appeared in the 1990 or the 1992 IPCC scientific assessments.

- *Revised values of Global Warming Potentials (GWPs)* — compared to GWPs listed in the 1992 IPCC report most values are larger by typically 10 to 30%. The uncertainties in the new GWPs are typically ±35%.

- *Revised methane GWP* — includes both direct and indirect effects. While the product of the revised GWP for methane and the current estimated annual anthropogenic emissions is significantly less than that of carbon dioxide over a 100-year time horizon, it is comparable over a 20-year time horizon.

- *Stabilisation of atmospheric carbon dioxide concentrations* — a range of carbon cycle models indicates that stabilisation of atmospheric carbon dioxide concentration at all considered levels between one and two times today's concentrations (that is to say, between 350 and 750 ppmv[1]) could be attained only with global anthropogenic emissions that eventually drop to substantially below 1990 levels.

- *Improved estimation of forcing by aerosols* — model calculations indicate that the negative radiative forcing from sulphate aerosols and aerosols from biomass burning, when globally-averaged, may be a significant fraction of the positive radiative forcing caused by anthropogenic greenhouse gases since the pre-industrial era. However, the estimates of the aerosol radiative forcing are highly uncertain, moreover the forcing is highly regional and cannot be regarded as a simple offset to greenhouse gas forcing.

- *Recent low growth rate of carbon dioxide concentration is not unusual*[2] — between 1991 and 1993 the rate of increase in the atmospheric

---

[1] 1 ppmv = 1 part per million by volume.

[2] In the sense that anomalies in the growth rate of atmospheric carbon dioxide are not unusual. The anomaly of the early 1990s does have some unusual features in terms of its magnitude, duration and its coincidence with the decrease in the growth rate of methane, but it remains too early to identify either the causes of the 1990s' anomaly or its significance to the long-term growth of carbon dioxide. See the box: "Variations in the growth rates of carbon dioxide and methane concentrations" to be found on page 24.

concentrations of carbon dioxide slowed substantially compared to the average rates of increase over the previous decade. However, the modern observational record for carbon dioxide since the 1950s contains other periods of similarly low growth rates. In the latter half of 1993, the carbon dioxide growth rates increased.

- *Sharp reduction in methane growth rate* — the rate of increase of the atmospheric abundance of methane has declined over the last decade, slowing dramatically in 1991 to 1992, though with an apparent increase in the growth rate in late 1993.

- *Climatic impact of Mt. Pinatubo* — the eruption of Mt. Pinatubo in June 1991 produced a large, transient increase of stratospheric aerosols which resulted in a surface cooling over about 2 years estimated from observations to be about 0.4 °C, consistent with model simulations which predicted a global mean cooling of 0.4 to 0.6 °C.

- *Global carbon budget* — New estimates of terrestrial carbon uptake during the 1980s have better quantified the known sinks, particularly forest regrowth in the Northern Hemisphere.

Sources of radiative forcing and their magnitude

Anthropogenic and natural factors cause radiative forcing of various magnitudes and of different signs. The concept of radiative forcing enables us to compare the potential effects of different factors, though care must be taken where these factors have large seasonal or regional variation.

First, we consider the gases carbon dioxide, methane, nitrous oxide and the halocarbons which have increased through human activities and which are well-mixed throughout the atmosphere.

- The increase in carbon dioxide (CO_2) since the pre-industrial era (from about 280 to 356 ppmv) makes the largest individual contribution to greenhouse gas radiative forcing: 1.56 Wm^{-2}, consistent with previous IPCC reports.

- The increase of methane (CH_4) since pre-industrial times (from 0.7 to 1.7 ppmv) contributes about 0.5 Wm^{-2} to radiative forcing.

- The increase in nitrous oxide (N_2O) since pre-industrial times (from about 275 to about 310 ppbv[1]) contributes about 0.1 Wm^{-2} to radiative forcing.

- The observed concentrations of halocarbons, including CFCs 11, 12, 113, 114, 115, methyl-chloroform and carbon tetrachloride, have resulted in a direct radiative forcing of about 0.3 Wm^{-2}.

- The atmospheric concentrations of a number of HCFCs and HFCs, which are being used as substitutes for halocarbons controlled under the Montreal Protocol have increased substantially. Their combined contribution to radiative forcing is, however, still less than 0.05 Wm^{-2} because of their low atmospheric concentrations.

Second, we consider changes in concentrations of ozone and aerosols which are believed to contribute significantly to radiative forcing. Patterns of historical change in these constituents are strongly regional in character, leading to two important consequences: (i) estimates of their globally-averaged radiative forcing are less certain than for the well-mixed gases (because the patterns of change are not well-quantified) and (ii) any negative forcing due to aerosols cannot be regarded as a simple offset to the effect of greenhouse gases (because the regional patterns of the forcings are different). Nevertheless, we report such estimates in order to provide a broad indication of their relative magnitude.

- Limited observations and model simulations suggest that tropospheric ozone in the Northern Hemisphere has increased since pre-industrial times resulting in a global average radiative forcing of 0.2 to 0.6 Wm^{-2}.

- Halocarbon-induced depletion of ozone in the stratosphere has resulted in a negative global average radiative forcing of about -0.1 Wm^{-2}. This has occurred mainly since the late 1970s over which period it has been of similar magnitude, but opposite sign, to the forcing caused by the halocarbons. Prior to the onset of significant ozone depletion the radiative forcing due to the halocarbons was between +0.1 and 0.2 Wm^{-2}.

- Anthropogenic particles in the atmosphere, derived from emissions of sulphur dioxide and from biomass burning, exert a net negative radiative forcing. The *direct* forcings, globally averaged, are probably in the ranges -0.25 to -0.9 Wm^{-2} for sulphate aerosols and -0.05 to -0.6 Wm^{-2} for aerosols from biomass burning. The *indirect* effect of aerosols, due to their effect on cloud properties, may cause a further negative forcing of a magnitude similar to the direct effect. The forcing shows large regional variations,

---

[1] 1 ppbv = 1 part per billion (thousand million) by volume.

with the largest values in industrialised regions in the Northern Hemisphere.

Third, we consider natural factors which can also exert positive or negative radiative forcings.

- Since about 1850 a change in the Sun's output may have resulted in a positive radiative forcing estimated at between 0.1 and 0.5 Wm^{-2}.

- Some volcanic eruptions, such as that of Mt. Pinatubo in June 1991, result in a short-lived (a few years) increase in aerosols in the stratosphere, causing a large (about -4 Wm^{-2} in the case of Mt. Pinatubo) but short-lived negative radiative forcing of climate. The effect of the Mt. Pinatubo eruption has been detected in the observed temperature record.

Trends in greenhouse gas and aerosol concentrations

- Over the decade 1980 to 89 the atmospheric abundance of CO_2 increased at an average rate of about 1.5 ppmv (0.4% or 3.2 billion tonnes of carbon) per year as a result of human activities, equivalent to approximately 50% of anthropogenic emissions over the same period.

- The rate of increase of the atmospheric abundance of methane has declined over the last decade, slowing dramatically in 1991 to 1992, though with an apparent increase in the growth rate in late 1993. The average trend over 1980 to 1990 is about 13 ppbv (0.8% or 37 million tonnes of methane) per year.

- The atmospheric abundance of nitrous oxide increased at an average annual rate (1980 to 1990) of about 0.75 ppbv (0.25% or 3.7 million tonnes of nitrogen) per year. The observations indicate that the growth rate varied during this period.

- The rates of increase of atmospheric concentrations of several major ozone-depleting halocarbons have fallen, demonstrating the impact of the Montreal Protocol and its amendments and adjustments. The total amount of organic chlorine in the troposphere increased by only 1.6% in 1992, about half of the rate of increase (2.9%) in 1989.

- The monitoring network for tropospheric ozone is sparse, making detection of global trends difficult. Since the 1960s concentrations of tropospheric ozone have almost certainly increased over large parts of the Northern Hemisphere but trends during the 1980s were small and of variable sign.

- Anthropogenic aerosol and precursor emissions have increased over the past 150 years, but while local trends (positive and negative) in concentrations are evident, no clear picture emerges of a contemporary *global* trend in atmospheric concentrations of anthropogenic aerosols in the size range important for radiative forcing.

The stabilisation of greenhouse gas concentrations

Several carbon cycle models have been used to study the implications for future atmospheric concentrations of carbon dioxide, of a range of global anthropogenic *emission* scenarios. The same models have been used to study the broad implications, in terms of *emissions*, of stabilising carbon dioxide *concentrations* in the range 350 ppmv (near current levels) to 750 ppmv. Differences in projected concentrations and emissions between models are typically ±15%; additional uncertainties arise from the various assumptions and simplifications used. The following results emerge:

- If carbon dioxide emissions were maintained at today's levels, they would lead to a nearly constant rate of increase in atmospheric concentrations for at least two centuries, reaching about 500 ppmv (approaching twice the pre-industrial concentration) by the end of the 21st century.

- A stable level of carbon dioxide concentration at values up to 750 ppmv can be maintained only with anthropogenic emissions that eventually drop below 1990 levels.

- To a first approximation the eventual stabilised concentration is governed more by the accumulated CO_2 emissions from now until the time of stabilisation, and less by the exact path taken to reach stabilisation. This means that, for example, for a given stabilisation scenario, higher emissions in early decades imply lower emissions later on. For the range of arbitrary stabilisation cases studied, accumulated emissions to the end of the 21st century were between 300 and 430 GtC[1] for stabilisation at 350 ppmv, between 880 and 1060 GtC for stabilisation at 550 ppmv, and between 1220 and 1420 GtC for stabilisation at 750 ppmv. For comparison the corresponding accumulated emissions for IPCC IS92[2] emission scenarios are 770 to 2190 GtC.

---

[1] 1GtC = 1 billion tonnes of carbon.
[2] In 1992 IPCC produced six scenarios, termed IS92a–f, for future emissions of greenhouse gases and their precursors.

If methane emissions were maintained at today's levels, atmospheric concentrations would effectively stabilise within 50 years at about 1900 ppbv, 11% higher than at present. Conversely, a reduction in annual methane emissions to levels about 35 million tonnes (roughly 10% of anthropogenic emissions) below current levels would stabilise concentrations at today's levels. (This calculation assumes that natural sources and atmospheric losses of methane are not affected by changing climate and atmospheric composition over the next century.)

If emissions of nitrous oxide were maintained at today's levels, atmospheric concentrations would effectively stabilise after several centuries at about 400 ppbv, 30% higher than at present and 50% above pre-industrial levels. Conversely a reduction of more than 50% of anthropogenic sources would stabilise concentrations at today's level of about 310 ppbv.

In contrast to the long-lived greenhouse gases, aerosols and tropospheric ozone are rapidly removed from the atmosphere and stabilisation of precursors would lead quickly to stable atmospheric concentrations.

The predictions of changes in atmospheric chlorine loading indicate that the depletion of stratospheric ozone should peak within the next decade and then slowly recover during the first half of the next century.

Global Warming Potential

Revised GWPs have been calculated. Furthermore, GWPs have been calculated for a number of new species, in particular hydrochlorofluorocarbons (HCFCs), hydrofluorocarbons (HFCs) and perfluorocarbons (PFCs).

The GWP concept is difficult to apply to short-lived species (for example, oxides of nitrogen, non-methane hydrocarbons and aerosols). New tools need to be developed to characterise their radiative forcing.

1 What is Radiative Forcing?

The ultimate energy source for all weather and climate is radiation from the Sun (called solar or short-wave radiation). Averaged globally and annually, about a third of incoming solar radiation is reflected back to space. Of the remainder, some is absorbed by the atmosphere, but most is absorbed by the land, ocean and ice surfaces. The solar radiation absorbed by the Earth's surface and atmosphere (which amounts to about 240 Wm^{-2}) is balanced at the top of the atmosphere by outgoing radiation at infrared wavelengths (Figure 1). Some of the outgoing infrared radiation is trapped by the naturally occurring greenhouse gases (principally water vapour, but also carbon dioxide (CO_2), ozone (O_3), methane (CH_4) and nitrous oxide (N_2O)) and by clouds, which keeps the surface and troposphere[1] about 33 °C. warmer than it would otherwise be. This is the *natural greenhouse effect*. In an unperturbed state, the net incoming solar radiation at the top of the atmosphere, averaged over the globe over long periods of time, must be balanced by net outgoing infrared radiation (Figure 1).

A *change* in average net radiation at the top of the troposphere (known as the tropopause), because of a change in either solar or infrared radiation, is defined for the purpose of this report as a *radiative forcing*. A radiative forcing perturbs the balance between incoming and outgoing radiation. Over time climate responds to the perturbation to re-establish the radiative balance. A positive radiative forcing tends on average to warm the surface; a negative radiative forcing on average tends to cool the surface. As defined here, the incoming solar radiation is not considered a radiative forcing, but a change in the amount of incoming solar radiation would be a radiative forcing.

---

[1] The troposphere is the lower part of the atmosphere from the surface to around 10–15 km.

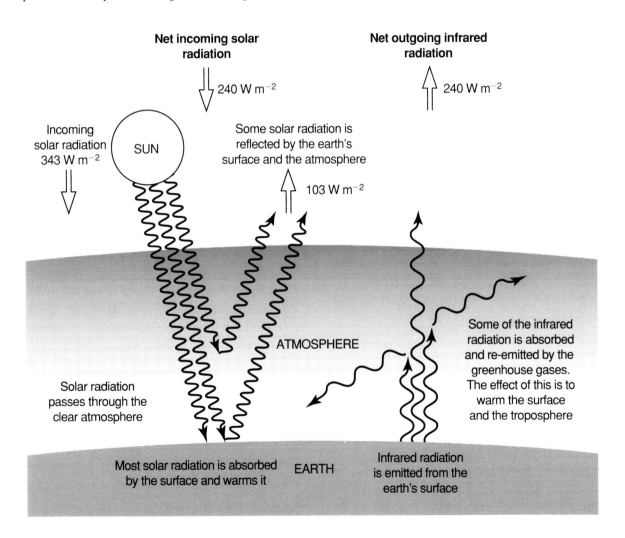

Figure 1: A simplified diagram illustrating the global long-term radiative balance of the atmosphere. Net input of solar radiation (240 Wm^{-2}) must be balanced by net output of infrared radiation. About a third (103 Wm^{-2}) of incoming solar radiation is reflected and the remainder is mostly absorbed by the surface. Outgoing infrared radiation is absorbed by greenhouse gases and by clouds keeping the surface about 33 °C warmer than it would otherwise be.

For example, an increase in atmospheric CO_2 concentration leads to a reduction in outgoing infrared radiation and a positive radiative forcing. For a doubling of the pre-industrial CO_2 concentration, in the absence of any other change, the global mean radiative forcing would be about 4 Wm^{-2}. For balance to be restored, the temperature of the troposphere and of the surface must increase, producing an increase in outgoing radiation. For a doubling of CO_2 concentration, the increase in surface temperature at equilibrium would be just over 1 °C, if other factors (e.g., clouds, tropospheric water vapour and aerosols) are held constant. Taking internal feedbacks into account, the 1990 IPCC report estimated that the increase in global average surface temperature at equilibrium resulting from a doubling of CO_2 would be likely to be between 1.5 and 4.5 °C, with a best estimate of 2.5 °C.

Other anthropogenically emitted gases which act in the same way as CO_2 and contribute to an *enhanced greenhouse effect* are CH_4, N_2O, and CFCs and other halocarbons. Some minor atmospheric constituents, such as the nitrogen oxides (NO_x) and carbon monoxide (CO), although not important greenhouse gases in their own right, can influence the concentration of some greenhouse gases (tropospheric ozone in particular) through atmospheric chemistry. Contributions of this kind are known as *indirect* radiative forcings.

Human activity has also led to an increase in the abundance of aerosols in the troposphere, mainly produced by oxidation of sulphur dioxide and from biomass burning, which cause a direct radiative forcing through their reflection and absorption of solar radiation. An indirect radiative forcing effect is believed to result from the influence of aerosols on the size of cloud droplets, and hence on cloud reflectivity. The radiative effects of aerosols are mainly negative and tend to cool the surface.

Natural factors, such as an increase in aerosols in the stratosphere produced by volcanic activity, or changes in the Sun's output, can also lead to radiative forcing.

The magnitude and timing of climate change due to human activities will depend on the ultimate concentrations of greenhouse gases and aerosols and their rates of growth and on the detailed response of the climate system.

Radiative forcing, averaged globally, has been used to compare the potential climatic effect of different climate change mechanisms. For a range of mechanisms there appears to be a similar relationship between global mean radiative forcing and global mean surface temperature change. However, the applicability of global mean radiative forcing to mechanisms such as changes in ozone or tropospheric aerosols which are spatially very inhomogeneous, is unclear. The degree of offset between the positive global mean radiative forcing from greenhouse gases and the negative forcing from aerosols may be an unreliable guide to the climatic consequences. For example, even a net global mean radiative forcing of zero could still lead to regional and possibly even global-scale climate changes, if the forcing mechanisms have different geographical distributions. This issue is beginning to be addressed through the use of climate models and a more detailed assessment can be expected in the 1995 IPCC report. Because of these problems, we avoid summing the various positive and negative contributions of the human-induced global mean radiative forcing to produce a net anthropogenic forcing.

It should also be noted that climate variations are believed to occur in the absence of any radiative forcing as a result of the complex interactions between the atmosphere and oceans and, possibly, the cryosphere, land surface and biosphere.

2 Carbon Dioxide (CO_2)

2.1 How Has the Atmospheric Concentration of CO_2 Changed in the Past?

CO_2 levels in the atmosphere have increased since the pre-industrial period[1] from about 280 to about 356 ppmv (Figure 2a). We know this from analysis of air trapped in ice cores and, since the late 1950s, from precise, direct measurements of atmospheric concentration. The radiative forcing due to this increase is 1.56 Wm^{-2} (Figure 3). Evidence that the observed increase in atmospheric CO_2 concentration is due to anthropogenic activity comes from the following facts.

- The long-term rise in atmospheric CO_2 closely follows the increase in anthropogenic CO_2 emissions (Figure 2a).

- Although CO_2 is well-mixed in the atmosphere, concentrations are slightly higher in the Northern Hemisphere (due to higher emissions). The increase in the inter-hemispheric gradient is growing in parallel with CO_2 emissions.

- Fossil fuel and biospheric carbon have a lower ratio of the carbon isotope ^{13}C to the isotope ^{12}C. Fossil fuels contain no ^{14}C because of their age. Decreases since pre-industrial times in the $^{13}C:^{12}C$ isotope ratio and in ^{14}C are fully consistent with the addition of fossil fuel and biospheric carbon by human activity.

Average rates of CO_2 concentration increase during the 1980s were 0.4% or 1.5 ppmv/yr. This is equivalent to 3.2 GtC/yr, approximately 50% of total anthropogenic CO_2

---

[1] The pre-industrial period is defined as the average over several centuries before 1750.

emissions. On decadal time-scales, the proportion of anthropogenic CO_2 emissions remaining in the atmosphere has stayed remarkably constant (at around 50%). The growth rate of atmospheric CO_2 concentration slowed during 1991 to mid-1993, although recently rates of growth have started to rise (Figure 2b). Short-term changes in growth rate are common in the past record of CO_2. See box on "Variations in the growth rates of CO_2 and CH_4 concentrations".

2.2 Sources and Sinks of CO_2 — Our Current Knowledge of the Carbon Budget

Our understanding of the carbon cycle has improved since previous IPCC reports, particularly in our knowledge of how the removal of CO_2 from the atmosphere is distributed between the sinks in the ocean and on land, and in the preliminary quantification of feedbacks. Considerable quantitative uncertainty remains regarding the processes which contribute to the sinks on land and in the ocean.

The main anthropogenic sources of CO_2 are the burning of fossil fuels (with additional contributions from cement production) and land-use changes. Over the period 1980 to 1989 the average emissions from fossil fuel burning and cement production were 5.5 ± 0.5 GtC/yr.

Land-use changes cause both release and uptake of CO_2. On average, CO_2 will be released to the atmosphere if the original ecosystem stored more carbon than the modified ecosystem which replaces it. Deforestation acts as a CO_2 source. In the current (1980 to 1989) budget, tropical deforestation is estimated to result in an average emission to the atmosphere of 1.6 ± 1.0 GtC/yr. However, in Northern Hemisphere mid- and high latitudes there are

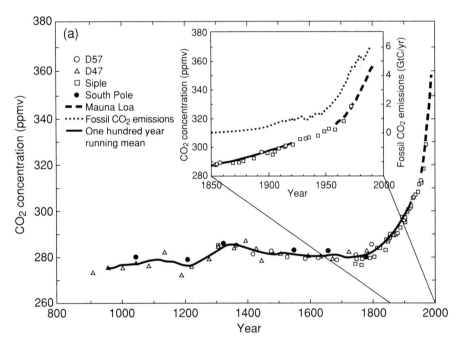

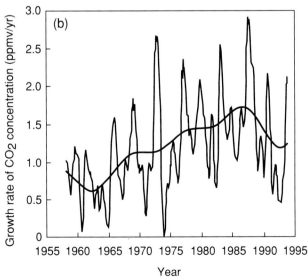

Figure 2: (a) CO_2 concentrations over the past 1000 years from ice-core records (D47, D57, Siple and South Pole) and (since 1958) from the Mauna Loa, Hawaii measurement site. The smooth curve is based on a hundred year running mean. The rapid increase in CO_2 concentration since the onset of industrialisation is evident and has followed closely the increase in CO_2 emissions from fossil fuels (see inset of period from 1850 onwards). (b) Growth rate of CO_2 concentration since 1958 in ppmv/yr at the Mauna Loa station showing the high growth rates of the late 1980s, the decrease in growth rates of the early 1990s, and the recent increase. The smooth curve shows the same data but filtered to suppress any variations on time-scales less than approximately 10 years.

> **ARE THE CO_2 ABSORPTION BANDS SATURATED?**
>
> The greenhouse effect of CO_2 is mainly due to its absorption bands between wavelengths of 14 μm[1] and 18 μm. There is a current misconception that, because there is already so much CO_2 in the atmosphere, absorption is saturated and addition of more CO_2 will not increase the greenhouse effect. Infrared absorption by CO_2 is well understood and over a small part of the spectrum, at the wavelength of strongest absorption (15 μm), increasing CO_2 causes little change in radiative forcing because absorption is indeed almost saturated there. However, at wavelengths greater and smaller than 15μm there is considerable capacity for increased absorption and an enhancement of the greenhouse effect.
>
> At present concentrations of CO_2, the relationship between concentration change and radiative forcing is strongly non-linear. For greenhouse gases with much smaller atmospheric concentrations the relationship is linear. This partly explains why changes in gases such as CFCs in the present atmosphere have a larger effect on radiative forcing, molecule for molecule, than does CO_2. The calculation of the radiative effects of CO_2 explicitly includes overlap with water vapour (a particularly important greenhouse gas) and clouds.
>
> ---
> [1] μm = 1 millionth of a metre.

areas where forests are regrowing after clearing in the past and where sequestration of CO_2 from the atmosphere is now occurring; we estimate a net sink of 0.5 ± 0.5 GtC/yr. The latest estimate of *net* CO_2 release due to global land-use changes is 1.1 ± 1.2 GtC/yr. This figure is lower than that quoted in IPCC 1990 (1.6 ± 1.0 GtC/yr) because of the inclusion of the estimate for Northern Hemispheric mid- and high latitude uptake in regrowing forests.

The oceans are a large sink of anthropogenic CO_2. At present net global oceanic uptake cannot be measured directly: it is estimated using models which describe the exchanges between the surface and deep ocean and the atmosphere. Such models can be tested by comparing the distribution of radiocarbon (released as a result of atom bomb tests in the 1950s and 1960s) from observations with model simulations.

Table 1 shows that summation of the best current estimates of CO_2 sources, sinks and atmospheric storage leads to an apparent unattributed terrestrial sink of 1.4 ± 1.5 GtC/yr. In previous IPCC reports this apparent imbalance in the carbon budget was referred to as a "missing sink", a term now felt to be inappropriate as sink mechanisms have been identified which could account for the imbalance.

CO_2 fertilisation

Photosynthesis can be stimulated by increased levels of CO_2. Studies carried out on small-scale experimental stands of vegetation, under optimal conditions of water and nutrient supply, suggest potential increases in photosynthesis of 20 to 40% when CO_2 is doubled. However, attempting to quantify the effect on a global scale is much more difficult. When the availability of water and nutrients is taken into account the fertilisation effect is likely to be reduced; several model results suggest reduction by around a half. The interaction of CO_2 fertilisation with the nitrogen cycle also has to be considered; model results of this effect are contradictory. During the 1980s CO_2 fertilisation may have accounted for a sink of 0.5 to 2.0 GtC/yr.

Table 1: Annual average anthropogenic carbon budget for 1980 to 1989. CO_2 sources, sinks and storage in the atmosphere are expressed in GtC/yr.

CO_2 sources	
(1) Emissions from fossil fuel and cement production	5.5 ± 0.5
(2) Net emissions from changes in tropical land-use	1.6 ± 1.0
(3) Total anthropogenic emissions = (1)+(2)	7.1 ± 1.1
Partitioning amongst reservoirs	
(4) Storage in the atmosphere	3.2 ± 0.2
(5) Ocean uptake	2.0 ± 0.8
(6) Uptake by Northern Hemisphere forest regrowth	0.5 ± 0.5
(7) Additional terrestrial sinks (CO_2 fertilisation, nitrogen fertilisation, climatic effects) = [(1)+(2)]-[(4)+(5)+(6)]	1.4 ± 1.5

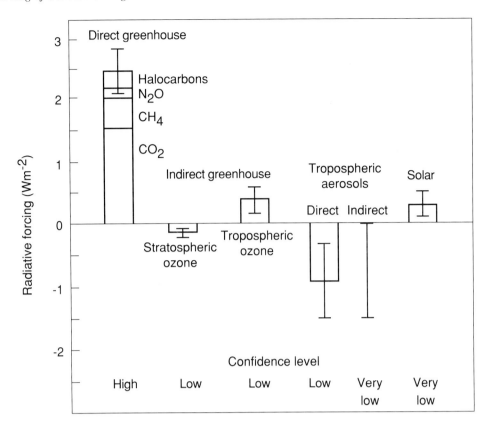

Figure 3: Estimates of the globally averaged radiative forcing due to changes in greenhouse gases and aerosols from pre-industrial times to the present day and changes in solar variability from 1850 to the present day. The height of the bar indicates a mid-range estimate of the forcing whilst the lines show the possible range of values. An indication of relative confidence in the estimates is given below each bar. The contributions of individual greenhouse gases are indicated on the first bar for direct greenhouse gas forcing. The major indirect effects are a depletion of stratospheric ozone (caused by the CFCs and other halocarbons) and an increase in the concentration of tropospheric ozone. The negative values for aerosols should not necessarily be regarded as an offset against the greenhouse gas forcing because of doubts over the applicability of global mean radiative forcing in the case of non-homogeneously distributed species such as aerosols and ozone (see Section 1 and Section 7).

Nitrogen fertilisation

Ecosystems receive substantial inputs of anthropogenic nitrogen, which in many areas can act as a fertiliser and could have increased terrestrial carbon storage by 0.2 to 1.0 GtC/yr in the 1980s. However, high levels of nitrogen addition are often associated with acidification and high surface ozone concentrations, which in the long term may damage ecosystems and possibly reduce carbon storage.

2.3 CO_2 Concentrations in the Future

Understanding how CO_2 concentrations will change in the future requires adequate knowledge of the relationship (including its quantification) between CO_2 emissions and atmospheric concentration using models of the carbon cycle (see the box "Modelling the carbon cycle").

Two questions are considered:

- For a given CO_2 emission scenario, how might CO_2 concentrations change in the future?

- For a given CO_2 concentration profile leading to stabilisation of the level of concentration, what anthropogenic emissions are implied?

Results from a range of different carbon cycle models are considered in order to assess the sensitivity of calculated emission and concentration profiles to model formulation. However, model intercomparison alone gives an underestimate of uncertainty because the calculations performed have the following limitations:

- The carbon cycle models were calibrated to balance the contemporary carbon budget according to earlier estimates (IPCC 1990 and 1992), rather than the budget shown in Table 1 which was not finalised until after the model calculations had been completed. The differences between the 1990 and 1992 budgets and the budget in Table 1 are: (i) a change in the estimate of atmospheric accumulation from 3.4 to 3.2 GtC/yr, and (ii) a reduction in the net

MODELLING THE CARBON CYCLE

Carbon is exchanged between the atmosphere, the oceans, the terrestrial biosphere (Figure 4), and, on geological time-scales, with sediments and sedimentary rocks. Fossil fuel burning, cement manufacture, and forest harvest and other changes of land-use transfer carbon (as CO_2) to the atmosphere. Although the anthropogenic flux of CO_2 is small compared with mean natural fluxes (Figure 4), it is sufficient to perturb the carbon cycle. The additional anthropogenic CO_2 cycles between the atmosphere, ocean, and marine and terrestrial biospheres. The net uptake of anthropogenic CO_2, particularly by the deep ocean, occurs slowly (for the oceans: on a time-scale of centuries), so addition of anthropogenic CO_2 has a long-lasting effect on atmospheric concentration. For example, if CO_2 emissions were held constant at present day levels, atmospheric concentrations would continue to rise for at least two centuries.

Key processes in the carbon cycle include:

- The exchange of CO_2 between the atmosphere and ocean.

- The exchange of CO_2 between the surface waters and long-term storage in the deep ocean.

- The net release or uptake of CO_2 from changes in land-use practices (e.g., deforestation).

- The photosynthetic uptake of CO_2 by land plants; the transfer of their carbon into long-term storage in wood and soils; the response of these processes to changing CO_2 and climate, and the release of CO_2 back to the atmosphere through plant and soil respiration.

To examine the relationship between CO_2 emissions and atmospheric concentration and to calculate future concentration levels we need a model of the carbon cycle which explicitly includes all of the above elements. However, most carbon cycle models include only simple representations of terrestrial biotic processes. The oceanic components vary in complexity from a few simplified equations to spatially explicit, detailed descriptions of ocean biology, chemistry, and transport processes. Attempts to model the effect of climate feedbacks on the carbon cycle are only just beginning and are not included in most carbon cycle models. The calculation of future CO_2 concentrations also requires assumptions regarding future anthropogenic CO_2 emissions.

source from changing land-use due to the inclusion of a sink (0.5 GtC/yr) in Northern Hemisphere forest regrowth. Atmospheric concentration changes calculated by models calibrated using the 1990 and 1992 budgets are lower by as much as 5 to 10% (for given emissions) and emissions higher by a similar amount (for given concentrations), compared with results of models calibrated using the budget in Table 1.

- The models include a sink term dependent on CO_2 concentration (i.e., acting like a simple CO_2 fertilisation effect) in order to balance the 1980 to 1989 carbon budget. This is an oversimplification because: (i) CO_2 fertilisation is much more complex than this: it depends on water and nutrient availability and on the state of the future biosphere (for example, deforestation of large areas would inevitably lead to a weaker CO_2 fertilisation effect) and (ii) other sink mechanisms exist which are currently not modelled, but in reality are likely to play a part in the carbon budget (e.g., nitrogen fertilisation).

- No attempt is made to model climate feedbacks on the carbon cycle (see Section 2.4).

2.3.1 For a given CO_2 emission scenario, how might CO_2 concentrations change in the future?

Six greenhouse gas emission scenarios were described in IPCC (1992), based on a wide range of future assumptions regarding economic, demographic and policy factors. The anthropogenic CO_2 emissions for these scenarios are shown in Figure 5a. IS92c, which has the lowest CO_2 emissions, assumes low population growth, low economic growth and severe constraints on fossil fuel supplies. The highest emission scenario (IS92e) assumes moderate population growth, high economic growth, high fossil fuel availability and a phase out of nuclear power. Figure 5b shows the resulting CO_2 concentrations. All show increases in concentration well above pre-industrial levels

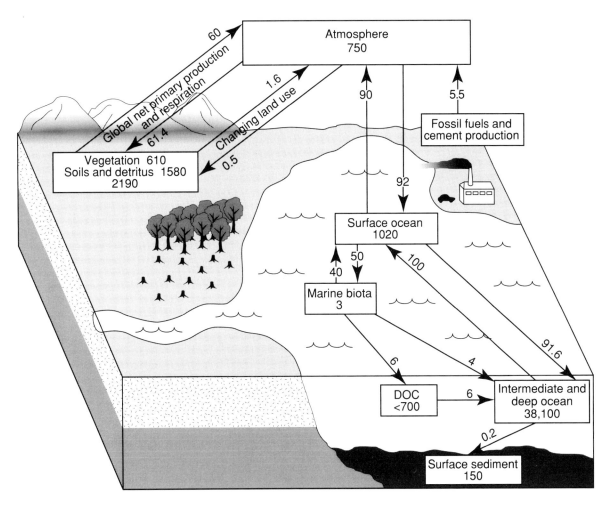

Figure 4: The global carbon cycle. The numbers in boxes indicate the size in GtC of each reservoir. On each arrow is indicated the magnitude of the flux in GtC/yr (DOC = dissolved organic carbon).

by 2100 (75 to 220% higher). None of the scenarios show a stabilisation of concentration before 2100, although IS92c produces slow growth in CO_2 concentration. IS92a, b, e and f all produce a doubling of the pre-industrial CO_2 concentration before 2070 with rapid rates of concentration growth. Neither IS92c nor d results in doubled pre-industrial CO_2 concentrations by 2100.

Stabilisation of current global emissions of CO_2 **does not** lead to stabilisation of CO_2 concentration by 2100. CO_2 concentrations reach about 500 ppmv by the end of the 21st century (Figure 5c) and calculations show that concentrations continue to increase slowly for several hundred years.

2.3.2 For a given CO_2 concentration profile leading to stabilisation, what anthropogenic emissions are implied?

In the context of the ultimate objective of the UN Framework Convention on Climate Change (quoted in the preface to Part I of this report), it is important to investigate, for all the greenhouse gases, the emission profiles which would lead to stabilisation of their concentration in the atmosphere. In this section CO_2 is considered; because of the complex nature of the lifetime of atmospheric CO_2, the calculations for CO_2 are relatively complex and therefore require considerable explanation. The stabilisation of other greenhouse gases, CH_4 and N_2O, for which the calculations are simpler, are considered in later sections.

Carbon cycle models have been used to calculate the emissions of CO_2 which would lead to stabilisation at a number of different concentration levels (i.e., the inverse of the type of calculation considered in Section 2.3.1). These calculations are designed to illustrate the relationship between CO_2 concentration and emissions. Concentration profiles have been devised (Figure 6) which stabilise at CO_2 concentrations from 350 to 750 ppmv (for comparison, the pre-industrial CO_2 concentration was close to 280 ppmv and the 1993 concentration was 356 ppmv).

The calculations which have been made so far are

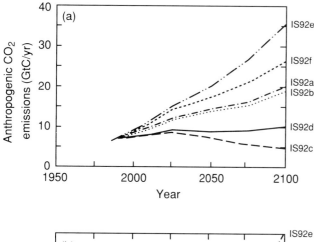

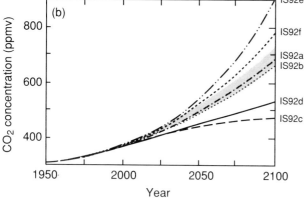

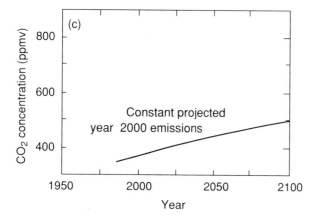

Figure 5: (a) Prescribed anthropogenic emissions of CO_2 (from fossil fuel use, deforestation and cement production) for the IS92 Scenarios, (b) CO_2 concentrations resulting from the IS92 emission scenarios calculated using the "Bern" model, a mid-range carbon cycle model (a range of results from different models is indicated by the shaded area of the IS92a curve) and (c) CO_2 concentrations resulting from constant projected year 2000 emissions (using the model of Wigley).

necessarily limited in their scope and ranges. They are designed to illustrate the relationship between CO_2 concentration and emissions. The selection of the range of concentrations from 350 ppmv to 750 ppmv was arbitrary and should not be construed as having any policy implications. Many different stabilisation levels, time-scales for achieving these levels, and routes to stabilisation could have been chosen. Those in Figure 6 give a smooth transition from the current rate of CO_2 concentration increase to stabilisation. As a result, the year of stabilisation differs with stabilisation level from around 2150 for 350 ppmv to 2250 for 750 ppmv.

Figure 7 shows the model-derived profiles of total anthropogenic emissions (from fossil fuel use, changes in land-use and cement production) that lead to stabilisation following the concentration profiles shown in Figure 6. Initially emissions rise, followed some decades later by quite rapid and large reductions. Stabilisation at any of the concentration levels studied (350 to 750 ppmv) is only possible if emissions are eventually reduced well below 1990 levels (Figure 7). For comparison the emissions from IS92a, c and e are shown up to 2100 in Figure 7. Emissions for all the stabilisation levels studied are lower than those for IS92a and e, even in the first few decades of the 21st century. Emissions for the IS92c Scenario lie between the emissions which in this study achieve stabilisation at 450 and 550 ppmv.

The concentration profiles here are illustrative. Stabilisation at the same level, via a different route, would produce different curves from those shown in Figure 7.

Table 2: Emissions of carbon accumulated from 1990 to the end of the 21st century leading to stabilisation of CO_2 concentration at 350, 450, 550, 650, and 750 ppmv. The range of uncertainty is derived from the spread of model results. For comparison the accumulated emissions are also shown for the IS92 emission scenarios.

	Accumulated emissions from 1990 to 2100 (GtC)
IS92 emission scenarios	
e	2190
f	1830
a	1500
b	1430
d	980
c	770
Stabilisation Case	
350 ppmv	300-430
450 ppmv	640-800
550 ppmv	880-1060
650 ppmv	1000-1240
750 ppmv	1220-1420

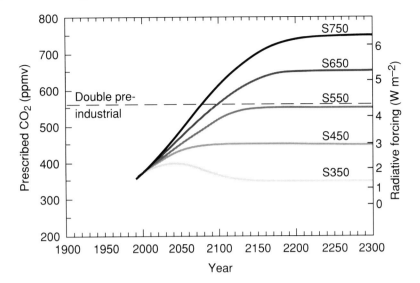

Figure 6: Profiles of atmospheric CO_2 concentration leading to stabilisation at 350, 450, 550, 650 and 750 ppmv. Doubled pre-industrial CO_2 concentration is 560 ppmv. The radiative forcing resulting from the increase in CO_2 relative to pre-industrial levels is marked on the right-hand axis. Note the non-linear nature of the relationship between CO_2 concentration change and radiative forcing.

However, to a first approximation, the total amount of emitted carbon accumulated over time (the area under the curves in Figure 7), is relatively insensitive to the concentration profile used. Stabilisation at a lower concentration implies lower accumulated emissions (Figure 8). Stabilisation of CO_2 concentration at or below 750 ppmv (the highest level studied) would require accumulated emissions from 1990 to 2100 lower than those occurring under the IS92a, b, e and f Scenarios (Table 2) and even lower in the next two centuries.

Figure 8 also shows the amount of carbon which accumulates in the atmosphere (known as the airborne fraction). On the century time-scale, the airborne fraction depends on the level of stabilisation, ranging from 15–25% (for 450 ppmv) to 30–40% (for 750 ppmv) of total anthropogenic emissions.

Although the range of results from different models is indicated in Figures 7 and 8, this is an underestimate of uncertainty. Changing the assumptions regarding the strength of the CO_2 fertilisation term indicated that future atmospheric concentration may vary by about ±15% from the stabilisation levels shown in Figure 6. The results in Figure 7 and 8 do not account for possible climate feedbacks on the carbon cycle (see Section 2.4). Different assumptions about land-use changes would give different results. For example, if large areas were deforested the

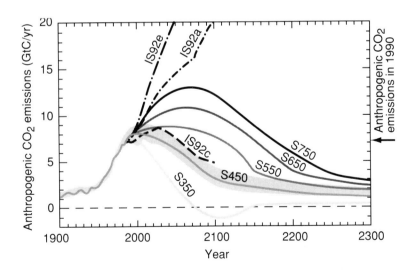

Figure 7: Illustrative anthropogenic emissions of CO_2 leading to stabilisation at concentrations of 350, 450, 550, 650 and 750 ppmv following the profiles shown in Figure 6 (using a mid-range carbon cycle model). The range of results from different models is indicated on the 450 ppmv profile. The emissions for the IS92a, c and e Scenarios are also shown on the figure. The negative emissions for stabilisation at 350 ppmv are an artefact of the particular concentration profile imposed.

> **VARIATIONS IN THE GROWTH RATES OF CARBON DIOXIDE AND METHANE CONCENTRATIONS**
>
> Observations of carbon dioxide (CO_2) since the 1950s show regular annual increases in both concentration and rate of concentration growth, albeit with year to year variations in the growth rate (Figure 2). During the period 1991 to 1993, the rate of increase of CO_2 per year slowed substantially (to as low as 0.5 ppmv/yr from more than 1.5 ppmv/yr) (Figure 2b). There are numerous examples in the record of short periods where growth rates are higher or lower than the long-term mean. The most recent observations indicate that growth rates of CO_2 are now increasing.
>
> The rate of methane (CH_4) concentration increase has declined over the last decade, slowing dramatically in 1991 to 1992, though with an apparent increase in the growth rate in late 1993 (Figure 9b). The slow down in growth in 1991 to 1992 was a maximum at high latitudes in the Northern Hemisphere, suggesting a drop in emissions as a possible explanation. Longer-period variations in the growth rate of CH_4 occurred in the 1920s and 1970s as observed from air trapped in ice cores.
>
> At present it is unclear what mechanisms account for these recent variations in the growth rate of CO_2 and CH_4. Although the decrease in growth rates of CO_2 and CH_4 occurred during roughly the same period, the decreases may be due to independent mechanisms. Further analysis of global observations during the most recent period should allow better assessment of the causes of the recent "anomalous" growth rates. Such studies should eventually improve our understanding of the global cycles of these important trace gases.
>
> Because of the variability at the year-to-year time-scale, growth rates averaged on a decadal time-scale should be considered when looking at anthropogenic trends.

capacity of the biosphere to act as a sink would be reduced, hence more CO_2 would remain in the atmosphere. Conversely, afforestation efforts could increase the capacity of the biosphere, and less CO_2 would remain in the atmosphere. While certain robust conclusions emerge from the model studies, uncertainties exist and these results must be considered a "first look".

2.4 Climate Feedbacks Associated with the Carbon Cycle

Higher temperatures and precipitation can increase photosynthesis and plant growth and hence increase carbon storage in living vegetation and litter (a negative feedback). The storage of carbon in soil tends to decrease with increasing temperature due to increased rates of

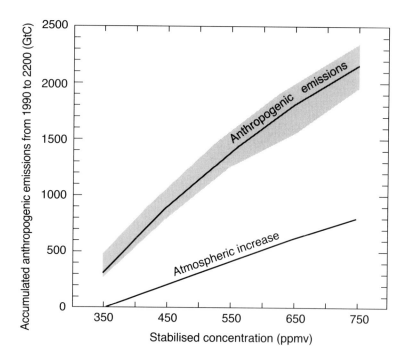

Figure 8: Anthropogenic CO_2 emissions shown in Figure 7 accumulated from 1990 to 2200 plotted against the final stabilised concentration. For example, accumulated emissions of between 1200 and 1600 GtC lead to stabilisation at a concentration of 550 ppmv. The figure also shows the amount of CO_2 (in GtC) remaining in the atmosphere at each stabilisation level. The difference between accumulated emissions and atmospheric increase represents the accumulated uptake by the ocean and the marine and terrestrial biospheres. The range of results from different models is indicated by the shaded area.

Table 3: A summary of key greenhouse gases affected by human activities.

	CO_2	CH_4	N_2O	CFC-12	HCFC-22 (a CFC substitute)	CF_4 (a perfluoro-carbon)
Pre-industrial concentration	280 ppmv	700 ppbv	275 ppbv	zero	zero	zero
Concentration in 1992	355 ppmv	1,714 ppbv	311 ppbv	503 pptv†	105 pptv	70 pptv
Recent rate of concentration change per year (over 1980s)	1.5 ppmv/yr 0.4%/yr	13 ppbv/yr 0.8%/yr	0.75 ppbv/yr 0.25%/yr	18-20 pptv/yr 4%/yr	7-8 pptv/yr 7%/yr	1.1-1.3 pptv/yr 2%/yr
Atmospheric lifetime (years)	(50–200)††	(12–17)†††	120	102	13.3	50,000

† 1 pptv = 1 part per trillion (million million) by volume.
†† No single lifetime for CO_2 can be defined because of the different rates of uptake by different sink processes.
††† This has been defined as an adjustment time which takes into account the indirect effect of methane on its own lifetime.

decomposition (a positive feedback). Soils which are flooded can store large amounts of carbon in the form of peat. Drying (due to changes in precipitation and/or evapotranspiration), and possibly warming, of such regions would release additional CO_2.

Global ecosystem models which include the CO_2 fertilisation effect and attempt to model the effect of changes in temperature and precipitation on plant growth and decomposition are now providing insight into the issue of future terrestrial carbon storage. Some models also allow for the effect of redistribution of vegetation in response to changes in climate. In response to climate change resulting from gradually increasing CO_2 concentration, model results suggest release of carbon due to die-back of vegetation of 1 to 4 GtC/yr for a period of decades to centuries, followed by carbon accumulation. Results suggest increased carbon storage of 100 to 200 GtC after several centuries. Assumptions about land-use affect these conclusions, with high rates of simulated deforestation leading to reduced storage or even net emission of CO_2.

Climate feedbacks also influence the storage of carbon in the ocean through changes in sea surface temperature, ocean circulation and the marine biological pump.

Additional insights into climatic feedbacks come from ice core records going back over many thousands of years (known as palaeo-records). A clear correlation between atmospheric CO_2 concentration and global temperature (especially during warming periods) is evident in much of the palaeo-record over long time-scales, with increases of about 80 ppmv occurring during deglaciations. This relationship between CO_2 concentration and temperature may carry forward into the future, possibly causing a significant positive climate feedback on CO_2 fluxes.

3 Methane (CH_4)

Methane is another naturally occurring greenhouse gas whose concentration in the atmosphere is growing as a result of human activities (rice paddies, animal husbandry, landfills, biomass burning and fossil fuel production and use) (Figure 9a). Globally, methane increased by 7% over the decade from 1983. However, the 1980s were characterised by declining growth rates, dropping from 16 ppbv/yr in 1980 to about 10 ppbv/yr by 1990. Growth rates slowed dramatically in 1991 and 1992, although very recent data suggest that growth rates started to increase in late 1993 (see the box on "Variations in the growth rates of CO_2 and CH_4 concentrations"). The magnitudes of sinks and, especially, individual sources of methane, are less well known than its atmospheric increase (Table 4). However, carbon isotope measurements indicate that about 20% of the total annual methane emissions are related to fossil fuel use (e.g., combustion, coal mines, natural gas production and distribution, and petroleum industry operations). In total, anthropogenic activities are responsible for 60–80% of current methane emissions (Table 4). The direct radiative forcing due to the increase in methane concentration since pre-industrial times is about 0.5 Wm^{-2} (Figure 3).

CH_4 has clearly identified chemical feedbacks. The main

ATMOSPHERIC CHEMISTRY

While chemical and physical reactions in the lower atmosphere have a small impact on carbon dioxide (CO_2) concentrations, they control the abundance and lifetime of other greenhouse gases. A key process to understand is the way these reactions influence the concentration and spatial distribution of the hydroxyl radical (OH) which has a major role in atmospheric chemistry as it is highly reactive and acts to remove a number of greenhouse gases from the atmosphere.

The major sink mechanism for methane (CH_4) is reaction with OH. Increasing tropospheric methane concentrations will lead to decreased OH concentrations and to an increased methane lifetime. Reduction in tropospheric OH can also indirectly increase the abundance of some other greenhouse gases and ozone-depleting substances.

CH_4 oxidation in the presence of oxides of nitrogen (NO_x) produces ozone (O_3), another greenhouse gas. When NO_x is present in low concentrations, the CH_4 oxidation destroys O_3. Overall, methane oxidation is a net ozone source. When CH_4 is destroyed in the stratosphere, the chlorine and nitrogen cycles that destroy O_3 are suppressed, thus increasing stratospheric O_3 abundances. The destruction process also produces stratospheric water vapour which acts as a greenhouse gas and plays a role in stratospheric chemistry, tending to destroy some O_3.

Non-methane hydrocarbons, carbon monoxide and NO_x, while not important greenhouse gases in their own right, can have important indirect effects on radiative forcing through the production and destruction of tropospheric ozone and their influence on OH and methane concentrations. There are not enough baseline global and regional monitoring data for these species to derive the contributions from the different emission sources to their respective global distributions.

Table 4: Estimated sources and sinks of methane for 1980 to 1990. All figures are in Tg†(CH_4)/yr. Current global burden is 4850 Tg(CH_4).

(a) Observed atmospheric increase, estimated sinks and sources derived to balance the budget.

	Individual estimates	Total
Atmospheric increase	37 (35–40)	
Sinks of atmospheric CH_4		
tropospheric OH	**445** (360–530)	
stratosphere	**40** (32–48)	
soils	**30** (15–45)	
Total atmospheric sinks		**515** (430–600)
Implied sources (sinks + atmospheric increase)		**552** (465–640)

(b) Inventory of identified sources.

	Individual estimates	Total
Natural sources		**160** (110–210)
Anthropogenic sources:		
Fossil fuel related	**100** (70–120)	
Total biospheric	**275** (200–350)	
Total anthropogenic sources		**375** (300–450)
Total identified sources		**535** (410–660)

† 1 Tg = 1 million million grammes which is equivalent to 1 million tonnes.

removal process for CH_4 is reaction with the hydroxyl radical (OH). Addition of CH_4 to the atmosphere reduces the concentration of tropospheric OH concentration which can subsequently feed back and reduce the rate of CH_4 removal. A recent analysis has shown that these chemical feedbacks result in an adjustment time for the addition of a pulse of CH_4 to the atmosphere of 14.5 ± 2.5 years, as compared with the "budget" lifetime of 10 ± 2 years used in previous IPCC reports, when no account was taken of the effect of CH_4 on its own lifetime. Use of this longer adjustment time increases the direct radiative effect, of a given emission of CH_4; so part of what was previously included under the indirect effect of methane is now accounted for in the direct effect. This new adjustment time has been used when calculating the GWP for methane and contributes to a slightly larger value compared with the 1990 IPCC report (see Section 9).

In addition, chemical feedbacks resulting from addition of CH_4 increase the concentrations of other greenhouse gases (in particular, tropospheric O_3 and stratospheric water vapour); these are the indirect effects of CH_4 (see the box on "Atmospheric chemistry").

At present the emission of CH_4 (from a combination of anthropogenic and natural sources) must exceed the removal because the atmospheric abundance is increasing. CH_4 has a short adjustment time compared with CO_2 and so stabilisation of emissions would lead relatively quickly to a stable concentration. If current emissions were held constant, CH_4 concentrations would stabilise in less than 50 years at about 1900 ppbv (11% higher than at present). A reduction in annual methane emissions to levels about 35 million tonnes (roughly 10% of anthropogenic emissions) below current levels would stabilise concentrations at today's levels. (This calculation assumes that natural sources and atmospheric losses of methane are

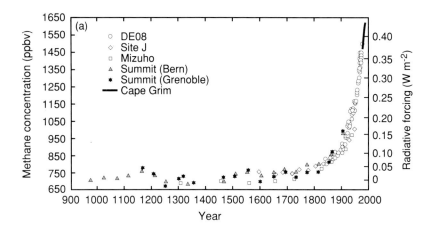

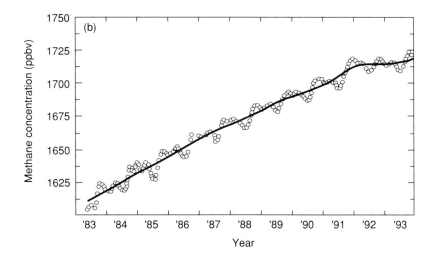

Figure 9: (a) CH_4 concentration derived from Antarctic ice cores over the past 1000 years. Direct observations of CH_4 concentration from Cape Grim, Tasmania, are included to demonstrate the smooth transition from ice core to atmospheric measurements. The radiative forcing resulting from increases in CH_4 relative to the pre-industrial period are indicated on the right-hand axis. The effect of overlap with N_2O is accounted for according to IPCC (1990). (b) Globally averaged CH_4 concentration for 1983 to 1993 showing the decline in growth rate during 1992 and 1993.

not affected by changing climate and atmospheric composition over the next century.)

As for CO_2, there are climate feedbacks which involve CH_4. In particular, methane emissions from northern wetlands, permafrost areas and decomposition of methane hydrates (clathrates) in continental shelf regions are sensitive to changes in temperature and precipitation. Measurements show that increases in either temperature or duration of water-logging increase methane emissions (a positive feedback). Conversely, a lowering of the water table in northern wetlands/peat lands may lead to a reduction in emissions of methane (a negative feedback).

Other insights into climate feedbacks come from palaeo-records which show a clear positive correlation between CH_4 concentration and global surface temperature on time-scales of order 10,000 years. CH_4 changes are typically about 300 ppbv during deglaciations. Recent results have also shown CH_4 changes of up to about 100 ppbv associated with more abrupt temperature fluctuations on the century time-scale. These shorter time-scale correlations are believed to result from changes in net fluxes from wetlands as a rapid response to climate change. Therefore the link between CH_4 concentration and temperature may carry forward into the future, with perhaps an additional effect due to clathrate decomposition.

4 Nitrous Oxide (N_2O)

There are many small sources of N_2O, both natural and anthropogenic, which are difficult to quantify. The main anthropogenic sources are from agriculture (especially the development of pasture in tropical regions), biomass burning, and a number of industrial processes (e.g., adipic

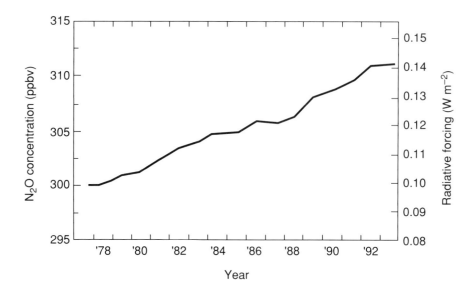

Figure 10: Globally averaged, annual mean N_2O concentration (ppbv) from 1977 to 1993 obtained from NOAA–CMDL flask sampling network. Pre-industrial N_2O concentration was around 275 ppbv. The radiative forcing resulting from increases in N_2O relative to the pre-industrial period is indicated on the right-hand axis. The effect of overlap with CH_4 is accounted for according to IPCC (1990).

acid and nitric acid production). A best estimate of the current (1980s) anthropogenic emission of N_2O is 3 to 8 Tg(N)/yr. Natural sources are probably twice as large as anthropogenic ones. Although sources/sinks cannot be well quantified, atmospheric measurements and evidence from ice cores show that the atmospheric abundance of N_2O has increased since the pre-industrial era, which is most likely due to human activities (Figure 10). The average growth rate over the past four decades is about 0.25%/yr (0.8 ppbv/yr). In 1992 atmospheric levels of N_2O were 311 ppbv; pre-industrial levels were about 275 ppbv. The radiative forcing due to the change in N_2O since pre-industrial times is about 0.1 Wm^{-2} (Figure 3).

N_2O, removed mainly by photolysis (breakdown by sunlight) in the stratosphere, has a long lifetime (about 120 years) which has implications for achieving stable concentrations. If emissions were held constant at today's level, the N_2O abundance would climb from 311 ppbv to about 400 ppbv over several hundred years. In order for N_2O abundances to be stabilised near current levels, anthropogenic sources would need to be reduced by more than 50%.

5 Halocarbons

Halocarbons containing fluorine, chlorine and bromine are significant greenhouse gases on a per-molecule basis. The chlorine and bromine containing species are also involved in the depletion of the ozone layer; the emissions of many such compounds are controlled by the Montreal Protocol and its subsequent amendments and adjustments.

The tropospheric growth rates of the major anthropogenic source species for stratospheric chlorine and bromine (CFCs, carbon tetrachloride, methylchloroform, halons) have slowed significantly (Figure 11), in response to substantially reduced emissions, as required by the Montreal Protocol and its subsequent amendments and adjustments. For example, the 1993 CFC-11 annual growth rate was 25 to 30% of that observed in the 1970s and 1980s. The total amount of organic chlorine in the troposphere increased by only 1.6% in 1992, about half of the rate of increase (2.9%) in 1989. Peak total chlorine/bromine loading in the troposphere is expected to occur in 1994, but the stratospheric peak will lag by about 3 to 5 years, so stratospheric abundances will continue to grow for a few more years before declining.

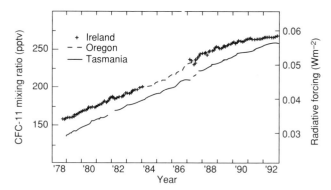

Figure 11: CFC-11 concentration from 1978 to 1992 from the ALE-GAGE global sampling network. Monthly mean clean air values are shown for three sites: Ireland, Oregon and Tasmania. CFCs are entirely of anthropogenic origin and did not exist in the atmosphere prior to the 1950s. The radiative forcing resulting from increases in CFC-11 relative to the pre-industrial period are indicated on the right-hand axis.

With phase out dates for the production and consumption of CFCs and other ozone-depleting substances now fixed by international agreement, several hydrochlorofluorocarbons (HCFCs) and hydrofluorocarbons (HFCs) are being manufactured and used as substitutes. The growth in the atmospheric concentration of HCFC-22 has been observed for several years and is currently about 7% per year. The direct global warming potentials of most HCFCs and HFCs are less than those of the compounds they replace, although some HFCs have substantial global warming potentials. Perfluorocarbons, which have been proposed as CFC substitutes in some applications and are by-products of some industrial processes, have very long atmospheric lifetimes (several thousand years) and are extremely powerful greenhouse gases.

The atmospheric residence times of CFC-11 and methylchloroform are now better known. Model studies simulating atmospheric abundances using known emissions suggest best-estimated lifetimes of 50 years for CFC-11 and 5.4 years for methylchloroform, with uncertainties of about 10%. These models, calibrated against CFC-11 and methylchloroform, are used to calculate the lifetimes, and hence GWPs, of other gases destroyed only in the stratosphere (other CFCs and nitrous oxide) and those which are highly reactive with tropospheric OH (HCFCs and HFCs).

The contribution to direct radiative forcing due to increases in the halocarbons since pre-industrial times is about 0.3 Wm^{-2} (the largest contributions come from CFC-11, CFC-12 and CFC-113). As well as this positive direct effect, the halocarbon induced loss of ozone in the lower stratosphere represents an indirect radiative forcing which is negative (see Section 6.1).

6 Ozone (O_3)

Ozone is an important greenhouse gas present in both the stratosphere and troposphere. Changes in ozone can cause radiative forcing by influencing both solar and infrared radiation. The net radiative forcing is strongly dependent on the vertical distribution of ozone change and is particularly sensitive to changes around the tropopause level, at which height trends are difficult to estimate due to a lack of reliable observations. Estimation of the radiative forcing due to changes in ozone is thus more complex than for the well-mixed greenhouse gases.

6.1 Stratospheric Ozone

Decreases in stratospheric ozone have occurred since the 1970s. The most obvious feature is the annual appearance of the Antarctic ozone hole in September and October. The October average total ozone values over Antarctica are 50–70% lower than those observed in the 1960s. The ozone loss occurs at altitudes between about 14 and 24 km and is caused by the chlorine and bromine compounds released in the stratospheric decomposition of halocarbons (principally CFCs and halons).

Statistically significant losses in total ozone have also been observed in the mid-latitudes of both hemispheres. In the Northern Hemisphere the trends in the 1980s are larger in local winter and spring than in summer or autumn, while in the Southern Hemisphere the trends show little seasonal variation. At Northern mid-latitudes in winter and spring the ozone losses amount to about a 10% decrease since around 1970. The weight of recent scientific evidence strengthens the conclusion that the mid-latitude ozone loss is due largely to anthropogenic chlorine and bromine compounds. Little or no downward trend in ozone has been observed in the tropics (20°N–20°S).

A variety of techniques indicates that ozone losses have occurred in the lower stratosphere (around 20 km altitude) where the bulk of the ozone resides. A statistically significant trend is also observed at 40 km, but this makes a small contribution to changes in total column ozone and has a negligible effect on radiative forcing.

Record low global ozone levels were measured over the past two years. Anomalous ozone decreases were observed in the mid-latitudes of both hemispheres, with Northern Hemispheric decreases being larger than those in the Southern Hemisphere. The Antarctic ozone depletions in 1992 and 1993 were the most severe on record, with ozone losses of more than 99% between 14 and 19 km in October 1993. The globally averaged ozone values were 1–2% lower than would be expected from an extrapolation of the trend prior to 1991 and allowing for the natural fluctuations resulting from the solar cycle and quasi-biennial oscillation. The 1994 global ozone levels are returning to values close to those expected from the long-term downward trend.

Since the stratospheric abundances of halocarbon compounds are expected to continue to grow for a few more years (see Section 5), continuing global ozone losses are expected for the remainder of the decade (other things being equal), with a gradual recovery throughout the first half of the 21st century.

A loss of ozone in the lower stratosphere over the past 15 to 20 years has led to a globally averaged radiative forcing of about -0.1 Wm^{-2}. This negative radiative forcing represents an indirect effect of CFCs and halons (the compounds thought to be largely responsible for the ozone loss) which, over the last 15 to 20 years, may have partially offset their direct warming effect. However, because the pattern of stratospheric ozone loss is not spatially uniform (it occurs mainly in mid-latitudes and polar regions), offsetting the direct and indirect global mean radiative forcing may not correctly reflect the climatic response (see Section 1). Prior to the onset of significant ozone depletion, the globally averaged radiative forcing caused by the increase in halocarbons was between +0.1 and +0.2 Wm^{-2}.

6.2 Tropospheric Ozone

In the troposphere ozone is produced from various short-lived precursor gases (carbon monoxide (CO), nitrogen oxides (NO_x) and non-methane hydrocarbons (NMHC)) and as a result of chemical feedbacks involving CH_4. Ozone can also be transported into the troposphere from the stratosphere. Changes in tropospheric ozone concentration are highly spatially variable, both regionally and vertically, making assessment of global long-term trends extremely difficult. Observations show that free tropospheric ozone has increased above many locations in the Northern Hemisphere over the last 30 years. There is also some evidence (from measurements made at high mountain sites in Europe) that levels have increased in the Northern Hemisphere since the early 1900s. Over the last decade, trends were small or non-existent. Model simulations and the limited observations together suggest that tropospheric ozone may have doubled in the Northern Hemisphere since pre-industrial times, an increase of around 25 ppbv. In the Southern Hemisphere, a decrease has been observed since the mid-1980s at the South Pole; in the hemisphere as a whole there are insufficient data to draw strong inferences.

Such changes in ozone have potentially important consequences for radiative forcing. Although detailed quantification is not possible, due to uncertainties in the size and distribution of the ozone change since pre-industrial times, estimates of the radiative forcing are of order a few tenths of a Wm^{-2} (tentatively put at 0.2 to 0.6 Wm^{-2}).

6.3 The Importance of NO_x

Our identification of the major physical and chemical processes affecting ozone in the troposphere is much further advanced than our ability to calculate or predict ozone concentrations. Uncertainties in the global budget of tropospheric ozone are associated primarily with our lack of knowledge of the distribution of O_3 and its short-lived precursors (NO_x, NMHC and CO). Observations of NO_x are just beginning to reveal the atmospheric distribution and large variability in these ozone-producing species, but even with the observed distributions we cannot define the importance of anthropogenic sources (e.g., transport of surface pollution out of the boundary layer, direct injection by aircraft) relative to natural sources (lightning, stratospheric air) in controlling the global NO_x distribution. Current estimates of anthropogenic NO_x sources attribute 24 Tg(N)/yr to fossil fuel combustion at the surface, 0.5 Tg(N)/yr to aircraft emissions, 8 Tg(N)/yr to biomass burning and an unknown, but significant fraction of the 12 Tg(N)/yr released from soils. Anthropogenic emissions dominate natural sources by far, but emissions from different sources cannot be directly compared and their impact is primarily local or regional.

Aircraft release NO_x directly into the free troposphere. Such emissions could increase ozone concentration at altitudes where ozone is at its most effective as a greenhouse gas. The radiative forcing resulting from such emissions could thus be more important than equivalent NO_x emissions at the surface. Until the various contributions to NO_x and O_3 levels in the free troposphere can be better quantified, the relative importance of the various NO_x sources in perturbing O_3 amounts cannot be reliably estimated. However, reasonable upper limits may be placed on the radiative forcing due to increased O_3 produced by the NO_x relative to that from the CO_2 emitted by aircraft. Aircraft emit about 3% of total CO_2 emissions from fossil fuel combustion and a similar fraction of anthropogenic NO_x. Our current best guess is that the positive radiative forcing due to the release of NO_x from aircraft could be of similar magnitude or smaller than the effect of CO_2 released from aircraft. These estimates are preliminary and may well change in future assessments.

Aircraft also emit carbon monoxide, water vapour, soot and other particles, sulphur gases and other trace constituents which have the potential to cause radiative forcing. The impact of such emissions has not yet been properly assessed.

7 The Effect of Tropospheric Aerosols

Aerosols are suspensions of particles (diameter range 10^{-3} to 10 μm) in the atmosphere. Tropospheric aerosols are formed by dispersal of material from the surface (e.g., soil dust), by direct emissions of material into the atmosphere (e.g., smoke) and by chemical reactions in the troposphere which convert gases (such as sulphur dioxide) into particles. The release of sulphur dioxide from fossil fuel combustion, and organic and elemental carbon, mainly from biomass burning, are the main anthropogenic sources of aerosols.

The addition of anthropogenic aerosols to the troposphere can influence the radiative balance of the Earth in two major ways: (i) through absorption and through scattering of solar radiation back to space, known as the *direct* effect and (ii) by acting as nuclei on which cloud droplets form, aerosols can influence the formation, lifetime and radiative properties of clouds (e.g., increase the amount of solar radiation they reflect): the *indirect* effect.

There are many uncertainties associated with estimating the climatic influence of aerosols. Aerosols are highly variable regionally, in both concentration and chemical composition, and observations of their spatial distribution in the atmosphere (currently and in the past) are lacking. Both the direct radiative effect of aerosols and their ability to modify cloud properties are strongly influenced by particle size and composition. As a result, the radiative effects of the anthropogenic component of aerosols cannot be related to aerosol mass loading in a simple way. The radiative effects of anthropogenic aerosols are relatively large compared with their mass contribution because most anthropogenic aerosols are in the size range which is most radiatively active.

New estimates of both the direct and indirect effect of anthropogenic aerosols in the troposphere have become available since IPCC 1992. In order to compare different aerosol effects it is useful to express them in terms of globally averaged values of radiative forcing. The direct radiative forcing due to increases in sulphate aerosol since 1850, averaged globally, is estimated to lie in the range -0.25 to -0.9 Wm^{-2}. The direct effect of aerosol from biomass burning is estimated to lie in the range -0.05 to -0.6 Wm^{-2}. Calculations of the indirect effect of aerosols are at an early stage. Preliminary results suggest that the radiative effect of aerosols on cloud radiative properties is probably a negative forcing and may be of similar magnitude to the direct effect. Note that in the global average the total aerosol induced radiative forcing is negative, but the absorption of solar radiation by carbonaceous aerosols may cause local positive radiative forcing. It is interesting to compare these estimates with the direct radiative forcing due to increases in greenhouse gases since pre-industrial times (+2.1 to +2.8 Wm^{-2}) (see Figure 3), although, as pointed out in Section 1, it is unlikely to be appropriate to add the negative global radiative forcing of aerosols to the positive global radiative forcing of greenhouse gases.

Figure 12 shows the spatial distribution of direct radiative forcing due to sulphate aerosols derived from a radiative transfer model forced with a model derived sulphate distribution. The largest forcing occurs over or close to regions of industrial activity. Over the eastern USA, central Europe, and eastern China, sulphate forcing may have offset much of, or in places been greater than, the greenhouse gas forcing. In other areas, particularly the Southern Hemisphere, the negative aerosol forcing is much weaker and greenhouse gas forcing dominates. However, care must be taken when comparing regional radiative forcing in this way. For example, a local net forcing of zero Wm^{-2} does not necessarily imply a lack of climate change, because climate change over a region can be affected by changes in circulation as the atmosphere responds to radiative forcing in another region (i.e., local forcing does not necessarily govern local response).

Tropospheric aerosols have a lifetime of only a few days, unlike most greenhouse gases which have lifetimes of tens to hundreds of years. So the atmospheric concentration of aerosols responds rapidly to changes in emissions. Control of sulphate emissions, motivated by other environmental considerations for instance, would immediately reduce the amount of aerosols in the atmosphere. In contrast, reductions in CO_2 emissions have a much slower effect (see Section 2.3). Emissions of aerosols and their precursors therefore create no long-term commitment to radiative forcing, in contrast to CO_2 and

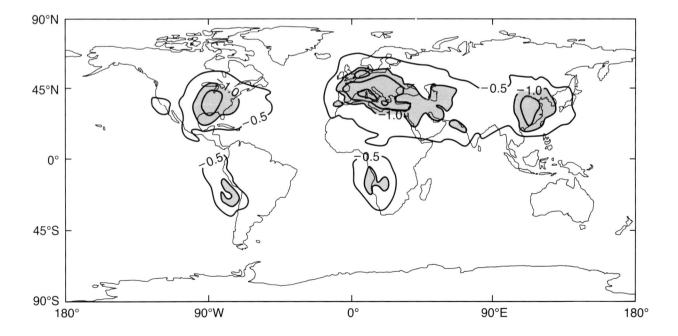

Figure 12: Modelled geographic distribution of annual mean direct radiative forcing (Wm^{-2}) from anthropogenic sulphate aerosols in the troposphere. The negative radiative forcing is largest over or close to regions of industrial activity.

other long-lived greenhouse gas emissions.

8 What Else Influences Radiative Forcing?

8.1 Solar Variability

We know that total solar irradiance varies with an 11-year cycle. Space-borne satellite measurements available since 1978 show that over the most recent sunspot cycle the changes in solar irradiance were equivalent to a radiative forcing of about 0.2 Wm^{-2}. This may initially seem significant, given that it is an appreciable fraction of the forcing due to greenhouse gases over the same period. However, these changes in solar irradiance are cyclical in nature and it is believed that, due to the thermal inertia in the climate system, only a small amount of the possible temperature change resulting from such transient changes in irradiance is realised. In contrast, the changes in greenhouse gases represent a sustained and cumulative effect over many decades.

Recent satellite observations show a relationship between total solar irradiance and other indicators of solar activity which allows a tentative reconstruction of past total solar irradiance. Although there is considerable uncertainty in estimating solar irradiance before direct measurements began, changes in solar irradiance since 1850 may have contributed a natural radiative forcing of around 0.3 Wm^{-2} (Figure 3). Future forcing due to changes in solar irradiance could be negative or positive.

8.2 Volcanic Activity

Volcanic eruptions can act to increase the amount of aerosol particles in the stratosphere. The dominant radiative effect is an increase in scattering of solar radiation which reduces the net radiation available to the surface/troposphere, thereby leading to cooling. Volcanoes have the potential to produce large radiative forcing, but the events are transitory.

The eruption of Mt. Pinatubo in the Philippines in June 1991 stands out from a climatic point of view as probably the most important eruption this century. The largest forcing is calculated to have been about -4 Wm^{-2} around one year after the eruption. This decayed to around -1 Wm^{-2} after 2 years. Thus, the radiative forcing resulting from Mt. Pinatubo for the first 2 years after the eruption was comparable, but of opposite sign, to the greenhouse gas forcing this century (+2.1 to +2.8 Wm^{-2}). A cooling of global surface temperature observed following the eruption reached a maximum of 0.3 to 0.5 °C during 1992. Simulation of the climatic effects of Mt. Pinatubo aerosols using general circulation models (GCMs) have produced results in good agreement with observations, with a maximum cooling of 0.4 to 0.6 °C. Such simulations increase confidence in the ability of GCMs to respond in a realistic way to transient, planetary-scale radiative forcings of large magnitude.

Clearly, individual volcanic eruptions can produce large radiative forcing effects, but these effects are transitory. An important issue here is whether changes in greenhouse gases and aerosols due to human activity are significant compared with natural factors. A key question is therefore whether there has been any trend in volcanic activity over the period since industrialisation; this is unlikely. However, variations in the occurrence of climatically significant eruptions may be a factor in explaining some interannual and interdecadal climate variations.

9 Global Warming Potential (GWP) — a Tool for Policymakers

Policymakers need some measure of possible future commitment to global warming resulting from current anthropogenic emissions. The GWP is an attempt to provide such a measure. The index is defined as the cumulative radiative forcing between the present and some chosen later time "horizon" caused by a unit mass of gas emitted now, expressed relative to some reference gas (here CO_2 is used). The future global warming commitment of a greenhouse gas over the reference time horizon is the appropriate GWP multiplied by the amount of gas emitted. For example, GWPs could be used to calculate the effect of reducing CO_2 emissions by a certain amount compared with reducing CH_4 emissions, for a specified time horizon.

Derivation of GWPs requires knowledge of the fate of the emitted gas (typically not well understood) and the radiative forcing due to the amount remaining in the atmosphere (reasonably well understood). Consequently, GWPs encompass the uncertainty associated with all the topics discussed in this report. Additionally the choice of time horizon will depend on policy considerations.

The latest estimates of GWPs are given in Table 5. Although the GWPs are quoted as single values, the typical uncertainty is ±35% relative to the carbon dioxide reference. The majority of GWPs are larger than those reported in IPCC (1992), typically by 10–30%. These increases are largely due to (i) an improved carbon dioxide reference and (ii) improved estimates of atmospheric lifetimes. Because GWPs are based on the radiative forcing concept, they are difficult to apply to radiatively important constituents that are unevenly distributed in the atmosphere (e.g., aerosols), for the reasons discussed in Section 1. No attempt is made to define a GWP for aerosols.

GWPs need to take account of any indirect effects of the emitted greenhouse gas, for example, the formation of another greenhouse gas, if they are correctly to reflect future warming potential. The calculation of many indirect GWP components is not currently possible because of

Table 5: *Global Warming Potentials, referenced to the absolute GWP for CO_2. The typical uncertainty is ±35% relative to the CO_2 reference.*

Species	Chemical Formula	Lifetime (yr)	Global Warming Potential (Time Horizon)		
			20 years	100 years	500 years
Methane†	CH_4	14.5±2.5††	62	24.5	7.5
Nitrous oxide	N_2O	120	290	320	180
CFCs					
CFC-11	$CFCl_3$	50±5	5000	4000	1400
CFC-12	CF_2Cl_2	102	7900	8500	4200
CFC-13	$CClF_3$	640	8100	11700	13600
CFC-113	$C_2F_3Cl_3$	85	5000	5000	2300
CFC-114	$C_2F_4Cl_2$	300	6900	9300	8300
CFC-115	C_2F_5Cl	1700	6200	9300	13000
HCFCs, etc.					
HCFC-22	CF_2HCl	13.3	4300	1700	520
HCFC-123	$C_2F_3HCl_2$	1.4	300	93	29
HCFC-124	C_2F_4HCl	5.9	1500	480	150
HCFC-141b	$C_2FH_3Cl_2$	9.4	1800	630	200
HCFC-142b	$C_2F_2H_3Cl$	19.5	4200	2000	630
HCFC-225ca	$C_3F_5HCl_2$	2.5	550	170	52
HCFC-225cb	$C_3F_5HCl_2$	6.6	1700	530	170
Carbon tetrachloride	CCl_4	42	2000	1400	500
Methylchloroform	CH_3CCl_3	5.4±0.6	360	110	35
Bromocarbons					
H-1301	CF_3Br	65	6200	5600	2200
Other					
HFC-23	CHF_3	250	9200	12100	9900
HFC-32	CH_2F_2	6	1800	580	180
HFC-43-10mee	$C_5H_2F_{10}$	20.8	3300	1600	520
HFC-125	C_2HF_5	36.0	4800	3200	1100
HFC-134	CHF_2CHF_2	11.9	3100	1200	370
HFC-134a	CH_2FCF_3	14	3300	1300	420
HFC-152a	$C_2H_4F_2$	1.5	460	140	44
HFC-143	CHF_2CH_2F	3.5	950	290	90
HFC-143a	CF_3CH_3	55	5200	4400	1600
HFC-227ea	C_3HF_7	41	4500	3300	1100
HFC-236fa	$C_3H_2F_6$	250	6100	8000	6600
HFC-245ca	$C_3H_3F_5$	7	1900	610	190
Chloroform	$CHCl_3$	0.55	15	5	1
Methylene chloride	CH_2Cl_2	0.41	28	9	3
Sulphur hexafluoride	SF_6	3200	16500	24900	36500
Perfluoromethane	CF_4	50000	4100	6300	9800
Perfluoroethane	C_2F_6	10000	8200	12500	19100
Perfluorocyclo-butane	$c\text{-}C_4F_8$	3200	6000	9100	13300
Perfluorohexane	C_6F_{14}	3200	4500	6800	9900

† The methane GWP includes the direct effect and those indirect effects due to the production of tropospheric ozone and stratospheric water vapour. The indirect effect due to the production of CO_2 is not included.
†† For methane the adjustment time is given, rather than the lifetime – see Section 3.

inadequate characterisation of many of the atmospheric processes involved.

The GWP value for methane in Table 5 includes both the direct and indirect components (e.g., the formation of tropospheric ozone), as well as the longer adjustment time discussed in Section 3. In the 1992 IPCC report, only the direct GWP for methane was quoted; no account was taken of the effect of methane on its own lifetime and indirect effects were not quantified. For the 100 year time horizon, the indirect contributions to the total GWP value are: tropospheric ozone change 19 ± 12% and stratospheric water vapour change about 4%. As the range indicates, there is substantial current uncertainty in the methane GWP.

The GWPs presented in Table 5 were calculated on the assumption that present background atmospheric composition remains constant indefinitely. An assumption of increasing CO_2 concentrations, which lowers the additional forcing of incremental CO_2 emissions, would increase the GWP of other gases relative to CO_2.

The indirect effect of CFCs and halons through stratospheric ozone depletion, which tends to reduce the GWPs for these gases, is not included in the values in Table 5. Some substitutes for CFCs have lower GWPs than the compounds they replace (e.g., short-lived gases like HFC-152a). On the other hand, other potential substitutes like perfluorocarbons (such as CF_4 and C_2F_6) have very long lifetimes and hence extremely large GWPs over long time horizons.

No GWPs are given for atmospheric constituents such as NO_x and CO, which are not important greenhouse gases in their own right (i.e., they have no significant direct effect), but which are covered by the IPCC Guidelines for National Greenhouse Gas Inventories. These gases indirectly influence the concentration of some other greenhouse gases (e.g., tropospheric ozone) through atmospheric chemistry. The indirect effects for these gases are complex and depend on when and where they are emitted. New techniques will need to be developed to assess their influence on radiative forcing.

Figure 13 shows the future warming potential of current emissions of various species (i.e., the GWP multiplied by estimated mean annual emissions typical of the 1980s). The radiative forcing due to contemporary global emissions of anthropogenic greenhouse gases over the next century is largest from CO_2. However, the GWP weighted emissions of CH_4 are also important; over a 20-year time horizon they are comparable to those of CO_2.

Current emissions of HFCs and perfluorocarbons are small, making the present commitment to future radiative forcing from these gases much less than for CO_2 and methane (Figure 13). However, if emissions were to increase in the future, their contribution to future radiative forcing would become more important.

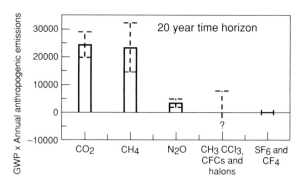
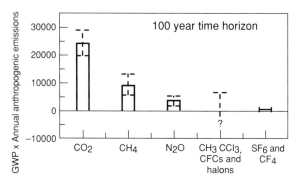
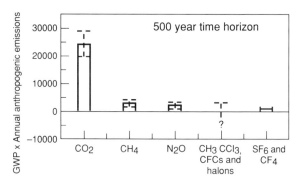

Figure 13: The product of GWP with estimated annual emissions typical of the 1980s for various gases, for time horizons of 20, 100 and 500 years. The dashed bars indicate the range of possible uncertainty considered here in estimating GWPs and total anthropogenic emissions. The estimated emissions are approximate. The indirect radiative forcing of CFCs and halons due to ozone depletion imply that the net GWPs for these gases could be negative, but this is presently uncertain.

1

CO$_2$ and the Carbon Cycle

D. SCHIMEL, I.G. ENTING, M. HEIMANN, T.M.L. WIGLEY,
D. RAYNAUD, D. ALVES, U. SIEGENTHALER

Contributors:
S. Brown, W. Emanuel, M. Fasham, C. Field, P. Friedlingstein, R. Gifford,
R. Houghton, A. Janetos, S. Kempe, R. Leemans, E. Maier-Reimer, G. Marland,
R. McMurtrie, J. Melillo, J-F. Minster, P. Monfray, M. Mousseau, D. Ojima, D. Peel,
D. Skole, E. Sulzman, P. Tans, I. Totterdell, P. Vitousek

Modellers:
J. Alcamo, B.H. Braswell, B.C. Cohen, W.R. Emanuel, I.G. Enting, G.D. Farquhar,
R.A. Goldstein, L.D.D. Harvey, M. Heimann, A. Jain, F. Joos, J. Kaduk, A.A. Keller,
M. Krol, K. Kurz, K.R. Lassey, C. Le Quere, J. Lloyd, E. Meier-Reimer, B. Moore III,
J. Orr, T.H. Peng, J. Sarmiento, U. Siegenthaler, J.A. Taylor, J. Viecelli, T.M.L. Wigley,
D. Wuebbles.

CONTENTS

Summary	39
1.1 Description of the Carbon Cycle	41
1.2 Past Record of Atmospheric CO_2	43
1.2.1 Atmospheric Measurements since 1958	43
1.2.2 Pre-1958 Atmospheric Measurements and CO_2-Ice Core Record over the Last Millennium	43
1.2.3 The CO_2 Record over the Last Climatic Cycle	45
1.3 The Anthropogenic Carbon Budget	46
1.3.1 Introduction	46
1.3.2 Methods for Calculating the Carbon Budget	46
1.3.2.1 Classical approaches	46
1.3.2.2 New approaches: budget assessment based on observations of $^{13}C/^{12}C$ and O_2/N_2 ratios	46
1.3.2.3 Constraints from spatial distributions	48
1.3.3 Sources and Sinks of Anthropogenic CO_2	48
1.3.3.1 Fossil carbon emissions	48
1.3.3.2 Atmospheric increase	49
1.3.3.3 Ocean exchanges	49
1.3.3.4 Terrestrial exchanges	51
1.3.3.4.1 Emissions from changing land use	52
1.3.3.4.2 Uptake of CO_2 by changing land use	53
1.3.3.4.3 Other terrestrial sink processes	53
1.4 The Influence of Climate and Other Feedbacks on the Carbon Cycle	55
1.4.1 Introduction	55
1.4.2 Feedbacks to Terrestrial Carbon Storage	56
1.4.3 Feedbacks on Oceanic Carbon Storage	57
1.5 Modelling Future Concentrations of Atmospheric CO_2	58
1.5.1 Introduction	58
1.5.2 Calculations of Concentrations for Specified Emissions	60
1.5.3 Stabilisation Calculations	60
1.5.4 Assessment of Uncertainties	62
References	66

SUMMARY

Interest in the carbon cycle has increased because of the observed increase in levels of atmospheric CO_2 (from ~280 ppmv in 1800 to ~315 ppmv in 1957 to ~356 ppmv in 1993) and because the signing of the UN Framework Convention on Climate Change has forced nations to assess their contributions to sources and sinks of CO_2, and to evaluate the processes that control CO_2 accumulation in the atmosphere. Over the last few years, our knowledge of the carbon cycle has increased, particularly in the quantification and identification of mechanisms for terrestrial exchanges, and in the preliminary quantification of feedbacks.

The increase in atmospheric CO_2 concentration since pre-industrial times

Atmospheric levels of CO_2 have been measured directly since 1957. The concentration and isotope records prior to that time consist of evidence from ice cores, moss cores, packrat middens, tree rings, and the isotopic measurements of planktonic and benthic foraminifera. Ice cores serve as the primary data source because they provide a fairly direct and continuous record of past atmospheric composition. The ice cores indicate that an increase in CO_2 level of about 80 ppmv paralleled the last interglacial warming. There is uncertainty over whether changes in CO_2 levels as rapid as those of the 20th century have occurred in the past. However, there is essentially no uncertainty that for approximately the last 18,000 years, CO_2 concentrations in the atmosphere have fluctuated around 280 ppmv, and that the recent increase to a concentration of ~356 ppmv, with a current rate of increase of ~1.5 ppmv/yr, is due to combustion of fossil fuel, cement production, and land use conversion.

The carbon budget

The major components of the anthropogenic perturbation to the atmospheric carbon budget are anthropogenic emissions, the atmospheric increase, ocean exchanges, and terrestrial exchanges. Emissions from fossil fuels and cement production averaged 5.5 ± 0.5 GtC/yr over the decade of the 1980s (estimated statistically). The measured average annual rate of atmospheric increase in the 1980s was 3.2 ± 0.2 GtC/yr. Average ocean uptake during the decade has been estimated by a combination of modelling and isotopic measurements to be 2.0 ± 0.8 GtC/yr.

Averaged over the 1980s, terrestrial exchanges include a tropical source of 1.6 ± 1.0 GtC/yr from ongoing changes in land use, based on land clearing rates, biomass inventories, and modelled forest regrowth. Recent satellite data have reduced uncertainties in the rate of deforestation for the Amazon, but rates for the rest of the tropics remain poorly quantified. For the tropics as a whole, there is incomplete information on initial biomass and rates of regrowth. Potential terrestrial sinks may be the result of several processes, including the regrowth of mid-latitude and high latitude Northern Hemisphere forests (0.5 ± 0.5 GtC/yr), enhanced forest growth due to CO_2 fertilisation (0.5-2.0 GtC/yr) and nitrogen deposition (0.2-1.0 GtC/yr), and, possibly, response to climatic anomalies (0-1.0 GtC/yr). Partitioning the sink among these processes is difficult, but it is likely that all components are involved. While the CO_2 fertilisation effect is the most commonly cited terrestrial uptake mechanism, existing model studies indicate that the magnitude of contributions from each process are comparable, within large ranges of uncertainty. For example, some model-based evidence suggests that the magnitude of the CO_2 fertilisation effect is limited by interactions with nutrients and other ecological processes. Experimental confirmation from ecosystem-level studies, however, is lacking. As a result, the role of the terrestrial biosphere in controlling future atmospheric CO_2 concentrations is difficult to predict.

Future atmospheric CO_2 concentrations

Modelling groups from many countries were asked to use published carbon cycle models to evaluate the degree to which CO_2 concentrations in the atmosphere might be expected to change over the next several centuries, given a standard set of emission scenarios (including changes in land use). Models were constrained to balance the carbon budget and match the atmospheric record of the 1980s via CO_2 fertilisation of the terrestrial biosphere.

Stabilisation of atmospheric CO_2 concentrations

Modelling groups carried out stabilisation analyses to explore the relationships between anthropogenic emissions and atmospheric concentrations. The analyses assumed arbitrary concentration profiles (i.e., routes to stabilisation) and final stable CO_2 concentration; the models were then used to perform a series of inverse calculations (i.e., to derive CO_2 emissions given CO_2 concentrations). These calculations did three things: (1) assessed the total amount of fossil carbon that has been released (because land use was prescribed), (2) determined the partitioning of this carbon between the ocean and the terrestrial biosphere, and (3) ascertained what the time course of carbon emissions from fossil fuel combustion must have been to arrive at the selected arbitrary atmospheric CO_2 concentrations while still matching the atmospheric record through the 1980s.

Results suggest that in order for atmospheric concentrations to stabilise below 750 ppmv, anthropogenic emissions must eventually decline relative to today's levels. All emissions curves derived from the inverse calculations show periods of increasing anthropogenic emissions, followed by reductions to about a third of today's levels (i.e., to ~2 GtC/yr) for stabilisation at 450 ppmv by the year 2100, and to about half of current levels (i.e., ~3 GtC/yr) for stabilisation at 650 ppmv by the year 2200. Additionally, the models indicated that if anthropogenic emissions are held constant at 1990 levels, modelled atmospheric concentrations of CO_2 will continue to increase over the next century.

Among models the range of emission levels that were estimated to result in the hypothesised stabilisation levels is about 30%. In addition, the range of uncertainty associated with the parametrization of CO_2 fertilisation (evaluated with one of the models) varied between ±10% for low stabilisation values and ±15% for higher stabilisation values. The use of CO_2 fertilisation to control terrestrial carbon storage, when in fact other ecological mechanisms are likely involved, results in an underestimate of concentrations (for given emissions) of 5 to 10% or an overestimate of emissions by a similar amount (for given concentrations).

Feedbacks to the carbon cycle

Climate and other feedbacks via the oceans and terrestrial biosphere have the potential to be significant in the future. The effects of temperature on chemical and biological processes in the ocean are thought to be small (tens of ppmv changes in the atmosphere), but the effects of climate on ocean circulation could be larger, with possible repercussions for atmospheric concentration of ±100-200 ppmv. Effects of changing precipitation, temperature and atmospheric CO_2 can also have effects on the terrestrial biosphere, resulting in feedbacks to the atmosphere. Models suggest transient losses of about 200 GtC from terrestrial ecosystems as temperatures warm, with a potential for long-term increases in carbon storage above present levels by a few hundred gigatons. Patterns of changing land-use will have a substantial effect on terrestrial carbon storage and decrease the potential of terrestrial systems to store carbon in response to CO_2 and climate. Representation of feedbacks on the carbon cycle through oceanic and terrestrial mechanisms need to be improved in subsequent analyses of future changes to CO_2.

1.1 Description of the Carbon Cycle

Atmospheric CO_2 provides a link between biological, physical, and anthropogenic processes. Carbon is exchanged between the atmosphere, the oceans, the terrestrial biosphere, and, more slowly, with sediments and sedimentary rocks. The faster components of the cycle are shown in Figure 1.1. In the absence of anthropogenic CO_2 inputs, the carbon cycle had periods of millennia in which large carbon exchanges were in near balance, implying nearly constant reservoir contents. Human activities have disturbed this balance through the use of fossil carbon and disruption of terrestrial ecosystems. The consequent accumulation of CO_2 in the atmosphere has caused a number of carbon cycle exchanges to become unbalanced. Fossil fuel burning and cement manufacture, together with forest harvest and other changes of land use, all transfer carbon (mainly as CO_2) to the atmosphere. This anthropogenic carbon then cycles between the atmosphere, oceans, and the terrestrial biosphere. Because the cycling of carbon in the terrestrial and ocean biosphere occurs slowly, on time-scales of decades to millennia, the effect of additional fossil and biomass carbon injected into the atmosphere is a long-lasting disturbance of the carbon cycle. The relationships between concentration changes and emissions of CO_2 are examined through use of models that simulate, in a simplified manner, the major processes of the carbon cycle. The terrestrial and oceanic components of carbon cycle models vary in complexity from a few key equations to spatially explicit, detailed descriptions of ocean and terrestrial biology, chemistry, and transport

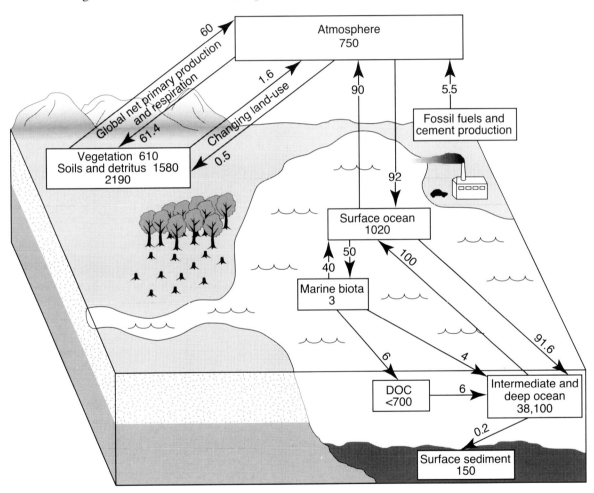

Figure 1.1: The global carbon cycle, showing the reservoirs (in GtC) and fluxes (GtC/yr) relevant to the anthropogenic perturbation as annual averages over the period 1980 to 1989 (Eswaran et al., 1993; Potter et al., 1993; Siegenthaler and Sarmiento, 1993). The component cycles are simplified and subject to considerable uncertainty. In addition, this figure presents average values. The riverine flux, particularly the anthropogenic portion, is currently very poorly quantified and so is not shown here (see text). While the surface sediment storage is approximately 150 Gt, the amount of sediment in the bioturbated and potentially active layer is of order 400 Gt. Evidence is accumulating that many of the key fluxes can fluctuate significantly from year to year (terrestrial sinks and sources: INPE, 1992; Ciais et al., submitted; export from the marine biota: Wong et al., 1993). In contrast to the static view conveyed by figures such as this one, the carbon system is clearly dynamic and coupled to the climate system on seasonal, interannual and decadal time-scales (Schimel and Sulzman, 1994; Keeling and Whorf, 1994).

> **RECENT ANOMALIES**
>
> The last few decades have been characterised by a number of observed changes in the carbon cycle:
>
> - The early 1980s were characterised by a period of relatively constant or slightly declining fossil carbon emissions. After 1985, emissions again exceeded the 1979 level; each year's release during the latter half of the 1980s was 0.1 to 0.2 GtC above that for the previous year (Boden *et al.*, 1991).
>
> - Direct measurements and ice core data have revealed a general decrease in atmospheric levels of ^{13}C relative to ^{12}C by about 1‰ over the last century (Friedli *et al.*, 1986; Keeling *et al.*, 1989a; Leuenberger *et al.*, 1992). This decrease is expected from the addition of fossil and/or terrestrial biospheric carbon, both of which are poor in ^{13}C relative to the atmosphere. In contrast, the atmospheric $^{13}C/^{12}C$ ratio remained nearly constant from 1988 to 1993. This constant ratio must reflect changes in the fluxes between the ocean, terrestrial biosphere, and atmosphere which currently are not quantified. The atmospheric record of ^{13}C shows a decrease of ~0.4‰ between the first measurements in 1978 and the present time. The ice core record provides information on the changing ^{13}C levels over the past few hundred years (Friedli *et al.*, 1986; Leuenberger *et al.*, 1992).
>
> - Relative to the long-term average rate of atmospheric CO_2 concentration increase (~1.5 ppmv/yr), the years 1988 to 1989 had relatively high CO_2 concentration growth rates (~2.0 ppmv/yr) while subsequent years (1991 to 1992) had very low growth rates (0.5 ppmv/yr) (Boden *et al.*, 1991). The magnitude of the 1988 to 1989 anomaly depends on what is defined as "normal" for the long-term trend. At Mauna Loa, Hawaii, the 1988 to 1989 increase was similar to a variation which occurred in 1973 to 1974 (see Figure 1.2). The subsequent decrease exceeds any previous anomaly since the Mauna Loa record began in 1958. Data for 1993 indicate a higher growth rate than that for 1991 to 1992.
>
> - The CO_2 record exhibits a seasonal cycle, with small peak-to-peak amplitude (about 1 ppmv) in the Southern Hemisphere but increasing northward to about 15 ppmv in the boreal forest zone (55-65°N). This cycle is mainly caused by the seasonal uptake and release of atmospheric CO_2 by terrestrial ecosystems. Part of the seasonal signal is driven by oceanic processes (Heimann *et al.*, 1989). The amplitude of the seasonal atmospheric CO_2 cycle varies with time. For instance, at Mauna Loa (see Figure 1.3; Keeling *et al.*, 1989a), it was roughly constant at 5.2 ppmv (peak-to-trough) from the beginning of the Mauna Loa measurements in 1958 until the mid-1970s. It then increased over the late 1970s to reach 5.8 ppmv for most of the 1980s. The most recent data indicate a further increase. Because the trend is not well-correlated with the CO_2 concentration increase (Thompson *et al.*, 1986; Enting, 1987; Manning, 1993), it provides at best weak evidence for CO_2 fertilisation of terrestrial vegetation, in contrast with interpretations that claimed strong evidence (e.g., Idso and Kimball, 1993). The variation in amplitude does indicate changes in terrestrial metabolism, but not necessarily increased photosynthesis or storage.

processes. The simpler models, in general, are designed to reproduce observed behaviour while the more complex models are aimed at incorporating the processes that cause the observed behaviour. The latter are thus potentially more likely to yield realistic projections of changes in storage under conditions different from the present (e.g., changing climate).

Time-scales

Because CO_2 added to the atmosphere by anthropogenic processes is exchanged between reservoirs having a range of turnover times, it is not possible to define a single atmospheric "lifetime". This is in contrast with other anthropogenic compounds such as N_2O and the halogens that are destroyed chemically in the atmosphere. Because the time-scales involved in the CO_2 exchanges range from annual to millennial (thousands of years), the consequences of anthropogenic perturbations will be long-lived. In this regard, the "turnover time" of about 5 years for atmospheric CO_2 deduced from the rate of bomb $^{14}CO_2$ removal is relevant to the initial response of the carbon system, but it does not characterise the much slower, long-term response of atmospheric concentrations to the anthropogenic perturbation.

The remainder of this chapter reviews what is known about the carbon cycle as a basis for understanding past changes and relationships between future emissions and

CO_2 and the Carbon Cycle

concentrations. We do not attempt to make specific predictions of likely future changes in CO_2 concentration – rather, we assess the sensitivity of the system to particular scenarios of future emissions and concentrations. We also analyse key areas in which quantitative understanding is deficient. The final section of the chapter presents the results of a set of calculations relating future CO_2 concentrations and future CO_2 emissions. These calculations, produced by modelling groups from many countries following an agreed set of specifications, explored various aspects of uncertainty.

1.2 Past Record of Atmospheric CO_2

1.2.1 Atmospheric Measurements since 1958

Precise, direct measurements of atmospheric CO_2 started in 1957 at the South Pole, and in 1958 at Mauna Loa, Hawaii. At this time the atmospheric concentration was about 315 ppmv and the rate of increase was ~0.6 ppmv/yr. The growth rate of atmospheric concentrations at Mauna Loa has generally been increasing since 1958. It averaged 0.83 ppmv/yr during the 1960s, 1.28 ppmv/yr during the 1970s, and 1.53 ppmv/yr during the 1980s. In 1992, the atmospheric level of CO_2 at Mauna Loa was 355 ppmv (Figure 1.3) and the growth rate fell to 0.5 ppmv/yr (see "Recent Anomalies" box). Data from the Mauna Loa station are close to, but not the same as, the global mean.

Atmospheric concentrations of CO_2 have been monitored for shorter periods at a large number of atmospheric stations around the world (e.g., Boden *et al.*, 1991). Measurement sites are distributed globally and include sites in Antarctica,

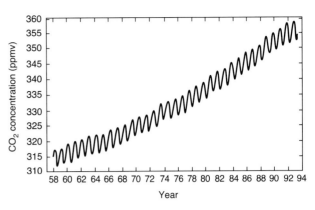

Figure 1.3: CO_2 concentrations measured at Mauna Loa, Hawaii since 1958 showing trends and seasonal cycle.

Australia, several maritime islands, and high northern latitude sites, but, at present, nowhere on the continents of Africa or South America. The reliability and high precision of the post-1957 record is guaranteed by comparing the measured concentration of CO_2 in air with the concentration of reference gas mixtures calibrated by a constant volume column manometer. The increase shown by the atmospheric record since 1957 can be attributed largely to anthropogenic emissions of CO_2, although considerable uncertainty exists as to the mechanisms involved. The record itself provides important insights that support anthropogenic emissions as a source of the observed increase. For example, when seasonal and short-term interannual variations in concentrations are neglected, the rise in atmospheric CO_2 is about 50% of anthropogenic emissions (Keeling *et al.*, 1989b) with the inter-hemispheric difference growing in parallel to the growth of fossil emissions (Keeling *et al.*, 1989a; Siegenthaler and Sarmiento, 1993; Figure 1.4).

1.2.2 Pre-1958 Atmospheric Measurements and CO_2-Ice Core Record over the Last Millennium

While several sets of relatively precise atmospheric measurements of CO_2 were carried out as early as the 1870s (e.g., Brown and Escombe, 1905), they did not

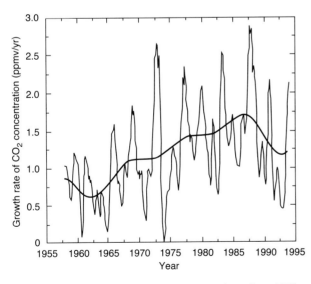

Figure 1.2: The growth rate of CO_2 concentrations since 1958 (from the Mauna Loa record). The high growth rates of the late 1980s, the extremely low growth rates of the early 1990s, and the recent increase in the growth rate are all evident. The smooth curve shows the same data but filtered to suppress any variations on time-scales less than approximately 10 years.

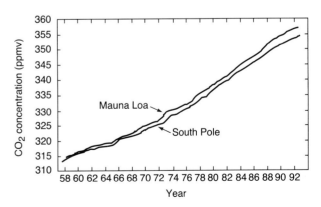

Figure 1.4: Trends in CO_2 concentration and the growing difference in concentration between the Northern and Southern Hemispheres.

allow assessment of concentration trends, as they were neither adequately calibrated nor temporally continuous.

Measurements of CO_2 concentration from air extracted from polar ice cores are presently the best means to extend the CO_2 record through the geologically recent past. The transformation of snow into ice traps air bubbles which are used to determine the CO_2 concentration. Providing certain conditions are met, which include no fracturing of the ice samples, absence of seasonal melting at the surface, no chemical alteration of the initial concentrations, and appropriate gas extraction methods, the ice record provides reliable information on past atmospheric CO_2 concentrations (Raynaud et al., 1993). It has been suggested that, under certain meteorological circumstances, the CO_2 data from Greenland ice cores may be contaminated, apparently influenced by varying levels of carbonate dust interacting with acid (Delmas, 1993) or organic matter deposition onto the ice sheet. Antarctic ice, however, has been uniformly acidic throughout the complete range of climate regimes of the last climatic cycle, and the available evidence suggests that these data are reliable throughout the entire record. Data from appropriate sites with unfractured ice are reliable to within ±3-5 ppmv (Raynaud et al., 1993). The recent ice core record is validated by comparison with direct atmospheric measurements (Neftel et al., 1985; Friedli et al., 1986; Keeling et al., 1989a).

Several high resolution Antarctic ice cores have recently become available in addition to the Siple core (Neftel et al., 1985; Friedli et al., 1986) for documenting both the "industrial era" CO_2 levels and the pre-industrial levels over the last millennium (Figure 1.5). The main results are:

- The ice core record can be used in combination with the direct atmospheric record to estimate, in conjunction with an oceanic model, the net changes in CO_2 flux between terrestrial ecosystems and the atmosphere (Siegenthaler and Oeschger, 1987).

- The pre-industrial level over the last 1000 years shows fluctuations up to 10 ppmv around an average value of 280 ppmv. The largest of these, which occurred roughly between AD 1200 and 1400, was small compared to the 75 ppmv increase during the industrial era (Barnola et al., in press, and Figure 1.5). Short-term climatic variability is believed to have caused the pre-industrial fluctuations through effects on oceanic and/or terrestrial ecosystems.

Finally, an important indicator of anthropogenically-induced atmospheric change is provided by the ^{14}C levels preserved in materials such as tree rings and corals. The ^{14}C concentration measured in tree rings decreased by about 2% during the period 1800 to 1950. This isotopic

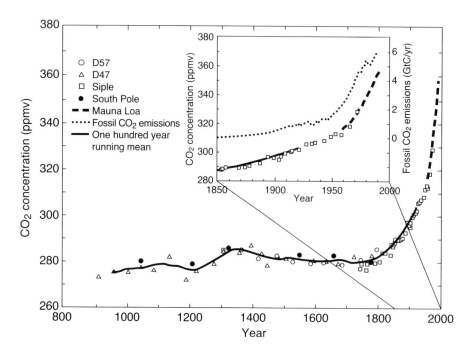

Figure 1.5: CO_2 concentrations over the past 1000 years from the recent ice core record and (since 1958) from the Mauna Loa measurement site. The inset shows the period from 1850 in more detail including CO_2 emissions from fossil fuel. Data sources: D47 and D57 (Barnola et al., in press); Siple (Neftel et al., 1985 and Friedli et al., 1986) and South Pole (Siegenthaler et al., 1988). The smooth curve is based on a 100yr running mean. All ice core measurements were taken in Antarctica.

decrease, known as the Suess effect (Suess, 1955), provides one of the most clear demonstrations that the increase in atmospheric CO_2 is due to fossil inputs.

1.2.3 The CO_2 Record over the Last Climatic Cycle

Although the magnitude and rate of climate changes observed in the palaeo-record covering the last climatic cycle may differ from those involved in any future greenhouse warming, these records provide an important perspective for recent and potential future changes. The glacial-interglacial amplitudes of temperature change are of similar order to the high estimate of equilibrium temperature shifts predicted for a doubling of CO_2 levels (Mitchell *et al.*, 1990), although the shifts of the past took thousands of years.

The close association between CO_2 and temperature changes during glacial-interglacial transitions was first revealed by data from the ice core record. Samples from Greenland and Antarctica representing the last glacial maximum (about 18,000 years before present) indicate that CO_2 concentrations at that time were 190-200 ppmv, i.e., about 70% of the pre-industrial level (Delmas *et al.*, 1980; Neftel *et al.*, 1982). Thus, an increase of about 80 ppmv occurred in parallel with the warming starting at the end of the glacial period, when the estimated glacial-interglacial rise of the mean surface temperature of the Earth rose by 4°C over about 10,000 years (Crowley and North, 1991). These discoveries have since been confirmed by detailed measurements of the Antarctic Byrd core for the 8,000 to 50,000 year BP period (Neftel *et al.*, 1988). Analyses of ice cores from Vostok, Antarctica, have provided new data on natural variations of CO_2 levels over the last 220,000 years (Barnola *et al.*, 1987; 1991; Jouzel *et al.*, 1993). The record shows a marked correlation between Antarctic temperature, as deduced from the isotopic composition of the ice, and the CO_2 profile (Figure 1.6).

Clear correlations between CO_2 and global mean temperature are evident in much of the glacial-interglacial palaeo-record. This relationship of CO_2 concentration and temperature may carry forward into the future, possibly causing a significant positive climatic feedback on CO_2 fluxes. Information about leads and lags between climatic variations and changes in radiatively active trace gas concentrations is contained in polar ice and deep sea sediment records (Raynaud and Siegenthaler, 1993). There is no evidence that CO_2 changes ever significantly (> 1 kyr) preceded the Antarctic temperature signal. In contrast, CO_2 changes clearly lag behind the Antarctic cooling at the end of the last interglacial. As temperature changes in the South generally preceded temperature changes in the North (CLIMAP Project Members, 1984), it cannot be assumed that CO_2 changes never led northern temperature changes. Comparison of atmospheric CO_2 concentration and continental ice volume suggests that CO_2 started to change ahead of any significant melting of continental ice. It is possible that CO_2 changes may have been caused by changes in climate, and that CO_2 and other trace gases acted to amplify palaeoclimatic changes.

Changes in climate on time-scales of decades to centuries have occurred in the past. The question remains whether these changes have been accompanied by changes in greenhouse trace gas concentrations. The Greenland ice cores (Johnsen *et al.*, 1992; Grootes *et al.*, 1993) show that during the last ice age and the last glacial-interglacial transition, there was a series of rapid (over decades to a century) and apparently large climatic changes in the North Atlantic region (~5 to 7°C in Central Greenland: Johnsen *et al.*, 1992; Dansgaard *et al.*, 1993). These changes may have been global in scale: the methane record suggests the potential for parallel changes in the tropics (Chappellaz *et al.*, 1993). Evidence for rapid climate oscillations during the last interglacial has also recently been reported (GRIP Project Members, 1993). However, because the details were not confirmed by a second core retrieved from the same area (Grootes *et al.*, 1993), the possibility that these features were caused by ice-flow perturbations cannot be discounted. The Dye 3 ice core

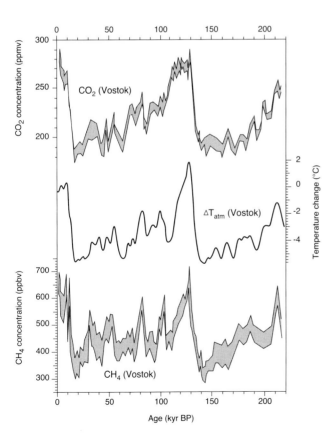

Figure 1.6: Temperature anomalies and methane and CO_2 concentrations over the past 220,000 years as derived from the ice core record at Vostok, Antarctica.

from Greenland indicates that CO_2 concentration shifts of ~50 ppmv occurred within less than 100 years during the last glacial period (Stauffer et al., 1984). These changes in CO_2 were paralleled by abrupt and drastic climatic events in this region. Such large and rapid CO_2 changes have not been identified in Antarctica, even after accounting for the lower time resolution of the Antarctic core (Neftel et al., 1988). It is possible that impurities have introduced artefacts into the Greenland CO_2 record (Delmas, 1993; Barnola et al., in press), and that these features do not represent real atmospheric CO_2 changes. Nevertheless, taken together with independent support from isotopic studies of mosses (White et al., 1994), rapid CO_2 events recorded during the past cannot be disregarded.

1.3 The Anthropogenic Carbon Budget

1.3.1 Introduction

The phrase "carbon budget" refers to the balance between sources and sinks of CO_2 in the atmosphere, expressed in terms of anthropogenic emissions and fluxes between the main reservoirs – the oceans, the atmosphere, the terrestrial carbon pool – and the build-up of CO_2 in the atmosphere. Because of the relative stability of atmospheric CO_2 concentrations over several thousand years prior to AD 1800, it is assumed that the net fluxes among carbon reservoirs were close to zero prior to anthropogenic disturbance. The data described in Section 1.2 provide the essential background for understanding the carbon budget's changes over time. Several approaches are used to quantify the components of this budget (other than atmospheric mass build-up, which can be measured directly), often in combination.

(a) direct determination of rates of change of the carbon content in atmospheric, oceanic, and terrestrial carbon pools, either by observations of local inventory changes or by local flux measurements, extrapolated globally;

(b) indirect assessment of the atmosphere-ocean and atmosphere-terrestrial biosphere fluxes by means of carbon cycle model simulations, either calibrated or partially validated using analogue tracers of CO_2, such as bomb radiocarbon or tritium, or with use of chlorofluorocarbons;

(c) interpretation of tracers or other substances that are coupled to the carbon cycle ($^{14}C/^{12}C$ and $^{13}C/^{12}C$ ratios and atmospheric oxygen).

The heterogeneity of some aspects of the oceanic and terrestrial carbon systems makes reliable extrapolation of flux measurements to the entire globe dependent on high resolution geographic information and accurate modelling of processes. Because of this, method (a) is used only for atmospheric carbon and for estimating effects of land use, and estimates of other carbon reservoirs are based primarily on methods (b) and (c).

1.3.2 Methods for Calculating the Carbon Budget

1.3.2.1 Classical approaches

The calculated ocean uptake rate, together with the estimated fossil emissions and the observed atmospheric inventory change, allows inference of the net terrestrial biospheric balance. Figure 1.7 shows the results of a time-series rate of change calculation for the atmospheric carbon mass, the oceanic component, fossil emissions, and the residual of the first three terms. In this calculation, the time history of fossil plus cement emissions was deduced from statistics (Keeling, 1973; Marland and Rotty, 1984; WEC, 1993; Andres et al., 1994), atmospheric accumulation was determined from the observational and ice core record (e.g., Barnola et al., 1991; Boden et al., 1991), and ocean uptake was modelled with the GFDL ocean general circulation model (Sarmiento et al., 1992). Ocean carbon uptake was determined by forcing the ocean chemistry with the time history of atmospheric concentrations to obtain uptake as a function of the non-linear chemistry of carbon dioxide in the ocean. This calculation illustrates the classic means by which the carbon budget is calculated. Figure 1.8 shows the balance of the terrestrial biosphere, now introducing the time history of land use effects together with a calculation of the inferred terrestrial sink over time. This calculation illustrates the classic estimation of a "missing sink", in which a residual sink arises because the sum of anthropogenic emissions (fossil, cement, changing land use) is greater than the sum of ocean uptake and atmospheric accumulation. Note that this approach is quite different from the estimation of sources and sinks from the spatial distribution of atmospheric concentrations and isotopic composition, discussed in Section 1.3.2.3 (Keeling et al., 1989b; Tans et al., 1990; Enting and Mansbridge, 1991).

1.3.2.2 New approaches: budget assessment based on observations of $^{13}C/^{12}C$ and O_2/N_2 ratios

Several promising approaches have recently been proposed to assess the current global carbon budget with less dependence on models. Included are observations of $^{13}C/^{12}C$ and O_2/N_2, both of which are strongly influenced by the anthropogenic perturbation of the carbon cycle.

The methods based on ^{13}C exploit the fact that the $^{13}C/^{12}C$ ratio in fossil fuels and terrestrial biomass is less than that in the atmosphere. There has been a decline in the $^{13}C/^{12}C$ ratio of atmospheric CO_2 over the last century (Keeling et al., 1989a). This atmospheric $^{13}C/^{12}C$ ratio

CO_2 and the Carbon Cycle

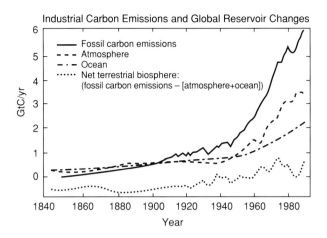

Figure 1.7: Fossil carbon emissions (based on statistics of fossil fuel and cement production), and representative calculations of global reservoir changes: atmosphere (deduced from direct observations and ice core measurements), ocean (calculated with the GFDL ocean carbon model), and net terrestrial biosphere (calculated as remaining imbalance). The calculation implies that the terrestrial biosphere represented a net source to the atmosphere prior to 1940 (negative values) and a net sink since about 1960.

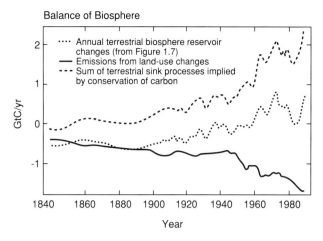

Figure 1.8: The carbon balance of the terrestrial biosphere. Annual terrestrial biosphere reservoir changes (from Figure 1.7), land use flux (plotted negative because it represents a loss of biospheric carbon) and the sum of the terrestrial sink processes (e.g., Northern Hemisphere regrowth, CO_2 and nitrogen fertilization, climate effects) as implied by conservation of carbon mass.

change propagates through the global carbon cycle, causing isotopic ratio changes in the ocean and the terrestrial carbon reservoirs.

Quay *et al.* (1992) proposed a method to determine the global budget from repeated measurements of vertical profiles of the $^{13}C/^{12}C$ ratio in oceanic dissolved inorganic carbon. In principle, these observations allow us to determine the rate of change of the oceanic $^{13}C/^{12}C$ ratio which, together with the observed changes in the atmospheric $^{13}C/^{12}C$ ratio, provide another constraint on the global carbon balance. The ratio permits discrimination between oceanic and terrestrial biospheric sinks because terrestrial uptake (photosynthesis) discriminates against the heavy isotope (^{13}C) much more strongly than does ocean uptake. A preliminary analysis based on a data set from seven stations in the Pacific Ocean (sampled in the early 1970s and again in 1990) yielded a mean oceanic sink of 2.1 ± 0.8 GtC/yr (Quay *et al.*, 1992).

Additional information for constraining the oceanic carbon budget may be provided by the $^{13}C/^{12}C$ isotopic disequilibrium between the air and the sea (Tans *et al.*, 1993). The disequilibrium reflects the isotopic adjustment of the ocean to the atmospheric perturbation and can be used to assess the atmospheric balances of ^{13}C and CO_2, thereby discriminating between oceanic and biospheric components. Tans *et al.* (1993) used this method to estimate the net air-sea flux of CO_2 for the period 1970 to 1990, which they found to be less than 0.4 GtC/yr, with unspecified error ranges.

Relative to achievable measurement precision, the anthropogenic perturbation has a greater effect on isotopic composition of dissolved organic carbon than on its concentration, because the former is not affected by chemical buffering reactions. The required analytical quality of the isotopic measurements is still high. The uncertainty range of the current oceanic sink, as estimated by the methods employing isotope measurements, may be reduced substantially if an extended measurement programme is vigorously pursued (Heimann and Maier-Reimer, submitted).

The trend in atmospheric oxygen, as revealed by measurements of the oxygen to nitrogen ratio, can also be used to assess the global carbon budget (Keeling and Shertz, 1992; Bender *et al.*, 1994). Oxygen in many respects is complementary to carbon. It is consumed during combustion and respiration and is released during biotic carbon uptake via photosynthesis. The crucial difference compared to carbon is that, owing to low solubility of O_2 in water, the magnitude of the oceanic oxygen pool is negligible and can be ignored in the global oxygen balance. Therefore, measurements of the temporal trend in atmospheric oxygen together with the known O:C stoichiometric relations during combustion, respiration, and photosynthesis, permit establishment of global oxygen and carbon balances, although interannual variations in marine photosynthesis and respiration may complicate the interpretation of O:C data (Keeling and Severinghaus, 1994). Data are limited, as measurements of the oxygen/nitrogen ratio have been reported for only two years. Within 10 years it is possible that an accurate, model-independent estimate of the oceanic carbon budget will be achieved by this method. The atmospheric oxygen trend during the last several decades can also be reconstructed from measurements of CO_2 and the O_2/N_2 ratio in glacial surface layers (firn) as a function of depth. Preliminary results indicate past ocean CO_2 uptake rates were large (> 3 GtC/yr), but associated uncertainty ranges are also large (Bender *et al.*, 1994).

Despite high uncertainty, these new approaches provide a different, model-independent means of assessing the global carbon budget. They provide only estimates of the oceanic uptake rate, however. The information gained can help reduce the uncertainty in the net terrestrial sink quantified by differencing, but to partition the uptake into land use fluxes and terrestrial sinks requires other approaches.

1.3.2.3 Constraints from spatial distributions

The different spatial patterns of sources and sinks of atmospheric CO_2 create gradients in the concentration that vary inversely with the strength of atmospheric transport. These spatial distributions of concentration can be interpreted quantitatively using models of atmospheric transport to match observed concentration distributions. This process is known as "inversion".

The problem of deducing sources and sinks of trace gases from observations of surface concentrations presents a considerable technical challenge. This is partly because the small-scale features of the spatial distribution of sources and sinks are blurred by atmospheric mixing. The "reconstruction" process amplifies the small-scale details, but the errors and uncertainties introduced along the way are similarly amplified. Hence, there is a trade-off between the spatial resolution sought and the accuracy of the estimates obtained. Consequently, only a small number of independent source/sink components can be reliably determined from the data, and, in general, only the largest spatial scales can be resolved (Enting, 1993). This explains, in part, the range of interpretations obtained in different studies. The uncertainty analysis of Enting *et al.* (1994a), indicates that such analyses cannot, on their own, estimate global totals of net fluxes more accurately than the classical approach to carbon budget analysis. Thus, at present, the constraints from the spatial distribution of CO_2 act mainly as consistency checks with budgets derived from analyses of individual budget components.

Synthesis inversions include additional information beyond atmospheric concentration and transport, and can suppress unwanted amplification of small-scale variations either by restricting the number of source components considered, or through the use of additional constraints such as incorporation of independent estimates of source strengths. Inversions using two-dimensional atmospheric transport models have been presented by Tans *et al.* (1989) and Enting and Mansbridge (1989; 1991). Three-dimensional modelling studies of source and sink distributions have been presented by Keeling *et al.* (1989b), Tans *et al.* (1990) and Enting *et al.* (1994a).

The main results that have been obtained from inverse calculations using atmospheric transport modelling are:

- *Northern Hemisphere source.* Inversion calculations reveal a strong Northern Hemisphere release, as expected from increased use of fossil fuels. This source is so strong that it tends to obscure other details of the source-sink distribution and is additional confirmation of the role of fossil fuel emissions in atmospheric concentration change.

- *Northern Hemisphere sink.* After subtracting the fossil source, there is a strong net sink in the Northern Hemisphere which may involve both marine and terrestrial components (e.g., Tans *et al.*, 1990). The partitioning remains controversial (Keeling *et al.*, 1989b; Ciais *et al.*, submitted), but recent isotopic analyses suggest an appreciable terrestrial component (Ciais *et al.*, submitted). A Northern Hemisphere sink, as estimated by inverse calculations, is consistent with results from terrestrial studies that suggest sinks are the result of changing land use (e.g., Dixon *et al.*, 1994) and nitrogen deposition (largely confined to the industrialised Northern Hemisphere) (e.g., Schindler and Bayley, 1993).

- *Indirect evidence of the existence of a tropical biotic sink.* The net equatorial source from oceanic outgassing and changes in tropical land use is smaller than expected based on other estimates (see Sections 1.3.3.3 and 1.3.3.4), consistent with a role for nutrient and CO_2 fertilisation and forest regrowth.

- *Southern Ocean source.* Most studies indicate a source (probably oceanic) at high southern latitudes.

- *Southern Hemisphere sink.* The net Southern Hemisphere sink (which is presumably oceanic, given the small proportion of Southern Hemisphere land) is weaker than expected in comparison with northern ocean sinks.

1.3.3 Sources and Sinks of Anthropogenic CO_2

1.3.3.1 Fossil carbon emissions

The dominant anthropogenic CO_2 source is that generated by the use of fossil fuels (coal, oil, natural gas, etc.) and production of cement. The total emissions of CO_2 from the use of fossil carbon can be estimated based on documented statistics of fossil fuel and cement production (Keeling, 1973; Marland and Rotty, 1984; WEC, 1993; Andres *et al.*, 1994). Average (1980 to 1989) fossil emissions were estimated to be 5.5 GtC/yr (Andres *et al.*, 1994). During 1991, the reported emissions totalled 6.2 GtC (Andres *et al.*, 1994). The cumulative input since the beginning of the industrial revolution (1751 to 1991) is estimated to be approximately 230 GtC (Andres *et al.*, 1994). Uncertainties associated with these estimates are less than 10% for the decade of the 1980s (at the 90% confidence level) based on the methods presented in Marland and Rotty (1984).

1.3.3.2 Atmospheric increase

The globally averaged CO_2 concentration, as determined through analysis of NOAA/CMDL data (Boden *et al.*, 1991; Conway *et al.*, 1994), increased by 1.53 ± 0.1 ppmv/yr over the period 1980 to 1989. This corresponds to an annual average rate of change in atmospheric carbon of 3.2 ± 0.2 GtC/yr. Other carbon-containing compounds like methane, carbon monoxide, and larger hydrocarbons, contain ~1% of the carbon stored in the atmosphere and can therefore be neglected in the atmospheric carbon budget.

1.3.3.3 Ocean exchanges

The ocean contains more than 50 times as much carbon as the atmosphere. Over 95% of the oceanic carbon is in the form of inorganic dissolved carbon (bicarbonate and carbonate ions); the remainder is comprised of various forms of organic carbon (living organic matter, particulate and dissolved organic carbon) (Druffel *et al.*, 1992).

The role of the oceans in the global carbon cycle is twofold: first, it represents a passive reservoir which absorbs excess atmospheric CO_2. It is this role that is discussed in this section. Second, changes in the physical state of the ocean (temperature, circulation) and the marine biota may affect the rate of air-sea exchange, and thus future atmospheric CO_2. This second role, subsumed under "feedbacks", is addressed in Section 1.4.3.

The oceanic uptake of excess CO_2 proceeds by (1) transfer of the CO_2 gas through the air-sea interface, (2) chemical interactions with the oceanic dissolved inorganic carbon, and (3) transport into the thermocline and deep waters by means of water mass transport and mixing processes. Though there are large geographical and seasonal variations of the surface ocean partial pressure of CO_2, averaged globally and annually, the surface water value is close to that for equilibrium with the atmosphere. Therefore, processes (2) and (3) are the main factors limiting the capacity of the ocean to serve as a sink on decadal and centennial time-scales. The chemical buffering reactions between dissolved CO_2 and the HCO_3^- and CO_3^{2-} ions reduce the rate of oceanic CO_2 uptake. At equilibrium, an atmospheric increase in CO_2 concentration of 10% is associated with an oceanic increase of dissolved inorganic carbon of merely 1%. This potential uptake occurs only in those parts of the ocean that are mixed with surface waters on decadal time-scales. Therefore, on time-scales of decades to centuries, the ocean is not as large a sink for excess CO_2 as it might seem from comparison of the relative sizes of the main carbon reservoirs (Figure 1.1). While carbon chemistry is known in sufficient detail to perform accurate calculations, oceanic transport and mixing processes remain the primary uncertainties in the determination of the oceanic uptake of excess CO_2.

The marine biota, if in steady state, are believed to play a minor role, if any at all, in the uptake of excess anthropogenic CO_2. The marine biota, however, play a crucial role in maintaining the steady-state level of atmospheric CO_2. About three-quarters of the vertical gradient in dissolved inorganic carbon is generated by the export of newly produced carbon from the surface ocean and its regeneration at depth (a process referred to as "biological pump"). In the open ocean, however, this process is believed to be limited by the availability of nutrients, light, or by phytoplankton population control via grazing, and not by the abundance of carbon (Falkowski and Wilson, 1992). Therefore, a direct effect of increased dissolved inorganic carbon (less than 2.5% since pre-industrial times) on carbon fixation and export is unlikely, although a recent study by Riebesell *et al.* (1993) suggested that under particular conditions the rate of photosynthesis and hence phytoplankton growth might indeed be limited by the availability of CO_2 as a dissolved gas. The global significance of this effect, however, remains to be assessed. (See Section 1.4.3 for a discussion of the potential indirect effects of increased dissolved inorganic carbon.)

Flux of carbon from the terrestrial biosphere to the oceans takes place via river transport. Global river discharge of carbon in organic and inorganic forms may be ~1.2-1.4 GtC/yr (Schlesinger and Melack, 1981; Degens *et al.*, 1991; Maybeck, 1993). A substantial fraction of this transport (up to 0.8 GtC/yr), however, reflects the natural geochemical cycling of carbon and thus does not affect the global budget of the anthropogenic CO_2 perturbation (Sarmiento and Sundquist, 1992). Furthermore, the anthropogenically induced river carbon fluxes reflect, to a large extent, increased soil erosion and not a removal of excess atmospheric CO_2.

The role of coastal seas in the global carbon budget is poorly understood. Up to 30% of total ocean productivity is attributed to marine productivity in the coastal seas, which comprise only ~8% of the oceanic surface area.

Here, discharge of excess nutrients by rivers might have significantly stimulated carbon fixation (up to 0.5-1.0 GtC/yr). At present, however, it is not known how much of this excess organic carbon is simply reoxidised, and how much is permanently sequestered by export to the deep ocean, or in sediments on the shelves and shallow seas. Because of the limited surface area, a burial rate significantly exceeding 0.5 GtC/yr is not very likely, as it would require all coastal seas to be under-saturated in partial pressure of CO_2 by more than 50 μatm on annual average, in order to supply the carbon from the atmosphere. Such under-saturations have been documented, e.g., in the North Sea (Kempe and Pegler, 1991), but these measurements are unlikely to be representative of all coastal oceans. Based on the above considerations, the role of the coastal ocean is judged most likely to be small, but, at present, cannot be accurately assessed and so is neglected in the budget presented in this chapter (Table 1.3).

Exchanges of carbon between the atmosphere and the oceans (net air-sea fluxes) can be deduced from measurements of the partial pressure difference of CO_2 between the air and surface waters. This calculation also requires knowledge of the local gas-exchange coefficient, which is a relatively poorly known quantity (Watson, 1993). Furthermore, while the globally and seasonally averaged partial pressure of surface waters is close to the value for equilibrium with the atmosphere, large geographical and seasonal variations exist, induced both by physical processes (upwelling, vertical mixing, sea surface temperature fluctuations) and by the activity of the marine biota. Representative estimates of the seasonally and regionally averaged net air-sea carbon transfer thus necessitates sampling with high spatial and temporal resolution (Garcon et al., 1992). Estimates of the regional net air-sea carbon fluxes have been obtained, albeit with considerable error margins. However, the global oceanic carbon balance is more difficult to deduce by this method, as it represents a small residual computed from summing up the relatively large emissions from super-saturated and uptake in under-saturated regions (Tans et al., 1990; Takahashi et al., 1993; Wong et al., 1993; Fung and Takahashi, 1994). The World Ocean Circulation Experiment (WOCE) survey has nevertheless demonstrated the feasibility of direct measurement programmes, and the importance of measurements as a cross-check for other approaches should not be underrated.

The oceanic contribution to the global carbon budget can also be assessed by direct observations of changes in the oceanic carbon content. This approach also suffers from the problem of determining a very small signal against large spatial and temporal background variability. Model estimates of the rate of change induced by the anthropogenic perturbation are of the order of 1 part in 2000/yr. Therefore, on a 10-year time-scale, variations in dissolved inorganic carbon would have to be measured with an accuracy of better than 1% in order to determine the carbon balance in a particular oceanic region. While such accuracy can be obtained, a substantial sampling effort is required (Keeling, 1993). Determination of the global oceanic budget by this approach does not appear feasible in the near future. However, repeated observational surveys might reveal regional carbon inventory changes, and thus might provide a cross-check for other approaches.

Model results

The present-day oceanic uptake of excess CO_2 is estimated using ocean carbon model simulation experiments, with observed atmospheric CO_2 concentration as a prescribed boundary condition. It is only necessary to model the transport and mixing of the excess carbon perturbation if the marine biota are assumed to remain on an annual average in a steady state, as excess CO_2 will disperse within the ocean like a passive tracer, independent of the background natural distribution of dissolved inorganic carbon. In particular, the perturbation excludes a natural cycle of carbon transported by rivers, outgassed by the oceans and then taken up by the terrestrial biota.

Until recently, most models of the oceanic uptake of anthropogenic CO_2 consisted of a series of well-mixed or diffusive reservoirs ("boxes") representing the major oceanic water masses, connected by exchange of water (Oeschger et al., 1975). Global transport characteristics of these models were obtained from simulation studies of the oceanic penetration of bomb radiocarbon (^{14}C) validated by comparison with observations (Broecker et al., 1985). Bomb radiocarbon, however, does not accurately track the oceanic invasion of anthropogenic carbon, because oceanic concentration changes of carbon isotopes depend on the full time history of those inputs and ^{14}C entered the ocean with an atmospheric history different from that of natural CO_2. Furthermore, CO_2 uptake depends on chemical interactions with dissolved inorganic C, while uptake of ^{14}C does not. Nevertheless, bomb radiocarbon provides a powerful constraint on ocean carbon models, a constraint which may be supplemented by analyses of other steady state and transient tracers, such as halocarbons, tritium from nuclear weapon testing, and possibly ^{13}C.

Recently, three-dimensional oceanic general circulation models (OGCMs) have been used for modelling CO_2 uptake by the oceans (Maier-Reimer and Hasselmann, 1987; Sarmiento et al., 1992; Orr, 1993). These models calculate oceanic circulation on the basis of the physics of fluid dynamics, and the few adjustable model parameters are primarily tuned to reproduce the relatively well-known large-scale patterns of ocean temperature and salinity. It is known that the OGCMs used in published global carbon cycle studies show significant and similar deficiencies (e.g., too weak a surface circulation, inaccurate deep convection, absence of high resolution features). Simulation of transient tracers, bomb radiocarbon in particular, provides an important validation test.

The average of modelled rates of oceanic carbon uptake is 2.0 GtC/yr over the decade 1980 to 1989 (Orr, 1993; Siegenthaler and Sarmiento, 1993). This value is corroborated by model experiments projecting future concentrations of atmospheric CO_2 (see Section 1.5, and Enting *et al.*, 1994b) and by a simulation with a newly-developed two-dimensional global ocean model (Table 1.1; Stocker *et al.*, 1994). The spread of the modelled uptake rates corresponds to a statistical uncertainty of only about ± 0.5 GtC/yr (at the 90% confidence level). However, recent assessments of the global radiocarbon balance (Broecker and Peng, 1994; Hesshaimer *et al.*, 1994) suggest that the oceanic carbon uptake estimates of box models tuned by bomb radiocarbon might have to be revised downwards by up to 25%. Furthermore, the spread of the OGCM results listed in Table 1.1 might be fortuitously small in view of the similar deficiencies in these models. Based on these considerations we do not change the estimate of the uncertainty of the ocean uptake from the value of ± 0.8 GtC/yr as given by IPCC in 1990 (Watson *et al.*, 1990) and in 1992 (Watson *et al.*, 1992).

1.3.3.4 Terrestrial exchanges

In previous assessments (IPCC, 1990; 1992), the remaining intact terrestrial biosphere has often been assumed to be a significant sink for carbon dioxide, balancing or exceeding emissions derived from changing land use. The calculated imbalance, which has been a persistent feature of global carbon cycle calculations (Broecker *et al.*, 1979), arises from the difference between measured atmospheric changes, statistically derived fossil fuel and cement emissions, modelled ocean uptake, and estimated emissions from changing land use. In this assessment, part of the previously calculated imbalance is accounted for by effects of changing land use in the middle and high latitudes (0.5 ± 0.5 GtC/yr: Table 1.3). Several processes may contribute to increased terrestrial carbon storage (Table 1.2), but the problem of detecting these increases is troublesome.

Difficulty in quantifying the role of the terrestrial biosphere in the global carbon cycle arises because of the complex biology underlying carbon storage, the great heterogeneity of vegetation and soils, and the effects of human land use and land management. We consider the following issues:

- *Deforestation and change in land use.* CO_2 is emitted to the atmosphere as a result of land use changes such as biomass burning and forest harvest through the oxidation of vegetation and soil carbon. These "emissions from changing land use" are currently largest in the tropics, but prior to the 1950s, the middle latitudes were a larger source than were the tropics.

- *Forest regrowth.* Carbon is absorbed by regrowing forests following harvest. This absorption is included as a factor in the calculation of net emissions from disturbed regions. In the tropics, emissions are

Table 1.1: Excess CO_2 uptake rates (average 1980 to 1989) calculated by various ocean carbon cycle models.

Model	Ocean uptake (GtC/yr)	Reference
Bomb radiocarbon-based box models		
Box-diffusion model	2.32	Siegenthaler & Sarmiento (1993)
HILDA model ("Bern model")	2.15	Siegenthaler & Joos (1992)
Three-dimensional ocean general circulation models		
Hamburg Ocean Carbon Cycle Model (HAMOCC-3)	1.47	Maier-Reimer (1993)
GFDL Ocean General Circulation Model	1.81	Sarmiento *et al.* (1992)
LODYC Ocean General Circulation Model	2.10	Orr (1993)
Two-dimensional ocean circulation models		
	2.1	Stocker *et al.* (1994)
AVERAGE of all models	**2.0**	

Table 1.2: Processes leading to increased terrestrial storage and their magnitudes (GtC/yr): average for the 1980s.

Processes	
Mid-/high latitude forest regrowth	0.5 ± 0.5
CO_2 fertilisation	0.5-2.0
Nitrogen deposition	0.2-1.0
Climatic effects	0-1.0

thought to exceed the uptake by secondary growth following forest harvest, but uptake from regrowth in the middle and high latitudes apparently results in a net sink for atmospheric CO_2 (see below).

- *Fertilisation by carbon dioxide.* As atmospheric CO_2 increases, plants increase their uptake of carbon, potentially increasing carbon storage in the terrestrial biosphere. This mechanism has often been hypothesised to account for the calculated imbalance in the global carbon budget (the "missing sink"). There is strong physiological evidence for photosynthetic increase with CO_2 fertilisation (e.g., Woodward, 1992), but interspecific differences in the magnitude of effects on photosynthesis and growth range from marked enhancement to negligible and even negative responses. Considerable uncertainty exists as to how to translate laboratory and field results into a global estimate of changing biospheric carbon storage, as ecosystem feedbacks and constraints are numerous (see Section 1.4.2).

- *Nitrogen fertilisation.* Most terrestrial ecosystems are limited by nitrogen, as evidenced by increased carbon storage after nitrogen fertilisation. Additions of nitrogen to the terrestrial biosphere through both intentional fertilisation of agricultural land and deposition of nitrogen arising from fossil fuel combustion and other anthropogenic processes can result in increased terrestrial storage of carbon.

- *Climate.* The processes of terrestrial carbon uptake (photosynthesis) and release (respiration of vegetation and soils) are influenced by climate. Decadal time-scale variations in climate may have caused natural changes in carbon storage by the terrestrial biosphere, acting in conjunction with or even counter to the anthropogenic effects described above. This effect on storage is separate from the potential future changes in carbon storage which might arise from greenhouse gas-induced changes of climate.

- *Interactions.* The processes mentioned above are not independent, and their magnitudes are not necessarily additive. For example, much of the anthropogenic nitrogen has been deposited by the atmosphere on regrowing forests of the middle and high latitudes, thereby potentially enhancing carbon storage in these regions. Recent measurements of mid-latitude and high latitude forest regrowth may therefore reflect the combined effects of nitrogen deposition, elevated CO_2 concentrations, and climate variability. Also, CO_2 fertilisation can affect the nitrogen cycle and vice versa (see Section 1.4.2). It is possible that nitrogen deposition will enhance the effectiveness of carbon storage by offsetting the increased plant demand for nitrogen caused by increased concentrations of CO_2. Interactive effects with air pollutants such as tropospheric ozone may also be important.

1.3.3.4.1 Emissions from changing land use

Deforestation and other changes in land use (including land management) cause significant exchanges of CO_2 between the land and atmosphere. Changes in land use from 1850 to 1990 resulted in cumulative emissions of 122 ± 40 GtC (Houghton, 1994a). From the last century through the 1940s, expansion of agriculture and forestry in the middle and high latitudes dominated carbon emissions from the terrestrial biosphere (Houghton and Skole, 1990). Since then, conversion of temperate forests for agricultural use has diminished and forests are regrowing in previously logged forests and on abandoned agricultural lands (Melillo *et al.*, 1988; Birdsey *et al.*, 1993; Dixon *et al.*, 1994). Emissions from the tropics have generally been increasing since the 1950s. Estimates for 1980 range from 0.4 to 2.5 GtC (Houghton *et al.*, 1987; Detwiler and Hall, 1988; Watson *et al.*, 1990).

While uncertainties in the estimated rate of tropical deforestation are large (FAO, 1993), estimation of deforestation rates and area of regrowing forests have improved recently through the use of remote sensing. For example, satellite imagery has reduced estimated rates of deforestation in the Brazilian Amazon (INPE, 1992; Skole and Tucker, 1993). The lack of satellite analyses for other tropical areas may mean that the flux from the tropics as a whole has been overestimated.

While there is considerable uncertainty in estimated emissions from changing land use, the most recent compilations (Dixon *et al.*, 1994; Houghton, 1994b) agree that tropical emissions averaged 1.6 ± 1.0 GtC/yr in the 1980s. These analyses take into account the changing estimates of deforestation rates (INPE, 1992; Skole and Tucker, 1993) as well as other new data. The uncertainty estimate reflects the controversy over quantity and distribution of biomass (e.g., Brown and Lugo, 1992; Fearnside, 1992) as well as the errors associated within and between individual studies. At this writing, considerable work is in progress; it is possible that the estimate of emissions from changing land use in the tropics may be modified in the near future. Of particular concern is the currently poor evaluation of effects of changing land use in the drier, open canopy forests of the seasonal tropics (FAO, 1993), and poor knowledge of carbon accumulation in regrowing forests.

1.3.3.4.2 Uptake of CO_2 by changing land use

Several recent analyses suggest a sink of carbon in regrowing Northern Hemisphere forests. Estimates for the magnitude of this term range from essentially zero to 0.74 GtC/yr (Melillo et al., 1988; Houghton, 1993; Dixon et al., 1994). It is difficult to assess the magnitude of this sink for two reasons. The first is the wide disparity in published regional estimates (e.g., a Russian sink of 0.01-0.06 GtC/yr (Melillo et al., 1988; Krankina and Dixon, 1994) versus 0.3-0.5 GtC/yr (Dixon et al., 1994)).

A more fundamental problem is the confusion of carbon uptake via forest growth versus forest *regrowth*. As a basis for understanding how fluxes might change in the future, we seek in this report to separate increases in forest carbon storage that result from the regrowth of previously harvested forests versus storage due to processes such as CO_2 fertilisation, nitrogen deposition, and changes in fire frequency, management practices, or climate, all of which can affect forest growth. Inventory-based estimates of forest carbon storage are made by multiplying the age distribution of forest stands (rates of carbon accumulation generally decrease as stands age) by age-specific rates of carbon accumulation, estimated from observations. However, as forest carbon accumulation has been measured during the period of changing CO_2 and nitrogen deposition, and possibly climate, forest inventories cannot distinguish the amount of accumulation attributable to individual processes.

The confusion of mechanisms is not important when inventory data are used to corroborate the results of inverse modelling (e.g., Tans et al., 1990; Enting and Mansbridge, 1991; Ciais et al., in press), but separation of carbon storage mechanisms is essential for projections of future changes in carbon storage. Estimates of carbon storage due to nitrogen deposition or CO_2 fertilisation (Peterson and Melillo, 1985; Schindler and Bayley, 1993; Gifford, 1994) cannot be added to inventory-based land use sinks (Dixon et al., 1994; Houghton, 1994b). Moreover, inventory data are ill-suited for the task of projecting future carbon uptake because the inventory-based method does not distinguish between demographic effects such as forest harvest, agricultural abandonment, and natural disturbance such as fire and wind (Kolchugina and Vinson, 1993). The problem of quantifying the contribution of individual mechanisms to total carbon storage is even greater in the tropics, where few demographic data exist. Thus, it is inappropriate to rely entirely on inventory-based measurements for initialisation of terrestrial carbon models.

In addition to changes on forested lands, the decline in soil carbon storage in agricultural lands may have been reversed in the middle latitudes. Best management practices may even lead to storage increases in the future (Metherell, 1992). While the world-wide extent of management changes and their effect on carbon balance is poorly known, a recent assessment of carbon storage in USA croplands suggests possible increases over the coming few decades of 0.02-0.05 GtC/yr; the authors note that similar changes are likely to have occurred over the past few decades (Donigian et al., 1994). In terms of the global C budget, increases in soil storage are small (Schlesinger, 1990), although they may be significant at the regional, or even national, scale.

1.3.3.4.3 Other terrestrial sink processes

While little has been done to increase understanding of potential sinks in tropical forests, analyses based on models of forest growth, age distribution data, and the age-growth relationship suggest carbon accumulation in Northern Hemisphere forests may be as high as ~1 GtC/yr (Dixon et al., 1994). Analyses based on observed atmospheric CO_2 and $^{13}CO_2$ suggest a strong Northern Hemisphere terrestrial sink. While various pieces of evidence support a substantial terrestrial sink in the Northern Hemisphere, direct observations to confirm the hypothesis and to establish the processes responsible for increasing carbon storage are lacking. For example, tree ring studies of contemporary and pre-industrial forest growth rates are contradictory. These studies have revealed enhanced growth in subalpine conifers (LaMarche et al., 1984), no growth enhancement beyond that explained by climatic variability (Graumlich, 1991; D'Arrigo and Jacoby, 1993), and both positive and negative changes in growth rate dependent upon species and location (Briffa et al., 1990). Models and observational data to confirm tropical sinks are even more deficient (e.g., Brown and Lugo, 1984; 1992; Fearnside, 1992; Skole and Tucker, 1993; Dixon et al., 1994). Improved observations for identifying terrestrial sinks is necessary for the improvement of models capable of projecting future states of the carbon cycle, and also for quantifying regional contributions to terrestrial sources and sinks. Processes that may contribute to a terrestrial sink are explored below (see Table 1.2).

CO_2 fertilisation

Experimental studies of agricultural and wild plant species have shown growth responses of typically 20-40% higher growth under doubled CO_2 conditions (ranging from negligible or negative responses in some wild plants to responses of 100% in some crop species (Körner and Arnone, 1992; Rochefort and Bazzaz, 1992; Coleman et al., 1993; Idso and Kimball, 1993; Owensby et al., 1993; Polley et al., 1993; Idso and Idso, 1994)). The majority of these studies were short-term and were conducted with potted plants (Idso and Idso, 1994). The effect of increased growth under elevated CO_2 conditions, known as the CO_2 fertilisation effect, is often assumed to be the primary mechanism underlying the imbalance in the global carbon budget. This assumption implies a terrestrial sink that increases as CO_2 increases, acting as a strong negative feedback. However, while the potential effect of CO_2 in experimental and some field studies is relatively strong (e.g., Drake, 1992; Idso and Kimball, 1993; see also review by Idso and Idso, 1994), natural ecosystems may be less responsive to increased levels of atmospheric CO_2. Evidence that the effects of CO_2 on long-term carbon storage may be less than is suggested by short-term pot studies of photosynthesis or plant growth includes:

(1) Field studies showing reduced responses over time (Oechel et al., 1993) and zero, small, or statistically insignificant responses (Norby et al., 1992; Jenkinson et al., 1994);

(2) Evidence that plants with low intrinsic growth rates, a common trait in native plants, are less responsive to CO_2 increases than are rapidly growing plants, such as most crop species (Poorter, 1993);

(3) Results of model simulations which suggest that increases in carbon storage eventually become nutrient-limited (Comins and McMurtrie, 1993; Melillo et al., 1994; reviewed in Schimel, 1995). (Long-term nutrient limitation is different from the short-term nutrient limitation of photosynthesis and plant growth observed in experimental studies; it reflects the need for nutrients in addition to carbon to increase organic matter production.)

There is also evidence that while water- or temperature-stressed plants are more responsive to CO_2 increase than are unstressed plants (because CO_2 increases water-use efficiency (Polley et al., 1993)), nitrogen-limited plants are less sensitive to CO_2 level (Bazzaz and Fajer, 1992; Comins and McMurtrie, 1993; Díaz et al., 1993; Melillo et al., 1993; Ojima et al., 1993; Idso and Idso, 1994). The reduction in response of short-term photosynthesis and plant growth to CO_2 caused by nitrogen limitation is highly variable and may on the whole be small (Idso and Idso, 1994), but as all organic matter contains nitrogen, carbon storage increases should eventually become limited by the stoichiometric relationships of carbon and other nutrients in organic matter (Comins and McMurtrie, 1993; Díaz et al., 1993; Schimel, 1995).

While current models assume that nitrogen inputs are not affected by CO_2 fertilisation, this may not be true (Thomas et al., 1991; Gifford, 1993; Idso and Idso, 1994). CO_2 fertilisation could stimulate biological nitrogen fixation, because nitrogen-fixing organisms have high energy (organic carbon) requirements. If nitrogen inputs are stimulated by increasing CO_2, acclimation of carbon storage to higher CO_2 levels could be temporarily offset. However, evidence from native ecosystems suggests that other nutrients, such as phosphorus, may be more limiting to nitrogen fixation than is energy (Cole and Heil, 1981; Eisele et al., 1989; Vitousek and Howarth, 1991).

As a sensitivity analysis, the effects of CO_2 enrichment was varied in one model over the plausible range of values (by the equivalent of 10, 25 and 40% increases in plant growth at doubled CO_2). The results indicate terrestrial carbon storage rates (increased net ecosystem production: NEP) due to CO_2 fertilisation were 0.5, 2.0 and 4.0 GtC/yr during the 1980s (Gifford, 1993). A modelling experiment by Rotmans and den Elzen (1993) indicated the strength of CO_2 fertilisation might be ~1.2 GtC/yr. Overall, it is likely that CO_2 fertilisation plays a role in the current terrestrial carbon budget, and may have amounted to storage of 0.5 to 2.0 GtC/yr during the 1980s.

Nitrogen fertilisation

Many terrestrial ecosystems are nitrogen limited: added fertiliser will produce a growth response and additional carbon storage (e.g., Vitousek and Howarth, 1991; Schimel, 1995). Nitrogen deposition from fertilisers and oxides of nitrogen released from the burning of fossil fuel during the 1980s is estimated to amount to a global total, but spatially concentrated, 0.05-0.06 GtN/yr (Peterson and Melillo, 1985; Duce et al., 1991). The carbon sequestration which results from this added nitrogen is estimated to be of the order 0.2-1.0 GtC/yr (Peterson and Melillo, 1985; Schindler and Bayley, 1993; Schlesinger, 1993), depending on assumptions about the proportion of nitrogen that remains in ecosystems. Estimates significantly higher than 1 GtC/yr are unrealistic because they assume that all of the N would be stored in forms with high carbon-to-nitrogen ratios, while much atmospheric nitrogen is in reality deposited on grasslands and agricultural lands where storage occurs in soils with low average carbon to nitrogen ratios. Uptake of carbon due to long-term increases in nitrogen deposition could increase as nitrogen pollution increases; but possibly to a threshold, after which additional nitrogen may result in ecosystem degradation (e.g., Aber *et al.*, 1989; Schulze *et al.*, 1989). Because most deposition of anthropogenic nitrogen occurs in the middle latitudes, some of the effect of added nitrogen may already be accounted for in measurement-based estimates of mid-latitude carbon accumulation (discussed above). However, existing analyses do not allow identification of the fraction of measured forest growth that is due to nitrogen addition or effects of CO_2.

Climate effects

Climate affects carbon storage in terrestrial ecosystems because temperature, moisture, and radiation influence both ecosystem carbon gain (photosynthesis) and loss (respiration) (Houghton and Woodwell, 1989; Schimel *et al.*, 1994). While a number of compensatory processes are possible, warming is thought to reduce carbon storage by increasing respiration, especially of soils (Houghton and Woodwell, 1989; Shaver *et al.*, 1992; Townsend *et al.*, 1992; Oechel *et al.*, 1993; Schimel *et al.*, 1994). Conversely, cooling would be expected to increase carbon storage. In nutrient-limited forests, however, warming may increase carbon storage by "mineralising" soil organic nutrients, which are generally stored in nutrient-rich, or low carbon-to-nutrient ratio forms, thereby allowing increased uptake by trees, which store nutrients in high carbon-to-nutrient forms (Shaver *et al.*, 1992). This is not true in nutrient-limited tundra ecosystems, where recent warming has resulted in a local CO_2 source (Oechel *et al.*, 1993). Increases in precipitation will generally increase carbon storage by increasing plant growth, although in some ecosystems compensatory increases in soil decomposition may reduce or offset this effect (Ojima *et al.*, 1993). While these effects have been discussed as components of future responses of ecosystems to climate change, climate variations during the past century may have influenced the terrestrial carbon budget. In a provocative paper, Dai and Fung (1993) suggested that climate variations over the past decades could have resulted in a substantial sink. Ciais *et al.* (in press) suggested that cooling arising from the effects of Mt. Pinatubo may have increased terrestrial carbon storage and contributed to the observed reduction in the atmospheric growth rate during the 1991 to 1992 period. Palaeoclimate modelling studies likewise suggest major changes in terrestrial carbon storage with climate (Prentice and Fung, 1990; Friedlingstein *et al.*, 1992). Modelling and observations of ecosystem responses to climate and climate anomalies will be important tools for validating predictions of future changes; meanwhile, investigations of the effects of climate on carbon storage remain suggestive rather than definitive.

1.3.4 Budget Summary

In Table 1.3 we present an estimated budget of carbon perturbations for the 1980s, shown as annual average values.

The budgetary inclusion of a Northern Hemisphere sink in forest regrowth reduces the unaccounted-for sink compared with earlier budgets (IPCC, 1990; 1992) by assigning a portion of this sink to forest regrowth. The remaining imbalance in this budget implies additional net terrestrial sinks of 1.4 ± 1.5 GtC/yr.

1.4 The Influence of Climate and Other Feedbacks on the Carbon Cycle

1.4.1 Introduction

The description of the carbon cycle in the previous sections addressed the budget of anthropogenic CO_2, without emphasising the possibility of more complex interactions within the system. There are, however, a number of processes that can produce feedback loops. One important distinction (see Enting, 1994) is between carbon cycle feedbacks and CO_2-climate feedbacks. The only direct carbon cycle feedback is the CO_2 fertilisation process described in Section 1.3.3.4.3. In contrast, CO_2-climate feedbacks involve the effect of climate change (potentially induced by CO_2 concentration changes) on the components of the carbon cycle. For terrestrial components, the most important effects are likely to be

Table 1.3: Average annual budget of CO_2 perturbations for 1980 to 1989. Fluxes and reservoir changes of carbon are expressed in GtC/yr, error limits correspond to an estimated 90% confidence interval.

CO_2 sources	
(1) Emissions from fossil fuel combustion and cement production	5.5 ± 0.5
(2) Net emissions from changes in tropical land use	1.6 ± 1.0
(3) Total anthropogenic emissions (1)+(2)	7.1 ± 1.1
Partitioning among reservoirs	
(4) Storage in the atmosphere	3.2 ± 0.2
(5) Oceanic uptake	2.0 ± 0.8
(6) Uptake by Northern Hemisphere forest regrowth	0.5 ± 0.5*
(7) Additional terrestrial sinks (CO_2 fertilisation, nitrogen fertilisation, climatic effects) [(1) + (2)] - [(4) + (5) + (6)]	1.4 ± 1.5

* from Table 1.2

those involving temperature, precipitation, and radiation changes (through changes in cloudiness) on net primary production and decomposition (including effects resulting from changes in species composition). For marine systems, the primary effects to be expected arise through climatic influences on ocean circulation and chemistry. Such changes would affect the physical and biological aspects of carbon distribution in the oceans including physical fluxes of inorganic carbon within the ocean and changes in nutrient cycling (Manabe and Stouffer, 1993). The feedbacks may also include changes in the species composition of the ocean biota, which determines the location and magnitude of oceanic CO_2 uptake.

Factors with a strong influence on the global carbon cycle that are similar to and/or modify feedback loops also exist. Among such factors are the effects of increased UV on terrestrial and marine ecosystems and the anthropogenic toxification and eutrophication of these ecosystems. While significant effects of possible changes in mid-latitude and high latitude UV radiation, tropospheric ozone and other pollutants have been documented in experimental studies, there is little basis for credible global extrapolation of these studies (Chameides et al., 1994). They will not be discussed further, but should be addressed in subsequent assessments as more data become available.

1.4.2 Feedbacks to Terrestrial Carbon Storage

The responses of terrestrial carbon to climate are complex, with rates of biological activity generally increasing with warmer temperatures and increasing moisture. Because photosynthesis and plant growth increase with warmer temperatures, longer growing seasons and more available water, storage of carbon in living vegetation generally increases as well (Melillo et al., 1993). Storage of carbon in soils generally increases along a gradient from low to high latitudes, reflecting slower decomposition of dead plant material in colder environments (Post et al., 1985; Schimel et al., 1994). Flooded soils, where oxygen becomes depleted, have extremely low rates of decomposition and may accumulate large amounts of organic matter as peat. Global ecosystem models based on an understanding of underlying mechanisms are designed to capture these patterns, and have been used to simulate the responses of terrestrial carbon storage to changing climate. Some models also include the effect of changes in land use. Model results suggest that future effects of changing climate, atmospheric CO_2, and changing land use on carbon storage may be quite large (Vloedbeld and Leemans, 1993).

Effects of temperature and CO_2 concentration

Ecosystem models have been used to simulate the response of the terrestrial biosphere to changes in climate, rate of change of climate ("transient" changes), effects of changing CO_2, and the effects of changing land use. Models may consider only the response of plant growth and decomposition, or they may also allow movement of vegetation "types" such as forests and grasslands. The simplest case, involving a general circulation model- (GCM-) simulated climate change and stationary vegetation patterns projected losses of terrestrial carbon of about 200 Gt over an implicit time period of a few hundred years (Melillo et al., 1993). When vegetation types were allowed to migrate, models projected a long-term increase in terrestrial carbon storage: 60-90 GtC over 100-200 years (Cramer and Solomon, 1993; Smith and Shugart, 1993). In Smith and Shugart's (1993) model, climate change and vegetation redistribution led to a transient release of CO_2 from die back before regrowth (~200 GtC over ~100 years), followed by an eventual accumulation of ~90 GtC. This study, like those mentioned above, did not incorporate the effects of CO_2 fertilisation. The TEM model of Melillo et al. (1993) was also used to assess the response of ecosystem carbon storage to CO_2 fertilisation under a scenario of climate change resulting from an instantaneous doubling of CO_2. The results of this experiment showed that, over the period of CO_2 doubling, uptake of carbon only occurred when the CO_2 fertilisation effect was included (loss of ~200 GtC without the fertilisation effect, and a gain of ~250 GtC when the effect was included: Melillo et al., 1994). Rotmans and den Elzen (1993) used the IMAGE model to assess the effects of including CO_2 fertilisation and temperature feedbacks on their modelled carbon budget. Estimated biospheric uptake increased by 1.2 GtC/yr over the decade of the 1980s when CO_2 and temperature feedbacks were included. The most complex assessment of the interactive effects on carbon storage are carried out with models of transient changes in climate that include CO_2 fertilisation. Esser (1990) simulated an increase of 170 GtC over 200 years with transient changes of climate and CO_2 fertilisation (this model does not take vegetation redistribution into account). Alcamo et al. (1994b) performed a similar experiment, but included redistribution of vegetation. Their results revealed eventual increased carbon storage of ~200-250 Gt, depending on assumptions of future land use. Results from ecosystem models suggest that changes in climate associated with CO_2 doubling are likely to lead to a significant transient release of carbon (1-4 GtC/yr) over a period of decades to more than a century (Smith and Shugart, 1993; Dixon et al., 1994).

Effects of land use

Several of these studies also included the effects of changing land use. For example, Esser (1990) simulated a gain of 170 GtC due to the effects of climate and CO_2 change, but losses of 322 GtC when he assumed areal reduction of global forests by 50% by the year 2300, emphasising the importance of maintaining forest ecosystems for terrestrial carbon storage. In the Cramer and Solomon (1993) study, inclusion of dense land clearing also substantially reduced simulated C storage (by up to 152 GtC). In the Alcamo et al. (1994b) study, global scenarios that required additional land for agriculture and biomass energy production resulted in lower carbon storage.

While the results from global terrestrial models range widely, all suggest that the terrestrial biosphere could eventually take up 100-300 GtC in response to warming, albeit after a significant transient loss (e.g., Smith and Shugart, 1993). While existing simulations have a number of shortcomings, including the lack of agreed-upon climate and land use scenarios that would allow rigorous comparison of results, they suggest possible exchanges between the atmosphere and the terrestrial biosphere of the order of 100s of GtC over decades to a few centuries (e.g., Alcamo et al., 1994a).

Soil feedbacks

Because soil carbon is released with increasing temperature, it has been suggested that global warming would result in a large positive feedback (Houghton and Woodwell, 1989; Townsend et al., 1992; Oechel et al., 1993). In order to evaluate this, the sensitivity of carbon storage to temperature was assessed using a number of models (Schimel et al., 1994). Inter-comparison of model results revealed rates of global soil carbon loss of 11-34 GtC per degree warming. Vegetation growth and C storage are stimulated in many of these models because as soil carbon is lost, soil nitrogen is made available to the modelled vegetation. This fertilisation effect can ameliorate or even reverse the overall loss of carbon (Shaver et al., 1992; Gifford, 1994; Schimel et al., 1994). In the Century ecosystem model, the nitrogen feedback reduces the effect of warming on soil carbon loss by 50% (Schimel et al., 1994).

Nutrient limitation of CO_2 fertilisation

The temperature feedback on nitrogen availability may interact with CO_2 fertilisation. CO_2-fertilised foliage is typically lower in nitrogen than foliage grown under current concentrations of CO_2 (e.g., Coleman and Bazzaz, 1992). As dead foliage from high-CO_2 conditions works its way through the decomposition process, soil decomposer organisms require additional nitrogen (Díaz et al., 1993), and, as a result, vegetation may become more nitrogen limited (attenuating but not eliminating the effects of CO_2 fertilisation (Comins and McMurtrie, 1993; Schimel, 1995)). Additional nitrogen released by warming, as described above, can alleviate the nitrogen stress induced by high-CO_2 foliage: this interaction between warming and CO_2 fertilisation is responsible for the large estimated effect of CO_2 on carbon storage in Melillo et al. (1993). Deposition of atmospheric nitrogen could also influence the effectiveness of CO_2 fertilisation (Gifford, 1994), although some modelling studies suggest the effect may be modest (Rastetter et al., 1992; Comins and McMurtrie, 1993). Empirical studies are few, but a recent analysis by Jenkinson et al. (1994) revealed no measurable change in hay yield over the past 100 years despite substantial increases in nitrogen inputs via precipitation and a 21% increase in atmospheric CO_2 concentration during the period of study. Nutrient feedbacks clearly influence the response of terrestrial ecosystems to the interactive effects of climate and CO_2.

1.4.3 Feedbacks on Oceanic Carbon Storage

Climate feedbacks influence the storage of carbon in the ocean through physical, chemical and biological processes. Changes in sea surface temperature affect oceanic CO_2 solubility and carbon chemistry. At equilibrium, a global increase in sea surface temperature of 1°C is associated with an increase of the partial pressure of atmospheric CO_2 of approximately 10 ppmv (Heinze et al., 1991), serving as a weak positive feedback between temperature and atmospheric CO_2. During a transient climate change on time-scales of decades to centuries this temperature feedback would be even weaker (MacIntyre, 1978).

Changes in temperature might also affect the remineralisation of dissolved organic carbon. Climatically driven changes in its turnover time are not expected to result in significant variations in atmospheric CO_2, as the quantity of dissolved organic carbon in the surface ocean is less than 700–750 Gt (Figure 1.1). Earlier suggestions (Sugimura and Suzuki, 1988) that this pool might be much larger than previously thought (i.e., about twice the value shown in Figure 1.1) are now generally discounted (Suzuki, 1993).

Changes in the oceanic circulation and their effects on the oceanic carbon cycle are more difficult to assess. Simulation studies of the transient behaviour of the ocean-atmosphere system using coupled GCMs (Cubasch et al., 1992; Manabe and Stouffer, 1993) indicate a strong reduction of the deep thermohaline circulation, which is driven by cooling and sinking of surface water at high latitudes, especially in the North Atlantic. Smaller effects are expected on the wind-driven features of the oceanic general circulation: the shallow upwelling cells at the equator and the eastern boundaries of the warm subtropical gyres and in the cyclonic areas in higher latitudes. A reduction of the vertical water exchange processes would impact the oceanic carbon cycle in several ways:

First, the downward transport of surface waters in contact with the atmosphere and thus enriched with excess CO_2 would be reduced, thereby leading to a smaller oceanic excess CO_2 uptake capacity.

Second, changes in the vertical water mass transports would affect the marine biological pump. This would reduce the flux of nutrients transported to the surface and, in nutrient-limited regions, result in reduced marine production. This effect would constitute a weakening of the marine biological pump (i.e., less export of carbon to depth), and could potentially lead to an increase in dissolved inorganic carbon and partial pressure of CO_2 at the surface. Conversely, smaller vertical water mass transports imply a reduced upward transport of deeper waters enriched in dissolved inorganic carbon. The two effects result from the same process, but affect the surface concentration of dissolved inorganic carbon, and hence the partial pressure and the air-sea flux of CO_2, in opposite directions. The latter of the two effects dominates in models that use constant carbon to nutrient ratios to describe the marine biosphere (Bacastow and Maier-Reimer, 1990; Keir, 1994). These models project a small increase in oceanic carbon storage as a result of the reduction of oceanic circulation and vertical mixing.

Initial model results suggest that the effects of predicted changes in circulation on the ocean carbon cycle are not large (10s rather than 100s of ppmv in the atmosphere). However, exploration of the long-term impacts of warming on circulation patterns has just begun; hence, analyses of impacts on the carbon cycle must be viewed as preliminary. For example, changes in the oceanic environment (temperature and circulation) have the potential to change the composition of marine ecosystems in ways not yet included in global models. This could lead to changes in carbon to nutrient ratios, which are assumed constant in present models, or to changes in the relationship between organic and inorganic carbon fixation, and/or change the efficiency by which marine organisms utilise available nutrients. Furthermore, the remineralisation depths of nutrients and carbon might be affected differently (Evans and Fasham, 1993). An extreme lower bound may be estimated by assuming complete utilisation of the oceanic surface nutrients which would reduce atmospheric CO_2 by 120 ppmv. Similarly, an upper bound is given by assuming extinction of all marine life, increasing atmospheric CO_2 by almost 170 ppmv (Bacastow and Maier-Reimer, 1990; Shaffer, 1993; Keir, 1994). Such large effects, however, are very unlikely to occur. Indeed, the oceanic carbon system appears to be rather stable as evidenced by the very small fluctuations of the atmospheric CO_2 concentration prior to anthropogenic perturbation. It has also proven very difficult to account for the lower atmospheric CO_2 concentration in glacial times (i.e., by 80 ppmv) merely by changing biospheric parameters within current three-dimensional ocean carbon models (Heinze et al., 1991; Archer and Maier-Reimer, 1994).

In areas of excess nutrients (e.g., in the equatorial and subarctic Pacific and in some parts of the Southern Ocean) the micronutrient iron appears to limit marine primary production (Martin, 1990). Changes in atmospheric iron loading thus potentially could affect the biological pump and impact atmospheric carbon dioxide levels. Modelling studies have shown, however, that even excessive "iron fertilisation" of these oceanic areas would have a relatively small impact on atmospheric CO_2 levels (Joos et al., 1991a; Peng and Broecker, 1991; Sarmiento and Orr, 1991; Kurz and Maier-Reimer, 1993). Therefore, at least during the past 200 years, it is very unlikely that iron loading had a significant impact on the present day carbon balance.

External impacts on the oceanic carbon system not directly related to global warming must also be considered. Increased ultraviolet radiation (UV-B, corresponding to light wavelengths of 280-320 nm) resulting from the depletion of stratospheric ozone could affect marine life and thus influence marine carbon storage. However, model simulation studies show that even a complete cessation of marine productivity in high latitudes, where increases in UV-B are expected to occur, would result in an atmospheric CO_2 increase of less than 40 ppmv (Sarmiento and Siegenthaler, 1992). Increased input of anthropogenic nitrogen or other limiting nutrients might also affect the oceanic biota. The global effects on atmospheric CO_2 are most likely much smaller than the extreme bounds discussed above.

1.5 Modelling Future Concentrations of Atmospheric CO_2

1.5.1 Introduction

The clear historical relationship between CO_2 emissions and changing atmospheric concentrations implies that continuing fossil fuel, cement, and land-use-related emissions of CO_2 at or above current rates will result in increasing atmospheric concentrations of this greenhouse gas. Understanding how CO_2 concentrations will change in the future requires quantification of the relationship between CO_2 emissions and atmospheric concentration using models of the carbon cycle. This section presents the results of a set of standardised calculations, carried out by modelling groups from many countries, that analyse the relationships between emissions and concentrations in a number of ways (see Enting et al., 1994b for full documentation of the modelling exercise).

Two questions are considered:

- For a given CO_2 emission scenario, how might CO_2 concentrations change in the future?

- For a given CO_2 concentration profile leading to stabilisation, what anthropogenic emissions are implied?

As an initial condition, it was required that all models have a balanced carbon budget in which the components matched, within satisfactory limits, a prescribed 1980s-mean budget based on Watson *et al.* (1992). Modelling groups were provided with prescribed future land use fluxes, and a variety of time-series of concentrations and fossil-plus-cement emissions. Because the model intercomparison was conducted simultaneously with the rest of the assessment, the prescribed budget differs from the budget presented in this chapter (Table 1.3) in that:

- (i) the 1980s mean concentration growth rate used (1.59 ppmv/yr or 3.4 GtC/yr) was higher than the current estimate (1.53 ppmv/yr or 3.2 GtC/yr);
- (ii) the net flux from changing land use was set at 1.6 GtC/yr rather than the current estimate of 1.1 GtC/yr;
- (iii) the only mechanism used in the models to simulate terrestrial uptake was CO_2 fertilisation, whereas we suggest that several other mechanisms may be important (Table 1.2).

Modellers were required to continue whatever processes were used to balance the 1980s budget into the future.

The approach used in balancing the budget probably biases the modelling exercise for this chapter towards lower concentrations when emission profiles were employed and higher anthropogenic emissions when concentrations were prescribed. There are two reasons for this. First, the effect of forest regrowth and nitrogen deposition result in changing terrestrial carbon storage which need not increase with increasing CO_2 concentrations as is required by the CO_2 fertilisation effect. Second, as noted in previous IPCC reports (Watson *et al.*, 1990; 1992) climate-related feedbacks may result in additional transient releases of CO_2 to the atmosphere. In contrast, most of the calculations employed in the 1990 IPCC assessment (Watson *et al.*, 1990) assumed constant terrestrial carbon content after 1990 and probably resulted in biases in the other direction. Revisions to the 1980s budget used in model initialisation for this report, however, are in a direction that would lead to higher concentrations by 5 to 10% (for given emissions) and lower emissions by a similar amount (for given concentrations).

Results from a range of different carbon cycle models are considered in order to assess the sensitivity of calculated emission and concentration profiles to model formulation. The complexity of the models employed varies considerably. The most detailed was a model coupling a three-dimensional ocean circulation and chemistry model to a terrestrial biosphere model incorporating a full geographical representation of ecological processes; but, as with other complex models, only a small set of calculations could be performed. The simplest models were designed to simulate only critical processes and represent terrestrial and oceanic carbon uptake with a minimum number of equations. Because most of the models employed incorporate some representation of terrestrial processes, they tended to have initially higher rates of CO_2 uptake from the atmosphere and produced lower estimates for the effective lifetime of CO_2 than the models used in the 1990 IPCC report (Moore and Braswell, 1994). Thus, the GWPs presented in Chapter 5 of this volume are higher for other trace gases than those of earlier assessments (Watson *et al.*, 1990; Chapter 5, this volume).

We chose one model, the "Bern model", for a number of important illustrative calculations, because its results were generally near the mid-point of the results obtained with all models, and because complete descriptions exist in the literature (Joos *et al.*, 1991a; Siegenthaler and Joos, 1992). Selected sensitivity analyses from the model of Wigley (1993) were also used as the configuration of that model allowed for ready modification to consider certain issues. Both the Bern and Wigley models have a balanced carbon cycle consisting of a well-mixed atmosphere linked to oceanic and terrestrial biospheric compartments. In the Bern model, the ocean is represented by the HILDA model, which is a box-diffusion model with an additional advective component. It was tuned to observed values of natural and bomb radiocarbon (Joos *et al.*, 1991a, b; Siegenthaler and Joos, 1992) and validated with CFCs and Argon-39 (Joos, 1992). The Wigley model uses a representation of the ocean based on the ocean general circulation carbon cycle model of Maier-Reimer and Hasselmann (1987). Both the models have similar terrestrial components, with representations of ground vegetation, wood, detritus, and soil (Siegenthaler and Oeschger, 1987; Wigley, 1993). A possible enhancement of plant growth due to elevated CO_2 levels is taken into account by a logarithmic dependency between additional photosynthesis and atmospheric CO_2, which, in the Bern model, probably overestimates biological storage at high CO_2 concentrations. The Bern model was chosen for illustrative purposes in discussions where presentation of multiple results would be confusing (and where all models produced similar patterns). This model was also used to define the reference case for the GWPs presented in

Chapter 5 of this volume. Neither model is recommended nor endorsed by the IPCC or the authors of this chapter as having higher credibility than other models.

1.5.2 Calculations of Concentrations for Specified Emissions

Six greenhouse gas emissions scenarios were described in the 1992 IPCC report (Leggett *et al.*, 1992), based on a wide range of assumptions regarding future economic, demographic, and policy factors. The anthropogenic CO_2 emissions for these scenarios are shown in Figure 1.9a. Scenario IS92c, which has the lowest CO_2 emissions, assumes an eventual decrease in population, low economic growth, and severe constraints on the availability of fossil fuel supplies. The highest emission scenario (IS92e) assumes moderate population growth, high economic growth, high fossil fuel availability, and a phase-out of nuclear power. Concentration estimates for these scenarios have previously been published (e.g., Wigley, 1993). Figure 1.9b shows concentration results from the Bern model that typify the responses of the wide range of models used for the full analysis. This and the other models show strong increases in concentration to well above pre-industrial levels by 2100 (75 to 220% higher). None of the six scenarios leads to stabilisation of concentration before 2100, although IS92c leads to a very slow growth in CO_2 concentration after 2050. IS92a, b, e, and f all produce a doubling of the pre-industrial CO_2 concentration before 2070, with rapid rates of concentration growth. Scenarios developed by the World Energy Council show a similar range of results (Figures 1.11a, b).

In addition to the IS92 emissions cases, three arbitrarily chosen "science" emissions profiles and the newly produced World Energy Council (WEC) Scenarios were also examined. In the former, fossil emissions followed IS92a to the year 2000 and then either stabilised (DEC0%) or decreased at 1% or 2%/yr (see Figure 1.10). For the WEC Scenarios, where only energy-related CO_2 emissions were originally given (WEC, 1993), estimates of gas-flaring and cement production emissions were added (M. Jefferson and G. Marland, personal communications) to ensure consistency with the IS92 Scenarios. Total fossil emissions are given in Figure 1.11a. In all cases net land use emissions were assumed to follow IS92a to 2075. For the WEC cases, IS92a was followed to 2100. For the science scenarios, land use emissions dropped to zero in 2100 and remained zero thereafter. Concentration results are shown in Figures 1.10 and 1.11. Perhaps the most important result (which could be anticipated from the IS92c case) is that stabilisation of emissions at 2000 levels does not lead to stabilisation of CO_2 concentration by 2100; in fact, the calculations show that concentrations

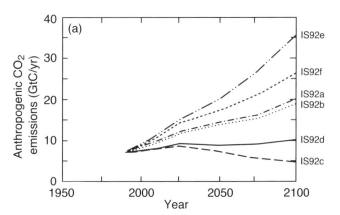

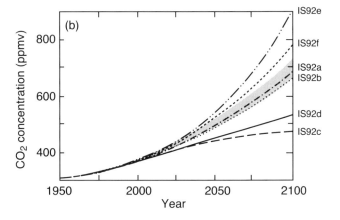

Figure 1.9: (a) Anthropogenic CO_2 emissions for the IS92 Scenarios. (b) Atmospheric CO_2 concentrations calculated from the scenarios IS92a-f (Leggett et al., 1992) using the Bern model (Siegenthaler and Joos; 1992). The typical range of results from different carbon cycle models is indicated by the shaded area.

continue to increase slowly for at least several hundred years. The lowest of the WEC Scenarios, where emissions were based on policies driven by "ecological" considerations (see WEC, 1993), gives an idea of the sort of emissions profile that could lead to concentration stabilisation.

1.5.3 Stabilisation Calculations

The calculations presented in this section illustrate additional aspects of what may be required to achieve stabilisation of atmospheric CO_2 concentrations. The exercise was motivated by the Framework Convention on Climate Change (United Nations, 1992), which states:

"The ultimate objective of this Convention and any related legal instruments that the Conference of the Parties may adopt is to achieve, in accordance with the relevant provisions of the Convention, stabilisation of greenhouse gas concentrations in the atmosphere at a

CO_2 and the Carbon Cycle

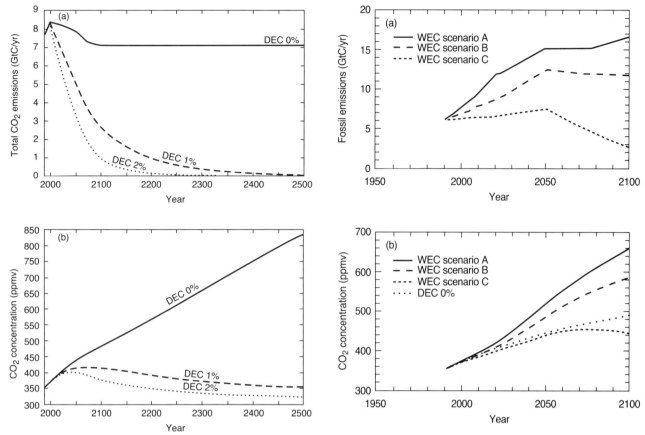

Figure 1.10: (a) Anthropogenic CO_2 emissions calculated by following the IS92a Scenario to the year 2000 and then either fixed fossil fuel emissions (DEC 0%) or emissions declining at 1%/yr (DEC 1%) or 2%/yr (DEC 2%). Land-use emissions followed the modified IS92a scenario (see Section 1.5.2). (b) Atmospheric CO_2 concentrations resulting from DEC 0%, DEC 1% and DEC 2% emissions. Curves are for the model of Wigley (1993).

Figure 1.11: (a) Emission scenarios from the World Energy Council (WEC) modified by including gas flaring and cement production. (b) Atmospheric CO_2 concentrations calculated from the WEC Scenarios using the model of Wigley (1993). Concentrations resulting from the fixed emissions case (DEC 0%) (see Figure 1.10) using the same model are included for comparison.

level that would prevent dangerous anthropogenic interference with the climate system. Such a level should be achieved within a time-frame sufficient to allow ecosystems to adapt naturally to climate change, to ensure that food production is not threatened and to enable economic development to proceed in a sustainable manner."

In the context of this objective it is important to investigate a range of emission profiles of greenhouse gases which might lead to atmospheric stabilisation. It is not our purpose here to consider the climate response (this will be done in the 1995 IPCC Scientific Assessment report), nor to define what might constitute "dangerous interference", nor to make any judgement about the rates of change that would meet the criteria of the objective. In this chapter only CO_2 is considered; the stabilisation of other greenhouse gas concentrations is discussed in Chapter 2.

Several carbon cycle models have been used to calculate the emissions of CO_2 that would lead to stabilisation at a range of different concentration levels. These calculations are designed to illustrate the relationship between CO_2 concentration and emissions. Concentration profiles have been devised (Figure 1.12) in which CO_2 concentrations stabilise at levels from 350 to 750 ppmv (for comparison, the pre-industrial CO_2 concentration was close to 280 ppmv and the 1990 concentration was ~355 ppmv). Many different stabilisation levels and routes to stabilisation could have been chosen. Those in Figure 1.12 give a smooth transition from the 1990 rate of CO_2 concentration increase to stabilisation. As a result, the year of stabilisation differs with stabilisation levels from around 2150 for 350 ppmv to 2250 for 750 ppmv. Further details on the concentration profiles, and the implied emissions results are given in Enting *et al.* (1994b).

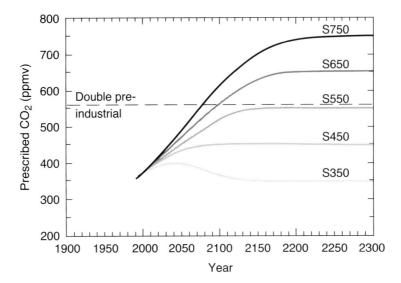

Figure 1.12: CO_2 concentration profiles leading to stabilisation at 350, 450, 550, 650 and 750 ppmv. These are the profiles prescribed for carbon cycle model calculations in which the corresponding emission pathways (shown in Figure 1.13) were determined.

Figure 1.13 shows the model-derived profiles of total anthropogenic emissions (i.e., the sum of fossil fuel use, changes in land use, and cement production) that lead to stabilisation following the concentration profiles shown in Figure 1.12. Initially emissions rise, followed some decades later by quite rapid and large reductions. Stabilisation at any of the concentration levels studied (350-750 ppmv) is only possible if emissions are eventually reduced well below 1990 levels (Figure 1.13). For comparison, the emissions corresponding to IS92a, c, and e are also shown up to 2100 in Figure 1.13. Emissions for all the stabilisation levels studied are lower than those for IS92a and e, even in the first few decades of the 21st century. For the IS92c Scenario, emissions lie between those which in this study eventually achieve stabilisation between 450 and 550 ppmv.

The concentration profiles here are illustrative. Stabilisation at the same level, via a different route, would produce different curves from those shown in Figure 1.13. However, the total amount of emitted carbon accumulated over time (the area under the curves in Figure 1.13), is less sensitive to the concentration profile than to the stabilisation level. Cumulative emissions and carbon storage amounts over the period 1990 to 2200 are shown in Figure 1.14. The separate model results are not shown, but the maximum and minimum estimates are indicated, along with the results from the Bern model, identified as a near-average model. Stabilisation at a lower concentration implies lower accumulated emissions (Figure 1.14).

The spread in results is large, in part because the problem of deducing sources given concentrations is inherently less stable than deducing concentrations from sources, analogous to the inverse problems described in Section 1.3.2.3. Any inferences regarding emissions strategies drawn from these results should take into account the inherent uncertainties in the projections and the limitations imposed by assumptions made to derive them (see Section 1.5.4 and Enting et al., 1994b), and allow response to knowledge gained through continuing observation of future trends in emissions and concentrations.

1.5.4 Assessment of Uncertainties

Two types of uncertainty analysis have been performed for the emission-concentration modelling studies. First, because the models used in the intercomparison incorporate model components of differing structure and levels of complexity to calculate terrestrial and ocean uptake, the range of results provides a view of uncertainties arising from different scientific approaches. While this is a useful view, it probably underestimates the overall uncertainty because the experiments were constrained to have a specific value for the flux arising from changing land use during the 1980s, and a terrestrial sink due solely to CO_2 fertilisation. Because ocean sink magnitudes were not constrained, models with smaller atmosphere-to-ocean fluxes have correspondingly (and possibly unrealistically) larger terrestrial sinks. None of the models used for the terrestrial sink process adequately accounted for the complexities of terrestrial uptake discussed in Section 1.3.3.4. Because the models were constrained to match a particular carbon budget during the

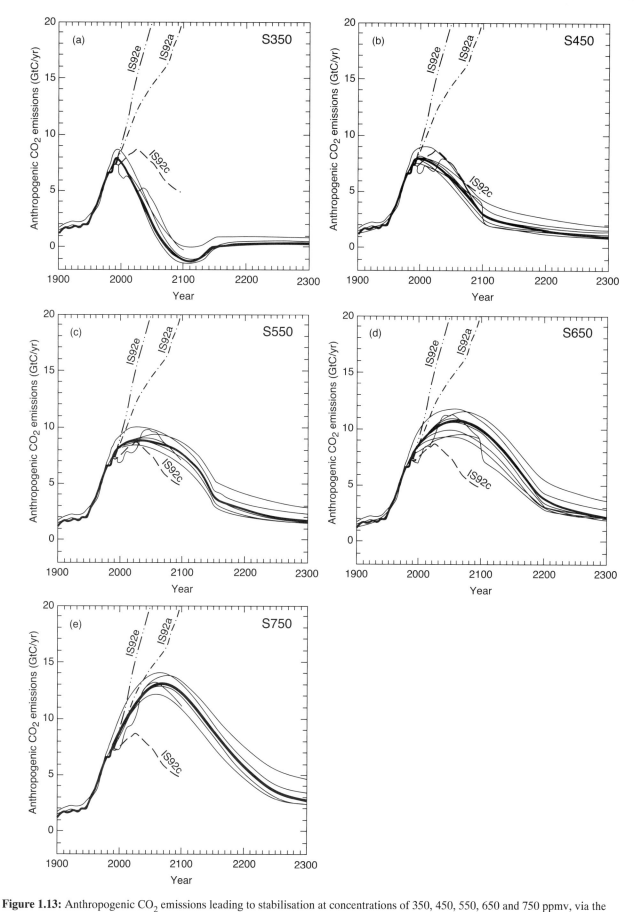

Figure 1.13: Anthropogenic CO_2 emissions leading to stabilisation at concentrations of 350, 450, 550, 650 and 750 ppmv, via the specified concentration profiles shown in Figure 1.12, for a range of different carbon cycle models. The bold line indicates emissions calculated using the Bern model.

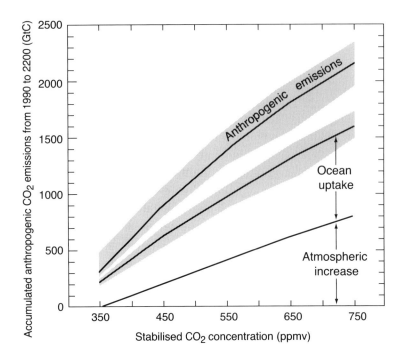

Figure 1.14: Accumulated anthropogenic CO_2 emissions over the period 1990 to 2200 (GtC) plotted against the final stabilised concentration level. Also shown are the accumulated ocean uptake and the increase of CO_2 in the atmosphere. The curves for accumulated anthropogenic emissions and ocean uptake were calculated using the model of Siegenthaler and Joos (1992). The shaded areas show the spread of results from a range of carbon cycle model calculations. The difference (i.e., the accumulated anthropogenic emissions minus the total of the atmospheric increase and the accumulated ocean uptake) gives the cumulative change in terrestrial biomass.

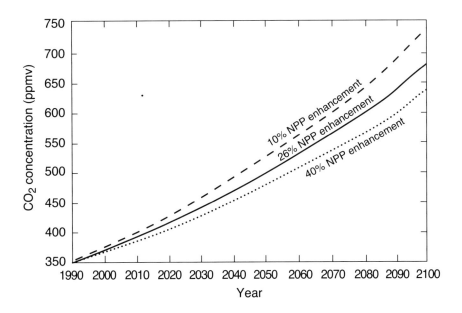

Figure 1.15: Concentration uncertainties associated with the CO_2 fertilisation effect for the IS92a emission scenario. The central curve shows the concentration projection using "best guess" model parameters using the model of Wigley (1993), which is similar to projections using other models. For the upper and lower curves the CO_2 fertilisation effect was decreased to 10% NPP enhancement (from 26% for the "best guess" simulation) and increased to 40% NPP enhancement for a CO_2 doubling from 340 ppmv to 680 ppmv. As explained in Wigley (1993), this assessment also captures some of the uncertainties arising from ocean flux uncertainties, through the need to balance the contemporary carbon budget within realistic limits.

1980s, the model's future projections were closer together than if the experiment had been less constrained. In addition, fossil fuel and cement emissions were prescribed for the 1980s; had the uncertainty in those numbers (about 10%) been allowed for, there would have been a wider range of model parameters for ocean and biospheric processes. Moreover, the results in Figures 1.13 and 1.14 do not account for possible climate feedbacks on the carbon cycle (see Section 1.4).

Although a complete assessment of model sensitivity and uncertainties (e.g., following Gardner and Trabalka, 1985) was outside the scope of the initial exercise carried out for this assessment, several specific model sensitivities have been studied. Figure 1.15 shows the sensitivity of future concentrations under emission scenario IS92a to varied strength of biospheric uptake. In this experiment (Wigley, 1993), the effectiveness of CO_2 fertilisation was varied across the range thought to be reasonable (10-40% enhancements of plant growth for a doubling of CO_2). Significant changes in projected concentrations occur depending upon the CO_2 fertilisation factor used, by approximately ± 14% over 1990 to 2100. Figure 1.16a shows an uncertainty assessment for the emissions corresponding to the S450 and S650 stabilisation cases, using the model of Wigley (1993). The effectiveness of CO_2 fertilisation was varied from 10% to 40% NPP enhancement for a CO_2 doubling (compared with the baseline case for this model of 26%), leading to lower and higher emissions, respectively, by up to ±1.4 GtC/yr for S650 and ±1.1 GtC/yr for S450. Figure 1.16b shows a similar assessment of uncertainty associated with concentrations corresponding to the S450 and S650 stabilisation profiles that result when the strength of terrestrial uptake of CO_2 (via fertilisation of plant growth) is varied.

While the sensitivity of the global carbon budget to the strength of CO_2 fertilisation does not appear great, if biospheric release were to exceed uptake in the future as a result of climate feedbacks (e.g., Smith and Shugart, 1993) or land use emissions (e.g., Esser, 1990), atmospheric concentrations could be significantly greater than those shown here: effects of changing land use could add 100s of GtC to the atmosphere (for comparison see Figure 1.14), over the next one to several centuries. Different assumptions about land use changes would produce different results. For example, if large areas were deforested, in the absence of substantial regrowth the capacity of the terrestrial biosphere to act as a sink would be reduced; hence, more CO_2 would remain in the atmosphere.

The effects of varying land use practices have not been assessed directly in the stabilisation exercise to date, but the changes in biospheric carbon storage from alternate land use scenarios, mitigation efforts, and climatic effects

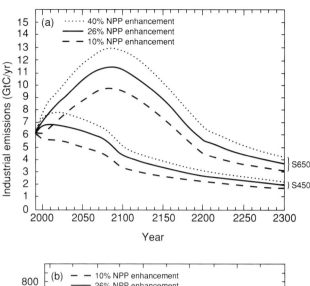

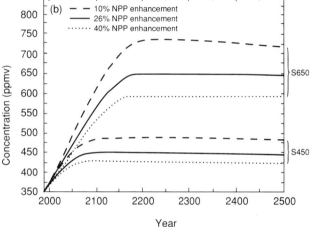

Figure 1.16: (a) Industrial emission uncertainties for concentration stabilisation profiles S450 and S650 associated with the CO_2 fertilisation effect, obtained using the model of Wigley (1993). The baseline emissions results are shown by the bold full lines. These correspond to a CO_2 fertilisation parameter equivalent to an increase in NPP of 26% for a CO_2 doubling. For the upper and lower (dashed) curves, the CO_2 fertilisation effect was increased to 40% NPP enhancement and decreased to 10% NPP enhancement, respectively. Note that it is the relative effects that are important here and that the baseline case is only one of the range of results presented in Figure 1.13. (b) Concentration uncertainties for stabilization profiles S450 and S650 associated with the CO_2 fertilisation effect. The baseline concentration profiles are shown as bold full lines. The dashed curves bounding these represent the range of uncertainty associated with uncertainties in terrestrial sink processes. They were constructed by first estimating the industrial emissions using "best guess" carbon cycle model parameters, which include a 1980s-mean ocean flux of 2.0 Gt C/yr and a CO_2 fertilization parameter equivalent to an increase in NPP of 26% for a CO_2 doubling. The fertilization parameter was then decreased to 10% and increased to 40% to obtain the higher and lower concentration projections. Essentially, this answers the question: what if we devise an industrial emissions policy based on a particular fertilization effect, and this happens to be either too high or too low? Model results were obtained using the model of Wigley (1993)

(e.g., Esser, 1990; Smith and Shugart, 1993; Vloedbeld and Leemans, 1993) discussed in Section 1.4.2 are large enough to influence integrated emissions for a given concentration profile (Figure 1.14) substantially.

References

Aber, J.D., J.K. Nadlehoffer, P.A. Steudler and J.M. Melillo, 1989: Nitrogen saturation in northern forest ecosystems - hypotheses and implications. *BioScience*, **39**, 378-386.

Alcamo, J., G.J.J. Kreileman, M. Krol and G. Zuidema, 1994a: Modeling the global society-biosphere-climate system. Part 1: model description and testing. *Water, Air, and Soil Pollution*, **76**, 1-35.

Alcamo, J., G.J. Van den Born, A.F. Bouwman, B. de Haan, K. Klein-Goldewijk, O. Klepper, R. Leemans, J.A. Oliver, B. de Vries, H. van der Woerd and R. van den Wijngaard, 1994b: Modelling the global society-biosphere-climate system. Part 2: computed scenarios. *Water, Air and Soil Pollution*, **76**, 37-78.

Andres, R.J., G. Marland, T. Boden and S. Bischoff, 1994: Carbon dioxide emissions from fossil fuel consumption and cement manufacture 1751 to 1991 and an estimate for their isotopic composition and latitudinal distribution. In: *The Carbon Cycle*. T.M.L. Wigley and D. Schimel (eds.), Cambridge University Press, Stanford, CA. (In press)

Archer, D. and E. Maier-Reimer, 1994: Effect of deep-sea sedimentary calcite preservation on atmospheric CO_2 concentration. *Nature*, **367**, 260-263.

Bacastow, R. and E. Maier-Reimer, 1990: Ocean-circulation model of the carbon cycle. *Climate Dynamics*, **4**, 95-125.

Barnola, J-M., D. Raynaud, Y.S. Korotkevitch and C. Lorius, 1987: Vostok ice core provides 160,000 year record of atmospheric CO_2. *Nature*, **329**, 408-414.

Barnola, J-M., P. Pimienta, D. Raynaud and T.S. Korotkevich, 1991: CO_2-climate relationship as deduced from the Vostok ice core: a re-examination based on new measurement and on a re-evaluation of the air dating. *Tellus*, **43B**, 83-90.

Barnola, J-M., M. Anklin, J. Porcheron, D. Raynaud, J. Schwander and B. Stauffer, 1994: CO_2 evolution during the last millenium as recorded by Antarctic and Greenland ice. *Tellus*. (In press)

Bazzaz, F.A. and E.D. Fajer, 1992: Plant life in a CO_2-rich world. *Scientific American*, **266**, 68-74.

Bender, M.L., T. Sowers, J-M. Barnola and J. Chappellaz, 1994: Changes in the O_2/N_2 ratio of the atmosphere during recent decades reflected in the composition of air in the firn at Vostok Station, Antarctica. *Geophys. Res. Lett.*, **21**, 189-192.

Birdsey, R.A., A.J. Plantiga and L.S. Heath, 1993: Past and prospective carbon storage in United States forests. *Forest Ecology Management*, **58**, 33-40.

Boden, T.A., R.J. Sepanski and F.W. Stoss, 1991: Trends '91: A compendium of data on global change. Oak Ridge National Laboratory, ORNL/CDIAC-46.

Briffa, K.R., T.S. Bartholin, D. Eckstein, P.D. Jones, W. Karlén, F.H. Schweingruber and P. Zetterberg, 1990: A 1,400 tree-ring record of summer temperatures in Fennoscandia. *Nature*, **346**, 434-439.

Broecker, W.S. and T-H. Peng, 1994: Stratospheric contribution to the global bomb radiocarbon inventory: Model versus observation. *Global Biogeochemical Cycles*, **8**, 377-384.

Broecker, W.S., T. Takahashi, H.J. Simpson and T-H. Peng, 1979: Fate of fossil fuel carbon dioxide and the global carbon budget. *Science*, **206**, 409-418.

Broecker, W.S., T-H. Peng, G. Ostlund and M. Stuiver, 1985: The distribution of bomb radiocarbon in the ocean. *J. Geophys. Res.* **90**, 6953-6970.

Brown, H.T. and F. Escombe, 1905: On the variations in the amount of carbon dioxide in the air of Kew during the years 1893-1901. *Proceedings of the Royal Society of London, Biology*, **76**, 118-121.

Brown, S. and A.E. Lugo, 1984: Biomass of tropical forests: a new estimate based on forest volumes. *Science*, **223**, 1290-1293.

Brown, S. and A.E. Lugo, 1992: Aboveground biomass estimates for tropical moist forests of the Brazilian Amazon. *Interciencia*, **17**, 8-18.

Chameides, W.L., P.S. Kasibhatla, J. Yienger and H. Levy II, 1994: Growth of continental-scale metro-agro-plexes, regional ozone production, and world food production. *Science*, **264**, 74-77.

Chappellaz, J., T. Blunier and D. Raynaud, 1993: Synchronous changes in atmospheric CH_4 and Greenland climate between 40 and 8 kyr BP. *Nature*, **366**, 443-445.

Ciais, P., P. Tans, J.W. White, M. Trolier, R. Francey, J. Berry, D. Randall, P. Sellers, J.G. Collatz and D. Schimel, 1994: Partitioning of ocean and land uptake of CO_2 as inferred by $\delta 13$ measurements from the NOAA/CMDL global air sampling network. *J. Geophys. Res.* (In press)

CLIMAP Project Members, 1984: The last interglacial ocean. *Quaternary Research*, **21**, 123-224.

Cole, C.V. and R.D. Heil, 1981: Phosphorus effects on terrestrial nitrogen cycling. In: *Terrestrial Nitrogen Cycles*, F.E. Clark and T. Rosswall (eds.), Ecological Bulletin, Swedish Natural Science Research Council, Stockholm, pp. 363-374.

Coleman, J.S. and F.A. Bazzaz, 1992: Effects of CO_2 and temperature on growth and resource use of co-occurring C_3 and C_4 annuals. *Ecology*, **73**, 1244-1259.

Coleman, J.S., K.D.M. McConnaughay and F.A. Bazzaz, 1993: Elevated CO_2 and plant nitrogen use: is reduced tissue nitrogen concentration size-dependent? *Oecologia*, **93**, 195-200.

Comins, H.N. and R.E. McMurtrie, 1993: Long-term response of nutrient limited forests to CO_2 enrichment: equilibrium behavior of plant-soil models. *Ecological Applications*, **3**, 666-681.

Conway, T.J., P. Tans, L.S. Waterman, K.W. Thoning, D.R. Buanerkitzis, K.A. Maserie and N. Zhang, 1994: Evidence for interannual variability of the carbon cycle from the NOAA/CMDL global air sampling network. *J. Geophys. Res.*, **99D**, 22, 831-22, 855.

Cramer, W.P. and A.M. Solomon, 1993: Climate classification and future global redistribution of agricultural land. *Climate Research*, **3**, 97-110.

Cubasch, U., K. Hasselmann, H. Hoeck, E. Maier-Reimer, U. Mikolajewicz, B.D. Santer and R. Sausen, 1992: Time-dependent greenhouse warming computations with a coupled ocean-atmosphere model. *Climate Dynamics*, **8**, 55-69.

Dai, A. and I.Y. Fung, 1993: Can climate variability contribute to the "missing" CO_2 sink? *Global Biogeochemical Cycles*, **7**, 599-609.

Dansgaard, W., S.J. Johnsen, H.B. Clausen, D. Dahl-Jensen, N.S. Gunderstrup, C.U. Hammer, J.P. Steffensen, A. Sveinbjörnsdoltir, J. Jouzel and G. Bond, 1993: Evidence for general instability of past climate from a 250-kyr ice-core record. *Nature*, **364**, 218-220.

D'Arrigo, R.D. and G.C. Jacoby, 1993: Tree growth-climate relationships at the northern boreal forest tree line of North America: evaluation of potential response to increasing carbon dioxide. *Global Biogeochemical Cycles*, **7**, 525-535.

Degens, E.T., S. Kempe and J.E. Richey, 1991: Summary: biogeochemistry of major world rivers. In: *Biogeochemistry of Major World Rivers*, E.T. Degens, S. Kempe and J.E. Richey (eds.), SCOPE Report 42, John Wiley and Sons, Chichester, pp.323-347.

Delmas, R.J., 1993: A natural artefact in Greenland ice-core CO_2 measurements. *Tellus,* **45B**, 391-396.

Delmas, R.J., J.M. Ascencio and M. Legrand, 1980: Polar ice evidence that atmospheric CO_2 20,000 yr B.P. was 50% of present. *Nature*, **284**, 155-157.

Detwiler, R.P. and C.A.S. Hall, 1988: Tropical forests and the global carbon cycle. *Science*, **239**, 42-47.

Díaz, S., J.P. Grime, J. Harris and E. McPherson, 1993: Evidence of a feedback mechanism limiting plant response to elevated carbon dioxide. *Nature*, **364**, 616-617.

Dixon, R.K., S.A. Brown, R.A. Houghton, A.M. Solomon, M.C. Trexler and J. Wisniewski, 1994: Carbon pools and flux of global forest ecosystems. *Science*, **263**, 185-190.

Donigian, A.S., Jr., T.O. Barnwell Jr., R.B. Jackson IV, A.S. Patwardhan, K.B. Weinrich, A.L. Rowell, R.V. Chinnaswamy and C.V. Cole, 1994: Assessment of alternative management practices and policies affecting soil carbon in agroecosystems of the Central United States. Environmental Protection Agency, *EPA/600/R-94/067.*

Drake, B.G., 1992: The impact of rising CO_2 on ecosystem production. *Water, Air, and Soil Pollution*, **64**, 25-44.

Druffel, E.R.M., P.M. Williams, J.E. Bauer and J.R. Ertel, 1992: Cycling of dissolved particulate organic matter in the open ocean. *J. Geophys. Res.*, **97C**, 15,639-15,659.

Duce, R.A., P.S. Liss, J.T. Merrill, E.L. Atlas, P. Buat-Ménard, B.B. Hicks, J.M. Miller, J.M. Prospero, R. Arimoto, T.M. Church, W. Ellis, J.N. Galloway, L. Hansen, T.D. Jickells, A.H. Knap, K.H. Reinhardt, B. Schneider, A. Soudine, J.J. Tokos, S. Tsunogai, R. Wollast and M. Zhou, 1991: The atmospheric input of trace species to the world ocean. *Global Biogeochemical Cycles*, **5**, 193-259.

Eisele, K.A., D.S. Schimel, L.A. Kapustka and W.J. Parton, 1989: Effects of available P and N:P ratios on non-symbiotic dinitrogen fixation in tallgrass prairie soils. *Oecologia*, **79**, 471-474.

Enting, I.G. 1987: The interannual variation in the seasonal cycle of carbon dioxide concentration at Mauna Loa. *J. Geophys. Res.*, **92(D)**, 5497-5504.

Enting, I.G., 1993: Inverse problems in atmospheric constituent studies. III. Estimating errors in surface sources. *Inverse Problems*, **9**, 649-665.

Enting, I.G., 1994: CO_2-climate feedbacks: aspects of detection. In: *Feedbacks in the Global Climate System*, G.M. Woodwell (ed.), Oxford University Press, New York. (In press)

Enting, I.G. and J.V. Mansbridge, 1989: Seasonal sources and sinks of atmospheric CO_2: direct inversion of filtered data. *Tellus*, **41**, 111-126.

Enting, I.G. and J.V. Mansbridge, 1991: Latitudinal distribution of sources and sinks of CO_2: results of an inversion study. *Tellus*, **43**, 156-170.

Enting, I.G., R.J. Francey and C.M. Trudinger, 1994a: Synthesis study of the atmospheric CO_2 and $^{13}CO_2$ budgets. *Tellus* (In press)

Enting, I.G., T.M.L. Wigley and M. Heimann, 1994b: Future emissions and concentrations of carbon dioxide: key ocean/atmosphere/land analyses. CSIRO Division of Atmospheric Research Technical Paper No. 31.

Esser, G., 1990: Modeling global terrestrial sources and sinks of CO_2 with special reference to soil organic matter. In: *Soils and the Greenhouse Effect*, A.F. Bouwman (ed.), John Wiley and Sons, New York, pp.247-262.

Eswaran, H., E. Van den Berg and P. Reich, 1993: Organic carbon in soils of the world. *Soil Science Society of America Journal*, **57**, 192-194.

Evans, G.T. and M.J.R. Fasham (eds.), 1993: *Towards a Model of Ocean Biogeochemical Processes*. NATO Workshop, Springer-Verlag, New York, 350 pp.

Falkowski, P.G. and C. Wilson, 1992: Phytoplankton productivity in the North Pacific ocean since 1990 and implications for absorption of anthropogenic CO_2. *Nature*, **358**, 741-743.

FAO (United Nations Food and Agriculture Organisation), 1993: Forest Resources Assessment 1990: Tropical Countries. FAO Forestry Paper No. 112, Rome, Italy.

Fearnside, P.M., 1992: Forest biomass in Brazilian Amazonia: comments on the estimate by Brown and Lugo. *Interciencia*, **17**, 19-27.

Friedli, H., H. Loetscher, H. Oeschger, U. Siegenthaler and B. Stauffer, 1986: Ice core record of the $^{13}C/^{12}C$ ratio of atmospheric CO_2 in the past two centuries. *Nature*, **324**, 237-238.

Friedlingstein, P., C. Delire, J-F. Müller and J-C. Gérard, 1992: The climate induced variation of the continental biosphere: a model simulation of the Last Glacial Maximum. *Geophys. Res. Lett.*, **19**, 897-900.

Fung, I.Y. and T. Takahashi, 1994: A re-evaluation of empirical estimates of the oceanic sink of anthropogenic CO_2. In: *The Carbon Cycle*, T.M.L. Wigley and D. Schimel (eds.), Cambridge University Press, Stanford, CA. (In press)

Garcon, V., F. Thomas, C.S. Wong and J.F. Minster, 1992: Gaining insight into the seasonal variability of CO_2 at O.W.S.P. using an upper ocean model. *Deep Sea Research*, **30**, 921-938.

Gardner, R.H. and J.R. Trabalka, 1985: *Methods of Uncertainty Analysis for a Global Carbon Dioxide Model*, US Department of Energy, Carbon Dioxide Research Division, Technical Report TR024, 41 pp.

Gifford, R.M., 1993: Implications of CO_2 effects on vegetation for the global carbon budget. In: *The Global Carbon Cycle,* M. Heimann (ed.), Proceedings of the NATO Advanced Study Institute, Il Ciocco, Italy, September 8-20, 1991, pp. 165-205.

Gifford, R.M., 1994: The global carbon cycle: a viewpoint on the missing sink. *Australian Journal of Plant Physiology*, **21**, 1-15.

Graumlich, L.J., 1991: Subalpine tree growth, climate, and increasing CO_2: an assessment of recent growth trends. *Ecology*, **72**, 1-11.

GRIP Project Members, 1993: Climatic instability during the last interglacial period revealed in the Greenland summit ice-core. *Nature*, **364**, 203-207.

Grootes, P.M., M. Stuvier, J.W.C. White, S. Johnsen and J. Jouzel, 1993: Comparison of oxygen isotope records from GISP2 and GRIP Greenland ice cores. *Nature*, **366**, 552-554.

Heimann, M. and E. Maier-Reimer, 1994: On the relations between the uptake of CO_2 and its isotopes by the ocean. *Global Biogeochemical Cycles*. (Submitted)

Heimann, M., C.D. Keeling and C.J. Tucker, 1989: A three dimensional model of atmospheric CO_2 transport based on observed winds: 3. Seasonal cycle and synoptic time-scale variations. In: *Aspects of Climate Variability in the Pacific and the Western Americas*, D.H. Peterson (ed.), American Geophysical Union, Washington, DC, pp. 277-303.

Heinze, C., E. Maier-Reimer and K. Winn, 1991: Glacial pCO_2 reduction by the world ocean: experiments with the Hamburg carbon cycle model. *Paleoceanography*, **6**, 395-430.

Hesshaimer, V., M. Heimann and I. Levin, 1994: Radiocarbon evidence suggesting a smaller oceanic CO_2 sink than hitherto assumed. *Nature*, **370**, 201-203.

Houghton, R.A., 1993: Is carbon accumulating in the northern temperate zone. *Global Biogeochemical Cycles*, **7**, 611-617.

Houghton, R.A., 1994a: Emissions of carbon from land-use change. In: *The Carbon Cycle*, T.M.L. Wigley and D. Schimel (eds.), Cambridge University Press, Stanford, CA. (In press)

Houghton, R.A., 1994b: Effects of land-use change, surface temperature and CO_2 concentration on terrestrial stores of carbon. In: *Biotic Feedbacks in the Global Climate System: Will the Warming Speed the Warming?*, G.M. Woodwell and F.T. Mackenzie (eds.), Oxford University Press, Oxford. (In press)

Houghton, R.A. and G.M. Woodwell, 1989: Global climatic change. *Scientific American*, **260**, 36-47.

Houghton, R.A. and D.L. Skole, 1990: Carbon. In: *The Earth as Transformed by Human Action*, B.L. Turner II, W.C. Clark, R.W. Kates, J.F. Richards, J.T. Mathews and W.B. Meyer (eds.), Cambridge University Press, New York, pp. 393-408.

Houghton, R.A., R.D. Boone, J.R. Fruci, J.E. Hobbie, J.M. Melillo, C.A. Palm, B.J. Peterson, G.R. Shaver, G.M. Woodwell, B. Moore, D.L. Skole and N. Myers, 1987: The flux of carbon from terrestrial ecosystems to the atmosphere in 1980 due to changes in land use: geographic distribution of the global flux. *Tellus*, **39B**, 122-139.

Idso, K.E. and S.B. Idso, 1994: Plant responses to atmospheric CO_2 enrichment in the face of environmental constraints: a review of the last 10 years' research. *Agricultural and Forest Meteorology*, **69**, 153-203.

Idso, S.B. and B.A. Kimball, 1993: Tree growth in carbon dioxide enriched air and its implications for global carbon cycling and maximum levels of atmospheric CO_2. *Global Biogeochemical Cycles*, **7**, 537-555.

INPE, 1992: *Deforestation in Brazilian Amazonia*. Instituto Nacional de Pesquisas Espaciais, São Paulo, Brazil.

IPCC (Intergovernmental Panel on Climate Change), 1990: *Climate Change: the IPCC Scientific Assessment*, J.T. Houghton, G.J. Jenkins and J.J. Ephraums (eds.). Cambridge University Press, Cambridge, UK. 365 pp.

IPCC, 1992: *Climate Change 1992: The Supplementary Report to the IPCC Scientific Assessment*, J.T. Houghton, B.A. Callander and S.K. Varney (eds.). Cambridge University Press, Cambridge, UK. 200 pp.

Jenkinson, D.S., J.M. Potts, J.N. Perry, V. Barnett, K. Coleman and A.E. Johnston, 1994: Trends in herbage yields over the last century on the Rothamsted long-term continuous hay experiment. *Journal of Agricultural Science*, **122**, 365-374.

Johnsen, S.J., H.B. Clausen, W. Dansgaard, K. Furher, N. Gundestrup, C.U. Hammer, P. Iverson, J. Jouzel, B. Stauffer and J.P. Steffensen, 1992: Irregular glacial interstadials recorded in a new Greenland ice core. *Nature*, **359**, 311-313.

Joos, F., 1992: Modellierung und verteilung von spurenstoffen im ozean und des globalen kohlenstoffkreislaufes. Ph.D. Thesis, University of Bern, Bern.

Joos, F., J.L. Sarmiento and U. Siegenthaler, 1991a: Estimates of the effect of Southern Ocean iron fertilization on atmospheric CO_2 concentrations. *Nature*, **349**, 772-774.

Joos, F., U. Siegenthaler and J.L. Sarmiento, 1991b: Possible effects of iron fertilization in the Southern Ocean on atmospheric CO_2 concentration. *Global Biogeochemical Cycles*, **5**, 135-150.

Jouzel, J., N.I. Barkov, J-M. Barnola, M. Bender, J. Chappellaz, C. Genthon, V.M. Kotlyakov, V. Lipenkov, C. Lorius, J.R. Petit, D. Raynaud, G. Raisbeck, C. Ritz, T. Sowers, M. Stievenard, F. Yiou and P. Yiou, 1993: Extending the Vostok ice-core record of paleoclimate to the penultimate glacial period. *Nature*, **364**, 407-412.

Keeling, C.D., 1973: Industrial production of carbon dioxide from fossil fuels and limestone. *Tellus*, **25**, 174-198.

Keeling, C.D., 1993: Surface ocean CO_2. In: *The Global Carbon Cycle*, M. Heimann (ed.), Springer-Verlag, Heidelberg, pp.413-429.

Keeling, C.D. and T.P. Whorf, 1994: Decadal oscillations in global temperature and atmospheric carbon dioxide. In: *Natural Variability of Climate on Decade-to-Century Time Scales*, W.A. Sprigg (ed.), National Academy of Sciences, Washington, DC. (In press)

Keeling, C.D., R.B. Bacastow, A.F. Carter, S.C. Piper, T.P. Whorf, M. Heimann, W.G. Mook and H. Roeloffzen, 1989a: A three-dimensional model of atmospheric CO_2 transport based on observed winds: 1. Analysis and observational data. In: *Aspects of Climate Variability in the Pacific and Western Americas. Geophysical Monograph 55*, D.H. Peterson (ed.), American Geophysical Union, Washington, DC, pp.165-236.

Keeling, C.D., S.C. Piper and M. Heimann, 1989b: A three-dimensional model of atmospheric CO_2 transport based on observed winds: 4. Mean annual gradients and interannual variations. In: *Aspects of Climate Variability in the Pacific and Western Americas. Geophysical Monograph 55*, D.H. Peterson (ed.), American Geophysical Union, Washington, DC, pp.305-363.

Keeling, R.F. and R. Shertz, 1992: Seasonal and interannual variations in atmospheric oxygen and implications for the global carbon cycle. *Nature*, **358**, 723-727.

Keeling, R.F. and J. Severinghaus, 1994: Atmospheric oxygen measurements and the carbon cycle. In: *The Carbon Cycle*, T.M.L. Wigley and D. Schimel (eds.), Cambridge University Press, Stanford, CA. (In press)

Keir, R.S., 1994: Effects of ocean circulation changes and their effects on CO_2. In: *The Carbon Cycle*. T.M.L. Wigley and D. Schimel (eds.), Cambridge University Press, Stanford, CA. (In press)

Kempe, S. and K. Pegler, 1991: Sinks and sources of CO_2 in coastal seas: the North Sea. *Tellus*, **43**, 224-235.

Kolchugina, T.P. and T.S. Vinson, 1993: Carbon sources and sinks in forest biomes of the former Soviet Union. *Global Biogeochemical Cycles*, **7**, 291-304.

Körner, C. and J.A. Arnone III, 1992: Responses to elevated carbon dioxide in artificial tropical ecosystems. *Science*, **257**, 1672-1675.

Krankina, O.N. and R.K. Dixon, 1994: Forest management options to conserve and sequester terrestrial carbon in the Russian Federation. *World Resources Review*, **6**, 88-101.

Kurz, K.D. and E. Maier-Reimer, 1993: Iron fertilization of the austral ocean – a Hamburg model assessment. *Global Biogeochemical Cycles*, **7**, 229-244.

LaMarche, V.C.J., D.A. Graybill and M.R. Rose, 1984: Increasing atmospheric carbon dioxide: tree ring evidence for growth enhancement in natural vegetation. *Science*, **225**, 1019-1021.

Leggett, J., W.J. Pepper and R.J. Swart, 1992: Emissions scenarios for IPCC: an update. In: *Climate Change 1992: The Supplementary Report to the IPCC Scientific Assessment*, J.T. Houghton, B.A. Callander and S.K. Varney (eds.),Cambridge University Press, Cambridge, UK. pp.69-95.

Leuenberger, M., U. Siegenthaler and C.C. Langway, 1992: Carbon isotope composition of atmospheric CO_2 during the last ice age from an Antarctic ice core. *Nature*, **357**, 488-490.

MacIntyre, F., 1978: On the temperature coefficient of pCO_2 in seawater. *Climatic Change*, **1**, 349-354.

Maier-Reimer, E., 1993: The biological pump in the greenhouse. *Global and Planetary Change*, **8**, 13-15.

Maier-Reimer, E. and K. Hasselmann, 1987: Transport and storage in the ocean - An inorganic ocean-circulation carbon cycle model. *Climate Dynamics*, **2**, 63-90.

Manabe, S. and R.J. Stouffer, 1993: Century-scale effects of increased atmospheric CO_2 on the ocean-atmosphere system. *Nature*, **364**, 215-218.

Manning, M.R., 1993: Seasonal cycles in atmospheric CO_2 concentrations. In: *The Global Carbon Cycle*, M. Heimann (ed.), Springer-Verlag, Heidelberg, pp.65-94.

Marland, G. and R.M. Rotty, 1984: Carbon dioxide emissions from fossil fuels: a procedure for estimation and results for 1950-1982. *Tellus*, **36B**, 232-261.

Martin, J.H., 1990: Glacial-interglacial CO_2 change: the iron hypothesis. *Paleoceanography*, **5**, 1-13.

Maybeck, M., 1993: Natural sources of C, N, P, and S. In: *Interactions of C, N, P, and S Biogeochemical Cycles and Global Change*, R. Wollast (ed.), Springer-Verlag, Berlin, pp.163-193.

Melillo, J.M., J.R. Fruci, R.A. Houghton, B. Moore III and D.L. Skole, 1988: Land-use change in the Soviet Union between 1850 and 1980: causes of a net release of CO_2 to the atmosphere. *Tellus*, **40B**, 166-128.

Melillo, J.M., A.D. McGuire, D.W. Kicklighter, B. Moore III, C.J. Vorosmarty and A.L. Schloss, 1993: Global climate change and terrestrial net primary production. *Nature*, **363**, 234-240.

Melillo, J.M., D.W. Kicklighter, A.D. McGuire, W.T. Peterjohn and K. Newkirk, 1994: Global change and its effects on soil organic carbon stocks. In: *Dahlem Conference Proceedings*, John Wiley and Sons, New York. (In press)

Metherell, A.K., 1992: Simulation of soil organic matter dynamics and nutrient cycling in agroecosystems. Ph.D. Dissertation, Colorado State University, Fort Collins.

Mitchell, J.F.B., S. Manabe, V. Meleshko and T. Tokioka, 1990: Equilibrium climate change - and its implications for the future. In: *Climate Change: the 1990 Scientific Assessment*, J.T. Houghton, G.J. Jenkins and J.J. Ephraums (eds.), Cambridge University Press, Cambridge, UK, pp.131-172.

Moore, B. III, and B.H. Braswell Jr., 1994: The lifetime of excess atmospheric carbon dioxide. *Global Biogeochemical Cycles*, **8**, 23-38.

Neftel, A., H. Oeschger, J. Schwander, B. Stauffer and R. Zumbrunn, 1982: Ice core sample measurements give atmospheric CO_2 content during the past 40,000 years. *Nature*, **295**, 220-223.

Neftel, A., E. Moor, H. Oeschger and B. Stauffer, 1985: Evidence from polar ice cores for the increase in atmospheric CO_2 in the past two centuries. *Nature*, **315**, 45-47.

Neftel, A., H. Oeschger, T. Staffelbach and B. Stauffer, 1988: CO_2 record in the Byrd ice core 50,000-5,000 years BP. *Nature*, **331**, 609-611.

Norby, R.J., C.A. Gunderson, S.D. Wullschleger, E.G. O'Neill and M.K. McCracken, 1992: Productivity and compensatory response of yellow-poplar trees in elevated CO_2. *Nature*, **357**, 322-324.

Oechel, W.C., S.J. Hastings, G. Vourlitis, M. Jenkins, G. Riechers and N. Grulke, 1993: Recent change of Arctic tundra ecosystems from a net carbon dioxide sink to a source. *Nature*, **361**, 520-523.

Oeschger, H., U. Siegenthaler and A. Guglemann, 1975: A box-diffusion model to study the carbon dioxide exchange in nature. *Tellus*, **27**, 168-192.

Ojima, D.S., W.J. Parton, D.S. Schimel, J.M.O. Scurlock and T.G.F. Kittel, 1993: Modelling the effects of climatic and CO_2 changes on grassland storage of soil carbon. *Water, Air, and Soil Pollution*, **70**, 643-657.

Orr, J.C., 1993: Accord between ocean models predicting uptake of anthropogenic CO_2. *Water, Air, and Soil Pollution*, **70**, 465-481.

Owensby, C.E., P.I. Coyne, J.M. Ham, L.M. Auen and A.K. Knapp, 1993: Biomass production in a tallgrass prairie ecosystem exposed to ambient and elevated CO_2. *Ecological Applications*, **3**, 644-653.

Peng, T-H. and W.S. Broecker, 1991: Dynamic limitations on the Antarctic iron fertilization strategy. *Nature*, **349**, 227-229.

Peterson, B.J. and J.M. Melillo, 1985: The potential storage of carbon caused by eutrophication of the biosphere. *Tellus*, **37B**, 117-127.

Polley, H.W., H.B. Johnson, B.D. Marino and H.S. Mayeux, 1993: Increase in C_3 plant water-use efficiency and biomass over Glacial to present CO_2 concentrations. *Nature*, **361**, 61-63.

Poorter, H., 1993: Interspecific variation in the growth response of plants to an elevated ambient CO_2 concentration. *Vegetatio*, **104/105**, 77-97.

Post, W.M., J. Pastor, P.J. Zinke and A.G. Stangenberger, 1985: Global patterns of soil nitrogen storage. *Nature*, **317**, 613-616.

Potter, C.S., J.T. Randerson, C.B. Field, P.A. Matson, P.M. Vitousek, H.A. Mooney and S.A. Klooster, 1993: Terrestrial ecosystem production: a process model based on global satellite and surface data. *Global Biogeochemical Cycles*, **7**, 811-841.

Prentice, K.C. and I.Y. Fung, 1990: The sensitivity of terrestrial carbon storage to climate change. *Nature*, **346**, 48-51.

Quay, P.D., B. Tilbrook and C.S. Wong, 1992: Oceanic uptake of fossil fuel CO_2: carbon-13 evidence. *Science*, **256**, 74-79.

Rastetter, E.B., R.B. McKane, G.R. Shaver and J.M. Melillo, 1992: Changes in C storage by terrestrial ecosystems: how C-N interactions restrict responses to CO_2 and temperature. *Water, Air, and Soil Pollution*, **64**, 327-344.

Raynaud, D. and U. Siegenthaler, 1993: Role of trace gases: the problem of lead and lag. In: *Global Changes in the Perspective of the Past*, J.A. Eddy and H. Oeschger (eds.), John Wiley and Sons, Chichester, pp173-188.

Raynaud, D., J. Jouzel, J-M.Barnola, J. Chappellaz, R.J. Delmas and C. Lorius, 1993: The ice record of greenhouse gases. *Science*, **259**, 926-934.

Riebesell, U., D.A. Wolf-Gladrow and V. Smetacek, 1993: Carbon dioxide limitation of marine phytoplankton growth rates. *Nature*, **361**, 249-251.

Rochefort, L. and F.A. Bazzaz, 1992: Growth response to elevated CO_2 in seedlings of four co-occurring birch species. *Canadian Journal of Forestry Research*, **22**, 1583-1587.

Rotmans, J. and M.G.J. den Elzen, 1993: Modelling feedback mechanisms in the carbon cycle: balancing the carbon budget. *Tellus*, **45B**, 301-320.

Sarmiento, J.L. and J.C. Orr, 1991: Three dimensional ocean model simulations of the impact of Southern Ocean nutrient depletion on atmospheric CO_2 and ocean chemistry. *Limnology and Oceanography*, **36**, 1928-1950.

Sarmiento, J.L. and U. Siegenthaler, 1992: New production and the global carbon cycle. In: *Primary Productivity and Biogeochemical Cycles in the Sea*, P.G. Falkowski and A.D. Woodhead (eds.), Plenum Press, New York, pp317-332.

Sarmiento, J.L. and E.T. Sundquist, 1992: Revised budget for the oceanic uptake of anthropogenic carbon dioxide. *Nature*, **356**, 589-593.

Sarmiento, J.L., J.C. Orr and U. Siegenthaler, 1992: A perturbation simulation of CO_2 uptake in an ocean general circulation model. *J. Geophys. Res.*, **97**, 3621-3645.

Schimel, D.S., 1995: Terrestrial biogeochemical cycles: global estimates with remote sensing. *Remote Sensing of Environment*. (In press)

Schimel, D.S., B.H. Braswell Jr., E.A. Holland, R. McKeown, D.S. Ojima, T.H. Painter, W.J. Parton and A.R. Townsend, 1994: Climatic, edaphic and biotic controls over storage and turnover of carbon in soils. *Global Biogeochemical Cycles*, **8**, 279-293.

Schimel, D.S. and E.W. Sulzman, 1994: Variability in the Earth-Climate system: decadal and longer timescales. In: *The U.S. National Report (1991-1994) to the International Union of Geophysics and Geodessey*, S.P. Nelson (ed.), American Geophysical Union, Washington, DC. (In press)

Schindler, D.W. and S.E. Bayley, 1993: The biosphere as an increasing sink for atmospheric carbon: estimates from increased nitrogen deposition. *Global Biogeochemical Cycles*, **7**, 717-734.

Schlesinger, W.H., 1990: Evidence from chronosequence studies for a low carbon-storage potential of soils. *Nature*, **348**, 232-234.

Schlesinger, W.H., 1993: Response of the terrestrial biosphere to global climate change and human perturbation. *Vegetatio*, **104/105**, 295-305.

Schlesinger, W.H. and J.M. Melack, 1981: Transport of organic carbon in the world's rivers. *Tellus*, **33**, 172-187.

Schulze, E.D., W. De Vries, M. Hauhs, K. Rosén, L. Rasmussen, O-C. Tann and J. Nilsson, 1989: Critical loads for nitrogen deposition in forest ecosystems. *Water, Air, and Soil Pollution*, **48**, 451-456.

Shaffer, G., 1993: Effects of the marine biota on global carbon cycling. In: *The Global Carbon Cycle*, M. Heimann (ed.), Springer-Verlag, Heidelberg, pp431-456.

Shaver, G.R., W.D. Billings, F.S. Chapin III, A.E. Giblin, K.J. Nadelhoffer, W.C. Oechel and E.B. Rastetter, 1992: Global change and the carbon balance of Arctic ecosystems. *BioScience*, **42**, 433-441.

Siegenthaler, U. and H. Oeschger, 1987: Biospheric CO_2 emissions during the past 200 years reconstructed by deconcolution of ice core data. *Tellus*, **39B**, 140-154.

Siegenthaler, U. and F. Joos, 1992: Use of a simple model for studying oceanic tracer distributions and the global carbon cycle. *Tellus*, **44B**, 186-207.

Siegenthaler, U. and J.L. Sarmiento, 1993: Atmospheric carbon dioxide and the ocean. *Nature*, **365**, 119-125.

Siegenthaler, U., H. Friedli, H. Loeeetscher, E. Moor, A. Neftel, H. Oeschger and B. Stauffer, 1988: Stable-isotope ratios and concentration of CO_2 in air from polar ice cores. *Ann. Glaciol.*, **10**, 151-156.

Skole, D. and C. Tucker, 1993: Tropical deforestation and habitat fragmentation in the Amazon: satellite data from 1978 to 1988. *Science*, **260**, 1905-1910.

Smith, T.M. and H.H. Shugart, 1993: The transient response of terrestrial carbon storage to a perturbed climate. *Nature*, **361**, 523-526.

Stauffer, B., H. Hofer, H. Oeschger, J. Schwander and U. Siegenthaler, 1984: Atmospheric CO_2 concentration during the last glaciation. *Ann. Glaciol.*, **5**, 160-164.

Stocker, T.F., W.S. Broecker and D.G. Wright, 1994: Carbon uptake experiments with a zonally-averaged global ocean circulation model. *Tellus*, **46B**, 103-122.

Suess, H.E., 1955: Radiocarbon concentration in modern wood. *Science*, **122**, 415-417.

Sugimura, Y. and Y. Suzuki, 1988: A high temperature catalytic oxidation method for non-volatile dissolved organic carbon in seawater by direct injection of a liquid sample. *Marine Chemistry*, **24**, 105-131.

Suzuki, Y., 1993: On the measurement of DOC and DON in seawater. *Marine Chemistry*, **41**, 287-288.

Takahashi, T., J. Olafsson, J.G. Goddard, D.W. Chipman and S.C. Sutherland, 1993: Seasonal variation of CO_2 and nutrients in the high-latitude surface oceans: a comparative study. *Global Biogeochemical Cycles*, **7**, 843-878.

Tans, P.P., T.J. Conway and T. Nakazawa, 1989: Latitudinal distribution of the sources and sinks of atmospheric carbon dioxide derived from surface observations and an atmospheric transport model. *J. Geophys. Res.*, **94**(D), 5151-5172.

Tans, P.P., I.Y. Fung and T. Takahashi, 1990: Observational constraints on the global atmospheric CO_2 budget. *Science*, **247**, 1431-1438.

Tans, P.P., J.A. Berry and R.F. Keeling, 1993: Oceanic $^{13}C/^{12}C$ observations: a new window on ocean CO_2 uptake. *Global Biogeochemical Cycles*, **7**, 353-368.

Thomas, R.B., D.D. Richter, H. Ye, P.R. Heine and B.R. Strain, 1991: Nitrogen dynamics and growth of seedlings of an N-fixing tree (*Gliricidia sepium* (Jacq.) Walp.) exposed to elevated atmospheric carbon dioxide. *Oecologia*, **88**, 415-421.

Thompson, M.L., I.G. Enting, G.I. Pearman and P. Hyson, 1986: Internal variations of atmospheric CO_2 concentration. *J. Atmos. Chem.*, **4**, 125-155.

Townsend, A.R., P.M. Vitousek and E.A. Holland, 1992: Tropical soils could dominate the short-term carbon cycle feedbacks to increased global temperatures. *Climatic Change*, **22**, 293-303.

United Nations, 1992: *Earth Summit Convention on Climate Change, 3-14 June 1992*, United Nations Conference on Environment and Development, Rio de Janeiro, Brazil.

Vitousek, P.M. and R.W. Howarth, 1991: Nitrogen limitation on land and in the sea: how can it occur? *Biogeochemistry*, **13**, 87-115.

Vloedbeld, M. and R. Leemans, 1993: Quantifying feedback processes in the response of the terrestrial carbon cycle to global change: the modeling approach of IMAGE-2. *Water, Air, and Soil Pollution*, **70**, 615-628.

Watson, A.J., 1993: Air-sea gas exchange and carbon dioxide. In: *The Global Carbon Cycle*, M. Heimann (ed.), Springer-Verlag, Heidelberg, pp397-411.

Watson, R.T., H. Rhode, H. Oeschger and U. Siegenthaler, 1990: Greenhouse gases and aerosols. In: *Climate Change: the IPCC Scientific Assessment*, J.T. Houghton, G.J. Jenkins and J.J. Ephraums (eds.), Cambridge University Press, Cambridge, UK, pp1-40.

Watson, R.T., L.G. Meira Filho, E. Sanhueza and A. Janetos, 1992: Sources and sinks. In: *Climate Change 1992: The Supplementary Report to the IPCC Scientific Assessment*, J.T. Houghton, B.A. Callander and S.K. Varney (eds.), Cambridge University Press, Cambridge, UK, pp25-46.

WEC (World Energy Council), 1993: *Energy for Tomorrow's World - the Realities, the Real Options and the Agenda for Achievement*, St. Martin's Press, New York, 320 pp.

White, J.W.C., P. Ciais, R.A. Figge, R. Kenny and V. Markgraf, 1994: A high resolution atmospheric pCO_2 record from carbon isotopes in peat. *Nature*, **367**, 153-156.

Wigley, T.M.L., 1993: Balancing the global carbon budget. Implications for projections of future carbon dioxide concentration changes. *Tellus*, **45B**, 409-425.

Wong, C.S., Y-H. Chan, J.S. Page, G.E. Smith and R.D. Bellegay, 1993: Changes in equatorial CO_2 flux and new production estimated from CO_2 and nutrient levels in Pacific surface waters during the 1986/87 El Niño. *Tellus*, **45B**, 64-79.

Woodward, F.I., 1992: Predicting plant responses to global environmental change. *New Phytologist*, **122**, 239-251.

2

Other Trace Gases and Atmospheric Chemistry

M. PRATHER, R. DERWENT, D. EHHALT, P. FRASER,
E. SANHUEZA, X. ZHOU

Contributors:
F. Alyea, T. Bradshaw, J. Butler, M.A. Carroll, D. Cunnold, E. Dlugokencky,
J. Elkins, D. Etheridge, D. Fisher, P. Guthrie, N. Harris, I. Isaksen, D.J. Jacob,
C.E. Johnson, J. Kaye, S. Liu, C.T. McElroy, P. Novelli, J. Penner, R. Prinn,
W. Reeburgh, J. Richardson, B. Ridley, T. Rudolph, P. Simmonds, L.P. Steele,
F. Stordal, R. Weiss, A. Volz-Thomas, A. Wahner, D. Wuebbles

Modelling Contributors: see tables

CONTENTS

Summary	77
2.1 Introduction	79
2.2 Atmospheric Chemistry	79
2.2.1 Chemical Processes and the Removal of Trace Gases	79
2.2.2 Atmospheric Adjustment Times of the Trace Gases	82
2.2.3 Current Tropospheric OH	83
2.2.4 Other Atmospheric and Surface Removal	84
2.2.5 Lifetimes from Stratospheric Removal	84
2.2.6 Examples of Chemical Feedbacks Affecting Greenhouse Gases	85
2.3 Methane	85
2.3.1 Methane Sources	85
2.3.2 Removal of Methane	87
2.3.3 Atmospheric Distribution	87
2.3.4 Trends and Sensitivities	87
2.4 Nitrous Oxide	89
2.4.1 Sources of Nitrous Oxide	89
2.4.2 Removal of Nitrous Oxide	91
2.4.3 Atmospheric Distribution	91
2.4.4 Trends and Sensitivities	91
2.5 Halocarbons	92
2.5.1 Atmospheric Distributions and Trends	92
2.5.1.1 CFCs and carbon tetrachloride	92
2.5.1.2 Methylchloroform and the HCFCs	93
2.5.1.3 Other chlorinated species	93
2.5.1.4 Methylbromide, halons and other brominated species	93
2.5.1.5 Other perhalogenated species	94
2.5.2 Industrial Production, Use and Emissions	94
2.5.3 Natural Sources	95
2.5.4 Halocarbon Removal Processes	95
2.6 Observed Ozone (O_3) and Tropospheric UV	95
2.6.1 Stratospheric Ozone	96
2.6.2 Tropospheric Ozone	98
2.6.3 Tropospheric UV	99
2.7 Tropospheric Nitrogen Oxides	99
2.7.1 Sources of Tropospheric NO_x	99
2.7.2 NO_x Removal Processes	100
2.7.3 Tropospheric Distribution of NO_x	100
2.7.4 Trends of NO_x	103
2.8 Carbon Monoxide and Volatile Organic Compounds	103
2.8.1 Sources and Removal Processes of CO	103
2.8.2 Atmospheric distribution and trends of CO	103
2.8.3 Volatile Organic Compounds	104
2.8.3.1 Introduction	104
2.8.3.2 Sources of volatile organic compounds	104
2.8.3.3 Sinks of the volatile organic compounds	105
2.8.3.4 The role of volatile organic compounds	105
2.9 Inter-Comparison of Tropospheric Chemistry/Transport Models	105
2.9.1 Intercomparison of Transport: A Case Study of ^{222}Radon	106
2.9.2 Intercomparison of Photochemistry: O_3 Production and Loss	109
2.9.3 Conclusions	111
2.10 Global Tropospheric Ozone Modelling	111
2.10.1 Tropospheric NO_x: Surface Combustion and Aircraft	112
2.10.2 CH_4 Increases: A Case Study	113
2.10.2.1 The current atmosphere	113
2.10.2.2 O_3 perturbations	113
2.10.2.3 Adjustment time of CH_4 emissions	115
2.10.3 Conclusions	116
2.11 Stabilisation of Atmospheric Chemical Composition	116
2.11.1 Methane	116
2.11.2 Nitrous Oxide	117
2.11.3 Halocarbons	117
2.11.4 Ozone	117
References	118

SUMMARY

Methane (CH_4)

Atmospheric CH_4 concentrations have increased from about 700 ppbv in pre-industrial times to a global mean of 1714 ppbv in 1992. The 1980s were characterised by declining methane growth rates which were approximately 10 ppbv/yr by the end of the decade. The average growth rate of 13 ppbv/yr corresponds to an imbalance between sources and sinks of about 37 $Tg(CH_4)$/yr. If emissions were frozen at this level, CH_4 would rise to about 1900 ppbv over the next 50 years. If emissions were cut by 37 $Tg(CH_4)$/yr, then CH_4 concentrations would remain at today's levels. Current estimates of the CH_4 budget assign 20-40% to natural sources, 20% to anthropogenic fossil fuel related sources and the remaining 40-60% to other anthropogenic sources.

The annual CH_4 increase from 1991 to 1992 was much smaller than in the previous decade, 1992 to 1993 levels were unchanged, and in late 1993 CH_4 apparently started to increase again. This anomaly was largest at high latitudes in the Northern Hemisphere, and could be explained by a rapid drop, of about 5%, in global annual emissions. Longer-period variations in the growth rate of CH_4 have been observed for the 1920s and 1970s from air trapped in ice cores.

Methane is the only long-lived gas that has clearly identified chemical feedbacks: increases in atmospheric CH_4 reduce the concentration of tropospheric hydroxyl radical (OH), increase the CH_4 lifetime and hence amplify the original CH_4 perturbation. A recent analysis has shown that these feedbacks result in an adjustment time for additional emissions of CH_4 equal to 14.5 ± 2.5 years based on a "budget" lifetime of about 10 years that reflects atmospheric chemical losses alone. It is uncertain how much the adjustment time would be affected by the small biological sink, and a range of 11 to 17 years encompasses this additional uncertainty. This lengthening of the effective duration of a CH_4 pulse applies also to any derived perturbations (e.g., in tropospheric O_3).

Several atmospheric chemistry models have calculated the impact of a 20% increase in CH_4 concentrations (from 1715 to 2058 ppbv). Two important results were extracted from these simulations: (1) the chemical feedback of CH_4 on OH chemistry results in a reduction of the CH_4 removal rate ranging from -0.17% to -0.35% for each 1% increase in CH_4 concentration; and (2) predicted increases in tropospheric O_3 varied by a factor of three or more across the models, averaging about 1.5 ppbv throughout most of the troposphere in both tropics and summertime mid-latitudes. The first result was used to derive the methane adjustment time (14.5 ± 2.5 yr or about 1.45 times the lifetime) and the latter to estimate the ratio, about 0.25, of radiative forcing from induced tropospheric O_3 increase to that from the CH_4 increase.

Nitrous oxide (N_2O)

Atmospheric N_2O concentrations have increased from about 275 ppbv in pre-industrial times to 311 ppbv in 1992. The trend during the 1980s was +0.25%/yr with substantial year-to-year variations. A growth rate of 0.8 ppbv/yr corresponds to an imbalance between sources and sinks of about 3.9 $Tg(N_2O)$/yr. If these emissions were frozen then N_2O levels would rise slowly to about 400 ppbv over the next two centuries.

Halocarbons

Tropospheric growth rates of the major anthropogenic source species for stratospheric chlorine and bromine (CFCs, CCl_4, CH_3CCl_3 and the halons) have slowed significantly in response to reduced emissions as required by the Montreal Protocol and its amendments. Total atmospheric chlorine from these gases grew by about 60 pptv (1.6%) in 1992 compared to 110 pptv (2.9%) in 1989. HCFC growth rates are accelerating as they are being used increasingly to substitute for CFCs. Tropospheric chlorine as HCFCs increased in 1992 by about 10 pptv compared to 5 pptv in 1989. Stratospheric chlorine levels are expected to peak in the next decade, and it is expected that stratospheric ozone depletion will follow this, recovering slowly in the first half of the next century.

For these long-lived, well-mixed gases, atmospheric lifetimes have been derived based on the recent re-evaluation of the lifetimes of CFC-11 (50 ± 5 yr) and CH_3CCl_3 (5.4 ± 0.6 yr). These new lifetimes are slightly smaller than those given in previous assessments and have significantly greater certainty. The global mean lifetimes for the well-mixed gases are used to infer sources and sinks, and to predict the increase in atmospheric concentration due to specified anthropogenic emissions.

Stratospheric ozone (O_3)

Trends in total ozone since 1979 have been updated to May 1994, and estimates of depletion since 1970, attributed to increases in halocarbons, have been made: (1) Northern Hemisphere mid-latitude loss is significantly negative in all seasons, with winter/spring cumulative losses of 10%; (2) tropical (20°S - 20°N) losses are small and not statistically significant; (3) southern mid-latitude losses are significant in all seasons (4-5%/decade since 1979). Unusually low values (lower than would be expected from an extrapolation of the 1980s trend) were observed in the 1991 to 1994 period, especially at Northern mid- and high latitudes. The Antarctic ozone "holes" of 1992 and 1993 were the most severe on record; for instance, parts of the lower stratosphere contained extremely low amounts of ozone corresponding to local depletions of more than 99%.

The eruption of Mt. Pinatubo in 1991 led to a massive increase in sulphate aerosol in the lower stratosphere. Observational evidence shows that this thirtyfold increase in aerosol surface area greatly enhanced the heterogeneous chemistry and accelerated photochemical loss of O_3. Further evidence points to an additional heating of the stratosphere by Mt. Pinatubo aerosols, resulting in circulation changes that altered the distribution of O_3 in the tropics immediately following the eruption, and possibly also in mid-latitudes.

Tropospheric ozone (O_3)

Tropospheric ozone appears to have increased in many regions of the Northern Hemisphere. Observations show that tropospheric ozone, which is formed by chemical reactions involving other hydrocarbons, carbon monoxide and some nitrogen oxides (NO_x), has increased above many locations in the Northern Hemisphere over the last 30 years. However, in the 1980s, the trends were variable, being small or non-existent. At the South Pole, a decrease has been observed; however in the Southern Hemisphere as a whole, there are insufficient data to draw strong inferences. Model simulations and limited observations together suggest that tropospheric ozone has increased, perhaps doubled, in the Northern Hemisphere since pre-industrial times.

NO_x and other short-lived tropospheric ozone precursors

Uncertainties in the global budget of tropospheric ozone are associated primarily with our lack of knowledge of the distribution of O_3, its short-lived precursors (NO_x, hydrocarbons, CO), and atmospheric transport. Observations of NO_x are just beginning to describe the global atmospheric distribution and they show the large variability in this ozone-producing species. Even with the observed distributions we cannot define the importance of anthropogenic sources (transport of surface pollution out of the boundary layer, direct injection by aircraft) relative to natural sources (lightning, stratospheric input) in controlling the global NO_x distribution. Current estimates of anthropogenic NO_x sources attribute 24 Tg(N)/yr to fossil fuel combustion at the surface, 0.5 Tg(N)/yr to aircraft emissions, 8 Tg(N)/yr to biomass burning and as much as 12 Tg(N)/yr release from soils, including fertilised fields. Anthropogenic emissions dominate natural sources in magnitude, but natural emissions may dominate a large fraction of the atmosphere remote from anthropogenic emissions.

Changes in ozone concentrations in the upper troposphere and lower stratosphere impact the radiative forcing. In addition to anthropogenic sources of O_3 precursors from the lower troposphere, aircraft currently represent a direct anthropogenic source of NO_x in that altitude range. Research evaluating the climatic effect of the current subsonic fleet is incomplete, but initial estimates of the possible changes to O_3 show that this radiative forcing is similar to, but not greater than, that of the CO_2 from the combustion of aviation fuel, about 3% of all fossil fuel combustion.

For compounds with lifetimes much shorter than 6 months, mixing within the troposphere is not rapid enough to average over variations in chemical loss. For these short-lived gases (e.g., NO_x, hydrocarbons, O_3 and CO), the mean concentration cannot be accurately calculated from the product of emissions or production and a global mean lifetime, but their distribution and impacts may be evaluated in future assessments by the more advanced three-dimensional atmospheric chemistry models.

Intercomparisons of atmospheric chemistry models

Two model intercomparison exercises have been conducted to test the ability of models to simulate (a) the transport of shortlived tracers and (b) the basic features of O_3 photochemistry. More than 20 models participated. A high degree of consistency was found in the global transport of a shortlived tracer within the 3D Chemistry /Transport Models (CTMs), but distinctly different results were found amongst 2-D models. General agreement was also found in the computation of photochemical rates affecting tropospheric O_3. These are the first extensive intercomparisons of global tropospheric models.

2.1 Introduction

The Earth's atmosphere is composed primarily of the gases N_2 (78% dry), O_2 (21%) and Ar (1%), whose abundances are controlled over geologic time-scales by uptake and release from crustal material, by degassing of the interior and by the biosphere. Water vapour (H_2O) is the next largest, though highly variable, constituent present mainly in the lower atmosphere at concentrations as high as 3%, where evaporation and precipitation control its abundance. The remaining gaseous constituents are considered trace gases, comprising less than 1% of the atmosphere, yet playing a disproportionately important role in the Earth's radiative balance. Among these, the greenhouse gases include ozone (O_3), methane (CH_4), nitrous oxide (N_2O), chlorofluorocarbons (CFCs), H_2O in the upper atmosphere, as well as carbon dioxide (CO_2). All but CO_2 are controlled in one way or another by atmospheric chemistry. Table 2.1 summarises the mean tropospheric abundance and atmospheric burdens of the more important of these gases.

The climatic impacts of the radiatively active trace gases increase as their global mean abundances increase. It is important to understand the atmospheric chemical cycles involving these gases in order to comprehend the changes in atmospheric composition observed to date and to predict future composition in response to alterations in natural and human-induced emissions. This chapter describes the life cycles of the more important gases, noting how chemical reactions in the atmosphere balance the emissions. A brief review of important chemical processes and of the concept of residence times is followed by a summary of the known sources, sinks and time-scales affecting CH_4, N_2O and the halocarbons (chlorine and bromine containing organic compounds such as the CFCs). Ozone is a special case emphasising the complexities of atmospheric chemistry: direct emissions of O_3 are not important; in situ production, loss and transport by the atmospheric circulation control its abundance. The net production in the lower atmosphere is controlled by a suite of short-lived trace gases (NO + NO_2 (= NO_x), non-methane hydrocarbons (NMHC) and carbon monoxide (CO)), as well as CH_4. These gases are not significant for their direct radiative effects, but rather for their ability to control the abundances of O_3 and other greenhouse gases such as CH_4.

The complexity of the ozone chemistry requires that our numerical models simulate not only the chemical reactions directly involving O_3, but also the global distributions of the short-lived species such as NO_x, and the related transport processes. The current models used to predict the chemistry of the lower atmosphere are, as a class, poorly analysed in comparison with those used to assess stratospheric ozone depletion. We examine some recent model studies, particularly those evaluating the meteorological transport. Because of current inadequacies in the atmospheric models and uncertainties in emissions of some gases, our assessment of stabilising the radiative impact of the trace greenhouse gases (i.e., the atmospheric composition) relies on a combination of models and recent observations.

2.2 Atmospheric Chemistry

2.2.1 Chemical Processes and the Removal of Trace Gases

The atmospheric residence times of many gases are determined primarily by chemical reactions. The bulk of the atmospheric mass, and of most greenhouse gases, is in the troposphere, which comprises on average the lowest 13 km of the atmosphere. Nearly all the the remaining atmosphere (~15%) is in the stratosphere from about 13 to 50 km altitude, the region containing most of the ozone (O_3). Tropospheric removal is dominated by reactions with the hydroxyl radical (OH); whereas stratospheric removal is dominated by ultraviolet photolysis. Tropospheric chemistry involving the production and destruction of O_3 is one of the most complex problems addressed by global atmospheric chemistry models to date.

The time-scales of atmospheric chemistry are carefully defined here, both in concept and usefulness. We examine the different loss mechanisms for the long-lived greenhouse gases and use theoretical models and empirical data to derive with a fair degree of confidence the residence times, i.e., the time for a perturbation to decay to 1/e (37%) of its original value. Less is known about the importance of the more complex feedback loops within atmospheric chemistry. An example shows the far reaching effects of increases in CH_4, but current understanding does not allow quantification of the complete range of such feedbacks.

Photochemistry and Losses

Most atmospheric chemistry is initiated by ultraviolet (UV) sunlight. Since only light with wavelengths greater than 290 nm reaches the troposphere, the number of compounds that can be directly photodissociated is limited. Most of the photochemical chains of interest begin with the dissociation of ozone (O_3) at wavelengths below 320 nm.

$$O_3 + \text{UV-sunlight} \rightarrow O_2 + O(^1D). \quad (2.1)$$

A fraction of this highly reactive form of atomic oxygen, $O(^1D)$, reacts with water vapour and is the primary source of tropospheric hydroxyl (OH),

$$O(^1D) + H_2O \rightarrow OH + OH. \quad (2.2)$$

In the troposphere, the OH radical is the most important chemical removal agent. It reacts with virtually all molecules containing hydrogen atoms (e.g., CH_4); in other

Table 2.1: Current atmospheric abundances and lifetimes of radiatively active trace gases.

Species	Mixing ratio (ppbv) 1992	1992 minus 1990	pre-ind.	Burden (Tg)	Lifetime (yr)
Strat H_2O	2,000-6,000				
Strat O_3	200 10,000			2,900	
Trop H_2O	10,000-20,000,000				
Trop O_3	10-200			300	
CO_2	356,000	2,000	278,000	760,000 (C)[††]	see Ch.1
CH_4	1,714	14	700	4,850	11-17[†]
N_2O	311	1.4	275	1,480 (N)[††]	120
CFC-11	0.268	0.005	0	6.2	50 (±5)
CFC-12	0.503	0.026	0	10.3	102
CFC-113	0.082	0.005	0	2.6	85
CFC-114	0.020	0.001	0		300
CFC-115	<0.01		0		1,700
CCl_4	0.132	-0.001	0	3.4	42
CH_3CCl_3	0.160	0.007	0	3.5	5.4 (±0.6)
CH_3Cl	0.600	?	0.600	5.0	1.5
HCFC-22	0.105	0.014	0	1.5	13.3
CH_3Br	0.012	0.0006	0.008	0.15	1.3
CF_2ClBr	0.0025	0.00014	0	0.08	20
CF_3Br	0.0020	0.0003	0	0.05	65
CF_4	0.070		0	0.9	50,000
C_2F_6	0.004		0		10,000
SF_6	0.0025		0		3,200
HCFC-123					1.4
HCFC-124					5.9
HCFC-141b					9.4
HCFC-142b	0.0035	0.001	0		19.5
HCFC-225ca					2.5
HCFC-225cb					6.6
HFC-125					36.
HFC-134a					17.7
HFC-143a					55.
HFC-152a					1.5
HFC-227ea					43

Note: All mixing ratios, except for stratospheric O_3 and H_2O, refer to mean tropospheric values.
† For CH_4 the adjustment time for decay of a perturbation is shown, rather than a "budget" lifetime - see text. The value of adjustment time conveyed to Chapter 5 for the calculation of GWPs, 14.5 ± 2.5 yrs, does not include possible losses of CH_4 in the soil.
†† The atmospheric burdens for CO_2 and N_2O are expressed in Tg(C) and Tg(N) respectively.

reactions it oxidises species such as CO, nitrogen dioxide (NO_2) and sulphur compounds (e.g., SO_2). In the unpolluted troposphere, the major reactions of OH are:

$$OH + CH_4 + O_2 \stackrel{net}{\rightarrow} CH_3O_2 + H_2O \qquad (2.3)$$
$$OH + CO + O_2 \stackrel{net}{\rightarrow} CO_2 + HO_2 \qquad (2.4)$$

The loss of OH is matched by the production of peroxy radicals (HO_2, CH_3O_2). When OH reacts with other, non-methane hydrocarbons (NMHC), such as ethane, analogous organic peroxy radicals are formed. Subsequent oxidation of CH_3O_2 and its NMHC analogues leads to production of one or more HO_2, which is recycled to OH by the reactions,

$$HO_2 + NO \rightarrow OH + NO_2 \qquad (2.5)$$
$$HO_2 + O_3 \rightarrow OH + O_2 + O_2. \qquad (2.6)$$

The reactions (2.3) or (2.4), followed by (2.5) or (2.6), form a catalytic chain which destroys methane and carbon monoxide but may regenerate OH. The cycling of OH and HO_2 is terminated by reactions involving OH, HO_2 and NO_2. These regenerate H_2O or form peroxides (H_2O_2, CH_3OOH, etc.) which are subsequently removed by clouds and precipitation. In a polluted environment with large concentrations of NO_2, the major loss of OH is through formation of nitric acid (HNO_3).

$$OH + NO_2 \rightarrow HNO_3 \qquad (2.7)$$

This reaction is also the major loss mechanism for NO_x (NO + NO_2) throughout the troposphere, since most of the HNO_3 is lost from the atmosphere by washout and dry deposition. Like the OH-HO_2 pair above, the NO_x pair is tightly linked through the reactions,

$$NO + O_3 \rightarrow NO_2 + O_2 \qquad (2.8)$$
$$NO_2 + sunlight + O_2 \stackrel{net}{\rightarrow} O_3 + NO \qquad (2.9)$$

In addition to the daytime photochemistry, there are some night-time reactions that affect O_3 and NO_x. The most important process is the reaction of NO_2 with O_3 to form a nitrate radical (NO_3), which further reacts, e.g., to form nitrate aerosol, so that the net effect is the removal of both O_3 and NO_x.

Molecules that escape tropospheric oxidation by OH, or removal by wet or dry deposition, will reach the stratosphere where they encounter sunlight with much shorter wavelengths, as low as 180 nm. These photons have sufficient energy to dissociate directly many of the compounds that could not be destroyed in the troposphere. This photolysis initiates the oxidation processes that, for example, turn CFCs into CO_2, HF and a mix of chlorine compounds. In addition, concentrations of O(^{1}D) from reaction (2.1) are higher in the stratosphere and contribute to the loss of the more stable gases. For gases destroyed primarily in the stratosphere, atmospheric lifetimes range from 40 to 200 yr and are limited at the lower range by the rate of transport into the stratosphere.

A very few greenhouse gases (e.g., CFC-115 and C_2F_6) are so stable that they are not efficiently photolysed in the stratosphere. Significant losses of these molecules occur above 60 km altitude through photolysis with Lyman-alpha sunlight (121.6 nm). The local rate of photolysis and the fractional mass of the atmosphere over which these losses occur are so small that the atmospheric lifetime of these gases exceeds 1000 yr (Ravishankara et al., 1993) and their atmospheric abundances equal accumulated emissions.

Ozone

The sunlight at 180-230 nm is also absorbed by molecular oxygen (O_2), which is photodissociated, generating O_3 in the process. This absorption is sufficiently strong to prevent these wavelengths reaching below about 20 km altitude in the atmosphere. The stratospheric amount of O_3 is controlled by a balance between this production and transport and the catalytic loss-cycles involving Cl and Br species, NO_x, OH and HO_2. The dominant source of stratospheric Cl today, 80%, is the family of industrial chlorocarbons, and human activity has also led to increases in the flux of Br into the stratosphere. Stratospheric NO_x is generated from N_2O. Stratospheric H_2O (the source of OH and HO_2) will increase along with any increases in atmospheric CH_4. Stratospheric O_3 is the major species absorbing solar UV between 220 nm and 320 nm. Hence, depletion of stratospheric O_3 is likely to change tropospheric chemistry through: (i) a lowered flux of O_3 into the troposphere, and (ii) increased tropospheric UV flux which enhances the primary production of OH.

While the flux of air from troposphere to stratosphere in the tropics represents a major loss process for many of the long-lived species, the return flux in the mid-latitudes brings significant sources of O_3 and NO_x into the upper troposphere. On a globally averaged basis this source of tropospheric O_3 is countered by removal through reactions with alkenes and NO near the Earth's surface and by dry deposition at the surface. In addition, comparable sources and sinks for O_3, and even larger sources for NO_x, occur throughout the troposphere.

The sequence consisting of the reactions (2.4), (2.5) and (2.9) illustrates how molecular oxygen is activated: an oxygen atom is transferred via HO_2 (or other peroxy radicals) to NO_2, and eventually leads to a net formation of O_3. In the destruction of CO, for example, both OH + HO_2 and NO_x are preserved and catalytically generate O_3. In the absence of NO_x, the reaction sequence (2.4) and (2.6) destroys both CO and O_3. NMHC also play an important role in producing O_3, the key being that they enhance overall the OH and HO_2 reactions ((2.3), (2.4), (2.6) and

(2.7)). The balance between in situ production and loss of O_3 depends on the NO concentration: production by reaction (2.5) followed by (2.9) is favoured over loss by reaction (2.6) when the NO/O_3 concentration ratio exceeds about (10 pptv)/(30 ppbv). Thus, in most parts of the troposphere, the addition of NO will induce additional production of O_3. In heavily polluted urban environments, where NO_x concentrations are already high, addition of NO can lead to a reduction in local O_3 concentrations, but will lead to increased O_3 levels farther downwind.

2.2.2 Atmospheric Adjustment Times of the Trace Gases

The budget of a trace gas is defined by its sources and sinks. The lifetime describes the rate at which the compound is removed from the atmosphere by chemical processes or by irreversible uptake by the land or ocean. If emissions are uniform, the concentration of a trace gas builds up to a steady-state value at which the product of the global mean abundance (Tg) with the inverse global mean lifetime (1/yr) balances the emissions (Tg/yr). The sum over the chemical rates described above defines the budget of a trace gas (i.e., the sources and sinks) and thus the time-scale for a small perturbation in its abundance to disappear. This time is defined formally as the "adjustment time", and is the essential quantity used in evaluating indices such as the Greenhouse Warming Potential (GWP). For a long-lived trace gas, the average atmospheric adjustment time is the e-folding time (63% reduction) of an additional kilogram added to the atmosphere to be removed by chemical processes. A second time-scale can be defined based on the integrated loss divided by the total abundance. This turn-over time, often referred to as the "lifetime" in atmospheric chemistry, relates the observed global mean abundance with the emissions needed just to sustain it.

These integrated time-scales require global summation of the local chemical loss processes acting on globally varying concentrations of the trace gas. For a compound such as CH_4, these vary widely throughout the atmosphere: the OH abundances depend on sunlight and other short-lived, highly variable constituents such as NO_x, H_2O and O_3; and the reaction rate of CH_4 with OH varies by a factor of 20 over the range of lower atmospheric temperatures. The atmospheric lifetime collapses all of these local chemical losses into a single, global-mean loss frequency. By integrating CH_4 losses over the globe for a year (i.e., $Tg(CH_4)/yr$) and dividing by the average CH_4 abundance (i.e., $Tg(CH_4)$), one derives the inverse of the atmospheric lifetime, τ. Because the lifetime is defined relative to the global atmospheric burden, we can add individual inverse lifetimes (i.e., losses) to get a total inverse lifetime, e.g.,

$$1/\tau = 1/\tau_{OH} + 1/\tau_{strat} + 1/\tau_{ocean} + 1/\tau_{land} + 1/\tau_{chem} + 1/\tau_{wash}$$

where the terms are reaction with tropospheric OH, stratospheric photolysis, irreversible oceanic uptake, net destruction at the land surface, additional chemical loss in the troposphere and removal by precipitation (washout), respectively. Calculation of these lifetimes is a difficult problem for global atmospheric chemistry models, but a good estimate helps constrain the total source of a compound. If observations indicate steady state, then sources are balanced by sinks. In the case of gases with measured trends in atmospheric abundance, the inferred source equals the loss calculated from the lifetime plus the annual increase in abundance derived from the trend.

In general, the adjustment time for a trace gas can be approximated to first-order by the lifetime, for which we often have a good estimate, either theoretical or empirical. Most trace gases have little impact on the chemistry of the atmosphere (e.g., the CH_3CCl_3 abundance is too low to affect OH concentrations), and for these gases the adjustment lifetime is equivalent to the lifetime. However two gases, CH_4 and N_2O, do interact strongly with the global atmospheric chemistry, and the exact calculation of their adjustment time requires inclusion of feedbacks and multiple reservoirs.

For example, it has been known that the CH_4-CO-OH system in the troposphere is highly coupled (Chameides *et al.*, 1976; Sze, 1977): increased CH_4 oxidation leads to additional CO formation; both suppress OH, the major sink for CH_4; and thus the time-scale for decay of the perturbation lengthens. This feedback was reported by the global models and treated in IPCC (1990) as an "indirect" GWP based on model calculations of the decrease in global OH for CH_4 increases (Isaksen and Hov, 1987). However, the concept of an extended adjustment time for perturbations has not been applied to GWPs until this assessment. The ratio of adjustment time to lifetime is greater than one and is independent of the magnitude of the methane perturbation for modest changes (Prather, 1994). Because of differing NO_x and NMHC fields, models of global tropospheric chemistry differ in the strength of their CH_4-CO-OH coupling (see Section 2.10.2). Thus the adopted adjustment time for CH_4, based on atmospheric chemistry alone, is 14 ± 2.5 yr using a chemical sink (OH and stratospheric loss) of 10 yr. If the biological soil sink were to respond to the atmosphere in the same way as the chemical sink, the response time would be reduced to a little under 14 yr. In this chapter we use the range 14 ± 3 (11-17) yr to cover this uncertainty.

The other major compound for which the lifetime and adjustment time are expected to be different is N_2O. The stratospheric NO_y derived from N_2O controls stratospheric O_3 and hence the ultraviolet radiation field in the stratosphere and finally the N_2O loss frequency. In contrast with CH_4, one would expect the adjustment time for N_2O to be shorter than its lifetime by about 10% because increases in N_2O lead to a greater photolytic destruction rate. For N_2O, however, there may be additional complications, similar to those for CO_2, due to the potentially important reservoir of N_2O dissolved in the oceans. If atmospheric N_2O concentrations were to change, such an oceanic reservoir would be likely to interact with the atmosphere on a time-scale longer than the 120-year adjustment time given in Table 2.1. Neither of these effects has been studied sufficiently to be included quantitatively here.

Short-lived trace gases such as NO_x (1-10 days) and CO (1-4 months) have not been included in this discussion because it is impossible to assign a useful global mean lifetime. Such gases have highly variable local loss rates, e.g., CO varies from 1 per month in the tropics, and 0.25 per month at mid-latitudes in spring and autumn, to very small values in high latitude winter. The atmospheric circulation does not mix CO from equator to pole within the troposphere in much less than a season. Thus emissions of CO in the tropics will decay in a month, whereas high latitude emissions in winter will accumulate for a season until they mix to lower latitudes. In the case of long-lived gases such as CH_4 with an adjustment time of about 11 to 17 years, the surface emissions become reasonably well-mixed throughout the troposphere before substantial losses occur, and thus the adjustment time does not depend on the location of the sources. For short-lived gases this is not true and they do not in general have well defined adjustment times, although a reasonable upper limit can be made by combining the chemistry with the atmospheric transport models. Given the finite rate of tropospheric mixing (about two weeks vertically, about three months from pole to tropics, and about one year between hemispheres) the concept of a global adjustment time as currently used in GWPs is of limited use for molecules such as NO_x, O_3 or even CO.

The discussion of adjustment time and turn-over time has assumed that we are dealing with small perturbations about the current atmosphere. We have observed changes in CH_4, N_2O and probably tropospheric O_3 since the pre-industrial atmosphere (see following discussion). These changes have been large, and we expect that the adjustment times for most gases have changed since 1850 and will continue to change into the future in response to changing emissions of a wide range of gases. Current trends in the lifetime of methane appear to be small and difficult to detect (e.g., Prinn *et al.*, 1992).

2.2.3 Current Tropospheric OH

The lifetimes of many greenhouse gases depend on the global OH field. Concentrations of OH respond almost instantly to variations in sunlight, O_3, NO_x, CO, CH_4 and NMHC; and therefore the OH field varies by orders of magnitude in space and time. Direct observations of OH can be used to test the photochemical models under specific circumstances, but are not capable of measuring the global OH field. Therefore we must rely on numerical models to estimate the global, seasonal distribution of OH; these models need to simulate the variations in sunlight caused by clouds, ozone column and time-of-day in addition to the local chemical fields. These calculations of global tropospheric OH and the consequent derivations of lifetimes have not advanced significantly and are still much the same as in WMO(1990): we cannot expect such calculations to achieve an accuracy much better than ±30%.

The modelled OH fields can be tested in an averaged sense by comparison with the budgets for some compounds which are removed mainly by OH. These compounds must be reasonably long-lived (and therefore well mixed), have well-known source strengths and only small, well-defined losses apart from OH removal. In addition, their atmospheric burdens must be well measured and well calibrated. However, this empirical derivation of a weighted, global-mean OH does not help test that part of the OH distribution that controls the NO_x lifetime and hence the net production of O_3 in the troposphere.

Empirical OH-Lifetimes based on CH_3CCl_3, ^{14}CO, and HCFC-22

CH_3CCl_3 fulfils all of the above requirements for calibrating tropospheric OH. A recent assessment (Kaye *et al.*, 1994) has reviewed and re-evaluated the lifetimes of two major industrial halocarbons, methylchloroform (CH_3CCl_3) and CFC-11. An optimal fit to the observed concentrations of CH_3CCl_3 from the five ALE/GAGE surface sites over the period 1978 to 1990 was done with two statistical/atmospheric models: the ALE/GAGE 12-box atmospheric model with optimal inversion (Prinn *et al.*, 1992); and the NCSU/UCI 3D-GISS model with autoregression statistics (Kaye *et al.*, 1994). There are well-defined differences in the two atmospheric models' simulations; while important for modelling CFC-11, these have no impact on the derived CH_3CCl_3 lifetimes. But the empirical lifetime and its possible trend are closely tied to the absolute calibration of CH_3CCl_3 measurements. With the ALE/GAGE calibration factor, a lifetime of 5.7± 0.3 yr in 1990 with a decrease in lifetime of 1± 0.8%/yr from 1978 to 1990 is obtained. For the calibration factor from CMDL (0.88 times the ALE/GAGE value, Fraser *et al*, 1994), a lifetime of 5.1 ± 0.3 with a significantly smaller trend is derived. The interpretation of such a trend as increasing tropospheric OH is unresolved, since other factors, including the known geographic change in CH_3CCl_3 emissions, may influence the observations. Given the uncertainties in absolute calibration, we choose

Table 2.2: CH_3CCl_3 Lifetime (yr) and the Range in OH-based Lifetimes

total	(range)	strat	(range)	ocean	(range)
5.4	(4.8 - 6.0)	45	(40 - 50)	85	(50 - infinite)
upper limit:	1/8.2 = 1/6.0 - 1/40 - 1/50				
central value:	1/**6.6** = 1/5.4 - 1/45 - 1/85				
lower limit:	1/5.3 = 1/4.8 - 1/50 - 1/infinity				
→ tropospheric lifetime due to OH: **6.6 yr (±25%)**					

Note: The lifetimes of many greenhouse gases, including CH_4, are determined primarily by the concentrations of tropospheric OH radicals. Our best calibration of the global average OH is based on the empirical total lifetime for CH_3CCl_3 (5.4 yr), which has three identified sinks: tropospheric OH, stratospheric photodissociation, and oceanic hydrolysis. The magnitude of the lifetime with respect to removal by OH (6.6 yr) is calculated by removing the stratospheric and oceanic components and increases the uncertainty range. The OH-based lifetime for CH_4 is 1.65 times that of CH_3CCl_3, based on the ratio of rate coefficients for reaction with OH. To that loss, a stratospheric lifetime corresponding to 120 yr must be added to get the atmospheric budget lifetime for CH_4 of 10 yr.

a CH_3CCl_3 lifetime of 5.4 yr with an uncertainty range, ±0.6 yr, to encompass the two calibrations. From this total atmospheric lifetime of CH_3CCl_3, the losses to the ocean and stratosphere are deducted to derive the tropospheric lifetime for reaction with OH of 6.6 yr (±25%) as shown in Table 2.2.

Analysis of the tropospheric budget of the radio-isotope ^{14}CO complements the CH_3CCl_3 test of tropospheric OH: the ^{14}CO is produced primarily in the stratosphere and upper troposphere by cosmic rays and has a much shorter lifetime, 1-3 months. Recent modelling (Derwent, 1994b) has supported the earlier results (Volz et al., 1981); however, new observations (Brenninkmeijer et al., 1992; Mak et al., 1992), particularly in the Southern Hemisphere, have not yet been reconciled with the latitudinal distribution of OH generated by global atmospheric chemistry models. The removal of ^{14}CO emphasises more the upper troposphere in the mid-latitudes, whereas the loss of CH_3CCl_3 and similar industrial halocarbons is dominated by the OH in the lower tropical troposphere (Prather and Spivakovsky, 1990).

Another candidate that might offer a test of model predicted OH is HCFC-22 (Montzka et al., 1993). Although HCFC-22 production is well documented (Midgley and Fisher, 1993), the atmospheric emissions are too uncertain to improve the accuracy of the global budget analysis for CH_3CCl_3.

For well-mixed gases destroyed by OH, the mean OH estimated from the budget of one species can be transferred to another by the ratio of their respective rate coefficients. The original AFEAS study, and subsequent assessments (WMO, 1992; IPCC, 1992) have derived lifetimes relative to that of CH_3CCl_3 by scaling their respective rate coefficients for reaction with OH at 277 K (Prather and Spivakovsky, 1990). The values of the derived lifetimes here rely on the precision of the empirical derivation of the CH_3CCl_3 lifetime (Prinn et al., 1992).

2.2.4 Other Atmospheric and Surface Removal

For short-lived gases that react rapidly with OH, the possible impact of other, unidentified removal processes is likely to be small. On the other hand, for long-lived compounds (e.g., CFCs, CCl_4, HCFCs 142b & 143a), small additional losses may noticeably reduce the atmospheric lifetimes calculated here.

The potential role of oceanic uptake and destruction as well as reactions in clouds of HCFC and HFCs was examined by Wine and Chameides (1990) and their conclusion that cloud processes appear unimportant still seems valid. More recently the role of the oceans in CH_3CCl_3 loss (Butler et al., 1991) and CH_3Br buffering (Butler, 1994) has been discussed. In general, oceanic loss is proportional to solubility (very high for CH_3Br) and is limited by the air-sea exchange. Only for these two gases has oceanic loss been included in the lifetimes (Table 2.3).

Chemical reactions with Cl atoms in the marine boundary layer, or possibly NO_3 radicals at night, have been identified as losses for NMHC and HCFC/HFCs. Currently, these losses are thought to be small compared with that from OH. Reactions on biologically or mineralogically active land surfaces have been identified for many long-lived greenhouse gases but none are thought to be important (see sections on individual gases below): these have not been included in the atmospheric lifetimes in Table 2.1.

2.2.5 Lifetimes from Stratospheric Removal

For compounds destroyed only in the stratosphere, a lower limit can be placed on the total lifetime. It can be derived simply from an estimate of how fast tropospheric air is circulated through the stratosphere, with the limiting assumption that all of the trace gas entering the stratosphere is destroyed. Using the mass flux through the 100 mbar surface in the tropics (Holton, 1990) yields a value of 25 yr. The gas which most closely approaches this limit is CCl_4, a rapidly photolysed species, with a lifetime of about 40 years, somewhat longer than the 25 year value, because many CCl_4 molecules re-enter the troposphere.

The derivation of the stratospheric-removal lifetimes in Table 2.1 is based on current two-dimensional stratospheric models which have a long history of development and testing (e.g., Ko *et al.*, 1991; Prather and Remsberg, 1993). These models calculate vertical profiles and relative variations of halocarbons, N_2O and CH_4 in the stratosphere that are consistent with observations. The model derived lifetimes for CFC-11, ranging from 40 to 60 yr, match the empirically derived value of 50 ± 5 years, based on reconciling emissions with the observed trend (see Kaye *et al.*, 1994; and recent update, Cunnold *et al.*, 1994).

2.2.6 Examples of Chemical Feedbacks Affecting Greenhouse Gases

The varied chemical and physical couplings in the atmospheric system mean that a perturbation in one trace gas propagates through a series of interactions into a perturbation of many other species. Methane has an especially large number of identified couplings, some of which are used as an example to illustrate this point. To first order, a kilogram of CH_4 released from the surface becomes well-mixed in troposphere, and a fraction of this CH_4, in proportion to its tropical abundance, is transported into the stratosphere. This added kilogram of CH_4 decays with an adjustment time of 11 to 17 years and not with the chemical sink time-scale of 10 years due to OH and stratospheric loss (see Section 2.2.2 above). This enhanced CH_4 concentration has a direct radiative effect. In the stratosphere, CH_4 directly affects the radical chemistry (Cl/HCl, OH) that controls O_3: by reducing the effectiveness of chlorine catalytic cycles, it would increase O_3. At the same time, oxidation of this CH_4 increases stratospheric H_2O, leading to enhanced efficiency of the HO_x-catalysed O_3 loss and potentially increasing heterogeneous chemistry in the stratosphere by providing more condensable water. Changes to stratospheric H_2O and O_3 have direct greenhouse impacts. Any induced changes to stratospheric O_3 will feed back on tropospheric chemistry through changes in UV and O_3 fluxes into the troposphere. The additional CH_4 also impacts tropospheric chemistry directly through reductions in OH and increased production of O_3. Lower OH increases the lifetimes and hence the concentrations of a whole host of atmospheric trace gases, many of which are direct greenhouse gases. Further, the concentrations of HCFCs and CH_3CCl_3 would increase with a secondary impact on stratospheric chlorine levels and hence O_3. Another secondary impact is in tropospheric chemistry where lower OH will reduce the oxidation of NO_x, which controls tropospheric O_3 production and the recycling of OH. To quantify these interactions, the numerical models must simulate realistically the atmospheric chemistry and transport of the troposphere and stratosphere.

2.3 Methane

After water vapour and carbon dioxide, methane (CH_4) is the most abundant greenhouse gas in the troposphere. It is also a reactive gas, which through various chemical interactions (detailed in Section 2.2) influences the concentrations of other greenhouse gases both in the troposphere and stratosphere, most notably that of ozone.

2.3.1 Methane Sources

CH_4 is emitted by a large number of sources. These are listed in Table 2.3 and total 410-660 Tg(CH_4)/yr globally. This range has shifted little from previous estimates in IPCC (1992). In fact, despite numerous publications of numerous new flux measurements and estimates, only few of the contributions from individual sources have been revised. These revisions are relatively small, usually concerned with the range rather than with the central values of the emissions estimates. Thus, recent flux data from the Amazon region suggest that a large fraction of CH_4 is emitted from tropical wetlands (20°N – 30°S), with a global estimate of ~60 Tg(CH_4)/yr (Bartlett *et al.*, 1990; Bartlett and Harriss, 1993). High latitude tundra studies indicate emissions ranging from 20 to 60 Tg/yr (Whalen and Reeburgh, 1992; Reeburgh *et al.*, 1993). A re-evaluation of the ocean source (Lambert and Schmidt, 1993) suggests only ~3.5 Tg/yr are emitted by the open ocean, but emissions from methane-rich areas could be higher, pushing the upper limit of the oceanic source to 50 Tg/yr. A recent estimate made by Martius *et al.* (1993) for the contribution of termites to the global budget agrees well with the central value of 20 Tg/yr given in the IPCC (1992). A new estimate has been made for methane originating from geological sources (including methane hydrates) of 5-15 Tg/yr (Judd *et al.*, 1993).

Studies of the carbon-14 content of atmospheric CH_4 indicate that 15 - 25% of total annual CH_4 emissions are from fossil carbon sources (IPCC, 1992). However, there are large uncertainties in the contribution of the various fossil sources: coal mines, natural gas and the petroleum industry. New global figures for emissions from coal mines range from 17 Tg/yr to 50 Tg/yr (CIAB, 1992; Müller, 1992; Beck, 1993; Beck *et al.*, 1993; Kirchgessner *et al.*, 1993). Müller (1992) estimated an emission from natural gas activities of 65 Tg/yr, higher than the range given in IPCC (1992) (2542 Tg/yr). Beck *et al.* (1993) give a value of 30 Tg/yr. Khalil *et al.* (1993a) proposed that low

Table 2.3: *Estimated sources and sinks of methane, $Tg(CH_4)/yr$.*
(a) Observed atmospheric increase, estimated sinks and sources derived to balance the budget

	Individual estimate	Total
Atmospheric increase	37 (35-40)	37 (35-40)
Atmospheric removal (lifetime = 9.4 yr)		
tropospheric OH	445 (360-530)	
stratosphere	40 (32-48)	
soils	30 (15-45)	
Total sinks		515 (430-600)
Implied total sources (atmospheric increase + total sinks)		552 (465-640)

(b) Inventory of identified sources

Identified sources	Individual estimate	Total
Natural		
Wetlands	115 (55-150)	
Termites	20 (10-50)	
Oceans	10 (5-50)	
Other	15 (10-40)	
Total identified natural sources		160 (110-210)[†]
Anthropogenic		
Total fossil fuel related		100 (70-120)[††]
Individual fossil fuel related sources		
Natural gas	40 (2550)	
Coal mines	30 (1545)	
Petroleum industry	15 (530)	
Coal combustion[†††]	? (130)	
Biospheric carbon		
Enteric fermentation	85 (65100)	
Rice paddies	60 (20-100)	
Biomass burning	40 (2080)	
Landfills	40 (2070)	
Animal waste	25 (2030)	
Domestic sewage	25 (15-80)	
Total biospheric		275 (200-350)
Total identified anthropogenic sources		375 (300-450)[†]
TOTAL IDENTIFIED SOURCES		535 (410-660)

TOTAL Global Burden: 4850 Tg(CH_4)

Note: The observed increases in methane show that sources exceed sinks by about 35 to 40 Tg each year. All data are rounded to the nearest 5 Tg.

[†] A pre-industrial level of 700 ppbv would have required a source of 210 Tg(CH_4)/yr if the lifetime has remained constant, and 280 Tg (CH_4)/yr if current tropospheric chemical feedbacks can be extrapolated back. The total anthropogenic emissions of CH_4 based on identifed sources, 375 (300-450), is slightly higher than the inferred range from pre-industrial levels, 270-340, but is well within the uncertainties.

[††] Fractional source from fossil carbon based on a measure of the atmospheric ratio of $^{14}CH_4$ to $^{12}CH_4$.

[†††] Judd *et al.* (1993), Khalil *et al.* (1993a)

temperature combustion of coal (not included previously) could be a significant source of methane, with a global emission of 16 Tg/yr and a range of 5-30 Tg/yr. Emissions of methane from the inefficient combustion of coal may be significant. However, there is considerable uncertainty in the magnitude of these emissions and further research is needed to refine preliminary estimates.

Anastasi and Simpson (1993) estimated a 1990 emission of 84 Tg/yr from enteric fermentation in cattle, sheep, and buffalo, which contributes 90% of the total emission from domestic animals, towards the upper part of the range given in IPCC (1992) (65-100 Tg/yr). In a complete re-evaluation of emissions from biomass burning, Andreae and Warneck (1994) report a total of 43 Tg/yr in good agreement with the global estimates made in IPCC (1992). Other recent estimates of CH_4 from biomass burning are 30 Tg/yr (Hao and Ward, 1993), 35 Tg/yr (Subak et al., 1993) and 50 Tg/yr (Levine et al., 1994).

Numerous studies of emissions from rice paddies (e.g., Bachelet and Neue, 1993; Delwiche and Cicerone, 1993; Lal et al., 1993; Subak et al., 1993; Wang et al., 1993a,b; Wassman et al., 1993; Shao et al., 1994; Shearer and Khalil, 1994) indicate a global source of 60 Tg/yr with a range of 20-100 Tg/yr.

New estimates of landfills suggest a global source of about 40 Tg/yr (Müller, 1992; Subak et al., 1993; Tohjima and Wakita, 1993).

As Table 2.3 indicates, about 70% of the CH_4 emissions (375 Tg/yr) result from human activity and approximately 20% of the total emissions are of fossil origin.

2.3.2 Removal of Methane

As detailed in Section 2.2, CH_4 is removed from the atmosphere through reaction with tropospheric OH, about 445 Tg/yr (Table 2.3) and stratospheric removal, 40 Tg/yr. Another significant sink is the microbial uptake in soils, about 15-45 $Tg(CH_4)$/yr. A growing number of studies (reviewed by Reeburgh et al., 1993) show that methane from the atmosphere, and also that diffusing upward from deep soils, is consumed by microbial communities in the upper soils. Methane fluxes at the terrestrial and marine surface are highly variable because they result from the difference of two large processes (in situ production and consumption). It is noted that this oxidation of CH_4 is also important in modulating methane emissions from rice paddies, wetlands and landfills. Disturbance (cultivation, fertilisation) has been shown to reduce the effectiveness of the soil sink in temperate soils (Keller et al., 1990; Mosier et al., 1991; Hutsch et al., 1993; Minami et al., 1993; Ojima et al., 1993).

The budget imbalance depends on the recent trends (see Section 2.3.4) but sources must have exceeded sinks by about 7-8% over the last decade.

2.3.3 Atmospheric Distribution

The fact that sources of CH_4 are essentially land based and therefore lie predominantly in the Northern Hemisphere can explain much of the latitudinal gradient observed in CH_4 with about 6% lower values in the Southern Hemisphere. This difference corresponds to an excess Northern Hemisphere source of about 280 $Tg(CH_4)$/yr. The observed seasonal cycle (±1.2% at mid-latitudes) can be explained in large part by the annual variation in tropospheric OH concentrations, and to a lesser extent by seasonally varying sources and atmospheric transport. The seasonal cycle in the Southern Hemisphere also contains a signal from transport of tropical biomass burning sources (Lassey et al., 1993; Law and Pyle, 1993).

2.3.4 Trends and Sensitivities

Atmospheric CH_4 concentrations have changed considerably over time. Most striking is the increase of more than a factor of two over the last two centuries, derived from air trapped in ice cores (Figure 2.1). During the past decade the rate of that increase has declined. Measurements from two global observing networks have derived globally averaged growth rates for methane of approximately 20 ppbv/yr in 1979-80, 13 ppbv/yr in 1983, 10 ppbv/yr in 1990 and about 5 ppbv/yr in 1992 (Steele et al., 1992; Khalil and Rasmussen, 1993; Dlugokencky et al., 1994a). The decline in growth rate was more rapid in the 30°-90°N semi-hemisphere than in the other semi-hemispheres (Figure 2.2). The change in CH_4 from 1991 to 1992 in the Northern Hemisphere was close to zero. The cause of this global decline in methane growth rates is unknown and still a matter of speculation. It could, for

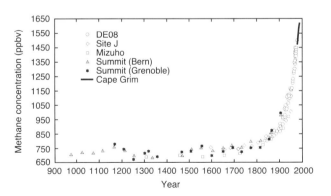

Figure 2.1: Methane mixing ratios from the air bubbles trapped in ice cores from Antarctica over the past 1,000 years (Law Dome (DE08): Etheridge et al., 1992; Mizuho: Nakazawa et al., 1993) and Greenland (Site J: Nakazawa et al., 1993; Summit: Blunier et al., 1993). Atmospheric measurements from Cape Grim, Tasmania, are included to demonstrate the smooth transition from ice core to atmospheric measurements.

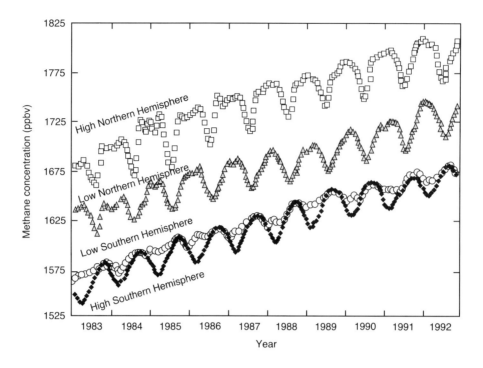

Figure 2.2: Seasonal cycles and trends in atmospheric methane over the decade 1983 to 1992. Results from the NOAA-CMDL global flask network (37 sites) have been gridded to 14-day intervals and averaged over four equal areas of the atmosphere: high northern (30°N - 90°N), low northern (0 - 30°N), low southern (0 - 30°S) and high southern (30°S - 90°S) semi-hemispheres (Steele et al., 1992; Dlugokencky et al., 1994c).

example, be explained by a rapid drop of about 5% in global annual emissions. One suggestion is a decrease in CH_4 emissions from the former Soviet Union (Dlugokencky et al., 1994b). Additional CH_4 data from Antarctica by Aoki et al. (1992) confirm the trends observed by the NOAA-CMDL stations in the same region.

A significant depletion in $^{13}CH_4$ has been observed in the Southern Hemisphere since mid-1991, coincident with the significant drop in the CH_4 growth rate between 1991 and 1992. The isotopic data are used to infer that the change in CH_4 growth rate is due to decreasing sources rather than increasing sinks and that it results from a combination of decreased tropical biomass burning and a lower release of fossil CH_4 in the Northern Hemisphere (Lowe et al., 1994).

Observed methane levels in the high Arctic (Alert, 83°N) in 1993 were actually lower than those observed in 1992. The data indicate that both Europe and the former Soviet Union are major CH_4 sources, but the cause of the change has not been identified (Worthy et al., 1994).

The historic record of methane concentrations through the industrial period has been improved by the analysis of four new ice cores included in Figure 2.1. An Antarctic ice core (Law Dome) covers the period from 1840 to 1980 and overlaps with the direct atmospheric measurements of the 1970s (Etheridge et al., 1992). It thus allows the trends of methane concentration during the industrial period to be studied in detail. An approximate doubling of methane is seen, as in previous ice core studies, but the growth rate is not monotonic, with apparent stabilisation around the 1920s and again during the 1970s (Etheridge et al., 1992; Dlugokencky et al., 1994c). From Greenland and Antarctic ice cores, Nakazawa et al. (1993) have extracted methane records that begin at about 1300 and 1600 respectively and continue to about 1950. The pre-industrial global mean and inter-hemispheric difference were found to be about 720-740 ppbv and 35-75 ppbv respectively, about one third of the present day difference, indicating that the pre-industrial natural sources in the Northern Hemisphere were larger than those in the Southern Hemisphere. A Greenland ice core has produced a further methane record over the past 1000 years (Blunier et al., 1993). A similar pre-industrial level was found (700 ppbv) and pre-industrial variations of about 70 ppbv were detected before AD 1500 which are attributed to the changed oxidising capacity of the atmosphere, climatic impacts on wetlands and possible agricultural sources of methane. Further back in time, large changes in the methane concentration in Greenland ice cores through the deglaciation and the last glacial period are observed to be in phase with inferred temperature (Figure 2.3). The record shows clearly the marked CH_4 variations associated with the younger Dryas event as well as the rapid CH_4 variations during the glacial periods associated with rapid Greenland climatic oscillations. Warm periods, each lasting hundreds of years within the

Other Trace Gases and Atmospheric Chemistry

last glacial period, are associated with methane increases of about 100 ppbv. New data from the Antarctic Vostok ice core have extended the methane record from 160 thousand years before present (kaBP) through the penultimate glaciation to the end of the previous interglacial, about 220 kaBP (Jouzel *et al.*, 1993). There is a clear positive correlation of methane with temperature. Methane minima (300-400 ppbv) occur at times of glacial maxima with approximately double this amount for the interglacials.

There are several potential climate feedbacks that could affect the atmospheric methane budget (IPCC, 1990). Recent attention has been focused on northern wetlands and permafrost, where changes in surface temperature and rainfall are predicted to occur in climate models. Traditionally, temperature increases are assumed to generate increases in CH_4 emissions (Hameed and Cess, 1983). Recent calculations suggest a moderate increase in CH_4 emissions in a $2 \times CO_2$ scenario (Harriss *et al.*, 1993). Climate models generally predict lower soil moisture in mid-continental regions in summer for a $2 \times CO_2$ climate (IPCC, 1990), although changes in soil moisture at high latitudes are model dependent. Lowering the water table increases the thickness of the layer over which methane oxidation can take place, so northern wetlands appear to be more sensitive to changes in moisture than temperature. Persistent water table lowering of 50-75 cm resulted in zero or even slightly negative fluxes (Roulet *et al.*, 1992, 1993), which suggests that a reduction in soil moisture may significantly reduce methane fluxes from northern peatlands and wetlands. However, response of high latitude wetlands to seasonal changes in soil moisture is not well understood (Reeburgh *et al.*, 1993).

Recent data on the methane concentration in permafrost (Kvenvolden and Sorenson, 1993; Rasmussen *et al.*, 1993) suggest a median concentration of 2-3 mg/kg. Estimates of future methane emissions from permafrost are less than 10 Tg/yr based on a one-dimensional heat conduction model that does not consider microbial oxidation (Moraes and Khalil, 1993).

2.4 Nitrous Oxide

Nitrous oxide (N_2O) is an important, long-lived greenhouse gas that is emitted predominantly by biological sources in soils and water, which are not well quantified in global terms (Table 2.4). It is removed mainly in the stratosphere by photolysis and reaction with excited oxygen atoms ($O(^1D)$). A small soil sink has been suggested. The loss of N_2O in the stratosphere yields NO, providing the major input of NO_x to the stratosphere, thus in part regulating stratospheric ozone and influencing the NO_x balance in the upper troposphere.

2.4.1 Sources of Nitrous Oxide

Tropical forest soils are probably the single most important source of nitrous oxide to the atmosphere. New data on tropical land-use changes and intensification of tropical agriculture indicate significant, growing sources of N_2O (Matson and Vitousek, 1990). The flux of N_2O depends on the age of the pasture, with young pasture (<10 years) emitting 3-10 times more N_2O than tropical forests, whereas older pastures emit less N_2O than tropical forest (Keller *et al.*, 1993). However, further research on tropical agricultural systems is required before conclusions can be reached concerning the relative importance of tropical agricultural systems as growing N_2O sources (Keller and Matson, 1994). The total N_2O source from tropical soils (forest, savannah) is estimated at 4 Tg(N)/yr (range 2.7 - 5.7). The magnitude of N_2O emissions from intensively fertilised tropical agricultural soils has not been quantified. No attempt has been made to speciate the tropical soil source into natural and anthropogenic components. More

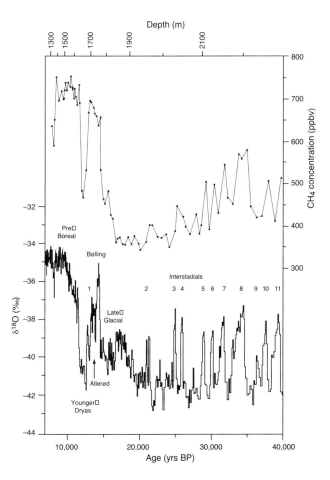

Figure 2.3: The methane record from a Greenland ice core over the period 7,000 to 40,000 years before the present is shown in parallel with $\delta^{18}O$ (a surrogate for temperature). The significant climatic events are named or numbered as interstadial episodes (Chappelaz *et al.*, 1993).

Table 2.4: *Estimated sources and sinks of N_2O typical of the last decade (Tg(N)/yr).*

	Range	Likely
Atmospheric increase	3.1 - 4.7	3.9†
Sinks		
stratosphere	9 - 16	12.3
soils	?	
Total Sinks	9 - 16	12.3
Implied total sources (atmospheric increase + total sinks)	13 - 20	16.2
Identified sources	Range	Likely
Natural		
oceans	1 - 5	3
tropical soils		
wet forests	2.2 - 3.7	3
dry savannas	0.5 - 2.0	1
temperate soils		
forests	0.1 - 2.0	1
grasslands	0.5 - 2.0	1
Total identified natural sources	6 - 12	9
Anthropogenic		
cultivated soils	1.8 - 5.3	3.5
biomass burning	0.2 - 1.0	0.5
industrial sources	0.7 - 1.8	1.3
cattle and feed lots	0.2 - 0.5	0.4
Total identified anthropogenic	3.7 - 7.7	5.7
TOTAL IDENTIFIED SOURCES	**10 - 17**	**14.7**

† The observed atmospheric increase implies that sources exceed sinks by 3.9 Tg(N)/yr.

reliable estimates require better models of N_2O emissions from soils and better data bases of underlying soil and ecosystem properties.

There is evidence to show that grasslands emit small amounts of N_2O (Mosier *et al.*, 1981, 1991; Minami *et al.*, 1993; Meyer *et al.*, 1994) and Krielman and Bouwman (1994) estimated a global grassland source of 1.4 Tg(N)/yr. A global value of approximately 1 Tg(N)/yr has been adopted in Table 2.4.

The Earth's oceans are significant N_2O sources. Nitrous oxide fluxes from upwelling regions of the Indian (Law and Owen, 1990) and Pacific (Codispoti *et al.*, 1992) Oceans suggest that oceans may be a larger source than previously estimated (1.4 - 2.6 Tg(N)/yr; IPCC 1992). The total pre-industrial N_2O source was approximately 9 Tg(N)/yr of which approximately 3 Tg(N)/yr was oceanic (Weiss, 1994). An isotopic study of atmospheric N_2O (Kim and Craig, 1993) suggests a large gross flux of N_2O between the atmosphere and the ocean, but the possible implications for net fluxes are not clear. The ocean flux estimate that has been adopted for this assessment is 3 Tg(N)/yr (range 1-5).

New research suggests that N_2O emissions from cropped, nitrogen-fertilised (by mineral, manure and legumes) agricultural systems are significant on a global scale. Such sources have been extensively evaluated by IPCC Working Group II (personal communication) and estimated at 3.5 Tg(N)/yr (range 1.8 - 5.3).

Nitrous oxide is also emitted by a large number of smaller sources, most of which are difficult to evaluate. These include soils of other natural ecosystems, biomass burning, degassing of ground water used for irrigation and specialised industrial processes. There are very few studies of these sources, particularly in the tropics, and uncertainties in their emission estimates are large. These sources are listed in Table 2.4 and are largely based on IPCC (1992). A few, previously unevaluated sources have been added including N_2O from cattle and feed lots (0.2-0.5 Tg(N)/yr; Khalil and Rasmussen, 1992) and N_2O from cars fitted with catalytic converters (0.1-0.6 Tg(N)/yr; Dasch, 1992; Khalil and Rasmussen, 1992; Berger *et al.*, 1993), the latter having been included in the figure for industrial sources, along with adipic acid production. The biomass burning source (0.2-1.0 Tg(N)/yr) may be underestimated due to unaccounted for enhanced post-burning biogenic emissions (Levine, 1994).

Table 2.4 indicates that about 40% of N_2O sources are anthropogenic. This figure could be an underestimate because tropical agricultural soil sources resulting from human activities have not been separated from natural tropical soil sources. Nitrous oxide from natural sources is estimated at 9 Tg(N)/yr (range 6-12), in agreement with Weiss (1994). These estimates of natural sources fall at the lower end of those needed to maintain pre-industrial concentrations of 275 ppbv, about 11 ± 3 Tg(N)/yr.

2.4.2 Removal of Nitrous Oxide

The major sink for N_2O is photodissociation by sunlight (wavelengths 180-230 nm) in the stratosphere; a secondary removal process, about 10%, occurs through reaction with $O(^1D)$. The best estimates for the lifetime of N_2O come from the 2D stratospheric chemical transport models that have been tested against observed distributions and tracer-tracer correlations (Prather & Remsberg, 1993) and that include the accurate modelling of transmission of ultraviolet sunlight (Minschwaner *et al.*, 1993). The current best estimate for stratospheric removal is 120 ± 30 yr. Mahlman *et al.* (1986) reached a similar conclusion based upon 3-D transport model considerations. There is some evidence that N_2O is consumed by some soils (cf. Donoso *et al.*, 1993), but there are not enough data to make a reasonable global estimate of this sink. The consequences of a significant ocean reservoir of nitrous oxide have been neglected here in the consideration of the above lifetime. An ocean reservoir could potentially extend the response time for N_2O beyond the above estimate.

The strengths of these sinks is summarised in Table 2.4. No estimate is made of the soil sink. The sum of the stratospheric loss and the atmospheric increase is about 16 Tg(N)/yr (range 12-20). Although not indicated by the uncertainty ranges, this is probably better known than the total N_2O source (approximately 15 Tg(N)/yr, range 10-17). Based on these estimates the identified sources exceed sinks by about 15%, but clearly the uncertainty ranges are such that there may not be any major, unidentified sources. Sources and sinks whose strengths have not been estimated, or may be underestimated, are tropical agriculture, biomass burning, temperate grasslands (sources) and soils (sinks).

2.4.3 Atmospheric Distribution

Owing to its long lifetime, N_2O exhibits only small spatial and temporal variations in the free troposphere. By empirically scaling with CFC interhemispheric gradients, the interhemispheric difference of about 1 ppbv (0.3%) corresponds to a source imbalance with an excess Northern Hemisphere source of 5 Tg(N)/yr. No seasonal variation has been observed.

2.4.4 Trends and Sensitivities

Analysis of available global nitrous oxide data indicates a clear continuous increase in N_2O since the pre-industrial era. The rate of growth over the last decade is small, about 0.25%/yr, and there is difficulty in measuring the global abundance of N_2O with this precision. For example, the decadal trend from Prinn *et al.* (1990) is 0.7 to 1.0 ppbv/yr and that from Khalil and Rasmussen (1992) is 0.5 to 1.2 ppbv/yr. In another recent analysis, Weiss (1994) showed that the global average abundance at the beginning of 1976 was 299 ppbv and had risen to 310 ppbv at the beginning of 1993. During 1976-82 the growth rate was about 0.55 ppbv/yr, which increased to a maximum of 0.8 ppbv/yr in 1988-89, declining to the current rate of 0.6 ppbv/yr. A similar result has been observed in the 15-year global record of NOAA-CMDL (Elkins *et al.*, 1993a, see Figure 2.4). Zander *et al.* (1994) have extended atmospheric measurements of N_2O from solar spectra collected at Jungfraujoch, and derived a mixing ratio of 275 ppbv for April 1951 together with an increase of about 10% over the past four decades.

Previously, ice core records of nitrous oxide showed an increase of about 8% over the industrial period (Etheridge *et.al.*, 1988; Khalil and Rasmussen, 1988a; Zardini *et al.*, 1989). A new record covering the last 45,000 years was obtained from two ice cores from Antarctica and Greenland (Leuenberger and Siegenthaler, 1992). The Greenland record covers the period 1780 to 1950 and suggests a pre-industrial nitrous oxide level of about 260 ppbv, 10 and 25 ppbv lower than other records. Results from a new Antarctic core (Machida *et al.*, 1994) indicate a pre-industrial level of about 273 ppbv. Thus estimates of pre-industrial levels of N_2O average about 275 ppbv with a range of 260 to 285 ppbv. Nitrous oxide levels have risen approximately 15% since pre-industrial times. The Antarctic core shows that nitrous oxide was about 30% lower than these pre-industrial levels (about 180-190 ppbv)

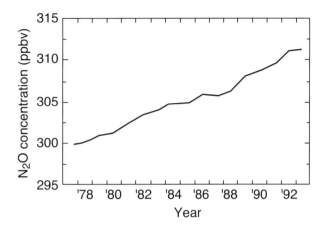

Figure 2.4: Nitrous oxide mixing ratios from 1977 to 1993. Globally averaged values are those reported from the NOAA-CMDL flask sampling network (Elkins *et al.*, 1993a).

at the Last Glacial Maximum (LGM) and between the pre-industrial and LGM levels (220-250 ppbv) between 30 and 50 kaBP, consistent with the hypothesis that soils are a major natural source of nitrous oxide.

2.5 Halocarbons

Perhalocarbon species such as the CFCs, carbon tetrachloride (CCl_4), carbon tetrafluoride (CF_4) and the halons are powerful greenhouse gases, because they strongly absorb terrestrial infrared radiation and have long (typically ≥ 50 years) atmospheric lifetimes. The hydrohalogenated species such as methylchloroform, HCFCs and CFCs are also significant infrared absorbers but their shorter atmospheric lifetimes (typically ≤ 15 years) reduce their climatic impact compared to the perhalocarbon species. All species containing chlorine and bromine play a role in lower stratospheric ozone depletion, and hence climate cooling, and this tends to offset their ability to cause surface warming. The emissions of the major anthropogenic chlorine- and bromine-containing species are now largely controlled (> 90%) by the requirements of the Montreal Protocol. A recent, comprehensive review (Kaye *et al.*, 1994) has provided extensive details on the global distributions, trends, emissions and lifetimes of halocarbons such as CFCs, halons and related species. This section provides a summary and update of that review.

2.5.1 Atmospheric Distributions and Trends

2.5.1.1 CFCs and carbon tetrachloride

CFCs-11, -12 and -113 and carbon tetrachloride have been measured in a number of global programmes and their tropospheric mixing ratios have increased steadily from the mid-1970s to 1990, with CFC-11 and -12 increasing by about 100%, CFC-113 by about 300% and CCl_4 by about 20% (Cunnold *et al.*, 1986; Rasmussen and Khalil, 1986; Makide *et al.*, 1987; Simmonds *et al.*, 1988; Fraser *et al.*, 1994 and references therein). There is now clear evidence that the growth rates of the CFCs and CCl_4 have slowed significantly. CFC-11 and -12 (Figure 2.5) trends in the late 1970s to late 1980s were about 9-11 pptv/yr and 17-20

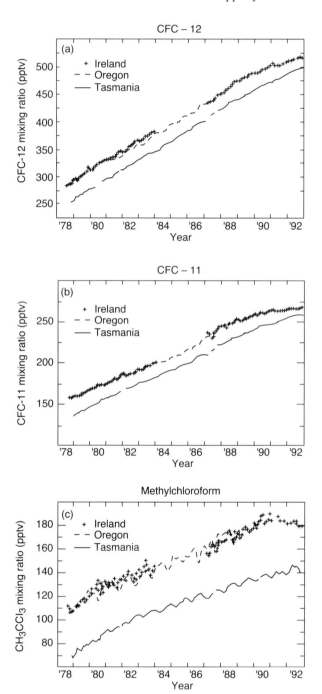

Figure 2.5: Atmospheric abundances of (a) CFC-12, (b) CFC-11, and (c) CH_3CCl_3 from 1978 to 1992 from the ALE-GAGE global sampling network (Prinn *et al.*, 1992, Cunnold *et al.*, 1994; Fraser *et al.*, 1994, Fraser and Derek, 1994 and unpublished data from the ALE/GAGE network). Monthly mean clean air values are shown for three sites: Tasmania, Oregon and Ireland.

pptv/yr respectively. These declined to about 7 and 16 pptv/yr respectively around 1990 and to about 3 and 11 pptv/yr by 1993 (Elkins *et al.*, 1993b; Simmonds *et al.*, 1993; Cunnold *et al.*, 1994; Rowland *et al.*, 1994).

Global CFC-113 and CCl_4 data to the end of 1990 were recently reviewed (Fraser *et al.*, 1994). A global average trend CFC-113 of about 6 pptv/yr was observed, with no sign of the slow-down observed for CFC-11 and -12, whereas carbon tetrachloride appeared to have stopped accumulating in the atmosphere. Global CFC-113 data extended to the end of 1992 now indicate that the growth rate has started to slow down, presumably in response to reduced emissions (Fisher *et al.*, 1994; Fraser *et al.*, 1994). Carbon tetrachloride data collected at Cape Grim, Tasmania, indicate that the levels of this trace gas have started to decline (Fraser and Derek, 1994).

2.5.1.2 Methylchloroform and the HCFCs

Global methylchloroform and HCFC-22 data to the end of 1990 have recently been reviewed (Prinn *et al.*, 1992; Montzka *et al.*, 1993; Fraser *et al.*, 1994), with growth rates in 1990 of 4-5 and 6-7 pptv/yr respectively. The methylchloroform data to the end of 1992 (Figure 2.5) indicate that the slowing of the growth rate observed in 1990 has continued, presumably due to the combination of atmospheric oxidation and reduced emissions in 1991-92 compared to 1990. Part of the declining methylchloroform trend has been ascribed to increasing OH levels (1.0 ± 0.8%/yr, Prinn *et al.*, 1992). The methylchloroform calibration problems detailed in Fraser *et al.* (1994) are important (see Section 2.2) and have yet to be resolved.

Global HCFC-22 data (Montzka *et al.*, 1993) indicate a mean mixing ratio in 1992 of 102 ± 1 pptv, an interhemispheric difference of 13 ± 1 pptv and a globally averaged growth rate of 7.4 ± 0.3 pptv/yr. Based on the latest industry estimates of HCFC-22 emissions (Midgley and Fisher, 1993), the data indicate an atmospheric lifetime for HCFC-22 of 13.3 (11.8-15.2) years, see Table 2.1. The latest HCFC-22 data indicate that the near linear trend observed in earlier data has continued. The possible HCFC-22 calibration problems addressed in Fraser *et al.* (1994) have not yet been resolved. HCFC-142b (CH_3CClF_2) and HCFC-141b (CH_3CCl_2F) have recently been introduced as CFC substitutes, replacing CFC-11 in foam-blowing processes. The study of their accumulation in the atmosphere has only just commenced, but hopefully will provide essential information about their atmospheric lifetimes and conversely about OH levels in the atmosphere. Pollock *et al.* (1992) detected upper tropospheric levels of HCFC-142b at about 1.1 pptv in 1989, growing at 7%/yr. Measurements of these two species has commenced using the NOAA-CMDL flask sampling network. The preliminary 1992 annual global mean HCFC-142b mixing ratio was 3.6 ± 0.1 pptv, growing at ~ 1 pptv/yr (Elkins *et al.*, 1993b). The preliminary 1992 and 1993 annual global mean HCFC-142b mixing ratios for 1992 and 1993 were 3.2 ± 0.1 and 4.3 ± 0.1 pptv respectively, growing by 1.1 pptv/yr. The 1993 mean mixing ratio of HCFC-141b was 0.7 ppbv, growing at 0.9 ppbv/yr into 1994 (Elkins *et al.*, 1993a; Montzka *et al.*, 1994).

2.5.1.3 Other chlorinated species

There are a number of other short-lived chlorinated species such as methylchloride, chloroform, dichloromethane and chlorinated ethenes that are present in the background atmosphere. Their atmospheric lifetimes are short and they do not make a significant contribution to radiative forcing, but methylchloride is a significant source of stratospheric chlorine. Available data on the abundance of these species have been reviewed (Fraser *et al.*, 1994). Global mean methylchloride (CH_3Cl, 600-640 pptv), chloroform ($CHCl_3$, 10-15 pptv), dichloromethane (CH_2Cl_2, 30 pptv), tetrachlorethylene (CCl_2CCl_2, 10 pptv) and trichloroethylene ($CHClCCl_2$, 2-3 pptv) abundances are indicated. No long-term trends for these species have been observed, although they all exhibit distinct annual cycles (summer minimum, winter maximum). Measurements made in 1989 in the Atlantic (45°N-30°S) showed average methylchloride levels of 540 pptv, with slightly higher levels in the SH (550 ± 12 pptv) than the NH (532 ± 8 pptv). Significant gradients in dichloromethane (366 pptv, NH; 18 ± 1 pptv, SH), tetrachlorethylene(1-10 pptv, NH; 1-3 pptv, SH) and trichloroethylene (1-15 pptv, NH; < 1pptv SH) were observed consistent with anthropogenic, largely Northern Hemispheric, sources for these species (Koppmann *et al.*, 1993). More recent measurements (1990) in the tropical Pacific Ocean (15°N to 10°S) showed that the average methylchloride level was 630 pptv; chloroform, 9 pptv; dichloromethane, 24 pptv; and tetrachlorethylene, 5 pptv. The methylchloride data showed no interhemispheric gradient, indicative of a largely oceanic source, whereas the other species showed clear gradients (higher in the Northern Hemisphere), indicative of Northern Hemispheric (presumably anthropogenic) sources (Atlas *et al.*, 1993).

2.5.1.4 Methylbromide, halons and other brominated species

Since the IPCC reports of 1990 and 1992, interest in and understanding of brominated species in the background atmosphere has expanded considerably, driven by the recognition of bromine's significant role in stratospheric ozone depletion.

Available measurements on the global distribution, sources and sinks of methylbromide have been reviewed

(Albritton and Watson, 1993; Penkett et al., 1994) and re-analysed (Reeves and Penkett, 1993; Singh and Kanakidou, 1993). The data suggest global mean levels of 10-15 pptv. Another data set (Khalil et al., 1993b) indicates an average level of about 10 pptv, which has increased over the past four years at about 3 ± 1%/yr. Six years of halon data (H-1211 and H-1301) show that the global background levels are about 2.5 pptv (H-1211) and 2.0 pptv (H-1301), currently growing at about 3 and 8%/yr respectively (Butler et al., 1992). These rates have slowed significantly in recent years, consistent with reduced emissions (McCulloch, 1992).

Data from the tropical Pacific Ocean (Atlas et al., 1993) indicate concentrations of methylbromide, dibromomethane (CH_2Br_2), bromoform ($CHBr_3$) and dibromochloromethane ($CHBr_2Cl$) of 14, 1.8, 1.8 and 0.2 pptv, respectively. The methylbromide data did not show an interhemispheric gradient, in contrast to the Atlantic Ocean data of Penkett et al. (1985), from which a large anthropogenic source was inferred. Clearly more research on the global distribution of methylbromide is required. Bromoform and dibromochloromethane show distinct equatorial maxima, indicating a tropical source related to biogenic activity.

2.5.1.5 Other perhalogenated species

Perhalogenated species such as tetrafluoromethane (CF_4), hexafluoroethane (CF_3CF_3) and sulphur hexafluoride (SF_6) strongly absorb infrared radiation, have very long atmospheric lifetimes and are therefore very powerful greenhouse gases. The global mean concentration of tetrafluoromethane was measured in 1979 at 70 ± 7 pptv (Penkett et al., 1981). Measurements at the South Pole between the late-1970s and mid-1980s indicate a mean concentration of 70 pptv growing at about 2%/yr (Khalil and Rasmussen, 1985). Fabian et al. (1987) have reported northern mid-latitude values of 70 pptv for tetrafluoromethane and 2 pptv for hexafluoroethane. Sulphur hexafluoride is a long-lived atmospheric trace gas whose current global background level is 2-3 pptv, increasing at about 9 ± 1%/yr, based on data observed in the lower stratosphere between 1981 and 1992 (Rinsland et al., 1993). Column measurements in Europe (1986 to 1990) and North America (1981 to 1990) both show increases of 6-7%/yr (Zander et al., 1991).

2.5.2 Industrial Production, Use and Emissions

World-wide consumption of CFCs has fallen significantly following enactment of the Montreal Protocol. From 1988 to 1992, production of CFCs-11, -12 and -113 fell by about 40% as a result of substitution and conservation measures taken in the refrigeration, foam-blowing, cleaning agent and aerosol propellant industries. During this same period, annual emissions fell by about 35%, somewhat less than production reductions, due to the time lag between production and emission. World-wide emissions of individual halocarbons are shown in Figure 2.6, based on industrial survey results (AFEAS, 1993; Fisher et al., 1994). Individual CFC compounds show substantial reductions in emissions over this period which vary among the compounds due to differences in application, ease of substitution/replacement and use-banking times. Allowances are included for production outside the region surveyed. Emissions of methylchloroform have also declined over this period, but to a lesser degree than the CFCs, as allowed by the provisions of the Protocol. HCFC-22 emissions rose continuously throughout this period since HCFC-22 has been used as a replacement for CFCs and its use has not yet been curtailed by the Protocol. Emissions of halons have been reduced substantially (McCulloch, 1992) (no data are available for 1991 and 1992). No emission data are available for carbon tetrachloride due to the fact that its production/use has not been surveyed.

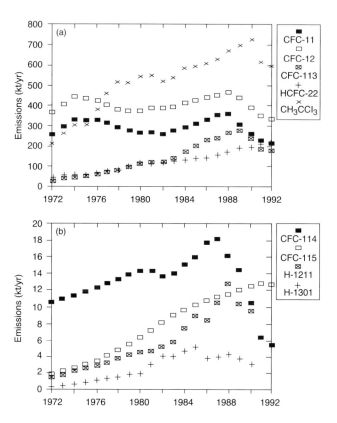

Figure 2.6: Annual emissions of some important industrial halocarbons from 1972 to 1992 based on production statistics (AFEAS, 1993; Fisher et al., 1994; Fisher, 1994 and unpublished work from D. Fisher and P. Midgley). For CFC-11, CFC-12, CFC-113, CH_3CCl_3 and the halons, the estimates are for global use; for HCFC-22, CFC-114 and CFC-115, the estimates are from reporting companies and may be somewhat less than the global emissions.

CH_2Cl_2 and $CHClCCl_2$ are used as industrial cleaning solvents. Sources of 0.9 and 0.6 Tg/yr have been recently estimated from observed atmospheric abundances (Koppmann *et al.*, 1993). The aluminium refining industry is believed to be the major source of CF_4 (0.018Tg/yr) and CF_3CF_3 (0.001 Tg/yr) although it still remains an open question as to what fraction of the atmospheric burden can be attributed to other sources, such as generation of F_2, reduction of UF_4 and UF_6, the use of fluorspar in steel making, etc. (Cicerone, 1979). No natural sources have been identified. 80% of SF_6 production (0.005 Tg in 1989) is used for insulation of electrical equipment, 5-10% for degassing molten reactive metals, and a small amount as an atmospheric tracer (Ko *et al.*, 1993).

2.5.3 Natural Sources

Methylhalides are produced primarily in the ocean, usually associated with algal growth (Sturges *et al.*, 1993; Moore and Tokarczyk, 1993), although a significant fraction may come from biomass burning. Methylchloride, with an atmospheric abundance of about 600 pptv, is the dominant halogenated methane species in the atmosphere. Maintaining this steady-state mixing ratio with an atmospheric lifetime on the order of two years requires a production of around 3.5 Tg/yr (Koppman *et al.*, 1993), most of which comes from the ocean. Sources from biomass burning have also been identified (Mano and Andreae, 1994; Rudolph *et al.*, 1994).

Total emissions of methylbromide have been estimated between 0.075 and 0.12 Tg/yr (Albritton and Watson, 1993; and other model evaluations noted there), but the identity of the sources is still in question. The ocean is believed to contribute between 30-70% of atmospheric methylbromide (Khalil *et al.*, 1993b; Reeves and Penkett, 1993; Singh *et al.*, 1983) and anthropogenic uses, such as fumigation of soils and fresh produce, are believed to contribute about 25%, with some arguments for a greater proportion. It has recently has been suggested that biomass burning could contribute 10-50% of the total flux (Mano and Andreae, 1994). Other halogenated methanes, such as bromoform and mixed halogens, are emitted mainly from coastal waters, but they do not accumulate significantly in the atmosphere. $CHBr_3$, $CHBr_2Cl$ and CH_2Br_2 are produced by macrophytic algae (seaweeds) in coastal regions (Manley *et al.*, 1992) and possibly by open ocean phyloplankton (Tokarczyk and Moore, 1994).

2.5.4 Halocarbon Removal Processes

Fully halogenated halocarbons are destroyed primarily by photodissociation and reactions with $O(^1D)$ in the mid to upper stratosphere. These gases have atmospheric lifetimes of decades to centuries (Table 2.1). Halocarbons containing at least one hydrogen atom, such as HCFC-22, chloroform, methylchloroform, the methyl halides and other HCFCs and HFCs, are removed from the troposphere mainly by reaction with OH. The atmospheric lifetimes of these gases range from months to decades, see Table 2.1. However, some of these gases also react with sea water. About 5-10% of the methylchloroform in the atmosphere is lost to the ocean, presumably by hydrolysis (Butler *et al.*, 1991; Wallace *et al.*, 1994). Only about 2% of atmospheric HCFC-22 is apparently destroyed in the ocean, mainly in tropical surface waters (Lobert *et al.*, 1993). Although the ocean is probably a net source of methyl bromide, it is removed from sea water at about 10-40% per day by hydrogen and halide exchange (Swain and Scott, 1953; Mabey and Mill, 1978; Elliot, 1984; Elliot and Rowland, 1993; Gentile *et al.*, 1989). The lifetime of methyl bromide with respect to tropospheric OH is about 2 yr, but the total lifetime may be significantly less. Butler (1994) estimated an effective lifetime of 1.2 yr with respect to increased or decreased emissions, owing to the role of the ocean as a "buffer" for atmospheric methyl bromide. Other, presently unquantified, land-based sinks may also contribute to a shorter atmospheric lifetime of methyl bromide (Rasche *et al.*, 1990; Oremland *et al.*, 1993a,b).

Recent studies show that carbon tetrachloride may also be destroyed in the ocean. Widespread, negative saturation anomalies (-6% to 0.8%) of carbon tetrachloride, consistent with a subsurface sink (Lobert *et al.*, 1993), have been reported (Butler *et al.*, 1993; Wallace *et al.*, 1994). Published hydrolysis rates for carbon tetrachloride are not sufficient to generate the observed saturation anomalies (Jeffers *et al.*, 1989), which, nevertheless, indicate that about 20% of the carbon tetrachloride in the atmosphere could be consumed by the oceans.

Recent investigation of the atmospheric lifetimes of perfluorinated species CF_4, CF_3CF_3 and SF_6 indicates lifetimes of >50,000, >10,000 and 3,200 years (Ravishankara *et al.*, 1993). Destruction processes considered include photolysis, reaction with $O(^1D)$, combustion and removal by lightning. Loss by uptake into the oceans is not significant for perfluorinated hydrocarbons.

2.6 Observed Ozone (O_3) and Tropospheric UV

Ozone is found throughout the atmosphere, with about 90% in the stratosphere and the remainder in the troposphere. In both regions it is continually formed and destroyed through photochemical processes. In the troposphere two other important processes are the transport of ozone rich air from the stratosphere and the destruction of ozone at the Earth's surface (see Section 2.2). Ozone is radiatively important in both the ultraviolet and infrared parts of the spectrum and, as described in Section 4.3 of Chapter 4, the radiative forcing due to ozone changes depends on whether the ozone is in the stratosphere or troposphere. This radiative forcing is greatest when the ozone changes occur near the tropopause and so the trends of most interest here are those observed in the upper troposphere and lower stratosphere. We consider the observed changes in ozone in these regions separately. Our understanding of and ability to model the processes controlling tropospheric ozone are discussed elsewhere in this chapter. Our current understanding of stratospheric ozone is described in the 1994 WMO/UNEP Ozone Assessment and is not discussed here. Similarly, a number of data quality issues which are discussed in some detail in the 1994 Ozone Assessment are not mentioned here.

2.6.1 Stratospheric Ozone

For some time it has been realised that stratospheric ozone levels over northern mid-latitudes (30°-60°N, where most of the ground-based Dobson spectrophotometers which monitor ozone are) have decreased since around 1970 (WMO, 1990). The losses are greatest in winter and spring. Satellite measurements from November 1979 to March 1991 confirmed this finding and showed that broadly similar mid-latitude losses have also occurred at equivalent latitudes in the Southern Hemisphere. Globally, the most obvious feature is the annual appearance of the Antarctic ozone hole in September and October. The October average values are 50-70% lower than those observed in the 1960s. The ozone loss occurs at altitudes between 14 and 20 km and is clearly caused by the chlorine and bromine compounds released in the stratospheric decomposition of halocarbons (principally CFCs, halons and methylbromide). The weight of evidence shows that the mid-latitude ozone loss is due largely to the same chlorine and bromine compounds. These results and the effect of more recent data are assessed in the 1994 Ozone Assessment.

The most obvious features in the recent record are the low values seen in the 1991 to 1993 period. Such effects occurred regionally, with the largest deviations (15% below historic levels) observed in the northern mid-latitudes in the springs of 1992 and 1993 (Bojkov *et al.*, 1993; Gleason *et al.*, 1993; Kerr *et al.*, 1993; Komhyr *et al.*, 1994). The globally averaged ozone values were 1-2% lower than would be expected from an extrapolation of the trend prior to 1991, allowing for the natural fluctuations resulting from the solar cycle and the quasi-biennial oscillation. In the spring of 1994 ozone values were back in line with the long-term trend seen through the 1980s.

The trends calculated up to May 1991 and May 1994 using data from the SBUV satellite instruments and ground-based Dobson instruments (February 1994) are shown in Figure 2.7. The first point to note is that SBUV shows a larger ozone decline, by about 2%/decade, than is shown by the Dobson network. The best agreement occurs in northern mid-latitudes where the Dobson network is strongest. The reason for this difference is being investigated. Overall the confidence in the trends from the ground based network is high, although the observed decreases in sulphur dioxide over Europe since the 1960s may have induced overestimates in the ozone decreases at polluted locations (De Muer and De Backer, 1992). The second point is that by continuing the analysis into 1994, the effect of the low ozone values in 1992-93 on the calculated trend is small and is largest in northern mid-latitudes where the largest ozone anomalies were observed. The ozone decreases, both long-term and in 1992-93, have occurred in the lower stratosphere at altitudes between about 15 and 25 km (e.g., London and Liu, 1992; Kerr *et al.* 1993; Hofmann *et al.*, 1993, 1994a; Logan, 1994). There is some disagreement between instrument types as to the magnitude of the changes at these altitudes, but at northern mid-latitudes the observed trends are in the range -5 to -15%/decade.

The trend in ozone in the tropics reported using data from the TOMS satellite instrument from November 1978 to March 1991 was effectively zero (WMO, 1991). That found using data from January 1979 to May 1991 the SBUV satellite instruments is -0.8 ± 2.1%/decade. Measurements made using Dobson instruments in this region show a zero trend, but the number of observing sites is small. It is impossible at the moment to decide which is the best trend estimate, but it is worth noting that the SBUV trends in the tropics are less than 95% significant and that the claimed stability of each system is about 1%/decade. While ozone levels in the tropics after 1991 were less obviously anomalous than those in the northern mid-latitudes, some effect on the trend is noticeable. When the data from 1979 to May 1994 from SBUV are analysed, the trend in the tropics is calculated to be -1.8 ± 1.4%/decade. The corresponding trend from the ground-based network is also more negative for the period ending in May 1994.

Large negative trends (-20 to -30 (± 18)%/decade at 16-17 km) have been reported for the tropics using the SAGE satellite instrument (McCormick *et al.*, 1993). Limited tropical ozone sonde records do not indicate significant trends between 15 and 20 km for this time period. With information currently available is difficult to evaluate the trends below 20 km in the tropics. The effect on the trend in the total column from any changes at these altitudes would be small because only small amounts of ozone are present.

Ozone values over southern mid-latitudes since 1991 have not been very different from those seen in the previous two years, apart from the continuing long-term

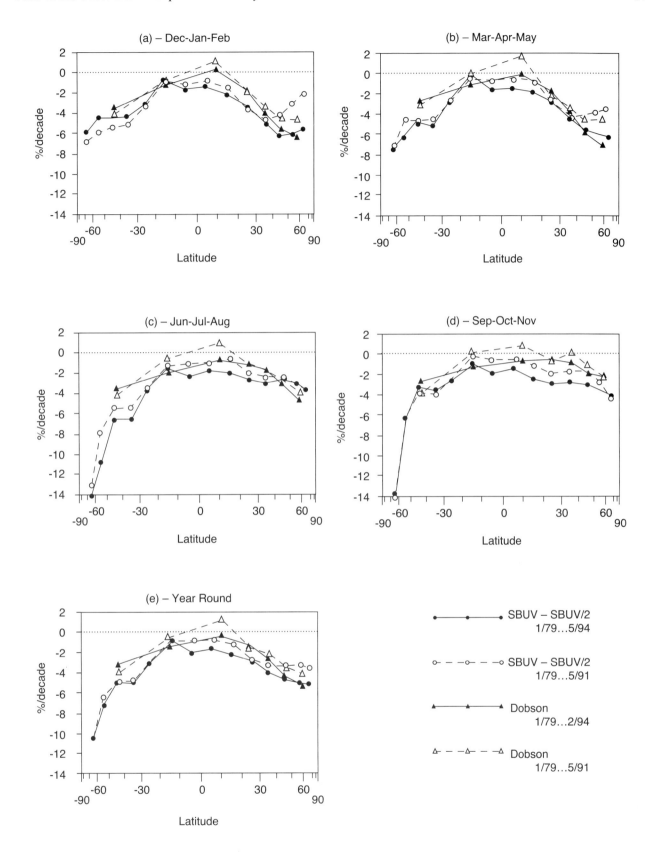

Figure 2.7: The seasonal and latitudinal variation of the total ozone trends calculated from the Dobson ground-based measurement record and the combined SBUV and SBUV-2 measurement records. The trends are calculated from January 1979 to May 1991 and to February 1994 (Dobson) and May 1994 (SBUV) so that the effect of the low total ozone values which were observed in the 1991 to 1994 period can be seen. The uncertainties associated with the combined SBUV records are 1-2% (95% confidence limits) except at the higher latitudes shown in their respective winters where they reach 3-4%. (Adapted from Harris *et al.*, 1994)

trend. In contrast, the springtime Antarctic depletions in 1992 and 1993 were greater than in earlier years with record low values observed. Measurements of the vertical profile over Antarctica show that large volumes of the lower stratosphere around 18 km altitude contained extremely low amounts of ozone (Hofmann et al., 1994b). In addition, depletion of ozone occurred at altitudes above and below the altitude range where it had occurred in previous years.

The reasons for the anomalous behaviour in ozone in the last three years are not known exactly, although a number of mechanisms have been suggested. One probable major influence was the eruption of Mt. Pinatubo in the Philippines in June 1991 which injected large amounts of gaseous SO_2 into the stratosphere which was then chemically transformed into sulphuric acid droplets (aerosols). A thirtyfold increase in the surface area of stratospheric aerosols was observed. The presence of aerosols in the stratosphere can change local heating rates which in turn affect the atmospheric dynamics and ozone distribution. For instance, tropical stratospheric temperatures rose in the second half of 1991, with maximum increases of 2-3°C at 30-50 hPa (Labitske and McCormick, 1992). At the same time negative ozone anomalies (about 6% of background values) were observed (Chandra, 1992; Schoeberl et al., 1992). The presence of aerosols also affects the chemical balance by providing surfaces on which chemical reactions involving chlorine and bromine compounds can proceed, an effect which is enhanced at lower temperatures. The increased altitude range of springtime ozone depletion over Antarctica in 1992 and 1993 may be related to the presence of aerosol at these altitudes (Hofmann et al., 1994b).

An updated analysis of the ground-based total ozone measurements through 1991 shows that the mid-latitude trends in the 1980s were significantly larger than those in the 1970s. This acceleration is consistent with the increasing halogen levels in the stratosphere and to the non-linearity between ozone loss and halogen levels. The total amount of organic chlorine in the troposphere increased by 1.6% in 1992, about half the rate of increase (2.9%) in 1989. Peak total of chlorine/bromine loading in the troposphere is expected to occur in 1994 and the stratospheric peak will lag by about 3 to 5 years. Thus in the next few years, stratospheric halogen amounts are expected to start to decline. In the interim, total ozone amounts should stop decreasing and then begin a recovery to pre-1970 values if future halocarbon emissions are controlled as prescribed in the current Montreal Protocol. This recovery will take place over decades, determined by the lifetimes of the CFCs.

While it is possible to reconcile the observed ozone losses with the increase in chlorinated and brominated compounds in the stratosphere, the predictive capabilities of the models still have significant uncertainties, in part because of the role of heterogeneous chemistry as demonstrated by the impact of Mt. Pinatubo. The depletion is not linear with additional chlorine and hence one cannot attribute a quantitative O_3 loss to a given CFC, since the greatest recovery of O_3 will occur with the first reduction in chlorine, whatever its source.

2.6.2 Tropospheric Ozone

The state of knowledge regarding ozone trends in the troposphere, especially in the free troposphere, is not as good as in the stratosphere. There are a limited number of ozone sonde stations with records suitable for long-term trend analysis, and even at these sites a number of questions about the data quality remain. The stations are mainly in the Northern Hemisphere in Europe, N. America and Japan. Even in these regions the data are somewhat sparse (particularly in Japan) so there is a fair degree of uncertainty involved in reaching conclusions. Elsewhere around the globe, especially in the tropics and Southern Hemisphere, insufficient measurements of good quality have been made with ozone sondes to allow credible, regionally-representative trends to be determined. Available records have been assessed recently by Akimoto et al. (1994), Furrer et al. (1992), Logan (1994), London and Liu (1992), Oltmans and Levy (1994) and Tarasick et al. (1994).

Over the last 20-30 years, the biggest change in free tropospheric ozone has occurred over Europe with an increase of perhaps 50% since the end of the 1960s. The change over N. America is less, possibly about 10-15%. The change over Japan is probably closer to that seen over Europe than that over N. America. The behaviour during the 1980s was different. Over N. America there seems to have been no increase, and over Canada levels have actually declined. The trend over Europe was also smaller than before with, for instance, little change seen at Hohenpeissenberg, Germany since the mid-1980s.

Information is contained in measurements of ozone made at the Earth's surface, although care has to be taken in the interpretation of these data as they are not directly representative of free tropospheric levels. Staehelin et al. (1994) reviewed measurements of ozone made at mountain sites in Europe in the 1930s, 1950s and the present day. They concluded that ozone concentrations over Europe (0-4 km, surface sites) are about a factor of two higher now than in the earlier period. A number of surface measurements of ozone have been made at remote and high altitude sites since the early 1970s (Wege et al., 1989; Scheel et al., 1990, 1993; Kley et al., 1994; Oltmans and Levy, 1994). All stations north of 20°N show a positive trend in ozone that is statistically significant over the whole period. Larger positive trends are observed at the high altitude European sites than elsewhere. In all records only a small increase, or no change, is seen in the 1980s. Thus the surface measurements of ozone are in qualitative agreement with the ozone sonde record. Again, there is only a limited amount of data in the tropics and Southern Hemisphere: surface observations at the South Pole show that ozone has decreased there over the last 10-20 years.

The apparent regional differences in the tropospheric ozone trends are not fully understood. As discussed elsewhere in this chapter, NO_x and NMHC amounts vary substantially through the troposphere, and their emissions have changed both temporally and regionally. Numerical simulation of tropospheric O_3, both global distribution and trends over the past century, is a research task for the three-dimensional chemical transport models (see Section 2.9).

2.6.3 Tropospheric UV

An important chemical coupling that affects the lifetimes of CO, CH_4, NMHC, HFCs and HCFCs is the change of solar UV radiation reaching the troposphere due to changes in stratospheric ozone column density, and tropospheric ozone (Liu and Trainer, 1988; Brühl and Crutzen, 1989; Fuglestvedt et al., 1994). Reductions in stratospheric ozone as observed since the late 1970s have been observed to lead to enhanced penetration of UV radiation into the troposphere (Smith et al., 1992; Kerr and McElroy, 1993) that in turn will yield increased production of $O(^1D)$ and OH. Tropospheric UV can also be altered on regional or local scales by changes in reflecting cloud cover, absorption by tropospheric O_3, or back-scattering by anthropogenic sulphate aerosols (Liu et al., 1991). Increased tropospheric UV will feed back into the abundances of CO and hydrogen-containing trace gases by increasing their sinks. In addition it will speed up the reactions that both create and destroy tropospheric ozone (see Section 2.2), which, depending on the NO_x abundance (Fuglestvedt et al., 1994), could lead to a net decrease or a net increase in tropospheric ozone.

2.7 Tropospheric Nitrogen Oxides

Although of little direct impact on the tropospheric radiation balance, NO_2 and NO (NO_x) have a large indirect effect owing to their importance in tropospheric chemistry. There, the role of NO_x is that of a catalyst promoting the formation of O_3 and controlling the concentration of OH, the most important oxidising agent of the troposphere (see Section 2.2). Through OH, emissions of NO_x influence the adjustment times and thus the abundances of many infrared absorbing gases.

2.7.1 Sources of Tropospheric NO_x

NO_x is emitted mainly as NO by a large variety of sources.

The major sources of tropospheric NO_x are summarised in Table 2.5. Fossil fuel combustion in the stationary power and transport sectors is the largest source of NO_x. Although emissions from fossil fuel use over North America and Europe, 6 Tg(N)/yr and 7.3 Tg(N)/yr respectively, still constitute the largest regional sources, they have hardly increased since 1979 owing to a levelling off in fuel consumption and to air quality abatement measures. The emissions over Asia, however, are continuing to increase at a rate of 4%/yr, reaching 4.7 Tg(N)/yr in 1987 with an estimated emission of 5.7 Tg(N)/yr in 1991 (Kato and Akimoto, 1992). According to Hameed and Dignon (1991) the global emissions from fossil fuel combustion increased from 18.1 Tg(N)/yr in 1970 to 24.3 Tg(N)/yr in 1986. Emissions from aircraft engines are listed separately from other fossil fuel sources because they are released predominantly in the free troposphere at cruise altitudes (8-12 km) rather than at the surface and are currently increasing at a mean rate of 4%/yr (Baughcum et al., 1993). Although only a small fraction of the total combustion source, they are potentially responsible for a large fraction of the NO_x found at those altitudes at northern mid-latitudes (Ehhalt et al., 1992). Microbial activity in soils also gives continental sources for NO (e.g., Sanhueza, 1992; Williams et al., 1992).

It is noted that the estimates of NO_x sources in Table 2.5 are the means of a range. In the cases of fossil fuel combustion, aircraft emissions and stratospheric input, ranges may be as narrow as ±30%; but in the case of natural sources, these are a factor of 2. NO_x produced by lightning is probably the most important natural source in the free troposphere and has an even larger uncertainty. Geographical distributions and seasonal variations of these emissions have been published and are needed to model properly the atmospheric distribution of NO_x (Penner et al., 1991; Kasibhatala, 1993).

Table 2.5: *Estimated global emissions of NO_x typical of the last decade (Tg(N)/yr).*

Sources	Magnitude (Tg(N)/yr)
Fossil fuel combustion	24
Soil release (natural and anthropogenic)	12
Biomass burning	8
Lightning	5
NH_3 oxidation	3
Aircraft	0.4
Transport from stratosphere	0.1 (0.6 total NO_y)

2.7.2 NO_x Removal Processes

Atmospheric oxidation of NO_2 to HNO_3 by OH in the daytime and to NO_3 by O_3 at night are the most important removal processes of NO_x from the troposphere. Dry deposition of NO_2 plays a lesser, but still significant, role. Some fraction of the NO_x so lost is regenerated from HNO_3 or NO_3 in the free troposphere by photodissociation. The tropospheric lifetime of NO_x is short and variable. It varies with altitude, for example, from less than a day in the boundary layer to about a week at the tropopause.

Although the removal processes of NO_x are reasonably well identified, the global distribution of NO_x is not well defined by observations and the loss rates are highly variable. Thus our estimates of the atmospheric budget of NO_x are based on model simulations using highly uncertain values for NO_x sources as noted above. Fortunately, the fact that, except for the small fraction which is dry deposited as NO_2, all of the NO_x is converted to nitrate and deposited as such by dry or wet processes allows an independent check of the removal of NO_x from the troposphere. The various analyses of global nitrate deposition based on the concentration in precipitation collected in a number of networks have been summarised by Warneck (1988). The latest such estimate was made by Logan (1983), giving a total deposition of (44 ± 20) Tg(N)/yr of which (17 ± 5) Tg(N)/yr were due to dry deposition of HNO_3, NO_3^- aerosol and NO_2. This estimate balances fairly well the emissions at that time, which have since grown to those given in Table 2.5.

2.7.3 Tropospheric Distribution of NO_x

Because of its complex geographical source pattern and its short lifetime, the spatial and temporal distribution of tropospheric NO_x is highly variable. That variability is

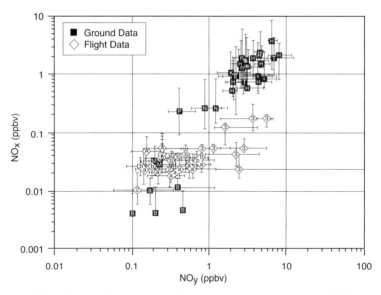

Figure 2.8: Measured NO_x and NO_y mixing ratios at the Earth's surface and in the lower and middle troposphere of North America (from data summarised by Carroll and Thompson, 1994). The location of the symbol indicates the mean of the measurements, the error bars the mean standard deviation. The NO_y mixing ratio given here is calculated from the mean measured NO_x ratio and the NO_x/NO_y ratio given by Carroll and Thompson (1994). The letters and numbers within the symbols refer to the following measurement campaigns:

a	ABLE-3a	Bethel, Barrow, AK, 0–6 km	summer	1988
A	AASE	North America 40°N, >59°N,10-12 km	winter	1988-89
b	ABLE-3b	Labrador, Hudson, 0–6 km	summer	1990
B		Barrow, AK		1990
H		Harvard Forest, MA	all year	1990–1992
K		Kinterbush, AL	summer	1990
M	MLOPEX	Mauna Loa Obs., HI	spring	1988
n	NACNEMS	Egbert, Ontario; Bondville, IL	summer	1988
N		Niwot Ridge, CO	summer	1984;1987
P		Point Arena, CA	spring	1985
s	SOS/SONIA	Candor, NC	summer	1991
S		Scotia, PA	summer	1986;1988
T	TOR	Schauinsland, Germany	winter,summer	1989–1990
2	CITE2	Pacific; Continental US, ~5 km	summer	1986
3	CITE3	Natal, Brazil; Wallops US, 0–6 km	autumn, summer	1989

demonstrated in Figure 2.8 which summarises measurements from rural and remote surface stations as well as airborne measurements in the lower and middle troposphere mostly over continental North America. NO_x is plotted against simultaneous measurements of NO_y, which is defined as $NO + NO_2 + NO_3 + 2 \times N_2O_5 + HNO_2 + HNO_3 + HNO_4 + PAN$ (peroxyacetyl nitrate) and other organic nitrates, and describes the sum of all reactive nitrogen compounds. NO_y results from the chemical ageing of NO_x, but is longer lived and therefore a more conservative tracer of NO_x sources.

Even in this data set which excludes the high NO_x concentrations around the urban centres and the very low concentrations in very remote areas, NO_x varies over 3 orders of magnitude, the higher mixing ratio being observed closer to the sources. The mean vertical profiles included in these data, which reached up to 6 km, showed little variation with altitude. The highest values were observed over the east coast of the United States which was the only profile with a significant vertical gradient. NO_y roughly correlates with NO_x, although not well enough to serve as a substitute for a direct NO_x measurement. As Figure 2.8 demonstrates, at a given NO_y mixing ratio NO_x can still vary over two orders of magnitude.

NO_x has now been measured up to the tropopause in a number of campaigns covering large geographical areas. Figure 2.9 summarises the means and variability of NO_x measured in the free troposphere from many aircraft campaigns. Unfortunately at the time of writing some of the most recent data, notably those from PEM West over the Pacific in September to October 1991, AASE II over northern mid-, subpolar and polar latitudes in January to March 1992, and TRACE A over the tropical Atlantic in October 1992, are not available for review. However, a number of broad conclusions can be drawn from data collected during the STRATOZ III campaign in June 1984, the AASE I effort in January to February 1989, the TROPOZ II campaign in January 1991, and the CITE efforts from 1984 to 1989 (Drummond *et al.*, 1988; Wahner *et al.*, 1994). The STRATOZ III and TROPOZ II results indicate that:

(i) mixing ratios of NO were lower in the Southern Hemisphere during both summer and winter;

(ii) NO mixing ratios increased with altitude in both hemispheres during summer and winter with high values (several hundred pptv) in the upper troposphere at the northern mid-latitudes. Due to the generally higher mixing ratios in the planetary boundary layer over the continents, a C-shape vertical profile is observed in continental air. In contrast, very low NO concentrations are seen in the surface layer in oceanic air, a finding that is shared by observations made, for example, during the CITE-2 campaign over the eastern Pacific in summer 1986 (Carroll *et al.*, 1992), the CITE-3 campaign over the western Atlantic in summer and autumn 1989 (Davis *et al.*, 1993), and the SAGA 3 campaign in the equatorial Pacific during February to March 1990 (Torres and Thompson, 1993);

(iii) the shape of the vertical profile at the northern mid-latitudes changes dramatically between summer and winter.

In summer the mid-troposphere has relatively low NO, with average values between 10 and 30 pptv, and the largest vertical gradient is located between 8 and 10 km altitude. In winter the troposphere is much more heavily loaded with NO and sharp gradients spread throughout the free troposphere. Some of the features observed in these distributions were due to the location of the flight track, which usually followed continental coast-lines. These are usually more heavily populated and therefore subject to human-made pollution, such that intense continental plumes would be sampled during times of off-shore winds. The zonal distribution of NO_x sources, in particular the difference between the oceanic lack and the continental abundance of surface sources, leads to large gradients in the tropospheric NO_x distribution. For example, large longitudinal gradients of NO mixing ratio were observed at all altitudes in the free troposphere across the North Atlantic from 55°N to 65°N latitude, presumably dominated by an outflow of polluted air from the European continent during STRATOZ III (Ehhalt *et al.*, 1992). This, together with the measurements shown in Figures 2.8 and 2.9, indicates that there is no sense in trying to define a typical, globally averaged, vertical NO (or NO_x) profile. The actual profile will always strongly and typically depend on location, season and local wind field (in the case of NO also on the time of day). In turn, such information is required for the interpretation of the observed profile and for a detailed comparison with model calculations. The AASE I results provide additional evidence for a C-shaped vertical profile in continental air and for the build-up of reactive nitrogen species in the upper troposphere at mid- to high latitudes in the Northern Hemisphere in winter (Carroll *et al.*, 1990). In addition to these campaigns where measurements were obtained throughout the entire vertical extent of the troposphere, a number of airborne campaigns have focused on lower and mid-tropospheric distributions of NO_x, including the ABLE-3A flights over Alaska in summer 1988 (Sandholm *et al.*, 1992) and the ABLE-3B campaign over northern Canada in summer 1990 (Sandholm *et al.*, 1994; Singh *et al.*, 1994; Talbot *et al.*, 1994). Altitude-specific results from these airborne campaigns are also shown in Figure

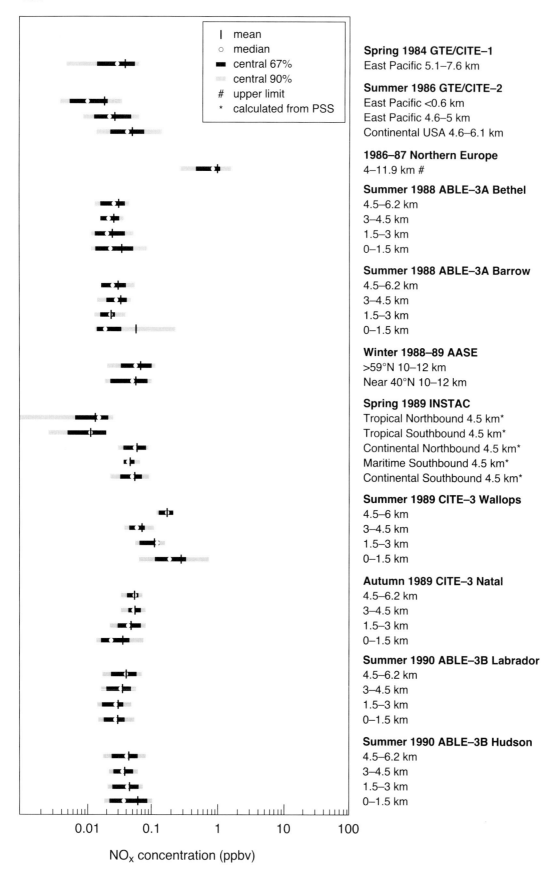

Figure 2.9: The statistical distribution of the NO_x mixing ratio (ppbv) observed at various altitudes during several aircraft campaigns. (Adapted from Carroll and Thompson, 1994)

2.9, but we have not included here a review of NO_x at remote surface sites such as the MLOPEX campaign (Carroll et al., 1992).

High mixing ratios of NO (300 to 500 pptv) in the upper troposphere, observed for instance during both STRATOZ III and TROPOZ II, indicate NO sources in addition to the stratospheric input (Ehhalt et al., 1992). Possible sources are: NO produced by lightning, aircraft emissions, and fast upward transport from a polluted boundary layer. There are a number of observations, where the vertical NO profile is strongly and unequivocally influenced by one or the other of these sources, e.g., lightning (Dickerson et al., 1987; Ridley, private comm.), aircraft emissions (Arnold et al., 1992), fast vertical transport (Drummond et al., 1988), which make it clear that all these sources can and do make a contribution to the NO_x in the upper troposphere. However, at present there are not enough data to derive their respective global contributions from atmospheric measurements alone.

2.7.4 Trends of NO_x

Because of the high variability in its tropospheric concentrations, it is not possible to establish a trend for NO_x from atmospheric measurements. Some of its sources, predominantly in the Northern Hemisphere, continue to increase (see Section 2.7.1), and the concentration of nitrate in Greenland ice cores shows an increase during the last century. This evidence suggests that tropospheric NO_x has been increasing, at least in the northern mid-latitudes. There is evidence that tropospheric O_3 has increased at northern mid-latitudes (see Section 2.6) which could be understood in terms of a combined increase in CH_4, CO and NO_x, but a quantitative reconciliation of all causes of tropospheric O_3 change has not been achieved.

2.8 Carbon Monoxide and Volatile Organic Compounds

Carbon monoxide (CO) and volatile organic compounds (VOCs) have little direct radiative impact on the atmosphere. However, as described in Section 2.2, these compounds can strongly influence tropospheric chemistry and, in particular, O_3 and OH concentrations, and so could have an important indirect impact. VOCs include non-methane hydrocarbons (NMHCs) and other organic compounds, containing additional elements such as oxygen.

2.8.1 Sources and Removal Processes of CO

The major sources of CO (Table 2.6) are technological (including transport, combustion, industrial processes and refuse incineration), biomass burning, and the oxidation of methane and non-methane hydrocarbons (IPCC, 1992). Ocean sources may also be important (Erikson and Taylor, 1992). Each of these sources has a large uncertainty. It is estimated that about two-thirds of CO currently results from anthropogenic activities, including that derived from anthropogenic CH_4. There is a large uncertainty associated with the magnitude of the CO source from the oxidation of VOC.

The most important removal mechanism for CO is reaction with OH radicals. Soil uptake and stratospheric removal are minor sinks of CO (WMO, 1986). In principle, the atmospheric removal rate for CO can be calculated from its observed distribution, the modelled distribution of OH, and the corresponding reaction rate coefficients. Based on measured CO distributions and an OH-field from model calculations, a removal rate of 2020 Tg(CO)/yr has been obtained. Studies of ^{14}CO indicate that OH levels in the Southern Hemisphere are significantly higher than in the Northern Hemisphere (Brenninkmeijer et al., 1992) and that global CO sinks may be larger than indicated in Table 2.6 (Mak et al., 1992). However, simulations of CH_3CCl_3 (e.g. Spivakovsky et al., 1990) do not require hemispheric differences in OH and place tight bounds on the CO sink consistent with the budgets adopted here.

2.8.2 Atmospheric distribution and trends of CO

Measurements made over the past 25 years have shown that CO mixing ratios in the troposphere range from

Table 2.6: Estimated sources and sinks of CO typical of the last decade (Tg(CO)/yr)

Sources	Range (Tg(CO)/yr)
Technological	300 - 550†
Biomass burning	300† - 700
Biogenics	60 - 160
Oceans	20 - 200
Methane oxidation	400 - 1000
NMHC oxidation	200 - 600
Total sources	1800 - 2700
Sinks	
OH reaction	1400 - 2600
Soil uptake	250 - 640
Stratospheric loss	~100
Total sinks	2100-3000

Note: Adapted from IPCC (1992) and †J. Logan (Harvard, pers. comm.).

approximately 40 to 200 ppbv (e.g., Novelli et al., 1992). Annual mean CO levels in the high latitudes of the Northern Hemisphere are about a factor of 3 greater than those at similar latitudes in the Southern Hemisphere (Figure 2.10). Mixing ratios vary seasonally: highest levels are observed during winter and lowest in summer. CO levels over continental locations are usually greater than over the oceans (Kirchhoff and Marinho, 1989; Poulida et al., 1991), and regional processes, such as biomass burning, can affect large areas distant from the emission source (Reichle et al., 1990; Fishman et al., 1991).

Long-term trends in atmospheric CO have been estimated at several locations world-wide using data collected over the past 30 years (Khalil and Rasmussen, 1988b; Brunke et al., 1990; Zander et al., 1989). From these studies a consensus emerged that up to 1990 CO mixing ratios in the Northern Hemisphere had probably increased at an average rate of 1%/yr (WMO, 1990). In the Southern Hemisphere, the rate of change in CO over time is less than in the Northern Hemisphere (Fraser et al., 1986; Dianov-Klokov et al., 1989). A review of available data concluded that there was no significant trend in the Southern Hemisphere (WMO, 1990).

More recent data (Novelli et al., 1994) indicate that global CO levels have fallen sharply since about 1990, at about 7%/yr. The largest decline has been observed at high latitudes of the Northern Hemisphere. In contrast, Khalil and Rasmussen (1994) for the period 1987–1992 reported decreases of $1.4 \pm 0.9\%$/yr in the Northern Hemisphere and $5.2 \pm 0.7\%$/yr in the Southern Hemisphere. There is no known explanation for this rapid decline although hypotheses include increasingly effective emission controls. The extent to which this trend is a truly global phenomenon is uncertain, because of the limited spatial coverage of the observing network for this short-lived gas.

2.8.3 Volatile Organic Compounds

2.8.3.1 Introduction

Non-methane hydrocarbons (NMHCs) constitute a large class of compounds containing only carbon and hydrogen atoms. They range from the simplest such as ethane (C_2H_6) and acetylene (C_2H_2) up to complexes with ten or more carbon atoms. Volatile organic compounds (VOCs) include the NMHCs and other organic compounds containing additional elements such as oxygen.

The oxidation of VOCs in the atmosphere has a general impact on atmospheric chemistry that depends only partly on the individual compound and thus allows us to consider them as a class. VOCs have a major influence on tropospheric chemistry (Derwent, 1994a) through three processes:

(i) as a source of tropospheric O_3 on the regional and global scales, by their irreversible oxidation in the presence of NO_x;
(ii) as a source of CO and other reactive VOCs such as aldehydes and ketones;
(iii) as a source of organic nitrogen compounds which act as temporary reservoirs for NO_x, allowing far greater global dispersion of the sources.

An accurate assessment of the role of VOCs in radiative forcing is complicated by the many individual compounds, whose sources and reactivities (e.g., the number of O_3 and CO molecules made per VOCs destroyed in the presence of NO_x) are not well known. This assessment recognises the potential importance of VOCs in controlling tropospheric O_3 and OH, but at this time can only provide the following synopsis, without any quantitative evaluation of their climate impacts.

2.8.3.2 Sources of volatile organic compounds

The estimation of VOC emissions is complicated by the wide variety of sources and by the large number of individual hydrocarbons. Sources of VOCs are associated with both natural biogenic processes and human activities. In some areas, natural sources dominate whereas in the industrialised regions of the Northern Hemisphere, human activities contribute the larger fraction.

The status of global emission inventories of VOCs has been extensively reviewed as part of the Global Emissions Inventory Activity (GEIA) of the International Global Atmospheric Chemistry Program (Graedel et al., 1993).

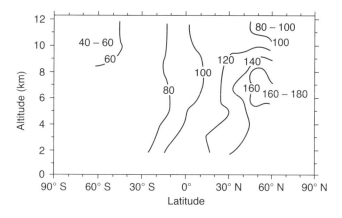

Figure 2.10: The CO mixing ratios (ppbv) as a function of latitude and altitude averaged over the four transects that were measured during the STRATOZ II and STRATOZ III flights in April/May 1980 and June 1984. During each of the missions, one distribution along the east coast of North America and the west coast of South America and another along the east coast of South America and the west coast of Africa were determined (Seiler and Fishman, 1981; Marenco et al., 1990).

Recent estimates indicate a global total for anthropogenic VOCs of about 140 Tg/yr with 25% due to road transport, 14% from solvent use, 13% fuel production and distribution, 34% fuel consumption and the rest uncontrolled burning and other minor sources (Bouwman, 1993).

VOC inventories without detailed species composition profiles offer little progress towards understanding the role of these compounds in tropospheric chemistry. Detailed speciated emission inventories are available only for North America and Europe (Placet et al., 1990; Lubkert and Zierock, 1989) with European NMHC emissions totalling 24.5 Tg/yr (Simpson, 1993). Some information on the long-term trends for man-made emissions is available for the USA (Gschwandtner et al., 1986; Placet et al., 1990). Recent estimates for the emission rates of NMHCs (Müller, 1992) indicate that this is the largest contribution of a single source to the budget of VOCs. However, the number of compounds and their source strengths which go up to make this total is poorly understood. A recent overview is found in Fehsenfeld et al. (1992).

Biomass burning is an important source of VOCs in subtropical and tropical regions, and includes forest and savannah fires, burning of agricultural wastes and the use of biomass fuels. The impact of biomass burning on global chemistry has been recently reviewed (Andreae, 1993). Many of the hydrocarbons liberated by biomass burning are alkenes, which in the presence of NO_x (also emitted by the fire) show a high propensity to ozone formation. Consequently, there are numerous observations of ozone production in biomass-burning plumes (Delany et al., 1985; Andreae et al., 1988; Cros et al., 1988; Kirchhoff et al., 1989).

2.8.3.3 Sinks of the volatile organic compounds

For most VOCs, reactions with hydroxyl radicals provide the major atmospheric loss. Adjustment times due to OH oxidation range from a fraction of a day for the biogenic hydrocarbons, to several months for ethane, one of the longest lived NMHCs. This is the only removal process which has been quantified for a significant number of VOCs.

Certain hydrocarbons have heightened reactivities with other atmospheric oxidants: ozone, chlorine atoms and NO_3 radicals. Ozone reacts rapidly only with olefinic NMHCs and then usually only significantly with certain specific natural biogenic terpenes. These reactions tend to be ozone sinks and may generate organic aerosols and other low volatility products such as organic carboxylic acids. Chlorine atoms react with a wide variety of VOCs but their importance in global hydrocarbon chemistry is not established. Nitrate (NO_3) radicals are present in the atmosphere to a significant amount only during night-time.

They are formed by the reaction of NO_2 with ozone and can react rapidly with certain VOCs. Night-time NO_3 chemistry may be significant under some circumstances but, again, its global importance in VOC chemistry is not established.

2.8.3.4 The role of volatile organic compounds

Because of their generally short atmospheric lifetimes and highly localised sources, organic compounds are not evenly distributed in space or time. There are now many measurements of the bewilderingly large number of organic compounds which contribute to global and regional scale ozone formation, carbon monoxide production and the formation of organic nitrogen compounds. However it is not possible to give at present an assessment of the major VOCs and their roles in tropospheric chemistry.

2.9 Inter-Comparison of Tropospheric Chemistry/Transport Models

A chemical response of the atmosphere is expected for the human-induced changes in the cycles of many trace gases. For example, we have accumulated evidence that tropospheric ozone in the northern mid-latitudes has increased substantially, on the order of 25 ppbv, since pre-industrial times. During this period, the global atmospheric concentration of CH_4 has increased regularly, and the emissions of NO_x and NMHC, at least over northern mid-latitudes, have also increased greatly. An accounting of the causes of the O_3 increases, and ascribing the induced climatic change to emissions of any particular gas, requires a global tropospheric 3-D Chemistry/Transport Model (CTM). A CTM provides the framework for coupling different chemical perturbations that are by definition indirect and thus cannot be evaluated simply with linear, empirical analyses.

In this report we use CTMs to calculate the total GWP for CH_4 (see Chapter 5), including the complex tropospheric chemical impacts on O_3. Thus we place an increasing responsibility on our CTM simulations of the atmosphere and should therefore ask how much confidence we have in these models. Models of tropospheric chemistry and transport have not been adequately tested in comparison with those stratospheric models used to assess ozone depletion associated with CFCs. The stratospheric assessment models are all two-dimensional (latitude by height), based on an established dynamical theory that allows inherently 3-D motions in the stratosphere to be mapped onto two dimensions (e.g., see WMO, 1986). There is no such corresponding theory for the troposphere, and it is expected that 3-D models will be needed to describe the tropospheric distribution of trace gases including phenomena such as convection, clouds and the different chemical regimes in continental and marine boundary layers. Furthermore, the stratospheric models have been extensively tested as a class against one another and against observations (e.g., WMO, 1994 and preceding reports; Prather and Remsberg, 1993). It has taken the stratospheric research community nearly a decade to develop assessments based on standard trace-gas scenarios that allow the results from different models to be compared. In contrast, most of the current tropospheric ozone models (2-D and a few 3-D) have published a few sample calculations, which differ from model to model and are not directly comparable.

There are numerous published examples of individual model predictions of the changes in tropospheric O_3 and OH in response to a perturbation (e.g., pre-industrial to present, doubling CH_4, aircraft or surface combustion NO_x, stratospheric O_3 depletion). Since these calculations in general used different assumptions about the perturbants and the background atmosphere, it is impractical, if not impossible, to use these results to derive an assessment. We need to understand how representative those models used in climate studies are, particularly the analyses of indirect radiative effects of CH_4 changes.

Thus, two quite different model intercomparisons are included as part of this report: (1) the transport of radon-222, a short-lived tracer, that highlights the difference between 2-D and 3-D models in the troposphere; and (2) a set of prescribed tropospheric photochemical simulations that test the approximations made in modelling chemical rates for O_3 production and loss. Both of these studies were initiated as blind intercomparisons, with model groups submitting results before seeing those of others. In the transport study, no obvious mistakes in performing the case studies were found, and detailed results will be published as a workshop report. In the photochemical study, some obvious errors in the set-up, diagnosis, or model formulation were identified and resubmitted by the contributors as discussed below. Both provide a first look at the consistency among current models. Participation in these two intercomparisons was a prerequisite for the use of any model's results in a third study (Section 2.10.2), which examines the chemical perturbations caused by CH_4 increases and provides the basis for calculating the indirect GWP of CH_4 in Chapter 5.

2.9.1 Intercomparison of Transport: A Case Study of ^{222}Radon

The model comparison that tested atmospheric transport was carried out in a World Climate Research Programme (WCRP) Workshop on short-range transport of greenhouse gases that followed a similar workshop on the long-range transport of CFC-11 (Dec 1991). The basic intercomparison examined the global distribution and variability predicted for the radon (^{222}Rn) emitted by radioactive decay of radium in soils. Radon is an ideal gas with a constant adjustment time of 5.5 days, and the daughter product, lead-210, is treated as a small aerosol. Although NO_x would seem a more relevant choice for these model comparisons, the large variations in the adjustment time for NO_x (e.g., <1 day in the boundary layer and >10 day in the upper troposphere) make it difficult to prescribe a meaningful experiment without running realistic chemistry — a task beyond the capability of most of the participating models. Furthermore, the non-linearities of the NO_x-OH chemistry would require that all major sources be included (see Section 2.7), which again is too difficult for this model comparison.

Twenty atmospheric models (both 3-D and 2-D) participated in the ^{222}Rn intercomparison for CTMs (see Table 2.7). Most of the participants were using established (i.e., published), synoptically varying (i.e., with daily weather) 3-D CTMs; several presented results from new

Table 2.7: Models participating in the Rn/Pb transport intercomparison

Code	Model	Contributor
CTMs established: 3-D synoptic		
1	CCM2	Rasch
2	ECHAM3	Feichter/Koehler
3	GFDL	Kasibhatla
4	GISS/H/I	Jacob/Prather
5	KNMI	Verver
6	LLNL/Lagrange	Penner/Dignon
7	LLNL/Euler	Bergman
8	LMD	Genthon/Balkanski
9	TM2/Z	Ramonet/Balkanski/Monfray
CTMs under development: 3-D synoptic		
10	CCC	Beagley
11	LaRC	Grose
12	LLNL/Impact	Rotman
13	MRI	Chiba
14	TOMCAT	Chipperfield
15	UGAMP	P. Brown
CTMs used assessments: 3-D/2-D monthly average		
16	Moguntia/3D	Zimmermann/Feichter
17	AER/2D	Shia
18	UCamb/2D	Law
19	Harwell/2D	Reeves
20	UWash/2D	M. Brown

models under development. Among these synoptic CTMs, the circulation patterns represented the entire range: gridpoint and spectral, first generation climate models (e.g., CCM1 and GISS), newly developed climate models (e.g., CCM2 and ECHAM3), and analysed wind fields from ECMWF (e.g., TM2Z and KNMI). One monthly averaged 3-D CTM and four longitudinally and monthly averaged 2-D models also participated.

We have a limited record of measurements of ^{222}Rn with which to test the model simulations. Some of these data are for the surface above the continental sources (e.g., Cincinnati, OH), and some are from islands far from land sources (e.g., Crozet I.). The former show a diurnal cycle with large values at the surface at night when vertical mixing is suppressed. The latter sites show a very low background level with large events lasting as long as a few days. An even more limited set of observations from aircraft over the Pacific (e.g., 300 mbar over Hawaii) shows large variations with small layers containing very high levels of radon, obviously of recent continental origin. A set of box plots in Figure 2.11 summarises the observations of radon at each of these three sites and compares with model predictions (see Table 2.7 for model codes). At Cincinnati the synoptic 3-D CTMs generally reproduce the mean afternoon concentrations in the boundary layer, although some have clear problems with excessive variability, possibly with sampling the boundary layer in the afternoon. At Crozet most of the synoptic models can reproduce the low background with occasional radon "storms." In the upper troposphere over Hawaii, the one set of aircraft observations shows occasional, extremely high values, unmatched by any model; but the median value is successfully simulated by several of the synoptic 3-D CTMs. The monthly averaged models could not, of course, simulate any of the time-varying observations.

The remarkable similarity of results from the synoptic CTMs for the free tropospheric concentrations of radon in

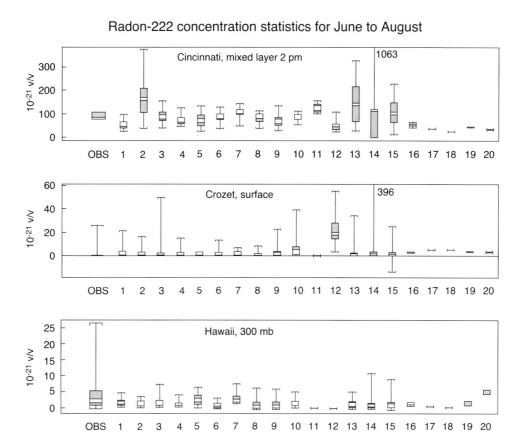

Figure 2.11: Radon-222 concentration statistics (OBS) for June, July and August at Cincinnati OH (40°N, 84°W, mixed layer at 2 p.m.), Crozet I. (46°S, 51°E, surface), and over Hawaii (20°N, 155°W, 300 mbar). Modelled time-series show minima and maxima, quartiles (shaded box), and medians (white band). Identification codes are given in Table 2.7. Observations at Hawaii (Balkanski *et al.*, 1992) show the same statistics; but for Cincinnati (Gold *et al.*, 1964) the shaded box gives the interannual range of June-August means; and for Crozet (Polian *et al.*, 1986), the lower bar gives background concentrations, with the upper bracket showing typical maximum events.

all three experiments was a surprise to most participants. All of the established CTMs produced patterns and amplitudes that agreed within a factor of two over a range in concentration of more than 100. As an example, the zonal mean radon from case (i) for December, January and February is shown for the ECHAM3 and CCM2 models in Figure 2.12a-b. The two toothlike structures result from the major tropical convergence and convective uplift south of the equator and the uplift over the Sahara in the north. This basic pattern is reproduced by all the other synoptic CTMs. In June, July and August (not shown) the 5×10^{-21} v/v contour shifts north of the equator, and again, the models produce similar patterns. In contrast, the 2-D model results, shown for the AER model in Figure 2.12c, have much smoother latitudinal structures, fail to match the expected zonally averaged concentrations, do not show the same seasonality, and, of course, cannot predict the large longitudinal gradients expected for ^{222}Rn (similar arguments hold for NO_x, see Kanakidou and Crutzen, 1993). Results from the Moguntia CTM (monthly average 3-D winds) fell in between these two extremes and could not represent the structures and variations predicted by the synoptic CTMs.

Such differences in transport are critical to this assessment. Both NO_x and O_3 in the upper troposphere have chemical time-scales comparable to the rate of vertical mixing, and the stratified layering seen in the monthly averaged models is likely to distort the importance of the relatively slow chemistry near the tropopause. Compared with the synoptic models, it is obvious that the monthly averaged models simulate the transport of surface-emitted NO_x into the free troposphere very differently, which may lead to inaccurate estimation of total NO_x concentrations. The 2-D models appear to have a clear systematic bias favouring high-altitude sources (e.g., stratosphere and aircraft) over surface sources (e.g., combustion) and may calculate a very different ozone response to the same NO_x perturbations.

The participating synoptic CTMs are derived from such a diverse range of circulation patterns and tracer models that the universal agreement is not likely to be fortuitous. It is unfortunate that we lack the observations to test these predictions. Nevertheless, it is clear that the currently tested 2-D models, and to a much lesser extent the monthly averaged 3-D models, have a fundamental flaw in transporting tracers predominantly by diffusion, and they cannot be viewed as reliable in simulating the global distribution of short-lived species. The currently tested synoptic 3-D CTMs are the only models which have the capability of simulating the global-scale transport of NO_x and O_3; however, this capability will not be realised until these models include improved simulations of the boundary layer, clouds and chemical processes.

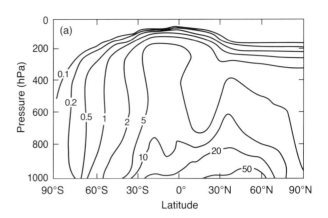

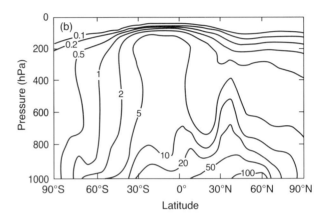

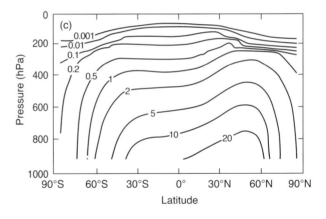

Figure 2.12: Latitude-by-altitude distribution of radon-222 as simulated in global chemical transport models. The contours are mixing ratios in units of 10^{-21} by volume and have been averaged over Dec-Jan-Feb. These results are examples taken from a WCRP workshop on atmospheric transport (December, 1993). The two three-dimensional models (panel a, CCM2; and panel b, ECHAM3) reported longitudinally averaged distributions that agreed in general with most of the other 3-D models in the workshop, but were dramatically different in magnitude and structure from the 2-D models, such as the AER model shown here (panel c).

2.9.2 Intercomparison of Photochemistry: O_3 Production and Loss

The photochemical evolution of NO_x and O_3 in a parcel of tropospheric air is as important as the transport in the CTM simulations of ozone. We need to evaluate the chemistry in these models separately. Unfortunately, there is no easy observational test of the rapid photochemistry of the troposphere that includes the net chemical tendency of O_3. Furthermore, uncertainties in the kinetic parameters would probably encompass a wide range of observations. Thus, we chose an engineering test in which all chemical mechanisms and data, along with the atmospheric conditions, were specified exactly in each case. This comparison becomes then a test of the photochemical schemes used by the different groups, and in general there is only one correct answer. For most of these results, many models give similar answers, resulting in a "band" of consensus, which we assume here to be the best answer. We are thus testing the consistency of the numerical solution of the photochemical reaction system under highly constrained conditions.

The original specifications for the comparison of photochemical schemes (PhotoComp) and the follow-on evaluation of CH_4's impact on global chemistry (delta-CH_4) brought results from a wide variety of research groups listed in Table 2.8. Twenty-three model groups submitted to PhotoComp, and seven groups to the delta-CH_4 study. Assessment of the initial results revealed numerous mistakes. Most of these obvious discrepancies occurred in model formulation, interpretation of instructions, or reporting of results. Assessment of tropospheric chemistry lacks experience when compared with that of stratospheric ozone, and one purpose in having such a blind intercomparison was to provide a measure of potential model discrepancies. (For more typical assessments such as the delta-CH_4 study, there remain so many options in these complex calculations that differences can be easily ascribed to many "justified" causes.) If these model assessments had been frozen in January 1994 as originally planned, then there would have been only two model results for the delta-CH_4 scenario. Therefore, the model intercomparison / assessment was continued with resubmissions up to June 1994. The current list of contributions is the same as for the parallel WMO Ozone Assessment. Removal or correction of obvious errors did not eliminate discrepancies among the models, and significant differences still remain and are presented here. Table 2.8 gives the participating models and their contributors along with letter codes used in the figures, codes for those submitted after the January 1994 meeting (§), and notation for those models contributing to delta-CH_4 (&).

Details of PhotoComp are given in Chapter 7 of the 1994 WMO Ozone Assessment. Atmospheric conditions were specified (1 July, US Standard Atmosphere with only molecular scattering and $O_2 + O_3$ absorption) and the air parcels with specified initial conditions were allowed to evolve in isolation for five days with diurnally varying photolysis rates ("J"s). The PhotoComp cases were selected as examples of different chemical environments in the troposphere. The wet boundary layer is the most extensive, chemically active region of the troposphere. Representative conditions for the low-NO_x concentrations over oceans (case: MARINE) and the high-NO_x concentrations over continents (case: LAND) were picked. In MARINE, ozone is lost rapidly (-1.4 ppbv/day), but in LAND the initial NO_x boosts O_3 levels. Over the continental boundary layer NO_x loss is rapid, and the high NO_x levels must be maintained by local emissions (e.g., Zhou et al., 1993). This region has the potential for export of O_3 (and its precursors) to the free troposphere (Pickering et al., 1992; Jacob et al., 1993). Rapid O_3 formation has been observed to occur in biomass burning plumes, and the rate is predicted to depend critically on whether hydrocarbons are present (PLUME/HC) or not (PLUME). In the dry upper troposphere (FREE), O_3 evolves very slowly, less than 1%/day, even at NO_x levels of 100 pptv.

The photolysis of O_3 yielding $O(^1D)$ (reaction (2.1)) is the first critical step in generating OH, and it controls the net production of O_3. In this case tropospheric values peak at about 4-8 km because of molecular scattering. Model predictions for this photolysis rate at noon, shown in Figure 2.13a, fall within a band, ±20% of the mean value, if a few outliers are not considered. These differences are still large considering that all models purport to be making the same calculation. This photolysis of O_3 and subsequent reaction with H_2O (reaction (2.2)) drives the major loss of O_3 in MARINE (0 km, 10 pptv NO_x) as shown in Figure 2.13b. The spread in results after 5 days, 21 to 23 ppbv, or ±12% in O_3 loss, does not seem to correlate with the O_3 photolysis rates in Figure 2.13a. Also shown in Figure 2.13b is the evolution of O_3 in LAND (0 km, 200 pptv NO_x). The additional NO_x boosts O_3 for a day or two, and doubles the discrepancy in the modelled ozone. The disagreement here is important since most tropospheric O_3 is destroyed by these reactions ((2.1), (2.2) and (2.6)) in the wet lower troposphere.

The production of O_3 in a NO_x-rich PLUME (4 km, 10 ppbv NO_x) without non-methane hydrocarbons is rapid and continues over 5 days as shown in Figure 2.13c. Model agreement is excellent on the initial increases from 30 to 60 ppbv O_3 in 48 hours, but starts to diverge as NO_x levels fall. When large amounts of NMHC are included in PLUME+HC (also Figure 2.13c), ozone is produced and NO_x depleted rapidly, in less than one day. Differences among models become much greater, in part because different chemical mechanisms for NMHC oxidation were

Table 2.8: *Models participating in the photochemical intercomparison and the delta-CH_4 simulation.*

Code		Model	Contributor[†]	Email
A	§	U. Mich.	S. Sillman	sillman@madlab.sprl.umich.edu
B	§	UKMO/UEA	R. Derwent	rgderwent@email.meto.govt.uk
B&	§	UEA-Harwell/2D	C. Reeves	c.reeves@uea.ac.uk
C		U.Iowa	G. Carmichael	gcarmich@icaen.uiowa.edu
D		UCIrvine	M. Prather	prather@halo.ps.uci.edu
E	§	NASA Langley	J. Richardson	richard@sparkle.larc.nasa.gov
F	§	AER (box)	R. Kotamarthi	rao@aer.com
G		Harvard	L. Horowitz	lwh@hera.harvard.edu
H		NASA Ames	B. Chatfield	chatfield@clio.arc.nasa.gov
I		NYU-Albany	S. Jin	jin@mayfly.asrc.albany.edu
J		KFA Jüelich	M. Kuhn	ICH304@zam001.zam.kfa-juelich.de
K		GFDL	L. Perliski	lmp@gfdl.gov
L		Ga.Tech.	P. Kasibhatla	psk@gfdl.gov
M&	§	U. Camb/2D	K. Law	kathy@atm.ch.cam.ac.uk
N	§	U. Camb (box)	O. Wild	oliver@atm.ch.cam.ac.uk
O		LLNL/2D	D. Kinnison	dkin@cal-bears.llnl.gov
P		LLNL/3D	J. Penner	penner1@llnl.gov
P&	§	"	C. Atherton	cyndi@tropos.llnl.gov
Q+	§	NASA Goddard	A. Thompson	thompson@gator1.gsfc.nasa.gov
R&	§	AER/2D	R. Kotomarthi	rao@aer.com
S	§	Cen. Faible Rad.	M. Kanakidou	mariak@asterix.saclay.cea.fr
T	§	U. Oslo/3D	T. Berntsen	terje.berntsen@geofysikk.uio.no
T&		"	I. Isaksen	–
U	§	NILU	F. Stordal	frode@nilu.no
Y/+	§	U. Wash.	H. Yang	yang@amath.washington.edu
Z/+		Ind. Inst. Tech.	M. Lal	mlal@netearth.ernet.in

Explanation of Codes: / submitted results only for the photolysis experiment, $J[O_3 \, O(^1D) + O_2]$.
    & submitted results for the methane perturbation study (second contributor if listed).
    + submitted results for the methane perturbation study, but could not be used.
    § results revised or submitted since the Irvine drafting session (January 1994), see text.

† Only a single point of contact is given here.

used. (The reaction pathways and rate coefficients for chemistry with CH_4 as the only hydrocarbon have become standardised, but different approaches are used for non-methane hydrocarbons.)

The 24-hour averaged OH concentrations are shown for LAND in Figure 2.13d. Values are high during the first day and demonstrate the dependence of OH on NO_x, which begins at 200 pptv and decays rapidly to about 10 pptv by day 4. Modelled OH values fall within a ±20% band. This variation in OH between models, however, does not correlate obviously with any other model differences such as the photolysis rate of O_3 or the abundance of NO. These results show that basic model-to-model differences of 30% or more exist in the calculations of O_3 change and OH concentrations. This spread is not a true scientific uncertainty, but presumably a result of different numerical methods that could potentially be resolved, although no single fix, such as O_3 photolysis rates, would appear to reduce the spread. A more significant uncertainty in the current calculations of O_3 tendencies is highlighted by the parallel experiments with and without NMHC: the sources, transport and oxidation, and in particular the correlation of NO_x emissions and NMHC emissions on a fine scale, may control the rate at which NO_x produces O_3.

Other Trace Gases and Atmospheric Chemistry

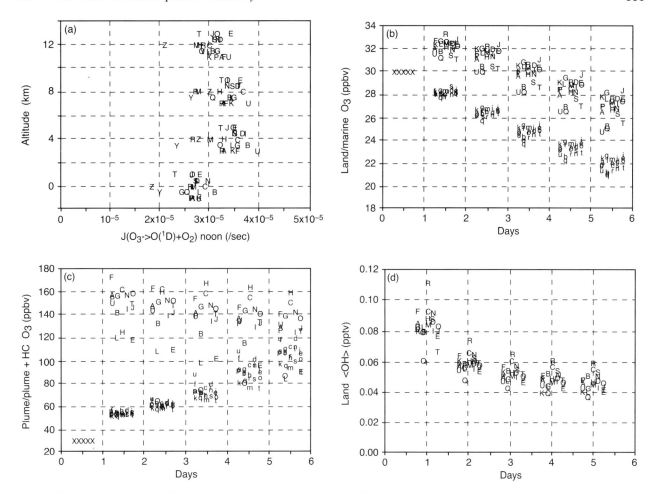

Figure 2.13: Intercomparison of 23 photochemical models, testing their ability to calculate photodissociation rates and chemical rates of ozone net production. All models agreed to adopt the same chemical rate coefficients and cross-sections; see Table 2.8 for the key to the letters.
(a) Photodissociation rate for O_3 into $O(^1D)$ and O_2; values refer to noon, July 1, 45° N, assuming the US Standard Atmosphere.
(b) The 5-day evolution of near-surface O_3, initialised at noon with 30 ppbv (XXXXX) and with high levels of NO_x (LAND, upper case key letters, see Table 2.8) or with low NO_x (MARINE, lower case keys).
(c) The 5-day evolution of O_3 in a simulated biomass burning plume at 4 km altitude, initialised at 30 ppbv of O_3 and extremely high levels of NO_x. Two cases of no NMHC (lower case keys) and equivalently high NMHC (upper case keys) are considered.
(d) The 5-day evolution of OH concentrations, calculated as 24-hour averages from noon to noon, are shown for the LAND simulation of panel (b).

2.9.3 Conclusions

3-D synoptic CTMs are needed for climate assessments that involve atmospheric chemistry. However, the currently available CTMs have been evaluated only for the transport of simple tracers; these models must develop the chemical mechanisms so that they can accurately simulate the scales of chemistry that remove NO_x and NMHC while making O_3. The chemistry in these models will continue to be tested and evaluated against observations. The growth in our understanding of tropospheric chemistry, and particularly that of O_3, has been and will continue to be driven by a combination of careful field observations of trace gases, laboratory investigations of chemical mechanisms and the theoretical development of CTMs and related models.

2.10 Global Tropospheric Ozone Modelling

Modelling tropospheric ozone is one of the more difficult tasks in atmospheric chemistry. The difficulty is due in part to the large number of processes that control tropospheric ozone and its precursors, and in part to the large range of spatial and temporal scales that must be resolved. Global tropospheric CTMs attempt to simulate the life cycles of many trace gases for which we have uncertain knowledge of their sources as well as the chemical mechanisms of their destruction.

The accurate description of the boundary conditions for a CTM simulation of tropospheric ozone is daunting. For stratospheric assessments we have been interested in long-lived gases which are well mixed in the troposphere, and whose concentrations are, therefore, independent of when or where they were emitted at the North pole or the South, in summer or winter. For tropospheric ozone, the key constituents, NO_x and NMHC, do not become well mixed in the troposphere, and their impact on ozone is likely to depend on when or where they were emitted. This greatly complicates any assessment since, unless great effort is made, the CTM simulations from different research groups will not be simulating the same atmosphere.

Such differences are apparent when comparing the published simulations of tropospheric ozone and its climatic impact. Most of the models published to date are 2-D (see discussion in Section 2.9.1) and have calculated changes in tropospheric O_3 since the pre-industrial atmosphere or made predictions of future ozone for assumed growth scenarios in CH_4, NO_x and other trace gases (Kanakidou *et al.*, 1991; Wuebbles *et al.*, 1992; Fuglestvedt *et al.*, 1993,1994a; Law and Pyle, 1993; Derwent, 1994b; Hauglustaine *et al.*, 1994; Strand and Hov, 1994). Two monthly averaged 3-D CTMs have included a fairly complete chemical model to study tropospheric O_3 (Müller, 1992; Lelieveld, 1994). The global synoptic 3-D CTMs (see Section 2.9.1) have been developed recently to the level where extensive publications are starting to appear (e.g., Penner *et al.*, 1991).

Few of the above studies of pre-industrial atmospheres can be compared directly because of the diversity in boundary conditions used. While most calculations will use similar boundary conditions for the long-lived gases (CH_4 and N_2O) based on the ice core record, the historical records of CO, NO_x and NMHC emissions are unknown. Furthermore, many of these models are solely tropospheric and do not include photochemical coupling with the stratosphere, and the stratospheric influxes of O_3 and NO_x are prescribed. Since the documented changes in CH_4 and N_2O are substantial, it is likely that stratospheric ozone has evolved. Thus CTMs simulating tropospheric O_3 must include the interactions with an evolving stratosphere. A final warning must be that these models cannot predict the possible changes in the circulation over the last centuries.

In spite of these obvious problems, the predicted changes in tropospheric O_3 in the Northern Hemisphere since pre-industrial times have been calculated (Hauglustaine *et al.*, 1994) as approximately a doubling. Combined with the historical evidence from observations (see Section 2.6), we may make a best guess that the increase in ozone since 1850 is about 25 ppbv throughout the troposphere at northern mid-latitudes, and possibly extending into the tropics. However, changes in the Southern Hemisphere, which is remote from human influence in terms of NO_x, CO and NMHC emissions, are not certain. The confidence in this number is no greater than a factor of 2 at best, and such "best guesses" do not substitute for critical assessment of the CTM simulations which await the ongoing development of the models and corresponding measurements.

2.10.1 Tropospheric NO_x: Surface Combustion and Aircraft

The sources of NO_x in the troposphere are numerous (see Section 2.7) and the large natural source from lightning is not well characterised. Thus the 3-D CTMs failure to simulate correctly the NO_x distributions and total nitrate observations in the troposphere, particularly in remote marine locations, is not surprising (Penner *et al.*, 1991; Kasibhatala *et al.*, 1993). These problems could be due to limitations in knowledge of emissions and to errors in the simulation of atmospheric chemistry and transport from intensive source regions.

Recent attention has focused on the role of aircraft relative to other anthropogenic sources of NO_x, as jet engine exhaust is placed directly into the upper troposphere and lower stratosphere. Aircraft are a small source of NO_x, producing about 1/80 of the NO_x released from other combustion sources at the Earth's surface. Similarly, aircraft fuel is about 3% of the total fossil fuel burned each year. However, upper tropospheric ozone exerts a greater radiative forcing than ozone at lower altitudes and so the importance of the various sources of upper tropospheric ozone need to be quantified.

Several recent studies have estimated peak increases of about 5% in tropospheric ozone over the northern mid-latitudes troposphere due to the current aircraft fleet (Beck *et al.*, 1992; Johnson *et al.*, 1992; Fuglestvedt *et al.*, 1993; Rohrer *et al.*, 1993). In general this work uses 2-D models, which have systematic discrepancies in the rate of vertical mixing between the surface and the mid troposphere (Section 2.9.1), although these studies are rapidly expanding to the best current 3-D transport models. Model tests also demonstrate that the ozone forming potential of NO_x emitted by aircraft depends upon transport formulation, injection height, removal by cloud processes, and the background level of NO_x from other sources. For example, if the lightning source of tropospheric NO_x is underestimated in the model, then the aircraft impact on ozone is overestimated (Fuglestvedt *et al.*, 1994).

If we select an upper limit for the effect of aircraft NO_x on ozone, then the short-term radiative forcing could be comparable to the radiative forcing due to the CO_2 from the burned fuel.

While aircraft emissions of NO_x are thought to have a greater radiative effect (through ozone formation in the upper troposphere) on a per kilogram basis, surface emissions of NO_x probably dominate the anthropogenically induced production of ozone within the troposphere. The 3-D synoptic CTMs show that similar short-lived trace species are rapidly mixed throughout the

Other Trace Gases and Atmospheric Chemistry

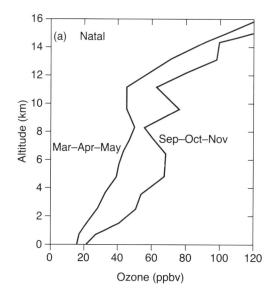
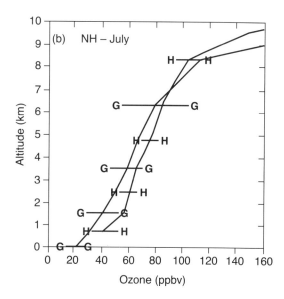

Figure 2.14: Profiles of O_3 observed in the tropical troposphere (Natal, panel a) and at northern mid-latitudes in July (G = Goose Bay and H = Hohenpeissenberg, panel b). The data from the two Northern hemisphere stations are averages over 1980 to 1991 (Logan, 1994). The tropical station shows measurements taken during the seasons of minimum ozone (March, April and May) and maximum ozone (September, October and November) (Kirchhoff *et al.*, 1989).

troposphere. Jacob *et al.* (1993) calculated the net export of O_3 and its precursors from the boundary layer over the North American continent and found it comparable to the mean stratospheric source of O_3 over the northern mid-latitudes.

2.10.2 CH_4 Increases: A Case Study

The impact of methane perturbations are felt throughout all of atmospheric chemistry from the surface to the exosphere, and most of these mechanisms are well understood. Quantification of these effects, however, is one of the classic problems in modelling atmospheric chemistry. Similar to the ozone studies noted above, the published methane-change studies have examined scenarios that range from 700 ppbv (pre-industrial) to 1700 ppbv (current) to a doubling by the year 2050 (e.g., WMO, 1992), but these scenarios are not consistent across models. The study of the effects of an increase in methane concentrations (delta-CH_4) was designed to provide a common framework for evaluating the multitude of indirect effects, especially changes in O_3 and OH, that are associated with an increase in CH_4 (see Section 2.2.6) and that provide the basis for the quantitative evaluation of radiative forcing from today's CH_4 emissions. The study centres on today's atmosphere; using each model's best simulation of the current atmosphere and then increasing the CH_4 concentration in the troposphere by 20%, from 1715 ppbv to 2058 ppbv. This increase is small enough that perturbations to current atmospheric chemistry are approximately linear. The list of participating research groups is given in Table 2.8, and the protocol and history are described above with PhotoComp.

2.10.2.1 The current atmosphere

Important diagnostics from delta-CH_4 include O_3 and NO_x profiles for the current atmosphere, providing a test of the realism of each model's simulation. Typical profiles observed for O_3 in the tropics and in northern mid-latitudes over America and Europe are shown in Figure 2.14. The corresponding calculated O_3 profiles, shown in Figure 2.15a-b, differ by almost a factor of 2, but encompass the observations. The clear divergence of results above 10 km altitude illustrates the difficulty in determining the transition between troposphere and stratosphere. This exercise is only the beginning of an objective evaluation of tropospheric ozone models. A more rigorous diagnosis is needed, including other latitudes and seasons.

The modelled zonal-mean NO_x profiles, shown in Figure 2.15c, differ by almost a factor of 10. Comparisons in the lowest 2 km altitude are not meaningful since the CTMs average regions of high urban pollution with clean marine boundary layer. The range of modelled NO_x values in the free troposphere often falls outside the range of typical observations, about 20 to 100 pptv, as shown in Figures 2.8 and 2.9.

2.10.2.2 O_3 perturbations

The calculated changes in tropospheric O_3 for June, July and August in northern mid-latitudes and the tropics are

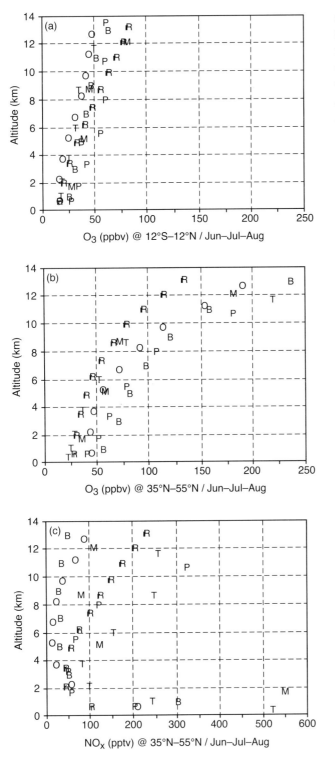

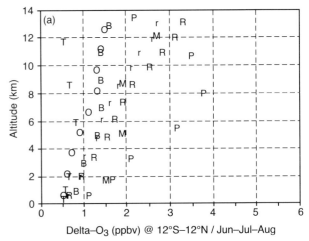

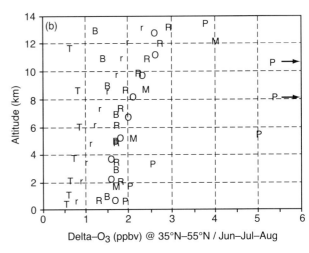

shown in Figure 2.16a-b for the delta-CH$_4$ study. Ozone increases everywhere in the troposphere, with values ranging from about 0.5 ppbv to more than 5 ppbv (the extremely high values for model P in the upper troposphere must be considered cautiously since this recent submission has not yet been scrutinised as much as the other results). In general the increase is larger at mid-latitudes, but not for all models. Results for the southern mid-latitudes in summer (December, January and February) (not shown) are similar to the northern.

The large spread in these results shows that our ability to predict changes in tropospheric O$_3$ induced by CH$_4$ perturbations is not very good. This conclusion is not unexpected given the large range in modelled NO$_x$ (Figure 2.15c), but the differences in O$_3$ perturbations do not seem

Figure 2.15: Profiles of O$_3$ modelled for the tropical troposphere (panel a, 12° S to 12° N), the northern mid-latitude troposphere (panel b, 35° N to 55° N), and those of NO$_x$ modelled for the northern mid-latitude troposphere (panel c, 35° N to 55° N). The models were simulating the current atmosphere, and results are averaged over northern summer (Jun-Jul-Aug). Results are from the delta-CH$_4$ study and the letter keys are given in Table 2.8.

Figure 2.16: Changes in tropospheric O$_3$ profiles predicted by models for a 20% increase in CH$_4$ relative to today's atmosphere (as shown in Figure 2.15). Results parallel those in Figure 2.15, averaged over northern summer and reported for (a) the tropics and (b) northern mid-latitudes.

to correlate with the model's NO_x. Nevertheless, a consistent pattern of increases in tropospheric O_3, ranging from 0.5 to 2.5 ppbv, occurs throughout most of the troposphere. Our best estimate is that a 20% increase in CH_4 would lead to an increase in ozone of about 1.5 ppbv throughout most of the troposphere in both tropics and summertime mid-latitudes. This indirect impact on the radiative forcing (see Chapter 4) is about 25 ± 15% of that due to the prescribed 343 ppbv increase in CH_4 alone.

2.10.2.3 Adjustment time of CH_4 emissions

Methane is the only long-lived gas that has clearly identified, important chemical feedbacks: increases in atmospheric CH_4 reduce tropospheric OH, increase the CH_4 lifetime, and hence amplify the climatic and chemical impacts of a CH_4 perturbation (Isaksen and Hov, 1987). The delta-CH_4 simulations from six different 2-D and 3-D models show that these chemical feedbacks change the relative loss rate for CH_4 by -0.17% to -0.35% for each 1% increase in CH_4 concentration as shown in Table 2.9. This range reflects differences in the modelled roles of CH_4, CO, and NMHC as sinks for OH. For example, model M, with the smallest feedback factor, has fixed the concentrations of CO; and model R has shown that calculating CO instead with a flux boundary condition (as most of the other models have done) results in a larger feedback. These differences cannot be resolved with this intercomparison, and this range underestimates our uncertainty in this factor.

Recent theoretical analysis has shown that the feedback factor (FF) defined in Table 2.9 can be used to derive an adjustment time that accurately describes the time-scale for the decay of a pulse of CH_4 added to the atmosphere.

Table 2.9: *Inferred CH_4 response time from the CH_4 perturbation simulations.*

Model code	Feedback factor	Adjustment time/lifetime
B	-0.20%	1.29
M	-0.17%	1.23†
O	-0.35%	1.62
P	-0.22%	1.32
R	-0.26%	1.39
(R)	-0.18%	1.26†
T	-0.34%	1.61

Note: Feedback factor = relative change (%) in the globally averaged CH_4 loss frequency (i.e., [OH]) for a +1% increase in CH_4 concentrations.
† These models use fixed CO concentrations and so underestimate this ratio.

Effectively, a pulse of CH_4, no matter how small, reduces the global OH levels by a similar amount (i.e. -0.3% per +1%). This leads to a relative build-up of a corresponding increase in the already existing atmospheric reservoir of CH_4, which cannot be distinguished from a longer adjustment time for the initial pulse. Thus the adjustment time (AT) is longer than the lifetime (LT) derived from the budget (i.e., total abundance divided by total losses). Prather (1994) has shown that the ratio, AT/LT, is equal to 1/(1 + FF) and that this adjustment time applies to all CH_4 perturbations, positive or negative, no matter how small or large, as long as the change in CH_4 concentration is not large enough to change the feedback factor. Based on model results, this assumption should apply at least over a ±30% change in current CH_4 concentrations. Two of the models with results in Table 2.9 have shown that small CH_4 perturbations decay with the predicted adjustment time.

Based on these limited results, we choose 1.45 as the best estimate for the ratio AT/LT, with an uncertainty bracket of 1.20 to 1.70. The budget lifetime of CH_4 is calculated to be about 10 yr, using the CH_3CCl_3 lifetime as a standard for OH and including stratospheric losses, and gives a best estimate for the CH_4 adjustment time of 14.5 ± 2.5 yr. If the biological soil sink were to respond to the atmosphere in the same way as the chemical sink, this would reduce the adjustment time to a little under 14 yr. In this chapter we use the range 14 ± 3 (11-17) yr to cover this uncertainty. This enhanced time-scale describes the effective duration for all current emissions of CH_4; it is independent of other emissions as long as current concentrations of CH_4, ±30%, are maintained. In the extreme case that all CH_4 emissions (including natural sources) ceased, the initial rate of decay (1/9.4 /yr) would become more rapid as OH levels increased, resulting in increasingly rapid decay, with an average adjustment time of about 8.5 yr. However, a small additional pulse on top of this decaying profile still produces chemical feedbacks that lengthen its adjustment time to 10.1 yr, a factor of 1.2 longer than the bulk of the CH_4.

Some of this effect was included in the previous assessment as an "indirect OH" enhancement to the size of the CH_4 perturbations. Here we recognise that the OH chemical feedbacks give an effective residence time for CH_4 emissions that is substantially longer than the lifetime used to derive the global budgets. This effective lengthening of a CH_4 pulse applies also to all induced chemical perturbations such as tropospheric O_3 and stratospheric H_2O.

Deriving natural sources based on observed pre-industrial concentrations is more uncertain for CH_4 than for N_2O because we do not know how tropospheric OH has changed. If the CH_4 lifetime has not changed, then a source of 210 $Tg(CH_4)$/yr would be needed to maintain

700 ppbv. If the current feedback factor is applicable to low values of CH_4, then 280 Tg(CH_4)/yr is required. However, industrial and agricultural emissions of CO, NMHC and NO_x also perturb tropospheric OH, and a clear picture of changes since the pre-industrial is not yet available (Thompson and Cicerone, 1986; Isaksen and Hov, 1987; Derwent, 1994b).

2.10.3 Conclusions

As noted above, there are numerous published studies examining the response of tropospheric ozone to changes in CH_4 and other trace gases since the pre-industrial era, particularly the emissions of NO_x from surface combustion and aircraft. It is difficult to evaluate the robustness of those results, as well as those from the delta-CH_4 study, without a more objective critique of the model performance, which should occur with time. This period will mark a significant transition in assessment models of the troposphere where we will rely increasingly on 3-D models.

A degree of caution is necessary in interpreting the results quoted here for tropospheric O_3 changes as a central value with a formal range of uncertainty (e.g., 95% confidence level). They represent our current best guesses; we must be aware that these values could change significantly.

2.11 Stabilisation of Atmospheric Chemical Composition

The atmospheric abundance of a chemical species is governed by the rates of production and removal, where these terms include processes occurring at the Earth's surface (emission and deposition) and in the atmosphere (chemical). When the gross production/emission rate is equal to the gross removal rate, the atmospheric concentration is in steady state and therefore remains constant over time. In practice, it is more straightforward to quantify the net change in atmospheric abundance (by direct measurement of the atmospheric concentration), than to estimate either of the gross terms with any accuracy. As discussed earlier in this chapter, a number of atmospheric feedback processes are known which have the potential to change atmospheric turn-over times (the removal rate), principally through perturbation of the tropospheric hydroxyl radical concentrations. How well current models of the atmosphere can quantify the effects of these feedbacks is unclear, and so in this discussion it is assumed (with the exception of CH_4 where the adjustment time is used) that the atmospheric lifetimes of the various gases remains constant.

The adjustment time determines the speed with which the atmospheric abundance responds to changes in the production rate - the faster the removal process (i.e. the shorter the adjustment time), the quicker the response. Thus if emissions for CH_4 (adjustment time about 14 years) and N_2O (about 120 years) were to be held constant, the atmospheric abundance of CH_4 would stabilise sooner. Thus we discuss stabilisation at different levels with particular emphasis on the implications for the anthropogenic component.

The major sources of most of the gases discussed in this section are at the Earth's surface. However some of the gases discussed in this chapter are predominantly formed in the atmosphere. Two of these are of particular importance: O_3 and CO. O_3 is formed exclusively through photochemical processes in the atmosphere, and its concentration is strongly influenced by chemicals of anthropogenic origin (e.g., halocarbons in the stratosphere, NO_x and hydrocarbons in the troposphere). A major source of CO is the oxidation of methane and other hydrocarbons in the troposphere. The link between surface emissions of NMHC and CO production in the atmosphere is more quantitatively established than the corresponding link between NO_x and NMHC emissions and the production of tropospheric ozone, although the mechanisms for both are well known.

2.11.1 Methane

The observed trend in the CH_4 atmospheric abundance over the past decade has been 35-40 Tg/yr (+0.7%/year); thus we know that gross production exceeded the gross removal by this rate. As shown in Table 2.3, the best estimate for the gross production of CH_4 is 535 Tg/yr. Of this about 375 Tg/yr is associated with anthropogenic activities and 160 Tg/yr with natural processes. Stabilisation of CH_4 concentrations will follow stabilisation of emissions, and a wide range of potential stabilisation profiles can be envisaged depending on how much control of anthropogenic emissions is possible. In this respect it should be noted that the continued increase in CH_4 observed since 1950 implies a sustained increase in emissions of about 1%/year. Four cases can be considered:

(a) If this (or any) increase in emissions were to continue, no stabilisation of the atmospheric CH_4 abundance would occur.

(b) If current emissions were held constant, CH_4 concentrations would stabilise in less than 50 years at about 1900 ppbv.

(c) If emissions could be cut by about 37 Tg/yr, the atmospheric CH_4 abundance would stabilise at current levels.

(d) If emissions could be reduced by more than 37 Tg/yr, the atmospheric abundance would stabilise below current levels. The 15-year adjustment time of CH_4

would allow such changes to occur on a decadal time-scale.

Very crudely, each reduction in emissions of 37 Tg/yr would lower the atmospheric concentration at stabilisation by about 170 ppbv (10% of current values). (Conversely each increase of 37 Tg/yr would raise the atmospheric concentration at stabilisation by about 170 ppbv.) It must be emphasised that these estimates should only be used as a guide because it is assumed that other changes in atmospheric chemistry that control the CH_4 adjustment time are not changing over this period.

2.11.2 Nitrous Oxide

The atmospheric N_2O abundance has, on average, increased by about 3.9 Tg(N)/yr (0.25%/yr) over the last 10-15 years. The anthropogenic emissions are estimated to be about 5.7 Tg(N)/yr, so a cut of about two-thirds of the anthropogenic emissions is needed to stabilise N_2O concentration at today's value. However, if the anthropogenic emissions were held constant at current values, the atmospheric N_2O abundance would climb from 310 ppbv to about 400 ppbv. The relatively long lifetime of N_2O (120 years) means that this change would be slow with the N_2O abundance reaching 370 ppbv by the year 2100. Further, any reductions in the atmospheric N_2O abundance would take a hundred years or so.

2.11.3 Halocarbons

Nearly all CFCs and other industrial halocarbons are out of balance in that emissions exceed atmospheric losses. For short-lived halocarbons such as CH_3CCl_3 with stable emissions over the past several years, the compound is nearly in steady-state, and the reduction in emissions expected under the Montreal Protocol would first stabilise, and then reduce atmospheric concentrations. For longer-lived compounds (CFCs), the primary production allowed under Article 5 of the Protocol and the continued release of the bank of current CFCs may sustain concentrations of the longer-lived CFCs over over the next decade or so. Overall, total tropospheric chlorine levels in the form of industrial halocarbons should peak in the next few years. In the longer term, atmospheric abundances will fall, the speed of decline depending on the adjustment time of the particular compound.

HCFCs and HFCs, used as substitutes for CFCs, are expected to grow as emissions increase. Under the terms of the Montreal Protocol, the HCFCs will be phased out in the early part of next century. There are no natural sources and so their atmospheric abundances will respond to the reductions in emissions. Once the emissions have been stopped, the atmospheric abundance will return to zero - the time needed for this to occur varies from chemical to chemical according to atmospheric adjustment time, but is shorter than for the longer-lived CFCs. Presently there are no controls over HFCs. The atmospheric abundances of these compounds would behave similarly to HCFCs if emissions were controlled.

Perfluorinated species (CF_4, C_2F_6, SF_6) are extremely long-lived. There are no known sinks for these compounds on the time-scales of centuries, and the only method of stabilising these compounds is complete cessation of emissions.

2.11.4 Ozone

Stratospheric ozone decreases over the last 25 years have occurred principally from the use of CFCs and halons. Future production of these and other ozone-depleting substances is limited under the Montreal Protocol. Depletion of stratospheric ozone is predicted to peak in the next decade, and ozone levels are expected to recover as the concentrations of these compounds fall, which will take several decades. Recovery of the Antarctic ozone hole is not expected until the middle of the 21st century. CFCs, halons and related compounds are not the only anthropogenic influences and other perturbations (NO_x from N_2O, changes in stratospheric temperatures from increased radiative forcing, etc.) will also influence future stratospheric ozone levels.

In most of the troposphere, the rate-limiting precursor of new ozone production is NO_x, although NMHCs play an important role in some regions. Accordingly, recommendations to stabilise tropospheric ozone focus on controlling emissions of NO_x. However, as shown in Section 2.10.2, some control of CH_4 (and CO and NMHC) is also necessary. The lifetime of ozone is very short (2-4 weeks), and so the effects of any reductions in precursor emissions should be felt quickly.

How to control these precursors, particularly NO_x and NMHC, is extremely difficult to assess, because their impact on tropospheric O_3 cannot be calculated by just summing their global emissions. The effect of these short-lived species depends on where and when they are emitted, as evidenced by the fact that aircraft emissions of NO_x have a proportionately greater impact than equal surface emissions on O_3 in the upper troposphere. Both this relative impact and the absolute source (i.e., anthropogenic surface emissions of NO_x are about 80 times those from aircraft) must be combined when assessing the role of short-lived species on radiative forcing.

References

AFEAS (Alternative Fluorocarbons Environmental Acceptability Study), 1993: *Production, Sales and Atmospheric Release of Fluorocarbons Through 1992*, SPA-AFEAS, Inc., Washington DC, USA.

Akimoto, H., N. Nakane and Y. Matsumoto, 1994: The chemistry of oxidant generation: tropospheric ozone increase in Japan. In: *The Chemistry of the Atmosphere: Its Impact on Global Change*, J.G. Calvert (ed), Blackwell Sci. Publ., pp.261-273.

Albritton, D. and R. Watson, 1993: Methyl bromide and the ozone layer: a summary of current understanding. In: *Methyl Bromide: Its Atmospheric Science, Technology and Economics. Montreal Protocol Assessment Supplement*, R. Watson, D. Albritton, S. Anderson and S. Lee-Bapty (eds), UNEP, Nairobi, Kenya.

Anastasi, C. and V.J. Simpson, 1993: Future methane emissions from animals. *J. Geophys. Res.*, **98**, 7181-7186.

Andreae, M.O., 1993: The influence of tropical biomass burning on climate and the atmospheric environment. In: *Biogeochemistry of Global Change: Radiatively Active Trace Gases*. R.S. Oremland (ed.), New York: Chapman & Hall, pp.113-150.

Andreae, M.O. and P. Warneck, 1994: Global methane emissions from biomass burning and comparison with other sources. *Pure and Applied Chemistry*, **66**, 162-169.

Andreae, M.O., E.V. Browell, G.L. Garstang, R.C. Gregory, R.C. Harriss, G.F. Hill, D.J. Jacob, M.C. Pereira, G.W. Sachse, A.W. Setzer, P.L. Silva Dias, R.W. Talbot, A.L. Torres and S.C. Wofsy, 1988: Biomass-burning emissions and associated haze layers over Amazonia. *J. Geophys. Res.* **93**, 1509-1527.

Aoki, S., T. Nakazawa, S. Murayama and S. Kawaguchi, 1992: Measurements of atmospheric methane at the Japanese Antarctic station, Syowa. *Tellus*, **44B**, 273-281.

Arnold, F., J. Scheid, Th. Stimp, H. Schlager and M.E. Reinhardt, 1992: Measurements of jet aircraft emissions at cruise altitude 1: The odd-nitrogen gases NO, NO_2, HNO_2 and HNO_3. *Geophys. Res. Lett.*, **12**, 2421-2424.

Atlas, E., W. Pollock, J. Greenberg, L. Heichat and A. Thompson, 1993: Alkyl nitrate, non methane hydrocarbons and halocarbon gases over the equatorial Pacific Ocean during SAGA 3. *J. Geophys. Res.*, **98**, 16933-16947.

Bachelet, D. and H. Neue, 1993: Methane emissions from wetland rice areas of Asia. *Chemosphere*, **26**, 219-237.

Balkanski, Y.J., D.J. Jacob, R. Arimoto and M.A. Kritz, 1992: Long-range transport of radon-222 over the North Pacific Ocean: implications for continental influences. *J. Atmos. Chem.*, **14**, 353-374.

Bartlett, K.B. and R.C. Harriss, 1993: Review and assessment of methane emissions from wetlands. *Chemosphere*, **26**, 261-320.

Bartlett, K.B., P.M. Crill, J.A. Bonassi, J.E. Richey and R.C. Harriss, 1990: Methane flux from the Amazon River floodplain: emissions during rising water. *J. Geophys. Res.*, **95**, 16773-16788.

Baughcum, S.L., D.M. Chan, S.M. Happenny, S.C. Henderson, P.S. Hertel, T. Higman, D.R. Maggiora and C.A. Oncina, 1993: Aircraft emissions scenarios. In: *The Atmospheric Effects of Stratospheric Aircraft: A 3rd Program Report*. NASA Ref. Publ. **1313**.pp87-208

Beck, L.L., 1993: A global methane emissions program for landfills, coal mines, and natural gas systems. *Chemosphere*, **26**, 447-452.

Beck, L.L., S. Piccot and D. Kirchgessner, 1993: Industrial sources. In: *Atmospheric Methane: Sources, Sinks and Role in Global Change*, M. Khalil (ed), NATO ASI series, pp.230-253.

Beck J.P., C.E. Reeves, F.A.A.M. de Leeuw and S.A. Penkett, 1992: The effect of aircraft emissions on tropospheric ozone in the Northern Hemisphere. *Atmos. Env.* **26A**, 17-29.

Berger, M.G.M., R.M. Hofmann, D. Scharffe and P.J. Crutzen, 1993: Nitrous oxide emissions from motor vehicles in tunnels and their global extrapolation. *J. Geophys Res.*, **98**(D10), 18527-18531.

Blunier, T., J. Chappellaz, J. Schwander, J. Barnola, T. Desperts, B. Stauffer and D. Raynaud, 1993: Atmospheric methane record from a Greenland ice core over the past 1000 years. *Geophys. Res. Lett.*, **20**, 2219-2222.

Bojkov, R.D., C.S. Zerefos, D.S. Balis, I.C. Ziomas and A.F. Bais, 1993: Record low total ozone during northern winters of 1992 and 1993. *Geophys. Res. Lett.*, **20**, 1351-1354.

Bouwman, A.F. 1993: Report of the third workshop of the Global Emissions Inventory Activity (GEIA). *RIVM Report* **481507002**, RIVM, Bilthoven, The Netherlands.

Brenninkmeijer, C.A.M., M.R. Manning, D.C. Lowe, G. Wallace, R.J. Sparks and A. Volz-Thomas, 1992: Interhemispheric asymmetry in OH abundance inferred from measurements of atmospheric ^{14}CO. *Nature*, **356**, 50-52.

Brühl, C., and P.J. Crutzen, 1989: On the disproportionate role of tropospheric ozone as a filter against solar UV-B radiation. *Geophys. Res. Lett.*, **17**, 703-706.

Brunke, E.G., M.E. Scheel and W. Seiler, 1990: Trends of tropospheric carbon monoxide, nitrous oxide and methane as observed at Cape Point, South Africa. *Atmos. Env.*, **24**(A), 585-595.

Butler, J.H., 1994: The potential role of the ocean in regulating atmospheric CH_3Br. *Geophys. Res. Lett.*, **21**, 185-188.

Butler, J.H., J.W. Elkins, T.M. Thompson, B.D. Hall, T.H. Swanson and V. Koropalov, 1991: Oceanic consumption of CH_3CCl_3: implications for tropospheric OH. *J. Geophys. Res.*, **96**, 22347-22355.

Butler, J.H., J.W. Elkins, B.D. Hall, S.O. Cummings and S.A. Montzka, 1992: A decrease in the growth rates of atmospheric halon concentrations. *Nature*, **359**, 403-405.

Butler, J.H., J.W. Elkins, T.M. Thompson, B.D. Hall, J.M. Lobert, T.H. Swanson and V. Koropalov, 1993: A significant oceanic sink for atmospheric CCl_4. In: *The Oceanography Society, Third Scientific Meeting, Apr 13-16, Seattle, Abstracts*, p.55.

Carroll, M.A. and Anne M. Thompson, 1994: NO_x in the Non-Urban Troposphere. In: *Current Problems and Progress in Atmospheric Chemistry*, John R. Barker (ed.), World Scientific Publishing Company. (In press)

Carroll, M.A., D.D. Montzka, G. Hubler, K.K. Kelly and G.L. Gregory, 1990: *In situ* measurements of NO_x in the Airborne Arctic Stratospheric Expedition. *Geophys. Res. Lett.*, **17** (4), 493-496.

Carroll, M.A., B.A. Ridley, D.D. Montzka, G. Hubler, J.G. Walega, R.B. Norton and B.J. Huebert, 1992: Measurements of nitric oxide and nitrogen dioxide during the Mauna Loa Observatory Photochemistry Experiment. *J. Geophys. Res.*, **97** (D10), 10361-10374.

Chameides, W.L., S.C. Liu and R.J. Cicerone, 1976: Possible variations in atmospheric methane. *J. Geophys. Res.*, **81**, 4997-5001.

Chandra, S., 1992: Changes in stratospheric ozone and temperature due to the eruption of Mt. Pinatubo. *Geophys. Res. Lett.*, **20**, 33-36.

Chappellaz, J., T. Blunier, D. Raynaud, J.M. Barnola, J. Schwander and B. Stauffer, 1993: Synchronous changes in atmospheric CH_4 and Greenland climate between 40 and 8 kyr BP. *Nature*, **366**, 443-445.

CIAB, 1992: Global methane emissions from the coal industry, Report of Global Climate Committee, Coal Industry Advisory Board.

Cicerone, R., 1979: Atmospheric carbon tetrafluoride : a nearly inert gas. *Science,* **206**, 59-60.

Codispoti, L., J. Elkins, T. Yoshinari, G. Frederich, C. Sakamoto and T. Packard, 1992: On the nitrous oxide flux from productive regions that contain low oxygen waters. In: *Oceanography of the Indian Ocean,* B. Desai (ed), Oxford & IBH Publishing Co., New Delhi, India, pp 271-284.

Cros, B., R. Delmas, A. Nganga, A. Minga, J. Fishman and V. Brackett, 1988: Seasonal trends of ozone in equatorial Africa: Experimental evidence of photochemical formation. *J. Geophys. Res.,* **93**, 8355-8366.

Cunnold, D., R. Prinn, R. Rasmussen, P. Simmonds, F. Alyea, C. Cardelino, A. Crawford, P. Fraser and R. Rosen, 1986: The atmospheric lifetime and annual release estimates for CCl_3F and CCl_2F_2 from 5 years of ALE data. *J. Geophys. Res.,* **91**, 10797-10817.

Cunnold, D., P. Fraser, R. Weiss, R. Prinn, P. Simmonds, B. Miller, F. Alyea and A. Crawford, 1994: Global trends and annual releases of CCl_3F and CCl_2F_2 estimated from ALE/GAGE and other measurements from July 1978 to June 1991. *J. Geophys. Res.,* **99**, 1107-1126.

Davis, D.D., G. Chen, W. Chameides, J. Bradshaw, S. Sandholm, M. Rodgers, J. Schendal, S. Madronich, G. Sachse, G. Gregory, B. Anderson, J. Barrick, M. Shipham, J. Collins, L. Wade and D. Blake, 1993: A photostationary state analysis of the NO_2-NO system based on airborne observations from the subtropical/tropical north and south Atlantic. *J. Geophys. Res.,* **98**, (D12), 23501-23523.

Dasch, J.M., 1992: Nitrous oxide emissions from vehicles. *J. Air Waste Manage. Assoc.,* **42**, 63-67.

Delany, A.C., S. Haagensen, A.F. Walters, A.F. Wartburg and P.J. Crutzen, 1985: Photochemically-produced ozone in the emission from large-scale tropical vegetation fires. *J. Geophys. Res.* **90**, 2425-2429.

Delwiche, C.C. and R.J. Cicerone, 1993: Factors affecting methane production under rice. *Global Biogeochem. Cycles.*, **7**, 143-155.

De Muer, D. and H. De Backer, 1992: Revision of 20 years of Dobson total ozone data at Uccle (Belgium): fictitious Dobson total ozone trends induced by sulphur dioxide trends. *J. Geophys. Res*, **97**, 5921-5937.

Derwent, R.G., 1994a: The estimation of global warming potentials for a range of radiatively active gases. In: *Non-CO_2 Greenhouse Gases,* J. Van Ham, L.J.H.M. Janssen,and R.J. Swart (eds.), Kluwer Academic Publishers, Netherlands, pp. 289-299.

Derwent, R,G, 1994b: The influence of human activities on the distribution of hydroxyl radicals in the troposphere. *Phil. Trans. Roy. Soc.* (Submitted)

Dianov-Klokov, V.I., L.N. Yurganov, E.I. Grechko and A.V. Dzhola, 1989: Spectroscopic measurements of carbon monoxide and methane 2: Seasonal variations and long-term trends. *J. Atmos. Chem.*, **8**, 153-164.

Dickerson, R.R., G.J. Huffman, W.T. Luke, K.E. Nunnermacker, K.E. Pickering, A.C.D. Leslie, C.G. Lindsey, W.G.N. Slinn, T.J. Kelly, P.H. Daum, A.C. Delany, P.J. Greenberg, P.R. Zimmerman, J.F. Boatman, J.D. Ray and D.H. Stedman, 1987: Thunderstorms: An important mechanism in the transport of air pollutions. *Science*, **235**, 460-465.

Dlugokencky, E.J., K.A. Masarie, P.M. Lang, P.P. Tans, L.P. Steele and E.G. Nisbet, 1994a: A dramatic decrease in the growth rate of atmospheric methane in the Northern Hemisphere during 1992. *Geophys. Res. Lett.*, **22**, 45-48.

Dlugokencky, E.J., J.M. Harris, Y.S. Chung, P.P. Tans and I. Fung, 1994b: The relationship between the methane seasonal cycle and regional sources and sinks at Tae-ahn Peninsula, Korea. *Atmos. Env.*, **27A**, 2115-2120.

Dlugokencky, E.J., L.P. Steele, P.M. Lang and K.A. Masarie, 1994c: The growth rate and distribution of atmospheric methane. *J. Geophys. Res*, **99**, 17021-17043.

Donoso, L., R. Santana and E. Sanhueza, 1993: Seasonal variation of N_2O fluxes at a tropical savannas site: Soil consumption of N_2O during the dry season. *Geophys. Res. Lett.*, **20**, 1379-1382.

Drummond, J.W., D.H. Ehhalt and A. Volz, 1988: Measurements of nitric oxide between 0-12 km altitude and 67°N to 60°S latitude obtained during STRATOZ III. *J. Geophys. Res.*, **93**, 15831-15849.

Ehhalt, D.H., F. Rohrer and A. Wahner, 1992: Sources and distribution of NO_x in the upper troposphere at northern mid-latitudes. *J. Geophys. Res.,* **97**, 3725-3738.

Elkins, J., J. Butler, S. Montzka, R. Myers, T. Thompson, T. Baring, S. Cummings, G. Dutton, A. Hayden, J. Lobert, G. Holcomb, W. Sturges and T. Gilpin, 1993a: Nitrous Oxide and Halocarbons Division. Section 5 in: *Climate Monitoring and Diagnostics Laboratory, Summary Report, 1992*, US Department of Commerce, National Oceanic and Atmospheric Administration, pp. 59-75.

Elkins, J., T. Thompson, T. Swanson, J. Butler, B. Hall, S. Cummings, D. Fisher and A. Raffo, 1993b : Decrease in the growth rates of atmospheric chlorofluorocarbons-11 and -12. *Nature*, **364**, 780-783.

Elliot, S.M., 1984: The chemistry of some atmospheric gases in the ocean. Ph.D. Thesis, University of California, Irvine. 274 pp.

Elliot, S.M. and F.S. Rowland, 1993: Nucleophilic substitution rates and solubilities for methyl halides in seawater. *Geophys. Res. Lett.*, **20**, 1043-1046.

Erikson, D. and J. Taylor, 1992: Three-dimensional tropospheric CO_2 modelling: the possible influence of the ocean. *Geophys. Res. Lett.*, **19**, 1955-1958.

Etheridge, D.M., G.I. Pearman and F. de Silva, 1988:

Atmospheric trace-gas variations as revealed by air trapped in an ice core from Law Dome, Antarctica. *Annals of Glaciology*, **10**, 28-33.

Etheridge, D.M., G.I. Pearman and P.J. Fraser, 1992: Changes in tropospheric methane between 1841 and 1978 from a high accumulation-rate Antarctic ice core. *Tellus*, **44B**, 282-294.

Fabian, P., R. Borchers, B. Kringer and S. Lal, 1987: CF_4 and C_2F_6 in the atmosphere. *J. Geophys. Res*, **92**, 9831-9835.

Fehsenfeld, F., J. Calvert, R. Fall, P. Goldan, A. Guenther, C.N. Hewitt, B. Lamb, S. Liu, M. Trainer, H. Westberg and P. Zimmerman, 1992: Emissions of volatile organic compounds from vegetation and the implications for atmospheric chemistry. *Global Biogeochem. Cycles*, **6**, 389-430.

Fisher, D., T. Duafala, P. Midgley and C. Niemi, 1994: Production and emissions of CFCs, halons and related molecules. In: *Concentrations, Lifetimes and Trends of CFCs, halons and related Species,* J. Kaye, S. Penkett and F. Ormond (eds), NASA Report no. 1339, 2.1-2.35.

Fishman, J., K. Fakhruzzaman, B. Cros and D. Nganga, 1991: Identification of widespread pollution in the Southern Hemisphere deduced from satellite analysis. *Science,* **252**, 1693-1696.

Fraser, P. and N. Derek, 1994: Halocarbons, nitrous oxide, methane and carbon monoxide - the GAGE program. In: *Baseline 91,* C. Dick and J. Gras (eds), Bureau of Meteorology-CSIRO, 70-81.

Fraser, P.J., P. Hyson, R.A. Rasmussen, A.J. Crawford and M.A.K. Khalil, 1986: Methane, carbon monoxide and methylchloroform in the Southern Hemisphere. *J. Atmos. Chem.*, **4**, 3-42.

Fraser, P., S. Penkett, M. Gunson, R. Weiss and F.S. Rowland, 1994: Measurements. In: *Concentrations, Lifetimes and Trends of CFCs, Halons and Related Species,* J. Kaye, S. Penkett and F. Ormond (eds), NASA Report no.1339, 1.1-1.68.

Fuglestvedt, J.S., T.K. Bentsen and I.S.A. Isaksen, 1993: Responses in tropospheric O_3, OH and CH_4 to changed emissions of important trace gases. *Centre for Climate and Energy Research, Oslo, Report 1993,* **4**.

Fuglestvedt, J.S., J.E. Jonson and I.S.A. Isaksen, 1994: Effects of reductions in stratospheric ozone on tropospheric chemistry through changes in photolysis rates. *Tellus,* **46B**, 172-192.

Furrer, R., W. Döhler, H. Kirsch, P. Plessing and U. Görsdorf, 1992: Evidence for vertical ozone redistribution since 1967. *J. Atmos. Terrest. Physics*, **11**, 1423-1445.

Gentile, I.A., L. Ferraris and S. Crespi, 1989: The degradation of methyl bromide in some natural freshwaters. Influence of temperature, pH and light. *Pestic. Sci.*, **25**, 261-272.

Gleason, J.F., P.K.Bhartia, J.R. Herman, R. McPeters, P. Newman, R.S. Stolarski, L. Flynn, G. Labow, D. Larko, C. Seftor, C. Wellemeyer, W.D. Komhyr, A.J. Miller and W. Planet, 1993: Record Low Global Ozone in 1992. *Science,* **260**, 523-526.

Gold, S., H. Barkhau, W. Shleien and B. Kahn, 1964: Measurement of naturally occurring radionuclides in air. In: *The Natural Radiation Environment,* J.A.S. Adams and W.M. Lowder, (eds), University of Chicago Press, pp.369-382.

Gschwandtner, G., K. Gschwandtner, K. Eldridge, C. Mann and D. Mobley, 1986: Historic emissions of sulphur and nitrogen oxides in the U.S. from 1980 to 1990. *J. Air Pollut. Contr. Assoc.* **36**, 19.

Graedel, T.E., T.S. Bates, A.F. Bouwman, D. Cunnold, J. Dignon, I. Fung, D.J. Jacob, B.K. Lamb, J.A. Logan, G. Marland, P. Middleton, J.M. Pacyna, M. Placet and C. Veldt, 1993: A compilation of inventories of emissions to the atmosphere. *Global Biogeochemical Cycles*, **7**,1-26.

Hameed, S. and R. Cess, 1983: Impact of global warming on biospheric sources of methane and its climate consequences. *Tellus,* **35B**, 1-7.

Hameed, S. and J. Dignon, 1991: Global emissions of nitrogen and sulphur oxides in fossil fuel combustion 1970-1986. *J. Air Waste Manage. Assoc.*, **42**, 159-163.

Hao, W. and D. Ward, 1993: Methane production from global biomass burning. *J. Geophys. Res.* **98**, 20657-20661.

Harris, N.R.P., G. Ancellet, L. Bishop, D.J. Hofmann, J.B. Kerr, R.D. McPeters, M. Préndez, W. Randel, J. Staehlin, B.H. Subbaraya, A.Volz-Thomas, J.M. Zawodny and C.S. Zerefos, 1994: Ozone Measurements. In: *Scientific Assessment of ozone depletion:1994, Rep. 37, WMO Global Ozone Res. and Mon. Project, Geneva.* (In press)

Harriss, R., K. Bartlett, S. Frolking and P. Crill, 1993: Methane emissions from northern high latitude wetlands. In: *Biogeochemistry of Global Change: Radiatively Active Trace Gases,* R.S. Oremland (ed.), Chapman and Hall, New York, pp. 449-486.

Hauglustaine, D.A., C. Granier, G.P. Brasseur and G. Mégie, 1994: The importance of atmospheric chemistry in the calculation of radiative forcing on the climate system. *J. Geophys. Res.*, **99**, 1173-1186.

Hofmann, D.J., S.J. Oltmans, J.M. Harris, W.D. Komhyr, J.A. Lathrop, T. DeFoor and D. Kuniyuki, 1993: Ozonesonde measurements at Hilo, Hawaii following the eruption of Mt. Pinatubo. *Geophys. Res. Lett.*, **20**, 1555-1558.

Hofmann, D.J., S.J. Oltmans, A.O. Langford, T. Deshler, B.J. Johnson, A. Torres and W.A. Matthews, 1994a: Ozone loss in the lower stratosphere over the United States in 1992-93: Evidence for heterogenous chemistry on the Mt. Pinatubo aerosol. *Geophys. Res. Lett.*, **21**, 65-68.

Hofmann, D.J., S.J. Oltmans, J.A. Lathrop, J.M. Harris and H. Voemel, 1994b: Record low ozone at the South Pole in the spring of 1993. *Geophys. Res. Lett.*, **21**, 421-424.

Holton, J.R., 1990: On the global exchange of mass between the stratospheric and troposphere, *J. Atmos. Sci.*, **47**, 392-395.

Hutsch, B,W., C.P. Webster and D.S. Powlson, 1993: Long-term effects of nitrogen fertilisation on methane oxidation in soil of the broadbalk wheat experiment, *Soil Biology and Biochemistry,* **25**, 10, 1307-1315.

IPCC (Intergovernmental Panel on Climate Change), 1990: *Climate Change: the IPCC Scientific Assessment*, J.T. Houghton, G.J. Jenkins and J.J. Ephraums (eds). Cambridge University Press, Cambridge, UK.

IPCC, 1992: *Climate Change 1992: The Supplementary Report to the IPCC Scientific Assessment,* J.T. Houghton, B.A. Callander and S.K. Varney (eds). Cambridge University Press, Cambridge, UK.

Isaksen, I.S.A. and O. Hov, 1987: Calculation of trends in the tropospheric concentration of O_3, OH, CH_4, and NO_x. *Tellus,* **39B**, 271-285.

Jacob, D.J., J.A. Logan, G.M. Gardner, R.M. Yevich, C.M. Spivakovsky, S.C. Wofsy, J.M. Munger, S. Sillman and M.J. Prather, 1993: Factors regulating ozone over the United States and its export to the global atmosphere. *J. Geophys. Res.*, **98**, 14817-14826.

Jeffers, P.M., L.M. Ward, L.M. Woytowitch and N.L. Wolfe, 1989: Homogeneous hydrolysis rate constants for selected chlorinated methanes, ethanes, ethenes, and propanes. *Environ. Sci. Technol.*, **23(8)**, 965-969.

Johnson, C.E., J. Henshaw and G. McInnes, 1992: Impact of aircraft and surface emissions of nitrogen oxides on tropospheric ozone and global warming. *Nature*, **355**, 69-71.

Jouzel, J., N.I. Barkov, J.M. Barnola, M. Bender, J. Chappellaz, C. Genthon, V.M. Kotlyakov, V. Lipenkov, C. Lorius, J.R. Petit, D. Raynaud, G. Raisbeck, C. Ritz, T. Sowers, M. Stievenard, F. Yiou and P. Yiou, 1993: Extending the Vostok ice-core record of palaeoclimate to the penultimate glacial period. *Nature*, **364**, 407-412.

Judd, A.G., R.H. Charlier, A. Lacroix, G. Lambert and C. Rowland, 1993: *Minor Sources of Methane; Sources, Sinks and Role in Global Change,* NATO ASI Series , Springer-Verlag, Berlin, pp 432-456.

Kanakidou, M. and P.J. Crutzen, 1993: Scale problems in global tropospheric chemistry modelling: comparison of results obtained with a 3-D model, adopting longitudinally uniform and varying emissions of NO_x and NMHC. *Chemosphere*, **26**, 787-801.

Kanakidou, M., H.B. Singh, K.M. Valentin and P.J. Crutzen, 1991: A two-dimensional study of ethane and propane oxidation in the troposphere. *J.Geophys. Res.*, **96**, 15,395-15,413.

Kasibhatala, P., 1993: NO_y from subsonic aircraft emissions: a global three-dimensional model study. *Geophys. Res. Lett.*, **20**, 1707-1710.

Kato, N. and H. Akimoto, 1992: Anthropogenic emissions of SO_2 and NO_x in Asia: Emission inventories. *Atmos. Env.*, **26A**, 2997-3017.

Kaye, J., S. Penkett and F. Ormond, 1994: Report on *Concentrations, Lifetimes and Trends of CFCs, Halons, and Related Species,* NASA Reference Publication 1339, NASA Office of Mission to Planet Earth, Science Division, Washington DC.

Keller, M. and P. Matson, 1994: Biosphere - atmosphere exchange of trace gases in the tropics : evaluating the effects of land-use changes. In: *Global Atmospheric-Biospheric Chemistry*, R. Prinn (ed), Plenum Press, New York and London, pp.103-118.

Keller, M., M.E. Mitre and R.F. Stallard, 1990: Consumption of atmospheric methane in soils of central Panama: effect of agricultural development. *Global Biogeochem. Cycles*, **4**, 21-27.

Keller, M., E. Veldkamp, A. Weltz and W. Reiners, 1993: Effect of pasture age on soil trace-gas emissions from a deforested area of Costa Rica. *Nature*, **365**, 244-246.

Kerr, J.B. and C.T. McElroy, 1993: Evidence for large upward trends of ultraviolet-B radiation linked to ozone depletion. *Science*, **262**, 1032-1034.

Kerr, J.B., D.I. Wardle and D.W. Tarasick, 1993: Record low ozone values over Canada in early 1993. *Geophys. Res. Lett.*, **20**, 1979-1982.

Khalil, M.A.K. and R.A. Rasmussen, 1985: Atmospheric carbon tetrafluoride (CF_4): sources and trends. *Geophys. Res. Lett.*, **12**, 671-672.

Khalil, M.A.K. and R.A. Rasmussen, 1988a: Nitrous oxide: trends and global mass balance over the last 3000 years. *Annals of Glaciology* **10**, 73-79.

Khalil, M.A.K. and R.A. Rasmussen, 1988b: Carbon monoxide in the Earth's atmosphere: Indications of a global increase. *Nature*, **332**, 242-245.

Khalil, M.A.K. and R.A. Rasmussen, 1992: The global sources of nitrous oxide. *J. Geophys. Res.*, **97**, 14,651-14,660.

Khalil, M.A.K. and R.A. Rasmussen, 1993: Decreasing trend of methane: unpredictability of future concentrations, *Chemosphere*, 26, 803-814.

Khalil, M.A.K. and R.A. Rasmussen, 1994: Global decrease of atmospheric carbon monoxide, *Nature*, **370**, 639-641.

Khalil, M.A.K., R.A. Rasmussen, M.J. Shearer, S. Ge and J.A. Rau, 1993a: Methane from coal burning. *Chemosphere*, 26, 473-477.

Khalil, M.A.K., R.A. Rasmussen and R. Gunawardena, 1993b: Atmospheric methyl bromide: Trends and global mass balance. *J. Geophys. Res.*, **98**, 2887-2896.

Kim, K-R. and H. Craig, 1993: Nitrogen-15 and oxygen-18 characteristics of nitrous oxide: A global perspective. *Science*, **262**, 1855-1857.

Kirchgessner, D.A., S.D. Piccot and J.D. Winkler, 1993: Estimate of global methane emissions from coal mines. *Chemosphere*, **26**, 453-472.

Kirchhoff, V.W. and E.V. Marinho, 1989: A survey of continental concentrations of atmospheric CO in the Southern Hemisphere. *Atmos. Env.*, **23**, 461-466.

Kirchhoff, V.W.J.H., A.W. Setzer and M.C. Pereira, 1989: Biomass burning in Amazonia: Seasonal effects on atmospheric O_3 and CO. *Geophys. Res. Lett.*, **16**, 469-472.

Kley, D., H. Geiss and V.A. Mohnen, 1994: Tropospheric ozone at elevated sites and precursor emissions in the United States and Europe. *Atmos. Env.*, **28**, 149-158.

Ko, M.K.W., N.D. Sze and D.K. Weisenstein, 1991: Use of satellite data to constrain the model-calculated atmospheric lifetime for N_2O: Implication for other gases. *J. Geophys. Res.*, **96**, 7547-7552.

Ko, M., N.D. Sze, W-C. Wang, G. Shia, A. Goldman, F. Murcray, D. Murcray and C. Rinsland, 1993: Atmospheric sulphur hexafluoride: sources, sinks and greenhouse warming. *J. Geophys. Res.*, **98**, 10499-10507.

Komhyr, W.D., R.D. Grass, R.D. Evans, R.K. Leonard, D.M. Quincy, D.J. Hofmann and G.L. Koenig, 1994: Unprecedented 1993 ozone decrease over the United States from Dobson spectrophotometer observations. *Geophys. Res. Lett.*, **21**, 210-214.

Koppman, R., F. Johnen, C. Plass-Dulner and J. Rudolph, 1993: Distribution of methylchloride, dichloromethane, trichlooroethane and tetrachloroethane over the North and South Atlantic. *J. Geophys. Res.*, **98**, 20,517-20,526.

Krieleman, G. and A. Bouwman, 1994: Computing land-use emissions of greenhouse gases. *J. Water, Air and Soil Pollution*, **76**, 231-258.

Kvenvolden, K.A. and T.D. Sorenson, 1993: Methane in

permafrost: Preliminary results from coring at Fairbanks, Alaska. *Chemosphere*, **26**, 609-616.

Labitske, K. and M.P. McCormick, 1992: Stratospheric temperature increases due to Mt. Pinatubo aerosols. *Geophys. Res. Lett.*, **19**, 207-210.

Lal, S., S. Venkataramani and B. Subbaraya, 1993: Methane flux measurements from paddy fields in the tropical Indian region. *Atmos. Env.*, **27A**, 1691-1694.

Lambert, G. and S. Schmidt, 1993: Re-evaluation of the oceanic flux of methane: Uncertainties and long-term variations. *Chemosphere*, **26**, 579-589.

Lassey, K.R., D.C. Lowe, C.A.M. Brenninkmeijer and A.J. Gomez, 1993: Atmospheric methane and its carbon isotopes in the Southern Hemisphere: their time-series and an instructive model. *Chemosphere*, **26**, 95-109.

Law, C.S. and N.J.P. Owen, 1990: Significant flux of atmospheric nitrous oxide from the north-west Indian Ocean. *Nature*, **346**, 826-828.

Law, K.S. and J.A. Pyle, 1993: Modeling trace gas budgets in the troposphere, 2: CH_4 and CO. *J. Geophys. Res.*, **98**, 18401-18412.

Lelieveld, J. and P.J. Crutzen, 1994: Role of deep cloud convection in the oxone budget of the troposphere. *Science*, **264**, 1759-1761.

Leuenberger, M. and U. Siegenthaler, 1992: Ice-age atmospheric concentration of nitrous oxide from an Antarctic ice core. *Nature*, **360**, 449-451.

Levine, J.S., 1994: Biomass burning and the production of greenhouse gases. In: *Climate Biosphere Interaction: Biogenic Emissions and Environmental Effects on Climate Change*, R.G. Zepp (ed.), John Wiley and Sons, Inc., New York, pp.139-159.

Levine, J., W. Cofer and J. Pinto, 1994: Biomass burning. Chapter 14 in *Atmospheric Methane: Sources, Sinks and Role in Global Change*, M. Khalil (ed.), NATO ASI series, 299-313.

Liu, S.C. and M. Trainer, 1988: Responses of tropospheric ozone and odd hydrogen radicals to column ozone change. *J. Atmos. Chem.*, **6**, 221-233.

Liu, S.C., S.A. McKeen and S. Madronich, 1991: Effect of anthropogenic aerosols on biologically active ultra violet radiation. *Geophys. Res. Lett.*, **18**, 2265-2268.

Lobert, J.M, T.J. Baring, J.H. Butler, S.A. Montzka, R.C. Myers and J.W. Elkins, 1993: Ocean-atmosphere exchange of trace compounds, 1992. *Final Report to AFEAS on Oceanic Measurements of HCFC-22, Methylchloroform, and Oher Selected Halocarbons*. NOAA/CMDL, Boulder, CO.

Logan, J.A., 1983: Nitrogen oxides in the troposphere: global and regional budgets. *J. Geophys. Res.*, **88**, 10785-10807.

Logan, J.A., 1994: Trends in the vertical distribution of ozone: An analysis of ozone sonde data. *J. Geophys. Res.* (In press)

London, J. and S. Liu, 1992: Long-term tropospheric and lower stratospheric ozone variations from ozonesondes observations. *J. Atmos. Terrest. Physics*, **5**, 599-625.

Lowe, D.C., C.A.M. Brenninkmeijer, G.W. Brailsford, K.R. Lassey, A.J. Gomez and E.G. Nisbet, 1994: Concentration and ^{13}C records of atmospheric methane in New Zealand and Antarctica: evidence for changes in methane sources. *J. Geophys. Res.*, **99**, 16913-16925.

Lubkert, B. and K.H. Zierock 1989: European emission inventories - a proposal of international worksharing. *Atmos. Env*, **23**, 37-48.

Mabey, W. and T. Mill, 1978: Critical review of hydrolysis of organic compounds in water under environmental conditions. *J. Phys. Chem. Ref. Data*, **7(2)**, 383-392.

Machida, T., T.Nakazawa, M. Tanaka, Y. Fufii, S. Aoki and O. Watanabe, 1994: Atmospheric methane and ntrous oxide concentrations during the last 250 years deduced from H15 Ice Core, Antarctica, *Proceedings International Symposium on Global Cycles of Atmospheric Greenhouse gases*, Sendai, Japan, 7-10 March, 1994, 113-116.

Mahlman, J., H. Levy and W. Moxim, 1986: Three-dimensional simulations of stratospheric N_2O: predictions for other trace constituents. *J. Geophys. Res.*, **91**, 2687-2707.

Mak, J.E., C.A.M. Brenninkmeijer and M.R. Manning, 1992: Evidence for a missing carbon monoxide sink based on tropospheric measurements of ^{14}CO. *Geophys. Res. Lett.*, **14**, 1467-1470.

Makide, Y., A. Yokohata, Y. Kubo and T. Tominaga, 1987: Atmospheric concentrations of halocarbons in Japan in 1979-1986. *Bull. Chem. Soc. Japan.*, **60**, 571-574.

Manley, S.L., K. Goodwin and W.J. North, 1992: Laboratory production of bromoform, methylene bromide and methyl iodide by macralgae and distribution in nearshore southern California waters. *Limnol. Oceanogr.*, **37**, 1652-1659.

Mano, S. and M.O. Andreae, 1994: Emission of methyl bromide from biomass burning. *Science*, **263**, 1255-1257.

Marenco, A., J.C. Medale and S. Prieur, 1990: Study of tropospheric ozone in the tropical belt (Africa, America) from STRATOZ and TROPOZ campaigns. *Atmos. Env.*, **24A**, 2823-2834.

Martius, C., R. Wassmann, U. Thein, A. Bandeira, H. Rennenberg, W. Junk and W. Seiler, 1993: Methane emission from wood-feeding termites in Amazonia. *Chemosphere*, **26**, 623-632.

Matson, P. and P. Vitousek, 1990: Ecosystems approach to a global nitrous oxide budget. *Bioscience*, **40**, 667-672.

McCormick, M.P., R.E. Veiga and W.P. Chu, 1993: Stratospheric ozone profile and total ozone trends derived from the SAGE I and SAGE II data. *Geophys. Res. Lett.*, **19**, 269-272.

McCulloch, A., 1992: Global production and emissions of bromochlorodifluoromethane and bromotrifluoromethane (Halons 1211 and 1301). *Atmos. Env.*, **26A**, 1325-1329.

Meyer, C., I. Galbally and Y. Wang, 1994: Recent advances in studies of regional greenhouse gas emissions: an Australian perspective. In: *Global Warming issues in Asia*, S. Bhattacharya *et al.*, (eds), Asian Institute of Technology, Bangkok, Thailand, pp.125-138.

Midgley, P. and D. Fisher, 1993: The production and release to the atmosphere of chlorodifluoromethane (HCFC-22). *Atmos. Env.*, **27A**, 2215-2223.

Minami, K., J. Goudriaan, E. Lantinga and T. Kimura, 1993: Significance of grasslands in emissions and absorption of greenhouse gases. *Proc. XVII International Grasslands Congress*, pp1231-1238.

Minschwaner, K., R.J. Salawitch and M.B. McElroy, 1993: Absorption of solar radiation by O_2: Implications for O_3 and lifetimes of N_2O, $CFCl_3$ and CF_2Cl_2. *J. Geophys. Res.*, **98**, 10,543-10,561.

Montzka, S.A., R.C. Myers, J.H. Butler and J.W. Elkins, 1993: Global tropospheric distribution and calibration scale of HCFC-22. *Geophys. Res. Lett.*, **20**, 703-706.

Montzka, S.A., R.C. Myers, J.H. Butler and J.W. Elkins, 1994: Early trends in the global atmospheric abundance of hydrofluorocarbon-141b and -142b. *Geophys. Res. Lett*. (In press)

Moore, R.M. and R. Tokarczyk, 1993: Volatile biogenic halocarbons in the north-west Atlantic. *Global Biogeochem. Cycles*, **7(1)**, 195-210.

Mosier, A., M. Stillwell, W. Parton and R. Woodmanse, 1981: Nitrous oxide emissions from native shortgrass prairie. *J. Soil Sci. Soc. Am.*, **45**, 617-619.

Mosier, A., D. Schimel, D. Valentine, K. Bronson and W. Parton, 1991: Methane and nitrous oxide emissions in native, fertilised and cultivated grasslands. *Nature*, **350**, 330-332.

Moraes, F. and M.A.K. Khalil, 1993: Permafrost methane content: 2. Modeling theory and results. *Chemosphere*, **26**, 595-607

Müller, J.F., 1992: Geographical distribution and seasonal variation of surface emissions and deposition velocities of atmospheric trace gases. *J. Geophys. Res.*, **97**, 3787-3804.

Nakazawa, T., T. Machida, M. Tanaka, Y. Fujii, S. Aoki and O. Watanabe, 1993: Differences of the atmospheric CH_4 concentration between the Arctic and Antarctic regions in pre-industrial/agricultural era. *Geophys. Res. Lett.*, **20**, 943-946.

Novelli, P.C., L.P. Steele and P.P. Tans, 1992: Mixing ratios of carbon monoxide in the troposphere. *J. Geophys. Res.*, **97**, 20731-20750.

Novelli, P.C., K. Masarie, P.P. Tans and P. Lang, 1994 : Recent changes in atmospheric carbon monoxide. *Science*, **263**, 1587-1590.

Ojima, D.S., D.W. Valentine, A.R. Mosier, W.J. Parton and D.S. Schimel, 1993: Effect of land-use change on methane oxidation in temperate forest and grassland soils. *Chemosphere*, **26**, 675-685.

Oltmans, S.J. and H. Levy II, 1994: Surface ozone measurements from a global network. *Atmos. Env.*(In press)

Oltmans, S.J., D.J. Hofmann, W.D. Komhyr and J.A. Lathrop, 1994: Ozone vertical profile changes over South Pole. *Proc. 1992 Quad. Ozone Symp., Charlottesville, VA,* 13 June 1992, NASA CP-3266, pp578-581.

Oremland, R.S., L.G. Miller, P.M. Blunden, S.W. Culbertson and M.D. Coulatkis, 1993a: Degradation of atmospheric halogenated methanes. I; Activity of soil methanotrophic bacteria. Submitted to *Environ. Sci. Technol.*

Oremland, R.S., L.G. Miller and F.E. Strohmaier, 1993b: Degradation of atmospheric halogenated methanes. II: Chemical and bacterial attack of methyl bromide in aneorobic sediments. Submitted to *Environ. Sci. Technol.*

Penkett, S., N. Prosser, R. Rasmussen and M. Khalil, 1981: Atmospheric measurements of CF_4 and other fluorocarbons containing the CF_3 group. *J. Geophys. Res.*, **86**, 5162-5178.

Penkett, S., B. Jones, B. Rycroft and D. Simmons, 1985: An interhemispheric comparison of the concentration of bromine compounds in the atmosphere. *Nature*, **318**, 550-553.

Penkett, S., J.H. Butler, M.J. Kurylo, C.E. Reeves, J.M. Rodriguez, H. Singh, D. Toohey, R. Weiss, M.O. Andreae, N.J. Blake, R.J. Cicerone, T. Duafala, A. Golombek, M.A.K. Khalil, J.S. Levine, M.J. Molina and S.M. Schauffler, 1994: Methyl Bromide. Chapter 10 of *Scientific Assessment of Ozone Depletion, 1994*, UNEP/WMO. (In press)

Penner, J.E., C.S. Atherton, J. Dignon, S.J. Ghan, J.J. Walton and S. Hameed, 1991: Tropospheric nitrogen: A three-dimensional study of sources, distributions, and deposition. *J. Geophys. Res.*, **96**, 959-990.

Pickering, K.E., A.M. Thompson, J.R. Scala, W-K. Tao, R.R. Dickerson and J. Simpson, 1992: Free tropospheric ozone production following entrainment of urban plumes into deep convection. *J. Geophys. Res.*, **97**, 17985-18000.

Placet, M., R.E. Battye, F.C. Fehsenfeld and G.W. Bassett, 1990: Emissions involved in acidic deposition processes. State-of-science/technology report 1. National Acid Precipitation Assessment Program. US Government Printing Office, Washington DC.

Polian, G., G. Lambert, B. Ardouin and A. Jegou, 1986: Long-range transport of radon in subantarctic and antarctic areas. *Tellus*, **38B**, 178-189.

Pollock, W., L. Heidt, R. Lueb, J. Vedder, M. Mills and S. Solomon, 1992: On the age of stratospheric air and ozone depletion potentials in polar regions. *J. Geophys. Res.*, **97**, 12993-12999.

Poulida, O., R.R. Dickerson, B.G. Doddrige, J.Z. Holland, R.G. Wardell and J.G. Watkins, 1991: Trace gas concentrations and meteorology in rural Virginia, 1. Ozone and carbon monoxide. *J. Geophys. Res.*, **96**, 22461-22476.

Prather, M.J., 1994: Lifetimes and eigenstates in atmospheric chemistry. *Geophys. Res. Lett.*, **21**, 801-804.

Prather, M.J. and C.M. Spivakovsky, 1990: Tropospheric OH and the lifetimes of hydrochlorofluorocarbons (HCFCs). *J. Geophys. Res.*, **95**, 18723-18729.

Prather, M.J. and E.E. Remsberg, 1993: (eds): The atmospheric effects of stratospheric aircraft: Report of the 1992 Models and Measurements Workshop, NASA Reference Publication,**1292.**

Prinn, R., D. Cunnold, R. Rasmussen, P. Simmonds, F. Alyea, A. Crawford, P. Fraser and R. Ramussen, 1990: Atmospheric emissions and trends of nitrous oxide deduced from 10 years of ALE-GAGE data. *J. Geophys. Res*, **95**, 18369-18385.

Prinn, R., D. Cunnold, P. Simmonds, F. Alyea, R. Boldi, A. Crawford, P. Fraser, D. Gutzler, D. Hartley, R. Rosen and R. Rasmussen, 1992: Global average concentration and trend for hydroxyl radicals deduced from ALE/GAGE trichloroethane (methylchloroform) data for 1978-1990. *J. Geophys. Res.*, **97**, 2445-2461.

Rasche, M.E., M.R. Hyman and D.J. Arp, 1990: Biodegradation of halogenated hydrocarbon fumigants by nitrifying bacteria. *Appl. Environ. Microbiol.*, **56(8)**, 2568-2571.

Rasmussen, R.A. and M.A.K. Khalil, 1986: Atmospheric trace gases: trends and distributions over the last decade. *Science*, **232**, 1623-1624.

Rasmussen, R.A., M.A.K. Khalil and F. Moraes, 1993: Permafrost methane content: 1. Experimental data from sites in Northern Alaska. *Chemosphere*, **26**, 591-594.

Ravishankara, A.R., S. Solomon, A.A. Turnipseed and R.F. Warren, 1993: Atmospheric lifetimes of long-lived halogenated species. *Science,* **259**, 194-199.

Reeburgh, W.S., N.T. Roulet and B.H. Svensson, 1993: Terrestrial biosphere-atmosphere exchange in high latitudes. In: *Global Atmospheric-Biospheric Chemistry*: R.G.Prinn(ed.), Plenum Press, New York and London, pp.165-178.

Reeves, C.E. and S.A. Penkett, 1993: An estimate of the anthropogenic contribution to atmospheric methyl bromide. *Geophys. Res. Lett.,* **20**, 1563-1566.

Reichle, H.G., V.S. Connors, J.A. Holland, R.T. Sherrill, H.A. Wallio, J.C. Casas, E.P. Condon, B.B. Gormsen and W. Seiler, 1990: The distribution of middle tropospheric carbon monoxide during early October 1984. *J. Geophys. Res.,* **95**, 9845-9856.

Rinsland, C., M. Gunson, M. Abrams, L. Lowes, R. Zanden and E. Mahieu, 1993: ATMOS/ATLAS 1 measurements of sulphur hexafluoride (SF_6) in the lower stratosphere and upper troposphere. *J. Geophys. Res.,* **98**, 20491-20497.

Rohrer, F., D.H. Ehhalt, E.S. Grobler, A.B. Kraus and A. Wahner, 1993: Model calculations of the impact of nitrogen oxides emitted by aircraft in the upper troposphere at northern mid-latitudes. *Proc of the 6th European Symposium on the Physico-chemical Behaviour of Atmospheric Pollutants. Vareses, Italy.*

Roulet, N., T. Moore, J. Bubier and P. Lafleur, 1992: Northern fens: methane flux and climatic change. *Tellus,* **44B**, 100-105.

Roulet, N.T., R. Ash, W. Quinton, and T. Moore, 1993: Methane flux from drained northern peatlands: effect of a persistent water table lowering on flux. *Global Biochemical Cycles,* **7**, 749-769.

Rowland, F.S., C. Wang and D. Blake, 1994: Reduction in the rates of growth of global atmospheric concentrations for CCl_3F(CFC-11) and CCl_2F_2(CFC-12) during 1988-1993. *Proceedings International Symposium on Global Cycles of Atmospheric Greenhouse gases, Sendai, Japan, 7-10 March, 1994,* pp38-51.

Rudolph, J., A. Khedim, R. Koppman and B. Bonsang, 1994: Field study of the emissions of methylchloride and other halocarbons from biomass burning in equatorial Africa. *J. Atmos. Chem.* (In press)

Sandholm, S.T, J.D. Bradshaw, G. Chen, H.B. Singh, R.W. Talbot, G.L. Gregory, D.R. Blake, G.W. Sachse, E.V. Browell, J.D. Barrick, M.A. Shipham, A.S. Bachmeier and D. Owen, 1992: Summertime tropospheric observations related to N_xO_y distributions and partitioning over Alaska: Arctic Boundary Layer Expedition (ABLE) 3A. *J. Geophys. Res.,* **97**, 16481-16509.

Sandholm, S., J. Olson, J. Bradshaw, R. Talbot, H. Singh, G. Gregory, D. Blake, B. Anderson, G. Sachse, J. Barrick, J. Collins, K. Klemm, B. Lefer, O. Klemm, K. Gorzelska, D. Herlth and D. O'Hara, 1994: Summertime partitioning and budget of NO_y compounds in the troposphere over Alaska and Canada: ABLE 3B. *J. Geophys. Res.,* **99**, 1837-1861.

Sanhueza, E., 1992: Biogenic emissions of NO and N_2O from tropical savanna soils. In: *Proceedings International Symposium on Global Change (IGBP), Tokyo, Japan, 27-29 March, 1992,* pp24-34.

Scheel, H.E., E.G. Brunke and W. Seiler, 1990: Trace gas measurements at the monitoring station Cape Point, South Africa, between 1978 and 1988. *J. Atm. Chem.,* **11**, 197-210.

Scheel, H.E., R. Sladovic and W. Seiler, 1993: Ozone-related species at the stations Wank and Zugspitze: trends, short-term variations and correlations with other parameters. In: *Photooxidants: Precursors and Products, Proc. EUROTRAC Symp.1992,* P.M. Borrell, P. Borrell, T. Cvitas and W. Seiler(eds), Acad. Publ. The Hague, pp.104-108.

Schoeberl, M.R., P.K. Bhartia and E. Hilsenrath, 1992: Tropical ozone loss following the eruption of Mt. Pinatubo, *Geophys. Res. Lett.,* **20**, 29-32.

Seiler, W. and J. Fishman, 1981: The distribution of carbon monoxide in the free troposphere. *J. Geophys. Res.,* **86**, 7255-7265.

Shao, Z., Xingsheng Li, Xiuji Zhou and Mulin Wang, 1994: Gradient measures of methane flux in Linan rice paddies, *Acta Meteorologica Sinica.* (In press)

Shearer, M. and M. Khalil, 1994: Rice agriculture: emissions. Chapter 12 in *Atmospheric Methane: Sources, Sinks and Role in Global Change,* ed. M. Khalil, NATO ASI series, pp230-253.

Simmonds, P., D. Cunnold, F. Alyea, C. Cardelino, A. Crawford, R. Prinn, P. Fraser, R. Rasmussen and R. Rosen, 1988: Carbon tetrachloride lifetimes and emissions determined from daily global measurements. *J. Atm. Chem.,* **7**, 35-58.

Simmonds, P., D. Cunnold, G. Dollard, T. Davies, A. McCulloch and R. Derwent, 1993: Evidence of the phase-out of CFC use in Europe over the period 1987-1990. *Atmos. Env.,* **27A**, 1397-1407.

Simpson, D., 1993: Photochemical model calculations over Europe for two extended summer periods: 1985 and 1989. Model results and comparison with observations. *Atmos.Env.,* **27A**, 921-943.

Singh, H.B. and M. Kanakidou, 1993: An investigation of the atmospheric sources and sinks of methyl bromide. *Geophys. Res. Lett.,* **20**, 133-136.

Singh, H.B., L.J. Salas and R.E. Stiles, 1983: Methyl halides in and over the eastern Pacific (40°N-32°S). *J. Geophys. Res.,* **88**, 3684-3690.

Singh, H.B., D. Herlth, D. O'Hara, K. Zahnle, J.D. Bradshaw, S.T. Sandholm, R. Talbot, G.L. Gregory, G.W. Sachse, D.R. Blake and S.C. Wofsy, 1994: Summertime distribution of PAN and other reactive nitrogen species in the northern high latitudes atmosphere of eastern Canada. *J. Geophys. Res.,* **99**, 1821-1835.

Smith, R.C., B.B. Prezlin, K.S. Baker, R.R. Bidigare, N.P. Boucher, T. Coley, D. Karentz, S. MacIntyre, H.A. Matlick, D. Menzies, M. Onderusek, Z. Wan and K.J. Waters, 1992: Ozone depletion: ultraviolet radiation and phytoplankton biology in Antarctic waters. *Science,* **255**, 952-959.

Spivakovsky, C.M., R. Yevich, J.A. Logan, S.C. Wofsy and M.B. McElroy, 1990: Tropospheric OH in a three-dimensional chemical tracer model: an assessment based on observation of CH_3CCl_3. *J.Geophys. Res.,* **95**, 18441-18471.

Staehelin, J., J. Thudium, R. Bühler, A. Volz-Thomas and W. Graber, 1994: Trends in surface ozone concentrations at Arosa (Switzerland). *Atmos. Env.,* **28**, 75-87.

Steele, L.P., E.J. Dlugokencky, P.M. Lang, P.P. Tans, R.C. Martin and K.A. Masarie, 1992: Slowing down of the global accumulation of atmospheric methane during the 1980s. *Nature,* **358**, 313-316.

Strand, A., and O. Hov, 1994: A two-dimensional global study of the tropospheric ozone production. *J. Geophys. Res.* (In press)

Sturges, W.T., C.W. Sullivan, R.C. Schnell, L.E. Heidt, W.H. Pollock and C.E. Quincy, 1993: Bromoalkane production by Antarctic ice algae: a possible link to surface ozone loss. *Tellus,* **45B**, 120-126.

Subak, S., P. Raskin and D. Hippel, 1993: National greenhouse gas accounts : current anthropogenic sources and sinks. *Climate Change,* **25**, 15-58.

Swain, C.G. and C.B. Scott, 1953: Quantitative correlation of relative rates. Comparison of hydroxide ion with other nucleophilic reagents toward alkyl halides, esters, epoxides and acyl halides. *J. Am. Chem. Soc.,* **75**, 141-147.

Sze, N.D., 1977: Anthropogenic CO emissions: implications for the atmospheric CO-OH-CH_4 cycle. *Science,* **195**, 673-675.

Talbot, R.W. J.D. Bradshaw, S.T. Sandholm, H.B. Singh, G.W. Sachse, J. Collins, G.L. Gregory, B. Anderson, D. Blake, J. Barrick, E.V. Browell, K.I. Klemm, B.L. Lefer, O. Klemm, K. Gorzelska, J. Olsen, D. Herlth and D. O'Hara, 1994: Summertime distribution and relations of reactive odd nitrogen species and NO_y in the troposphere over Canada. *J. Geophys. Res.,* **99**, 1863-1885.

Tarasick, D.W., D.I. Wardle, J.B. Kerr, J.J. Bellefleur and J. Davies, 1994: Tropospheric ozone trends over Canada: 1980-1993. *Geophys. Res. Let.* (Submitted)

Thompson, A.M. and R.J. Cicerone, 1986: Possible perturbations to atmospheric CO, CH_4 and OH. *J. Geophys. Res.,* **91**, 10853-10864.

Tohjima, Y. and H. Wakita, 1993: Estimation of methane discharge from a plume: a case of landfill. *Geophys. Res. Lett.,* **20**, 2067-2070.

Tokarczyk, R. and R. Moore, 1994: Production of volatile organohalogens by phytoplankton cultures. *Geophys. Res. Lett.,* **21**, 285-288.

Torres, A.L. and A.M. Thompson, 1993: Nitric oxide in the equatorial Pacific boundary layer: SAGA 3 measurements. *J. Geophys. Res.,* **98**, 16949-16954.

Volz, A., D.H. Ehhalt and R.G. Derwent, 1981: Seasonal and latitudinal variation of ^{14}CO and tropospheric concentration of OH radicals. *J. Geophys. Res.,* **86**, 5163-5171.

Wahner, A., F. Rohrer, D.H. Ehhalt, B. Ridley and E. Atlas, 1994: Global measurements of photochemically active compounds, *Proceedings of 1st IGAC Scientific Conference: Global Atmospheric-Biospheric Chemistry,* Eilat/Israel, pp205-222.

Wallace, D.W.R., P.Beinung and A.Putzka, 1994: Carbon tetrachloride and chlorofluorocarbons in the South Atlantic Ocean. *J. Geophys. Res.,* **99**, 7803-7819.

Wang, M-X., Shangguan Xingjian, Shen Renxing, R. Wassman and W. Seiler, 1993a: Methane production, emission and possible control measures in the rice agriculture. *Advances in Atmospheric Sciences,* **10**, 307-314.

Wang M-X., Dai Aiguo, Huang Jun, Ren Lixin, Shen Renxing, H. Schutz, H. Rennenberg, W. Seiler, R.A. Rasmussen and M.A.K. Khalil, 1993b: Estimate on methane emission from China. *Chinese Journal of Atmospheric Sciences,* **17**, 49-62.

Warneck, P., 1988: *Chemistry of the Natural Atmosphere.* International Geophysics Series, Vol. 41, Academic Press, Inc., San Diego, 757pp.

Wassmann, R., H. Papen and H. Rennenberg, 1993: Methane emission from rice paddies and possible mitigation strategies. *Chemosphere,* **26**, 201-217.

Wege, K., H. Claude and R. Hartmannsgruber, 1989: Several results from the 20 years of ozone observations at Hohenpeissenberg. In: *Ozone in the Atmosphere,* R.D. Bojkov and P. Fabian (eds.), Deepak publ., Hampton, VA., pp109-112.

Weiss, R. F.,1994: Changing global concentration of atmospheric nitrous oxide. *Proceedings International Symposium on Global Cycles of Atmospheric Greenhouse gases, Sendai, Japan, 7-10 March, 1994,* pp78-80.

Whalen, S.C. and W. Reeburgh, 1992: Interannual variations in tundra methane emission: A 4-year time-series at fixed sites. *Global Biogeochem. Cycles,* **6**, 139-159.

Williams, E.J., G.H. Hutchinson and F.C. Fehsenfeld, 1992: NO_x and N_2O emissions from soils. *Global Biogeochem. Cycles,* **6**, 351-388.

Wine, P.H. and W.L. Chameides, 1990: Possible atmospheric lifetimes and chemical reaction mechanisms for selected HCFCs, HFCs, CH_3CCl_3 and their degradation products against dissolution and/or degradation in seawater and cloudwater, pp. 271-295 in WMO/UNEP report 20, vol.2, appendix.

World Meteorological Organisation (WMO), 1986: Scientific Assessment of Atmospheric Ozone:1985, Rep. 16, *WMO Global Ozone Res. and Mon. Project, Geneva.*

World Meteorological Organisation (WMO), 1990: Scientific Assessment of Stratospheric Ozone:1989, Rep. 20, *WMO Global Ozone Res. and Mon. Project, Geneva.*

World Meteorological Organisation (WMO), 1992: Scientific Assessment of Ozone Depletion : 1991, Rep. 25, *WMO Global Ozone Res. and Mon. Project, Geneva.*

World Meteorological Organisation (WMO), 1994: Scientific Assessment of Ozone Depletion:1994, Rep. 37, *WMO Global Ozone Res. and Mon. Project, Geneva.* (in press).

Worthy, D., N. Trivett, K. Higuchi, J. Hopper, E. Dlugokencky, M. Ernst and M. Glumpak, 1994: Six years of *in situ* atmospheric methane measurements at Alert, N.W.T., Canada, 1988-1993; seasonal variations, decreasing growth rate and long-range transport. *Proceedings International Symposium on Global Cycles of Atmospheric Greenhouse gases, Sendai, Japan, 7-10 March, 1994,* pp130-133.

Wuebbles, D.J., D.E. Kinnison, K.E. Grant and J. Lean, 1992: The effect of solar flux variations and trace gas emissions on recent trends in stratospheric ozone and temperature. *J. Geomagnetism and Geoelectricity,* **43**, 709-718.

Zander, R., Ph. Demoulin, D.H. Ehhalt, U. Schmidt and C.P. Rinsland, 1989: Secular increases in the total vertical abundance of carbon monoxide above central Europe since 1950. *J. Geophys. Res.,* **94**, 11021-11028

Zander, R., C.P. Rinsland and P. Demoulin, 1991: Infrared spectrascopic measurements of the vertical column abundance of sulphur hexafluoride from the ground. *J. Geophys. Res.,* **96**, 15447-15454.

Zander, R., D.H. Ehhalt, C.P. Rinsland, U. Schmidt, E. Mahieu, J. Rudolph, P. Demoulin, G. Roland, L. Delbouille and A. Sauval, 1994: Secular trend and seasonal variability of the column abundance of N_2O above the Jungfraujoch station determined by IR solar spectra, *J. Geophys. Res,* **99**, 16745-16756.

Zardini, D., D. Raynaud, D. Scharffe and W. Seiler, 1989: N_2O of air extracted from Antarctic ice cores: implication on atmospheric N_2O back to the last glacial-interglacial transition. *J. Atmos. Chem.,* **8**, 189-201.

Zhou, X., C. Luo, G. Ding, J. Tang and Q. Liu, 1993: Preliminary analysis of the variations of surface ozone and nitric oxides in Lin'an. *Acta Meteo. Sinica,* **7**, 287-294.

3

Aerosols

P.R. JONAS, R.J. CHARLSON, H. RODHE

Contributors:
*T.L. Anderson, M.O. Andreae, E. Dutton, H. Graf, Y. Fouquart, H. Grassl,
J. Heintzenberg, P.V. Hobbs, D. Hofmann, B. Huebert, R. Jaenicke, M. Jietai,
J. Lelieveld, M. Mazurek, M.P. McCormick, J. Ogren, J. Penner, F. Raes, L. Schütz,
S. Schwartz, G. Slinn, H. ten Brink.*

CONTENTS

Summary 131

3.1 Introduction 133
   3.1.1 Radiative Forcing by Aerosols 133

3.2 Aerosol Sources and Types 134
   3.2.1 Soil Dust 134
   3.2.2 Sea Salt Aerosols 135
   3.2.3 Volcanic Dust 136
   3.2.4 Primary Organic Aerosols 136
   3.2.5 Industrial Dust 136
   3.2.6 Soot 136
   3.2.7 Biomass Burning 136
   3.2.8 Oxidation of Precursor Gases 136

3.3 Aerosol Transformation Processes 137
   3.3.1 Processes Relevant to Radiative Forcing 137
   3.3.2 Why Particle Size is Important 138
   3.3.3 Accumulation Mode Particles as the Repository of Most Anthropogenic Aerosol Substances 139

3.4 Aerosol Sinks and Lifetimes 139

3.5 Properties, Measurements and Budgets of Atmospheric Aerosols 140
   3.5.1 Aircraft, Satellite and Remote Sensing Measurements 140
   3.5.2 Regional and Global Models and Budgets of Anthropogenic Aerosol 140
   3.5.3 Measured Trends 141

3.6 Stratospheric Aerosols 142
   3.6.1 Sources, Sinks and Lifetimes 142
   3.6.2 Size Distributions, Compositions and Optical Depths 143
   3.6.3 Interaction with the Troposphere 143

3.7 Optical Properties of Aerosols Determining the Direct Radiative Effect 144
   3.7.1 Spatial Scales 144
   3.7.2 Local Optical Properties (Two Approaches) 144
      3.7.2.1 Dependence of local integral optical properties on size and composition 146
   3.7.3 Column Properties 146

3.8 Causes of Uncertainty in Calculating Direct Forcing by Aerosols 147
   3.8.1 Local vs. Regional vs. Global Models and Effects 147
   3.8.2 Coupled Chemical/Radiative Transfer Models 148
   3.8.3 Causes of Uncertainty in Calculating the Direct Forcing by Anthropogenic Aerosol 149

3.9 Influence of Anthropogenic Aerosol on Clouds (Indirect Effect) 150
   3.9.1 Cloud Formation 150
   3.9.2 Cloud Concentration Nuclei (CCN) Population and Aerosol Sources 151
      3.9.2.1 Sulphates and nitrates 151
      3.9.2.2 Organics 152
      3.9.2.3 Soot 152
      3.9.2.4 Mineral dust 152
      3.9.2.5 Sea salt 152
   3.9.3 Cloud Radiative Properties 152
      3.9.3.1 Water clouds 152
      3.9.3.2 Ice clouds 152
   3.9.4 The Indirect Effect of Anthropogenic CCN on Cloud Albedo 153
   3.9.5 Precipitation Formation and Cloud Lifetime 154

3.10 Future Trends 154
   3.10.1 Industrialisation 154
   3.10.2 Feedbacks 155

3.11 Likely Developments 155
   3.11.1 Local Measurements 156
   3.11.2 Ground-based and Satellite Monitoring 156
   3.11.3 Modelling 157

References 157

SUMMARY

- Atmospheric aerosol in the troposphere influences climate in two ways, directly through the reflection and absorption of solar radiation, and indirectly through modifying the optical properties and lifetime of clouds.
- Estimation of aerosol radiative forcing is more complex and hence more uncertain than radiative forcing due to the well-mixed greenhouse gases (see Chapter 4) for several reasons:
    (i) Both the direct and indirect radiative effect of aerosol particles are strongly dependent on particle size and chemical composition and cannot be related to aerosol mass source strengths in a simple manner.
    (ii) The indirect radiative effect of aerosols depends on complex processes involving aerosol particles and the nucleation and growth of cloud droplets.
    (iii) Aerosols in the troposphere have short lifetimes (around a week) and therefore their spatial distribution is highly inhomogeneous and strongly correlated with their sources.
- The net (direct and indirect) global mean radiative forcing due to the increase in anthropogenic aerosol since pre-industrial times is negative and the magnitude is significant; while forcing due to changes in sulphate and organic aerosols is negative, that due to soot carbon is probably positive although present estimates suggest the latter is relatively small (see Chapter 4 for more details).

Direct radiative effect

- There are many lines of evidence suggesting that anthropogenic aerosol has increased the optical depth over and downwind of industrial regions and that this increase is very large compared with the natural background in these regions.
- The major contributions to the anthropogenic component of the aerosol optical depth arise from sulphates (produced from sulphur dioxide released as a result of fossil fuel combustion) and from organics released by biomass burning.
- Owing to the effect of size and chemical composition, anthropogenic aerosol contributes approximately 50% to the global mean aerosol optical depth but only around 20% to the mass burden.

Indirect radiative effect

- The impact of aerosol on the optical properties of low level clouds has been demonstrated in localised observations but the global impact has yet to be quantified. The effect is expected to be smaller for additional aerosol introduced into already polluted air.
- There is some observational evidence suggesting that mean sizes of cloud droplets are larger in the Southern Hemisphere (where tropospheric aerosol concentration is generally lower) than in the Northern Hemisphere (where aerosol concentration is generally higher).

Stratospheric aerosols

- Large volcanic eruptions, such as Mt. Pinatubo, significantly influence the aerosol in the stratosphere and the effects persist over several years.
- Stratospheric aerosol has a longer lifetime (of order 1 year) than aerosol in the troposphere and is therefore more uniformly distributed. The amount of sunlight reflected by aerosols over industrial regions is comparable with the peak effects of volcanic aerosols.

Stabilisation of aerosol concentration

- Future concentrations of anthropogenic sulphate aerosols will depend on both fossil fuel use and emission controls. Even if the globally averaged concentration were stabilised – through stabilisation of total global emissions– the geographical distribution of the SO_2 emissions, and hence the aerosol concentration, would be likely to exhibit major changes.

3.1 Introduction

Aerosols in general, and anthropogenic aerosols in particular, are thought to exert a radiative influence on climate directly, through the reflection and absorption of solar radiation, and indirectly, through modifying the optical properties and lifetimes of clouds. The atmospheric loading of anthropogenic aerosols and the resultant radiative influence have increased over the industrial period roughly in parallel with that of the greenhouse gases and these aerosol particles could have exerted a significant radiative forcing over this period.

This chapter reviews the relevant physics and chemistry of atmospheric aerosol in order to provide the context for describing the climatic influence of anthropogenic aerosols (discussed in Chapter 4). It is demonstrated that:

(a) there is sound physical and chemical evidence supporting the existence of anthropogenic sulphate and organic aerosol radiative forcing which is negative in sign;
(b) uncertainty in calculating the magnitude of the forcing arises from a lack of observations of the chemical and physical properties of aerosol particles as well as their spatial and temporal distribution.

Aerosols present a large source of uncertainty in calculating radiative forcing over the industrial period. The effect of anthropogenic aerosol may help explain the differences between observed climate change over the industrial era and the best estimates of such change from model calculations which are based only on the effects of greenhouse gases. However, at the present time, the climatic effects of aerosols can only be estimated with considerable uncertainty (+Hansen *et al.*, 1990) due to the lack of sufficient detailed observations. A quantitative estimate of the uncertainty can be made only for direct forcing by sulphate and organic aerosols; other aerosol types, e.g., soil dust and soot, cannot yet be quantified.

3.1.1 Radiative Forcing by Aerosols

Atmospheric aerosol particles are conventionally defined as those particles suspended in air having diameters in the range 10^{-3} to 10 μm.[1] They are formed by the reaction of gases in the atmosphere (gas to particle conversion), or by the dispersal of material at the surface. Although forming a small part of the mass of the atmosphere, around 1 part in 10^9, they have the potential significantly to influence both radiative transfer through the atmosphere and the atmospheric water cycle. The best understood mechanism by which aerosol particles can influence climate is by scattering incoming solar radiation, thereby increasing the Earth's albedo (the direct effect). Absorption of solar radiation, for example by soot, may cool the surface while heating the atmosphere. Direct aerosol forcing is primarily a short-wave forcing process, and is therefore expected to dominate in the daytime and summer months (e.g., Karl *et al.*, 1991; Engardt and Rodhe, 1992; Hunter *et al.*, 1993). Although aerosol particles may absorb long-wave radiation, this effect is believed to be small for anthropogenic aerosols other than soot. Anthropogenic aerosol particles acting as cloud condensation nuclei (CCN) are believed to increase the number concentration of cloud droplets, increasing cloud albedo (the indirect effect) e.g., SMIC, 1971; Twomey, 1977; Twomey *et al.*, 1984. The corresponding decrease in cloud droplet size may inhibit precipitation formation, extending cloud lifetime. Long-wave absorption by wet aerosol particles (Marley *et al.*, 1993) may be a significant factor in the overall radiation balance but estimates of the magnitude of this effect are only preliminary.

In contrast to greenhouse gases, atmospheric aerosol particles are short-lived in the troposphere with lifetimes of around a week, while they have lifetimes of a year or more in the stratosphere. The short lifetimes, together with the highly non-uniform geographical distribution of aerosol sources, results in a highly non-uniform geographical distribution of anthropogenic aerosols. This non-uniformity results in very different patterns of radiative forcing and may result in different climate response patterns compared with greenhouse gases which are well-mixed.

Figure 3.1 is a schematic representation of typical aerosol particle mass and number distributions above 0.01 μm diameter as functions of particle diameter, with peaks corresponding to aerosol produced by gas-to-particle conversion and by disruption of the Earth's surface respectively. Although the fine and coarse modes account for most of the aerosol mass, there are many smaller particles and a graph of number concentration shows a further peak around 0.02 μm, the nucleation mode. The spectrum also shows a peak at a diameter smaller than the lower limit of the graph but particles of this size are not believed to have climatic effects.

The estimation of radiative forcing caused by tropospheric aerosol is fundamentally different from that caused by well-mixed greenhouse gases because:

(i) The aerosol mass and size distribution show considerable temporal and spatial variability due to the relatively short lifetime and localised nature of the sources and sinks. There are substantial differences in the distributions of both anthropogenic and natural aerosols between the Northern and Southern Hemispheres owing to the different anthropogenic sources and different distribution of land masses.

---

[1] 1 μm = 1 micrometre = 1 millionth of a metre.

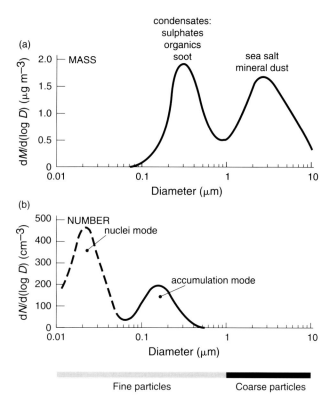

Figure 3.1: Schematic diagram showing the main features of the aerosol size distribution as a function of (a) mass and (b) number. The distributions have been plotted such that the area under the curve corresponds to the total mass and number concentrations respectively. Although coarse particles contribute substantially to the aerosol mass, they form a very small part of the number of aerosol particles.

(ii) Aerosols are not chemically homogeneous and are produced by a variety of processes including chemical reactions between gases in the air. Such heterogeneity makes it difficult to prescribe particle size distributions and optical properties of given types of aerosol particles, for example in climate models.

(iii) The radiative effects of aerosol are critically dependent on the size distribution, rather than simply on the total mass loading alone. Different sources have different aerosol size characteristics and the size distribution and chemical composition change as the aerosol ages.

3.2 Aerosol Sources and Types

Atmospheric aerosol particles may be emitted as particles (primary sources) or formed in the atmosphere from gaseous precursors (secondary sources). Some of the sources, e.g., volcanic emissions and sea spray from the oceans, are clearly natural; others, for example industrial emissions, are purely anthropogenic. There are also sources, e.g., biomass burning and soil dust emissions, where, although the distinction between natural and anthropogenic processes may be clearly defined, there is often insufficient information to apportion particular sources.

Table 3.1 (modified from Andreae, 1995) summarises the estimated annual emissions into the troposphere or stratosphere from the major sources of atmospheric aerosol. Emission of mineral aerosols from soil dust has been classified as a natural source, although a certain influence from anthropogenic processes, e.g., agriculture, is likely to exist. Likewise, biomass burning is not purely anthropogenic. The large upper estimates for sea salt and volcanic dust include large particles with very short atmospheric lifetimes. It should be noted that particle mass flux is not a good measure of either the mass loading or the instantaneous effect of the aerosol particles, for reasons to be discussed later. Further, the quantities shown are highly disparate in magnitude and uncertainty so that the totals are only intended to demonstrate that anthropogenic aerosols are a significant contribution to the total aerosol burden.

Owing to the short lifetime of aerosol particles in the troposphere and the non-uniform distribution of sources, their geographical distribution is highly non-uniform. As a consequence, the relative importance of the various sources shown in Table 3.1 varies considerably over the globe. For example, within and around the most industrialised regions in Europe and North America the industrial sources are relatively much more important. The nature of the various emissions is briefly summarised below.

3.2.1 Soil Dust

Atmospheric mineral dust is found all around the globe. This is due to long-range transport over thousands of kilometres within the general circulation of the atmosphere. Major sources of the particulate matter are the arid and semi-arid regions, where the material is continuously produced by bulk to particle conversion caused by physical and chemical weathering of soils and rock. The diameter of windblown soil dust ranges from less than 1 μm to 100 μm or more. The largest particles fall out rapidly and are of no concern in this context. The mass median diameter of that fraction which is transported over appreciable distances (several 100 km) is about 2-4 μm (Buat-Ménard *et al.*, 1983; Dulac *et al.*, 1989, 1992), i.e. in the coarse range. The contribution of fine particles is enhanced during intense dust events compared to moderate dust rising conditions.

The main sources of soil dust are surfaces with unconsolidated material and areas with active weathering

Table 3.1: Global emission estimates for major aerosol types in the 1980s. Sulphates and nitrates are assumed to occur as ammonium salts. Modified from Andreae (1994).
Units: Tg/yr (dry mass)

Source	Estimated flux			Particle size category (coarse: 1μm diameter fine: 1μm diameter)
	low	high	"best"	
Natural				
Primary				
Soil dust (mineral aerosol)	1,000	3,000	1,500	mainly coarse
Sea salt	1,000	10,000	1,300	coarse
Volcanic dust	4	10,000	33	coarse
Primary organic aerosols	26	80	50	coarse
Secondary				
Sulphates from biogenic gases	60	110	90	fine
Sulphates from volcanic SO_2	4	45	12	fine
Organic matter from biogenic VOC*	40	200	55	fine
Nitrates from NO_x	10	40	22	mainly coarse
Sum of Natural Sources	2,144	23,475	3,062	
Anthropogenic				
Primary				
Industrial dust (except soot)	40	130	100	coarse and fine
Soot (includes biomass burning)	5	25	10	mainly fine
Biomass burning (except soot)+	50	140	80	fine
Secondary				
Sulphates from SO_2	120	180	140	fine
Nitrates from NO_x	20	50	40	mainly coarse
Organic matter from biogenic VOC*	5	25	10	fine
Sum of Anthropogenic Sources	240	550	380	
TOTAL	**2,384**	**24,025**	**3,442**	

* VOC = Volatile Organic Carbon
+ Also forms secondary particles

(Pye, 1987). These are alluvial fans, outwashes, wadis, etc. Only a minor fraction of the total desert area can be considered as a source for airborne material. The annual global emission amounts to about 1500 Tg, which is of the order of the global sea salt production. Mineral dust particles originate mainly from the great desert areas of Northern Africa and Asia (Prospero, 1990). The largest source is probably the Sahara. Despite the inefficient light scattering by coarse particles, mineral dust does have an appreciable effect on the planetary radiation balance due to the large amounts involved (see section 3.7.3). For example, desert dust is clearly seen from space in satellite imagery (Durkee *et al.*, 1991; Jankowiak and Tanre, 1992). Dust originating from the Asian continent causes a decrease in visibility over large regions of Japan and China (Kai *et al.*, 1988).

Anthropogenic forcing due to mineral dust emissions by industrial, agricultural and vehicle activities has not yet been well quantified but is believed to be relatively small.

3.2.2 Sea Salt Aerosols

Sea salt, resulting from the evaporation of sea-spray droplets, is a major component of natural aerosol. The largest droplets, containing most of the mass injected into the air, are so large that they return to the ocean almost immediately. Smaller droplets may stay airborne long enough to dry out and shrink, whereby they then can remain airborne even longer. This inverse relationship between mass and atmospheric lifetime has resulted in estimates of the amount of these aerosols injected into the atmosphere which range from 1,000 to 10,000 Tg/yr (SMIC, 1971; Blanchard, 1983). The mass median diameter of sea salt aerosol near the sea surface is of the order of 8 μm and, because of their short lifetime and inefficient light-scattering, the largest particles are of little importance to radiative forcing of climate change. Therefore, in Table 3.1 a value of 1,300 Tg/yr is suggested as representative for that fraction of the sea salt aerosol which can be transported throughout the lower marine troposphere (Petrenchuk, 1980; Andreae, 1986).

3.2.3 Volcanic Dust

Volcanic emissions consist of solid particles (ash) and gases (mainly water vapour (H_2O), sulphur dioxide (SO_2) and carbon dioxide (CO_2)) in very variable concentrations. SO_2 is a precursor to aerosol formation by gas to particle conversion. The ash particles are mainly in the coarse particle range. Most of the known active volcanoes are in the Northern Hemisphere, and only 18% are between 10°S and the South Pole (Simkin et al., 1981; Simkin and Sibert, 1984). In regions with frequent volcanic activity, the emission of sulphur and other aerosol components may be important (e.g., Hobbs et al., 1982). For example, Allard et al. (1991) estimate the sulphur emission of Etna/Sicily to be 0.7 TgS (sulphur)/yr averaged over the years 1975 to 1987.

The global volcanic emission of SO_2 has been estimated to be 3.5 TgS/yr from effusive degassing and, on average, about 6 TgS/yr from erupting volcanoes (Stoiber et al., 1987). On average, less than 10% of this amount reaches the stratosphere. This number may vary by an order of magnitude in individual years, for example following explosive eruptions like Mt. Pinatubo (9 TgS in 1991) and El Chichon (3.5 TgS in 1982).

3.2.4 Primary Organic Aerosols

Natural emissions of particulate organic carbon (POC) are produced by marine and continental sources. Global emissions of POC from ocean regions are estimated to be slightly larger than from continental natural sources (Duce, 1978). Particles with nominal diameters > 1 μm account for 90% of the POC emitted from both source categories. The ocean is a large source of POC owing to the injection of naturally derived marine surfactants from bubble bursting processes (Monahan, 1986). Terrestrial sources of primary POC include natural products emitted from vegetation such as C_{10} to C_{30}, terpenes, plant waxes, and macromolecular plant fragments (Graedel 1979; Simoneit and Mazurek 1981). POC may also be emitted by combustion processes.

3.2.5 Industrial Dust

Primary aerosol particles originate from the incombustible material present as inorganic impurities in the fuel which passes through the combustion process and from incomplete fuel combustion. The amount of aerosol material produced from coal is much larger than from oil. The result of more efficient particle removal systems is that emissions from modern coal-fired installations are an order of magnitude lower than those of a few decades ago. However, for many developing countries the conditions appropriate to the older installations probably apply. In addition to combustion sources, industrial dust may be produced by other processes. The size of the emitted industrial dust particles depends on, among other things, the efficiency of the filtering process. Most of the 100 Tg/yr listed in Table 3.1 is likely to be in the coarse mode.

3.2.6 Soot

Because of the unique properties (light absorbing efficiency) of the soot fraction of industrial aerosol, this fraction is given a separate estimate in Table 3.1. Soot particles absorb solar radiation very efficiently. Most of the soot particles lie in the fine category. The estimates are based on Ogren and Charlson (1983), Turco et al. (1983) and Penner et al. (1992).

3.2.7 Biomass Burning

Biomass burning is likely to be the next largest source of fine mode anthropogenic aerosols, after sulphate (Andreae, 1991). Most of the precursors are carbon, sulphur and nitrogen compounds which form organic sulphate and nitrate aerosols. Another important component is soot and tar condensates (Cachier et al., 1989). The types of aerosols formed depend to a great extent on the combustion characteristics and have not been fully characterised (Levine, 1991).

3.2.8 Oxidation of Precursor Gases

Natural gases which are precursors of aerosol particles include sulphur compounds (mainly dimethyl sulphide (DMS) and SO_2), hydrocarbons (HC) and oxides of nitrogen (NO and NO_2, known collectively as NO_x), and HNO3. Oxidation of these precursors occurs both in the gas phase (mainly through oxidation processes initiated by the OH and other atmospheric radicals) and in liquid droplets. In this oxidation process low volatility products are formed which either condense onto pre-existing aerosol particles (including droplets) or form new particles. Most of the aerosol mass produced by this process is in the fine size range. In-cloud processes have a significant effect on the aerosol size distribution.

The dominant part of the sulphate from biogenic gases is derived from DMS emitted from the oceans where it is formed by biological processes (Andreae, 1995). Natural sources of NO_x include soil exhalation (about 10 TgN/yr) (Galbally and Roy, 1978), and lightning (2-20 TgN/yr) (Warneck, 1988). It is estimated that about half of the NO_x is oxidised to nitric acid (HNO_3) which is present in both the gas and the condensed phase (in aerosols and droplets). Part of the gas phase HNO_3 is deposited directly at the Earth's surface or scavenged by precipitation. The fraction of the HNO_3 that contributes to the aerosol mass (or number) is uncertain. There is a tendency for the aerosol nitrate to occur as coarse particles, which are not so important from a climatic point of view except possibly as CCN.

The contribution of natural non-methane hydrocarbons (NMHC) to the aerosol mass and number concentration is likely to be very significant, especially in coniferous forests where emissions of terpenes and other hydrocarbons are large. The global emission of natural NMHC is estimated to be about 700 Tg/yr out of which about 10% may be converted to aerosols (Andreae, 1995). Carbonaceous aerosol particles either emitted directly from natural and anthropogenic sources or produced from precursor gases have highly complex chemical compositions that contain varying proportions of elemental and organic carbon. The scattering and absorption characteristics depend strongly on the proportion of elemental and organic carbon. Natural aerosols contain essentially no elemental carbon except for smoke aerosols from wild fires. Anthropogenic particles produced from combustion processes contain substantial quantities of elemental carbon relative to organic carbon, depending on the source type (Hildemann *et al.*, 1991; Cachier *et al.*, 1989).

Today, industrial and other anthropogenic sources of precursor gases represent a very substantial addition to the natural precursors (see Table 3.1). The largest contribution comes from industrial SO_2. For a few decades, this source of gaseous sulphur has surpassed the natural emission of such gases (mainly DMS from the oceans and SO_2 from volcanoes). The current industrial SO_2 emission is about 70 - 90 TgS/yr (see e.g., Muller, 1992; Spiro *et al.*, 1992), roughly half of which is oxidised to aerosol sulphate before being deposited (Langner and Rodhe, 1991). Approximately 90% of the emissions are from industrial regions in the Northern Hemisphere, and little of this SO_2 or of the resulting sulphate aerosol is transported to the Southern Hemisphere.

Whereas most of the natural sources of aerosols can be expected to have changed only moderately during past centuries, anthropogenic emissions have grown dramatically especially during the 20th century. While industrial sulphur aerosol loadings have doubled since 1950, following a slow increase over the preceding 100 years, loadings from biomass burning have undergone a steady increase over the last 150 years (Andreae, 1991). Figure 3.2 shows the rapid increase in anthropogenic SO_2 emissions over this period, although as will be shown later, the rate of increase in emissions shows large regional variations, resulting in changes in the geographical distribution of aerosol. Emissions associated with eruptive volcanoes represent a special case with highly variable source strength.

3.3 Aerosol Transformation Processes

3.3.1 Processes Relevant to Radiative Forcing

Aerosol transformation processes are of particular importance to the problem of climatic effects, because these processes are the major determinants of aerosol optical and cloud droplet-nucleating properties. The size-

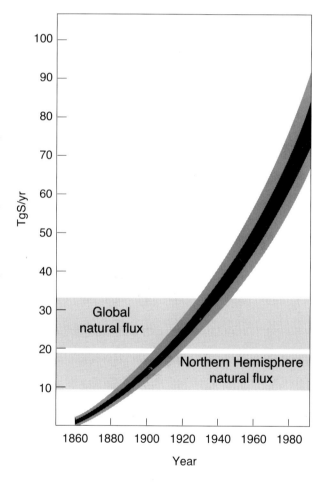

Figure 3.2: Time history of global emission of SO_2 (in TgS/yr) and estimates of the global and Northern Hemisphere natural flux (from Charlson *et al.*,1992). Anthropogenic sulphur is emitted mainly (~90%) in the Northern Hemisphere and emissions greatly exceed the natural emissions. Width of shading represents the uncertainty.

dependent properties of the particles that govern the direct and indirect radiative effects are:

- number, surface area and mass concentration;
- refractive index (determined by chemical composition);
- particle shape and morphology;
- water solubility;
- surface chemical properties;
- location in the atmosphere.

Because the radiative effects of aerosol vary strongly as a function of particle size, it is necessary to focus on the processes that determine the above variables as functions of size.

3.3.2 Why Particle Size is Important

Major changes of aerosol production and transformation processes occur at particle diameters around 0.1 and 1.0 μm. Table 3.2 compares a variety of aerosol properties above and below 1 μm diameter, and illustrates the very large differences that exist across this size range. Below 0.1 μm, particle behaviour is dominated by Brownian motion and by large ratios of surface area to volume (>30 m^2 cm^{-3}). Nuclei mode (smaller than ~ 0.02 μm) particles are freshly produced new particles from condensation at low temperature (atmospheric reactions) or high temperature (fire), and their presence in air is transitory (minutes to days). The accumulation mode (0.1 to 1.0 μm diameter) is so named because mass accumulates by Brownian coagulation, condensation of products of gas-phase reactions, and aqueous phase reactions. Nuclei mode and accumulation mode particles are together referred to as *fine particles* (see Figure 3.1). The bulk of the aerosol mass resides in accumulation and coarse modes. Because Brownian motion becomes unimportant as particle size grows above about 0.1 μm diameter, accumulation mode particles do not readily become attached to coarse mode particles. This prevents the chemical substances of the accumulation mode (e.g., H_2SO_4) from reacting with those in the coarse mode (e.g., basic soil dust). This mixture of coarse and fine particle aerosol is chemically heterogeneous and exhibits features that are the consequence of size dependent source, sink and transformation processes.

The general form of size distribution in Figure 3.1 (adapted from Andreae, 1995) is based on large amounts of data from both continental and marine locations. Chemical species in the accumulation mode particles thus must dominate the CCN concentration and, most significantly, anthropogenic changes in it. Because the number concentration is dominated by particles less than about 0.1 μm diameter, the chemical substances in fine mode particles must dominate the process of cloud nucleation even though the coarse mode particles contribute more to the aerosol mass. Similarly, but not quite so thoroughly, fine mode substances must dominate the direct optical effects as will be discussed in Section 3.7.

This complex but constant general form of size distribution is both the consequence and the controller of the chemical and physical processes that occur in the atmosphere. Several of these processes can be noted as of

Table 3.2: Comparison of properties of coarse and fine aerosol particles

Property/Attribute	Coarse (greater than 1 μm diameter)	Fine (accumulation plus nuclei) (less than 1 μm diameter)
Production mechanism	Mechanical	Nucleation, condensation; multiphase chemical processes; small amounts mechanically produced
Number concentration	Small; usually less than 1 cm^{-3}	Large; ranges from 10s to 1000s cm^{-3}
Motion	Dominated by sedimentation of larger particles; easily impacted, e.g., by falling raindrops	Brownian motion; no sedimentation or impaction
Cloud-nucleating characteristic	Growth at low supersaturation but concentrations are small	Requires higher (up to ~1%) supersaturation but usually dominates the number population of CCN
In-situ modification mechanism	Mixing with other aerosol solutes via cloud coalescence and multiphase chemical reactions; uptake of acidic gases	Condensation of products of gas phase reactions; multiphase reactions in particles and in clouds and mixing via Brownian coagulation; uptake of alkaline gases
pH	Generally alkaline	Acidic

central importance to those aerosol properties that are relevant for radiative forcing:

- chemical production of water soluble condensates such as H_2SO_4 and $(NH_4)_2SO_4$. These particles can act as cloud condensation nuclei, are hygroscopic or deliquescent, and have large light scattering efficiency due to particle size and to hygroscopic growth.
- accumulation of mass by numerous processes in the size range 0.1 and 1 μm. This is the size range for CCN *and* for maximum efficiency for scattering solar radiation.
- modification of the shape of the size distribution by multiple-passages of aerosol particles through clouds in which aqueous-phase chemical reactions occur. This increases the light-scattering efficiency (Lelieveld and Heintzenberg, 1992).
- rapid removal of coarse particles. As a result, a large mass source strength does not necessarily result in large climatic effects (Andreae, 1995).

3.3.3 Accumulation Mode Particles as the Repository of Most Anthropogenic Aerosol Substances

In Section 3.2, five key types of contemporary aerosol substances were identified as having a substantial anthropogenic component:

- sulphates from the oxidation of sulphur-containing gases;
- nitrates from gaseous nitrogen species;
- organic materials from biomass combustion and oxidation of reactive volatile hydrocarbons;
- soot from combustion;
- mineral dust from aeolian processes.

Because these substances constitute major additions to both the number and mass concentrations of the pre-existing natural aerosol, they are the cause of both a secular trend and temporal and geographical variability in aerosol properties, transformations and effects. Sulphate ion typically comprises around 25-50% of the anthropogenic accumulation mode aerosol mass, when it has been measured, and is, probably, the best understood (Heintzenberg, 1989). Much of the subsequent discussion will therefore concentrate on the effects of sulphate aerosol.

It is of particular importance to reiterate that the processes of particle formation, growth by deposition of vapour and by Brownian motion, and processes in clouds result in the bulk of anthropogenic sulphates, organic and soot aerosol *mass* being concentrated in the diameter range between 0.1 and 1.0 μm. The maximum increase in particle *number* concentrations due to anthropogenic effects occurs at a diameter range roughly an order of magnitude below that. The reason for the relevance of this peculiar size dependence lies in two factors which will be discussed in Sections 3.7 and 3.9 respectively:

- Particles in the 0.1 to 1.0 μm diameter range have the highest efficiency per unit mass for optical interactions with sunlight (due to the similarity of particle size and wavelength of light);
- Particles of sizes around 0.1 μm diameter composed of water soluble substances, like sulphates, are highly effective as cloud condensation nuclei.

3.4 Aerosol Sinks and Lifetimes

Whereas the number of aerosol particles and their surface area can be reduced by coagulation processes within the atmosphere, removal of aerosol mass is mainly achieved by transfer to the Earth's surface or by volatilisation. Such transfer is brought about by precipitation ("wet deposition") and by direct uptake at the surface ("dry deposition"). The efficiency of both these deposition processes is strongly dependent on particle size, especially in the diameter range 0.1-10 μm (Slinn, 1983). For particles in the accumulation mode, wet deposition is the major removal process. The lifetime of nuclei mode particles is determined by coagulation which reduces the number of particles but does not reduce the total particle mass.

The time spent in the atmosphere by an aerosol particle is a complex function of its physical and chemical characteristics (e.g., size, hygroscopic properties, etc.) and the time and location of its release. For sulphate particles in the diameter range 0.01-1.0 μm (fine mode) released or formed in the mid-latitude boundary layer, an average lifetime is typically of the order of several days (Junge, 1963; Rodhe, 1978; Chamberlain, 1991). This time scale is dominated mainly by the frequency of recurrence of precipitation. Particles transported into, or formed in the upper troposphere are likely to have a longer lifetime (weeks to months) because of less efficient precipitation scavenging (Balkanski *et al.*, 1993). Many estimates of atmospheric lifetimes or aerosol particles have been based on measurements of radio nuclides bound to aerosols; for example Cambray *et al.* (1987) estimated a lifetime of nine days for ^{137}Cs from the Chernobyl accident. The global model simulation of Langner and Rodhe (1991) gave a global average lifetime of sulphate aerosol of about five days. It should be noted that the atmospheric lifetime of aerosol particles in the troposphere is much smaller than the lifetimes of the main greenhouse gases (see Chapters 1, 2 and 5). The implications of this difference are discussed further in Sections 3.8 and 3.10. Aerosol particles injected into, or formed in, the stratosphere have lifetimes of order one year, much longer than those of tropospheric particles. This is largely due to the absence of precipitation scavenging and limited vertical turbulent transport.

3.5 Properties, Measurements and Budgets of Atmospheric Aerosols

In order to evaluate how aerosols influence the climate, it is necessary to know the cloud nucleating and optical properties of those types of anthropogenic and natural aerosols expected to have a significant climatic effect, in addition to the microphysical and chemical properties discussed earlier.

3.5.1 Aircraft, Satellite and Remote Sensing Measurements

A large number of aircraft measurements is available that provides information on aerosol loading and chemical and microphysical properties. However, difficulties in the sampling of aerosols from aircraft (e.g., Huebert et al., 1990) raise concerns over the accuracy of these measurements. Moreover, the lack of integrated data sets including all, or most, of the important quantities identified later in Table 3.3, makes it difficult to use such observations in radiative forcing calculations.

Remote-sensing of aerosol properties poses challenges that to date have been addressed only to a very limited extent (Penner et al., 1994). A change in the radiation field caused by aerosols can be of a magnitude that is significant for climate perturbation but is barely visible in typical remote sensing images.

While a number of techniques for retrieval of aerosol properties from satellite measurements have been investigated, the only instruments thus far flown which were specifically designed to provide aerosol information are SAGE I, SAGE II and SAM II. These experiments use photometers to observe the solar occultation at each satellite sunrise-sunset and thus determine the vertical profile of aerosol extinction at 1 μm wavelength in the stratosphere, and, in the absence of high-altitude clouds, in the upper troposphere (McCormick et al., 1979; Yue et al., 1989; Kent et al., 1991). Occultation observations are a particularly sensitive method for aerosol measurement because of the enhanced aerosol effects with the large path length at the limb and the self-calibrating nature of such measurements. However, the restriction to satellite sunrise-sunset locations leads to a sparse distribution of aerosol profiles and restricted ability to probe the atmosphere below about 7 km altitude although realisation of the full potential of satellite monitoring of aerosol optical depths, particle size shape and refractive index will depend on concurrent monitoring of aerosol profiles from fixed ground sites to provide calibration and to allow detailed process studies.

Except during periods following volcanic eruptions that significantly increase aerosol loading in the stratosphere, the largest contribution to the total column aerosol amount is from the lower troposphere, below the lowest altitude that the occultation method can probe easily even in cloud-free conditions. Most techniques proposed for retrieving the tropospheric aerosol from satellite observations rely upon the increase in solar radiation reflected by the Earth's surface and atmosphere caused by the additional backscattering from the aerosol. Since this increase may often be modest, such an approach is best suited for conditions where the surface has a low albedo that can be accurately estimated *a priori*, for example, observations over the ocean well away from sun glint (Durkee et al., 1991; Kaufman and Nakajima, 1993; Kaufman et al., 1990; Kaufman, 1993). Lidar holds great promise for sensing tropospheric aerosols and providing vertical soundings. Unfortunately little analysis of such data has yet been published.

3.5.2 Regional and Global Models and Budgets of Anthropogenic Aerosol

Models of the distribution and budget of sulphur and nitrogen species within regions of a few thousand square kilometres have been formulated in several parts of the world, mainly in the polluted regions in the Northern Hemisphere (for a recent review see Galloway and Rodhe, 1991). Such budgets indicate that dry and wet deposition, in roughly similar proportions, account for 50-75% of the removal of the anthropogenic emissions within the polluted regions (a few thousand km across), the rest being exported further away. About half of the anthropogenic SO_2 is estimated to be transformed to aerosol sulphate in the atmosphere.

Hemispheric or global scale models of the distribution of aerosol sulphate (and its precursors) have been formulated during the past few years (Erickson et al., 1991; Langner and Rodhe, 1991; Luecken et al., 1991; Tarrason and Iversen, 1992; Taylor and Penner, 1994). Figure 3.3 shows a model calculation of the annual mean distribution of the vertically integrated anthropogenic burden of sulphate aerosol, from Langner and Rodhe (1991). Such models indicate that industrial emissions are having a dramatic impact on the sulphate aerosol concentrations, not only within the industrialised regions, but over a large part of the Northern Hemisphere.

Figure 3.4 depicts the estimated division of global sulphur fluxes through the atmosphere between the major oxidation and deposition pathways (Langner et al., 1992). Note the important role played by oxidation in clouds for producing aerosol sulphate.

To our knowledge no attempt has yet been made to model the global distribution of aerosols in general, including sources and sinks of primary and secondary aerosols and their precursors. A different approach was used by d'Almeida et al. (1991) who estimated, directly from observations, a global distribution of 11 different

Aerosols

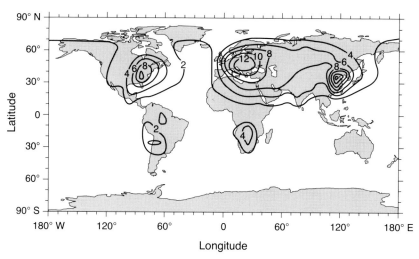

Figure 3.3: Calculated distribution of the vertically integrated amount of sulphate aerosol due to anthropogenic emissions only (from Langner and Rodhe, 1991: slow oxidation case). Values are annual averages expressed as mg sulphate per square metre and the maxima are associated with main industrialised regions. No comparable results are available for aerosol from biomass combustion.

types of aerosols based on a combination of observations and modelling, although owing to the limited database the representativeness of the observations is doubtful.

3.5.3 Measured Trends

The growth of anthropogenic aerosol *sources* over the past 150 years is indisputable, but direct evidence for an increase in aerosol *concentrations* in the atmosphere is scarce since, owing to the large variability of aerosol concentrations in space and time, secular time trends are much more difficult to observe than for the long-lived and relatively evenly distributed greenhouse gases.

Measurements of optical depth at some remote mountain top sites over several decades have not shown any significant trends (Ellis and Pueschel, 1971; Roosen et al., 1973; Charlson, 1988) due to the limited period of the record. No obvious trend in atmospheric turbidity was observed by Helmes and Jaenicke (1988) in their analysis of sunshine records and cloudiness over the past two decades. There is, however, a clear change evident in some statistical analyses of aerosol or haze occurrence in more directly polluted areas (Yamamoto et al., 1971; Husar et al., 1979, 1981). For example, Husar and Patterson (1987) document a considerable increase over the past two decades. A trend of increasing aerosol concentration has

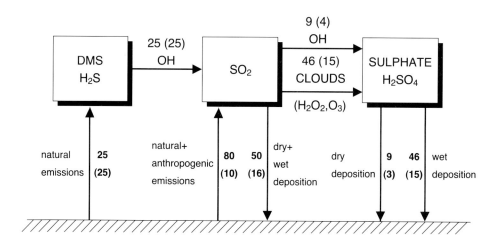

Figure 3.4: Schematic representation of fluxes of atmospheric sulphur species (excluding sea salts and soil dust) in different parts of the sulphur cycle. Numbers represent fluxes in TgS/yr. Natural (pre-industrial) fluxes are shown in parentheses (from Langner et al., 1992). Much of the sulphate aerosol results from sulphur dioxide emissions which have been oxidised by in-cloud processes.

also been demonstrated from balloon measurements by Hofmann (1993). However, Kaminski and Winkler (1988) show that as a result of control measures, the contribution of sulphate to the aerosol mass is decreasing in some parts of Europe although the total mass is increasing.

Field measurements of aerosol composition in remote regions also show the influence of anthropogenic emissions in areas far from continental sources (Lawson and Winchester, 1979; Andreae *et al.*, 1984; Warneck, 1988; Savoie and Prospero, 1989; Church *et al.*, 1991; Duce *et al.*, 1991; Losno *et al.*, 1992). Satellite measurements of light-scattering over the North Atlantic have shown that haze-forming aerosol particles can be transported for great distances over the oceans and still produce an easily detectable impact on the scattering properties of the atmosphere (Fraser *et al.*, 1984). Based on aircraft measurements, Andreae *et al.* (1988) showed that mid-tropospheric sulphate concentrations over the North Pacific off the west coast of the United States were dominated by long-range transport from Asia. In addition, ground-based observations suggest that long-range transport of Asian dust is seasonally dependent.

The increase in CCN above natural levels due to anthropogenic emissions is well documented in industrialised regions (Warner and Twomey, 1967; Hobbs *et al.*, 1970; Braham, 1974; Schmidt, 1974; Twomey *et al.*, 1978). That elevated sulphate levels over the North Atlantic are correlated with an increase in CCN is suggested by the data of Hoppel (1979), who finds that CCN concentrations over the North Atlantic are typically three or more times greater than over the Southern Hemisphere oceans (Twomey *et al.*, 1984). A pronounced upward trend (~10%/yr) in condensation nuclei (particles larger than 0.01 µm) over the past eight years at a site in Antarctica has been reported, although the cause has not been determined and the effect on CCN concentrations is unclear. It is possible that these changes, and small decreases in aerosol concentration at the South Pole (Samson *et al.*, 1990) may reflect changing dynamical factors. At Cape Grim, Tasmania (41°S), the only site with an extended record, Gras (1994) reported a significant decrease in CCN concentrations of around 3%/yr during 1981-91 while the concentration of condensation nuclei (CN) increased by 1.2%/yr. Hobbs *et al.* (1970) have also shown that ice nuclei may be transported long distances from their sources. This demonstrates the complexity of interactions between aerosol sources, sinks and transport processes.

Soot, a tracer of combustion-derived aerosols, has also been found at significant concentrations over many remote marine and continental areas (Heintzenberg, 1982; Andreae, 1983; Andreae *et al.*, 1984; Clarke, 1989; Ogren and Charlson, 1983; Warren and Clarke, 1990; *Hansen *et al.*, 1990). Clarke and Charlson (1985) cited the presence of light-absorbing material in the mid-tropospheric aerosol at Mauna Loa, Hawaii, as evidence for the long-range transport of soot and other anthropogenic aerosols, which results in a pervasive anthropogenic haze in the Northern Hemisphere. The transport of carbon and sulphate aerosols from mid-latitude sources in Eurasia results in the formation of Arctic haze observed at high latitudes in winter (Stonehouse, 1986).

Strong evidence for a large-scale increase in the atmospheric burden of anthropogenic aerosols can also be found in the record of atmospheric composition preserved in glacier ice cores. In the Greenland ice sheet, non-sea salt sulphate (and nitrate) show a pronounced increase (from about 20 to over 100 ng/g in the case of sulphate) over the past century (Figure 3.5; Mayewski *et al.*, 1986, 1990). Similar data from Antarctica show no such increase (Legrand and Delmas, 1987). Lead found in snow layers of the Greenland ice sheet is a good tracer of anthropogenic aerosol and records a 200-fold increase in lead concentrations since the beginning of the industrial era. However, there has been a reduction by a factor of 7.5 over the last 20 years (Boutron *et al.*, 1991) reflecting both the introduction of lead-free fuels and the rapid response of aerosol to emission controls.

3.6 Stratospheric Aerosols

3.6.1 Sources, Sinks and Lifetimes

The stratospheric aerosol record reveals at least three components: episodic volcanic enhancements; polar

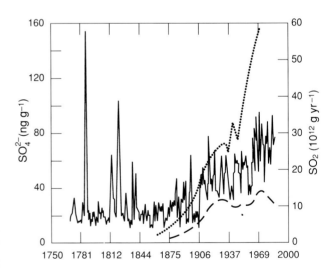

Figure 3.5: Concentration of sulphate in Greenland ice corresponding to the past 200 years (from Mayewski *et al.*, 1990). The dotted and dashed curves show the anthropogenic SO_2 emissions worldwide and over the USA, respectively (from Andreae 1993). While the secular trend is of anthropogenic origin, the larger spikes result from natural (volcanic) emissions.

stratospheric clouds (PSCs) and clouds just above the tropical tropopause. A background concentration of sulphuric acid droplets also appears to be present. At normal stratospheric temperatures, aerosols are most likely supercooled solution droplets of $H_2SO_4.H_2O$, with an acid weight fraction of 55 to 80%. The primary source of stratospheric aerosols is volcanic eruptions that are strong enough to inject SO_2 (sometimes H_2S) buoyantly into the stratosphere. After an eruption, and following conversion to H_2SO_4 (gas) and subsequently to sulphuric acid aerosols, aerosol loading decreases with a half life of 6 to 9 months, although this appears quite variable with altitude and latitude. The decay is most likely due to a combination of sedimentation, subsidence and exchange through tropopause folds. The net effect of this post-volcanic dispersion and natural cleansing is a greatly enhanced aerosol concentration in the upper troposphere after a major eruption, especially poleward of about 30° latitude. Except immediately after an eruption, stratospheric aerosol droplets tend to be concentrated into three distinct latitudinal bands, one over the equatorial region (between 30°N and 30°S) and the others over each high latitude region (50° to 90°N and S). Following a low latitude eruption, aerosol is dispersed into both hemispheres, whereas following a mid-to-high latitude eruption, aerosols tend to stay primarily in the hemisphere of the eruption. Potential sources of a background aerosol component include carbonyl sulphide from the oceans, low level SO_2 emissions from volcanoes, and various anthropogenic sources, including industrial and aircraft emissions. Measurements of aerosols in the upper troposphere and lower stratosphere using balloon sondes, commencing in 1960 (Hofmann, 1991) suggest a 5%/year increase in non-sea-salt sulphate that might be attributed to jet aircraft emissions. If the concentrations increase with anticipated increases in air traffic they could have significant climatic implications, and furthermore water vapour from aircraft emissions may increase the numbers of atmospheric ice particles.

Stratospheric aerosol loading in 1979 was approximately 0.5 Tg of sulphate, thought to be representative of background aerosol conditions. The present status of the aerosol is one of enhancement due to the June 1991 eruption of Mt. Pinatubo (15.1°N, 120.4°E), which produced on the order of 30 Tg of new aerosol in the stratosphere, about three times that of the 1982 eruption of El Chichon. The Mt. Pinatubo perturbation appears to be the largest of the century, perhaps the largest since the 1883 eruption of Krakatoa. By early 1993, stratospheric loading had decreased to approximately 13 Tg, about equal to the peak loading values after El Chichon, although the aerosol optical depth associated with the eruption of Mt. Pinatubo was comparable with that following El Chichon due to the differing aerosol distribution.

3.6.2 Size Distributions, Compositions and Optical Depths

Stratospheric aerosol sizes range from hundredths of a micrometre to several micrometres. Although there is some variability, especially just after a volcanic eruption, two log normal size distributions of spherical, submicrometre particles appears to be an apt description of the aerosol. Just after an eruption, the previously monomodal size distribution becomes bimodal, and some particles are non-spherical because of the addition of crustal material. While there are relatively few measurements of the size distribution, the volcanically perturbed/climatically important aerosol is currently described as having a bimodal number distribution. The smaller mode has a geometric mean diameter by number of 0.1 μm and a concentration of 10 cm^{-3}, and the larger mode is characterised by a mean diameter of 1 μm, and a concentration of 1 cm^{-3} (Hofmann, 1993, personal communication). Given this size distribution, the larger mode accounts for almost all of the mass as well as the optical effects. In terms of optical extinction, stratospheric aerosols of this sort demonstrate very little light absorption at solar wavelengths, with a scattering coefficient around 1-2 × 10^{-5} m^{-1} at 550 nm. For aerosol layers of several to perhaps 10 km depth, this results in optical depths of 0.1 to 0.2, as indeed was observed in 1992 over much of the globe. This is a large excursion compared with < 0.01 for the volcanically unperturbed case. Figure 3.6 from Dutton and Christy (1992) shows the observed optical depth at Mauna Loa, Hawaii illustrating the magnitude and systematic decay following both the El Chichon (1982) and Mt. Pinatubo (1991) eruptions. As is indicated in Chapter 4, there is evidence of significant warming of the stratosphere following both eruptions.

3.6.3 Interaction with the Troposphere

Just as stratospheric aerosol is derived from substances transported upward through the troposphere, the sink for it is transport back into the troposphere via sedimentation, subsidence and exchange through tropopause folds and turbulence in the vicinity of jet streams. This flux of aerosol material, particularly sulphates, directly into the upper troposphere represents a substantial local source of aerosol. The above estimate for the Mt. Pinatubo aerosol of 30 Tg SO_4^{2-}, decaying with a half life of 6-9 months, can be compared to estimates of the current natural non-volcanic and anthropogenic global annual fluxes of 30 and 140 Tg of SO_4^{2-}, respectively. Thus, the extra flux of sulphate mass represents a short-lived addition to the tropospheric sulphate burden that is substantial on a global basis and may be a major contribution at locations remote from anthropogenic sources. On the other hand, the large particle size of most of the mass, increased substantially by

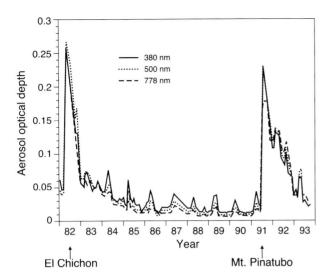

Figure 3.6: Variation of aerosol optical depth following the Mt. Pinatubo and El Chichon eruptions (from Dutton and Christy, 1992), showing the removal of aerosol over several years following the eruptions.

the higher humidity of the troposphere makes these particles substantially less important per unit mass to optical effects than ordinary accumulation mode sulphate, by around a factor of five. As a result, this sulphate aerosol does not substantially add to the global mean negative radiative forcing of the tropospheric aerosol and its effects are limited mainly to the time when it is in the stratosphere.

Of potentially larger importance are the local effects of this aerosol in the upper troposphere which normally has very low aerosol number and mass concentrations. The possibility exists that stratospheric aerosol particles injected at very low temperatures near the tropopause could act as nuclei of formation of cirrus cloud ice particles. More nuclei in the upper troposphere owing to volcanic eruptions and effects of aircraft would probably mean more but smaller cirrus particles, a higher solar albedo and, likely, decreased infrared absorption but this has not been quantified (see Chapter 2 for a discussion of the chemical effects on stratospheric ozone). In addition to aerosol from the stratosphere, the upper troposphere may be influenced significantly by particles and water vapour from high altitude commercial aircraft, although this has not been properly quantified.

3.7 Optical Properties of Aerosols Determining the Direct Radiative Effect

The simplest set of parameters describing the direct interaction of aerosol particles with solar radiation includes the aerosol optical depth (δ), single-scatter albedo (Ω), and asymmetry parameter (g), all as a function of wavelength, λ in the solar wavelength range. The aerosol optical depth is the vertical integral of the aerosol extinction coefficient, σ_e (= $\sigma_{sp} + \sigma_{ap}$, where σ_{sp} is the aerosol light scattering coefficient and σ_{ap} is the aerosol light absorption coefficient). The single-scatter albedo is a measure of the relative importance of scattering and absorption by the particles, and is defined as $\Omega = \sigma_{sp}/\sigma_e$. The extinction coefficient and its components are often approximated as being proportional to $\lambda^{\text{å}}$ where å is the Ångstrom exponent. The amount of light scattered through some angle ϕ from the incident beam is described by the angular scattering phase function $\beta(\phi)$. Simple parametrizations of $\beta(\phi)$ are the ratio of the hemispheric back scatter to the total scatter, R, or the asymmetry parameter which ranges from -1 (complete backscatter) to +1 (complete forward scatter).

3.7.1 Spatial Scales

To quantify the direct radiative forcing due to anthropogenic aerosol, it is necessary to define the optical properties on two spatial scales:

- local, microphysical properties,
- column integrated properties.

In-situ observations at the surface are potentially influenced strongly by source and removal processes at or near the surface, and measurements of the entire air column above are needed to evaluate the representativeness of the surface-based observations. Aircraft observations play an important role here, but remote sensing of the aerosol from ground level also provides important information. Table 3.3 summarises the range of optical properties observed in the troposphere along with other relevant integral properties. While the values that are currently available can be, and have been, used to estimate the magnitude of radiative forcing by aerosol particles, uncertainties will remain without an integrated set of simultaneous observations of the optical, chemical, and microphysical properties of the particles in key locations.

3.7.2 Local Optical Properties (Two Approaches)

Two complementary approaches exist for determining the local optical properties that are needed for calculating radiative forcing:

- Measure the key properties directly at representative locations and extrapolate to other locations (e.g., Charlson *et al.*, 1991);
- Measure the size distribution and the chemical composition, make certain assumptions regarding particle shape, morphology and distribution of properties with size, then calculate the optical properties (e.g., Kiehl and Briegleb, 1993).

Table 3.3: Typical range of observed integral properties of lower tropospheric aerosols (optical properties for 500-550 nm wavelength at low relative humidity). Entries indicate the range of observations.

Parameter	Polluted continental air (non-urban)	Clean continental air (minimal anthro-pogenic perturbation)	Clean marine air (negligible anthro-pogenic perturbation)
Optical depth, δ	0.2-0.8	0.02-0.1	0.05-0.1
Single-scatter albedo, Ω	0.8-0.95	0.9-0.95	close to 1
Back/total scatter, R	0.1-0.2	Not available	0.15
Light scattering coefficient, σ_{sp} (Mm^{-1})	50-300	5-30	5-20
Light absorption coefficient, σ_{ap} (Mm^{-1})	5-50	1-10	0.01-0.05
Submicrometre mass concentration (μg m^{-3})	5-50	1-10	1-5
CCN number concentration, 0.7-1% supersaturation (cm^{-3})	1,000-5000	100-1,000	10 - 200
Ångstrom exponent, å	1-2	1-2	1.5 - 2.1

Note: Mm = 10^6 m

For purposes of coupling radiative transfer models to chemical models (which calculate aerosol species mass concentrations) it is necessary to relate the scattering and absorption coefficients to measures of the aerosol concentration and composition. Three quantities are in common use to relate measurements of the scattering and absorption characteristics of an aerosol to its fine particle component (subscript f denotes fine particles, subscript i denotes chemical species).

(i) the fine particle scattering coefficient,

$$\sigma_{spf} \cong \alpha_f \cdot m_f \quad (3.1)$$

where m_f is the mass concentration of fine particles (defined as having diameters less than 1 μm) and α_f is the scattering efficiency. Typically, $\alpha_f \cong 3$ m^{2}g^{-1} at relative humidities less than 40% but is dependent on the size distribution and molecular chemical composition.

(ii) the absorption coefficient for visible light (usually dominated by elemental or black carbon, C(0))

$$\sigma_{ap} \cong \alpha_{aC} \cdot m_{C(0)} \quad (3.2)$$

where $m_{C(0)}$ is the mass of elemental or black carbon and α_{aC} is its absorption efficiency.

(iii) the scattering coefficient for all particles summed over all chemical species present in the aerosol,

$$\sigma_{sp} \cong \Sigma \alpha_i m_i \text{ or } \sigma_{spf} = \Sigma \alpha_{if} m_{if} \quad (3.3)$$

Here, the assumption is made that σ_{spf} and σ_{ap} are independent and that the scattering by different chemical species also are independent and additive. In all instances, α_f and α_{aC} are evaluated by simultaneous measurement of σ_{sp}, σ_{ap}, m_f and the fine particle mass concentration of individual chemical species i, m_{if}. Very substantial amounts of data of this sort have already been acquired, but not on a global scale. Most data were acquired at the Earth's surface and few exist from airborne platforms.

The alternative to direct measurement of the optical properties and scattering and absorption efficiencies is to calculate the optical properties from measured or assumed particle size distributions, compositions and concentrations. Results obtained using these approaches, however, are almost wholly dependent on assumptions, many of which are not based on observations. Nonetheless, it is necessary to show that both of the approaches yield the same answer in order to demonstrate internal consistency and to define the sensitivity of the measured or calculated optical properties to input quantities, whether they are measured or assumed. Blanchet (1989) has demonstrated this consistency for a limited number of cases. Much of the progress in estimating radiative forcing by anthropogenic aerosol has resulted from the use of both approaches simultaneously.

It is necessary to recognise that all of the factors controlling the optical properties are functions of particle size and wavelength and therefore so are the optical properties themselves. Figure 3.7 is an example of the distribution over particle size of mass scattering efficiency at 500 nm. Having measured the size distribution and knowing the refractive index of the aerosol one can thus determine the distribution of scattering efficiency by particle size, the integral of which is the scattering efficiency used in radiative transfer calculations.

3.7.2.1 Dependence of local integral optical properties on size and composition

It is convenient to subdivide the aerosol properties by region, size class and by chemical species, insofar as that is possible, and also to compare calculated and measured properties. While very substantial amounts of data exist, primarily at polluted or industrial-region surface locations, there have been almost no efforts to measure simultaneously all of the quantities. Thus, it is not yet possible to examine the data sets for internal consistency. Also, because of a lack of standardisation, for example, of selection of particle-size ranges, it is difficult to compare data sets.

White (1990) determined the contributions of individual compounds to σ_{sp} and σ_{ap} in various urban and non-urban locations in the USA. Table 3.4 summarises the contributions of scattering by fine and coarse particles and absorption by particles to the aerosol component of the extinction coefficient. While relative humidity (RH) was not controlled or measured in those measurements, it is likely that a temperature increase of a few degrees occurred during the measurement, yielding RH values in the instruments that were less than ambient. Note that the single scattering albedo values are in the range 0.8 to 0.95 in industrialised regions; the small amount of absorption is mainly attributed to soot. Table 3.4 generally supports the notion that scattering by fine particles dominates the aerosol contribution to extinction. It further suggests that scattering dominates over absorption, especially when the increased scattering at ambient RH is taken into account, and that the main effect of the combined aerosol is therefore cooling.

As expected, the dominant chemical species in the industrialised eastern USA is sulphate which, along with "organic" material, comprises about 84% of the mass when the sulphate compound is assumed to be $(NH_4)_2SO_4$. The residual material includes fine soil dust, water, nitrates and other, unidentified, materials. Caution is needed in interpreting these data because the chemical analyses were not often complete. In the western USA, SO_4^{2-} accounts for a smaller fraction (about 35%) and organics and soil dust are relatively more important. White (1990) also used data from simultaneous observations of sulphur in a multiple regression analysis against measured light scattering at low relative humidity which gave a squared correlation coefficient of 0.948. This clearly implicates sulphates and organics as consistently dominant contributors to σ_{sp}, at least over the USA.

While the similarity of chemical inputs and physical processes in other locations (e.g., Europe and Asia) and the high correlation coefficient suggest that these USA data may be applicable for models over the globe, there are currently few data from outside the USA against which to verify this idea. It is important to discover if source factors are sufficiently different in other locations or aloft to cause significant departures from White's (1990) results. Recent results reported by Nyeki et al. (1994) tend to support the wider application of White's results.

Table 3.5 briefly summarises the present estimates of σ_{sp}, hemispheric backscatter fraction and fractional increase of σ_{sp}, due to hygroscopic growth to 80% RH, $f(RH\ 80)$.

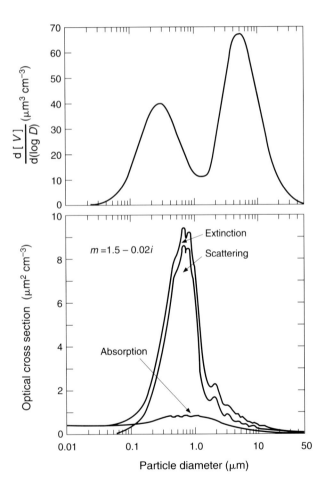

Figure 3.7: Aerosol volume size distribution and the calculated extinction, scattering and absorption cross sections per unit aerosol volume at a wavelength of 500 nm, calculated by Covert et al. (1980), using Mie theory and an assumed refractive index of m = 1.5 - 0.02i. Note that the major contribution to aerosol scattering is not associated with the part of the size distribution contributing most to the aerosol mass loading.

3.7.3 Column Properties

Integration of the local aerosol properties over altitude yields the key column properties used in radiative transfer models. The most important of these properties is the aerosol optical depth. Global mean optical depth (Table

Table 3.4: Percentage contributions to aerosol extinction coefficient in rural areas of the Eastern and Western USA, from White (1990).

Locale	$\dfrac{\sigma_{sp}\,(fine)}{\sigma_e}$ %	$\dfrac{\sigma_{sp}\,(coarse)}{\sigma_e}$ %	$\dfrac{\sigma_{ap}}{\sigma_e}$ %	Single scattering albedo of fine particles (RH < 40%)
Eastern USA	91	3	6	0.94
Western USA	69	17	14	0.86

Notes: a   Fine/coarse separation at diameter ≈ 2.5 μm.
        b   RH low, but not measured.
        c   All data corrected to λ ≈ 525 nm.

Table 3.5: Observed integral optical properties of anthropogenic sulphate, organic carbon and soot and for the fine aerosol mass at 525 nm (adapted from Penner et al., 1994). See Section 3.7 for a definition of the terms used.

Integral optical property	SO_4^{2-}	C_{ORG}	Soot	Fine aerosol mass
σ_{sp} m^{2}g^{-1}	5 (3.6 - 7)	5 (3.0 - 7)	3	3 (2 - 4)
σ_{sp} m^{2}g^{-1}	0	0	10 (8 - 12)	– (0 - 10)
R	0.15 (0.12 - 0.19)	0.15 (0.11 - 0.2)	–	0.15 (0.11 - 0.2)
f(RH = 80%)	1.7 (1.4 - 4.0)	1.7 (1.4 - 4.0)	(0-1.7)	1.7 (0 - 4.0)
å	–	–	–	1 - 2

Note: The indicated range (in brackets) represents the approximate standard deviation of the available data. For sulphate (SO_4^{2-}), and organic carbon, C_{ORG}, σ_{sp} is the effective scattering efficiency given as the cross section per unit mass of the indicated species. This is higher than the ratio of scattering to fine aerosol mass (last column), because gas-to-particle conversion of anthropogenic SO_2 combines sulphur with oxygen, hydrogen and/or ammonia to form an acidic and hygroscopic aerosol. The radiative effect of adding particulate sulphate mass to the atmosphere thus is a combination of scattering by the sulphate mass itself plus the additional mass of ammonia and/or water. Hegg et al. (1993) suggest that a value of 3 m^2 g^{-1} is appropriate for sulphate alone, based on assumptions that all submicron mass scatters with equal efficiency and that the sulphate ionic mass concentration is independent of the mass of other aerosol substances.

3.6) can be estimated using simple geochemical mass balances (e.g., Andreae, 1995). The largest contributions to the optical depth from natural sources are soil dust, sulphates and organic matter from volatile organic compounds. These values are calculated using best estimates in Table 3.1, the uncertainty associated with each parameter reflects the spread of fluxes shown in Table 3.1. Anthropogenic sulphate and material from biomass burning make similar contributions to the optical depth even though the source strengths and burdens are smaller than those of the natural aerosol.

Large amounts of data exist on measured optical depths, some of which (having been largely acquired by the use of hand-held sun photometers) is of unknown or dubious quality. Figure 3.8, showing atmospheric optical depths over the USA, derived from turbidity measurements, clearly demonstrates the region of high anthropogenic aerosol content. Such observations can also show the impact of emission control measures at specific locations (Kaminski and Winkler, 1988). Satellite observation can provide information on the global distribution of upwelling radiation but the derivation of aerosol optical depth from such information is complex and requires many, unproven, assumptions.

3.8 Causes of Uncertainty in Calculating Direct Forcing by Aerosols

3.8.1 Local vs. Regional vs. Global Models and Effects

Prior to about 1990, several disparate views prevailed as to the direct radiative effects of aerosol. Anthropogenic physical effects were viewed as local perturbations on a global background aerosol. Descriptions of the latter were given as temporally static (see for example Toon and Pollack, 1976; Coakley et al., 1983). On the other hand, global chemical models, e.g., of the sulphur cycle, that included the secular trend of anthropogenic aerosol, did produce useful simulations of acidic precipitation but failed to include a coupling to aerosol physics or physical properties (see e.g., Rodhe and Isaksen, 1980). One effort

Table 3.6: *Source strength, atmospheric burden and optical depth at 550 nm wavelength due to the various types of aerosols hydrated at 70 - 80% RH (after Andreae, 1994).*

Source	Emissions Tg/yr	Lifetime days	Column burden mg/m²	Mass extinction efficiency m²/g	Optical depth
Natural					
Primary					
Soil dust (mineral aerosol)	1500	4	32.2	0.7	0.023
Sea salt	1300	1	7.0	0.4	0.003
Volcanic dust	33	4	0.7	2.0	0.001
Biological debris	50	4	1.1	2.0	0.002
Secondary					
Sulphates from biogenic gases	90	5	2.4	5.1	
Sulphates from volcanic SO_2	12	5	0.3	5.1	0.014
Organic matter from biogenic NMHC[1]	55	7	2.1	5.1	0.011
Nitrates from NO_x	22	4	0.5	2.0	0.001
Total natural	3060		46		0.055
Anthropogenic					
Primary					
Industrial dust etc.	100	4	2.1	2.0	0.004
Black carbon (soot and charcoal)	20	6	0.6	10.0	0.006
Secondary					
Sulphates from SO_2	140	5	3.8	7.1	0.019
Biomass burning (w/o black carbon)	80	8	3.4	8.0	0.017
Nitrates from NO_x	36	4	0.8	2.0	0.002
Organic from anthropogenic NMHC[1]	10	7	0.4	8.0	0.002
Total anthropogenic	390		11.1		0.050
TOTAL	3450		57		0.105
Anthropogenic fraction	11%		19%		48%

[1] NMHC: Non-methane hydrocarbons

did include a rudimentary regional scale chemical model along with regional scale calculation of loss of solar irradiance (Ball and Robinson, 1982). Global two-dimensional models treating aerosol formation from sulphate of anthropogenic origin, as well as the modification of cloud properties (Rodhe and Isaksen, 1980: Rehkopf, 1984) showed a strong anthropogenic impact on fine particles.

The introduction around 1990 of a three-dimensional global model of the sulphur cycle (Zimmermann, 1987; Langner and Rodhe, 1991) along with great improvements in the quantification of both natural and anthropogenic gaseous sulphate aerosol precursors made it possible to develop rudimentary models that coupled the geographical and altitudinal chemical fields to local, column, hemispheric and global radiative effects. For estimation of the direct forcing, the key variable for achieving this coupling is the specific scattering efficiency, $\alpha(SO_4^{2-})$ Table 3.5 shows values of the key parameters for different aerosol species and for all fine particles.

3.8.2 Coupled Chemical/Radiative Transfer Models

A hierarchy of models has evolved recently which includes the necessary coupling of chemical processes and properties with radiative effects. The simplest of these serves to illustrate and define the controlling variables. Local chemical concentrations and integral properties are related to column integral properties over altitude, and then to the column burden of fine-particle SO_4^{2-} (e.g., derived from a chemical model, or integrated from a measured vertical profile of SO_4^{2-}). The scattering coefficient is treated similarly. In the first crude estimates (Charlson *et al.*, 1990), the Northern Hemisphere mean burden of anthropogenic SO_4^{2-} was estimated to be 6.6×10^{-3} g m^{-2}; from measured values of $\alpha(SO_4^{2-})$ and f(RH), the Northern Hemisphere mean optical depth due to anthropogenic SO_4^{2-} (δSO_4^{2-}) was estimated to be 0.066.

To connect this optical depth estimate to radiative forcing, a simple radiative transfer model was used, deliberately disregarding the influence of aerosol in cloudy

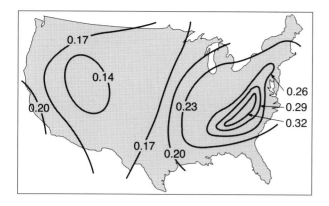

Figure 3.8: Annual average values of atmospheric optical depth over the USA. Redrawn from Flowers *et al.* (1969) and showing a pronounced maximum in the industrialised regions.

areas. For the simplest case, the loss of solar irradiance from the ground, L, becomes

$$L \sim (1 - A_c) R (2 \delta_{SO_4^{2-}}) \sim 0.8\% \qquad (3.4)$$

where A_c is the global average fractional cloud cover and $R \approx 0.15$ is the measured fraction of σ_{sp} in the backward hemisphere. The factor of two accounts for the average of the secant of zenith angle over the sunlit hemisphere. Thus, if solar irradiance incident upon the aerosol layer is 200 Wm^{-2}, a hypothetical loss to space of 1.6 Wm^{-2} would exist (Charlson *et al.*, 1990). This value is rather high for a variety of reasons and values based on more detailed model calculations are described in Chapter 4.

Much more sophisticated models have evolved; however, all of them depend on a geochemical mass balance model for the column burden of sulphate, $B_{SO_4^{2-}}$. Some of the models still disregard microphysical considerations and base the values of α, R and Ω on measurements (Charlson *et al.*, 1991). Others take a different approach and calculate the optical properties from assumed size distributions (Kiehl and Briegleb, 1993). The two approaches give significant differences in the value for the direct forcing. All of these models fail to account for correlations that influence the forcing, such as that between $B_{SO_4^{2-}}$ and cloud cover. Current, more sophisticated models, some of which are still in the developmental stage, can address such correlations with varying degrees of realism.

3.8.3 Causes of Uncertainty in Calculating the Direct Forcing by Anthropogenic Aerosol

It is necessary to recognise that besides the problem of correlations among controlling variables, there are both chemical and physical factors that contribute uncertainty to the estimates of forcing by anthropogenic aerosol. Again, using anthropogenic sulphate as an example (because it is better understood than other types of aerosol) the geographically averaged mean forcing can be used to identify the key variables (Charlson *et al.*, 1992) using a combination of a simple radiative transfer equation and a chemical box-model describing material balance:

$$\overline{\Delta F_R} \cong -0.5 F_T T^2 (1-A_c)(1-R_s)^2 \beta \alpha_{(SO_4^{2-})} f(RH)$$
$$\times Q_{SO_2} Y_{SO_4^{2-}} \tau_{SO_4^{2-}} / A \qquad (3.5)$$

$\overline{\Delta F_R}$ is the areal mean short-wave radiative forcing due to the aerosol in Wm^{-2}.

$\tfrac{1}{4} F_T$ is the global mean top of the atmosphere radiative flux in Wm^{-2}.

A_c is the fractional cloud cover.

T is the fraction of incident light transmitted by the atmosphere above the aerosol.

R_S is the albedo of the underlying surface.

β is the upward fraction of the radiation scattered by the aerosol, calculated from an aerosol model or measured values of hemispheric backscatter ratio, R.

$\alpha_{(SO_4^{2-})}$ is the scattering efficiency of fine-particle sulphate at a reference low (e.g., 30%) relative humidity, m^{2}g^{-1}.

$f(RH)$ accounts for the relative increase in scattering due to relative humidity above the reference value.

Q_{SO_2} is the source strength of anthropogenic SO$_2$ in gS/yr.

$Y_{SO_4^{2-}}$ is the fractional yield of emitted SO$_2$ that reacts to produce sulphate aerosol.

$\tau_{SO_4^{2-}}$ is the sulphate lifetime in the atmosphere in yrs.

A is the area of the geographical region under consideration in m^2.

This equation represents the simplest way of estimating uncertainty; it neglects, for example, direct radiative forcing in regions containing cloud. Table 3.7 lists the uncertainties in quantities used in the example rudimentary model calculations (the uncertainty factor is used here to define the upper and lower limits of the range by multiplication/division of the most probable value). For anthropogenic SO$_4^{2-}$ the total resultant uncertainty factor is estimated to be 2.2, not including consideration of spatial or temporal correlations. Thus the overall uncertainty is as yet not fully defined. The uncertainty factor for aerosol from biomass burning has been estimated by Penner *et al.* (1994), using similar techniques, as 2.7.

In contrast to greenhouse gases, atmospheric aerosol particles are relatively short-lived in the atmosphere. This property, together with the highly non-uniform geographical distribution of sources, results in a highly non-uniform geographical distribution of anthropogenic aerosols, which results in very different geographical patterns of radiative forcing by aerosols and greenhouse gases (see Chapter 4). Likewise, the aerosol radiative

Table 3.7: *Evaluation of uncertainties of global mean direct radiative forcing due to anthropogenic sulphate.*

Quantity	$1-A_c$	T	$1-R_s$	R	$\alpha_{(SO_4^{2-})}$	$f(RH)$	Q_{SO_2}	$Y_{SO_4^{2-}}$	$\tau_{SO_4^{2-}}$	$\overline{\Delta FR}$
Uncertainty factor	1.1	1.15	1.05	1.3	1.4	1.2	1.15	1.5	1.4	2.2

Note: Total uncertainty factor evaluated as $f_t = \exp[\Sigma(\log f_i)^2]^{1/2} = 2.2$.

forcing is largely a short-wave forcing dependent on sunlight and therefore exhibits different diurnal and seasonal patterns from long-wave greenhouse gas forcing. The pattern is described in more detail in Chapter 4 and raises the question as to whether global mean forcing which has proved useful when comparing the radiative forcing of different greenhouse gases is equally useful in describing the direct forcing due to anthropogenic aerosol.

3.9 Influence of Anthropogenic Aerosol on Clouds (Indirect Effect)

In addition to the direct aerosol radiative forcing effect, aerosol particles modify the properties of clouds which are known to be a significant component in the climate system and hence there exists a potential indirect aerosol effect. While the physical basis for such an effect is well established, attempts to quantify the effect are at present limited. It is demonstrated in this section that:

- The change in cloud droplet concentration is related to aerosol particle concentration and mass in a non-linear manner which is also dependent on cloud type and the chemical composition and age of the aerosol.
- Natural aerosol sources have a substantial impact on clouds and some of these sources are sensitive to climate, so allowing a climate feedback.
- The sensitivity of ice clouds to aerosol has not been quantified.
- The impact of aerosol on cloud cover and lifetime has been demonstrated although the climatic impact cannot be assessed at present.

3.9.1 Cloud Formation

In clouds, condensation occurs on particles which act as cloud condensation nuclei (CCN). The critical supersaturation at which a particle will act as a nucleus for drop formation depends on the size and chemical composition of the particle. As the air is cooled and the supersaturation increases, more nuclei are activated until there are sufficient droplets to achieve a balance between water which is condensed, and that made available by cooling. Subsequent cooling then results in increased droplet sizes, although the concentration remains more or less constant. Since the peak supersaturation in clouds seldom exceeds 0.05%, at which only aerosol particles with dry diameters larger than around 0.1 μm become activated, it is the concentration of aerosol particles larger than this radius that is important for the determination of the cloud microphysical properties.

The fact that increasing aerosol particle concentrations can affect the microphysical and radiative properties of clouds is seen by the formation of ship tracks in extensive sheets of layer cloud (e.g., Radke *et al.*, 1989, King *et al.*, 1993). The increased aerosol loadings due to the passage of the ship give rise to increases in the droplet concentration and to reduced droplet sizes, as seen in Figure 3.9. The relation between droplet concentration and aerosol concentration is complex, since it depends on the updrafts in the clouds, the degree of mixing between the cloud and its environment and the ability of the aerosol particles to act as CCN (for a more complete review, see Fouquart and Isaka, 1992). Initial cloud droplet concentration is determined by the CCN contained in the air prior to cloud formation and by the vertical velocity of the air; this process is well understood and confirmed by observations (Hudson and Rogers, 1986; Leaitch *et al.*, 1986). Further evolution of cloud droplet spectra depends on mixing and entrainment processes in the cloud layer. Observations (Austin *et al.*, 1985; Cooper, 1989; Brenguier,1990) show that some CCN contained in the entrained air can become activated, leading to complex droplet spectra. Since cloud reflectance is most sensitive to the droplet population near cloud top, mixing processes considerably degrade the ability of simple models to predict the influence of changes in aerosol concentration or cloud optical properties.

As a result of many observational studies (e.g., Leaitch *et al.*, 1986; Leaitch and Isaac, 1994; Martin *et al.*, 1994 and Raga and Jonas, 1993), empirical relationships have been derived between droplet concentration (and hence size) and aerosol number concentration below cloud base. These results show a correlation between aerosol particle number concentration and the cloud droplet concentration, although the form of the relationship is dependent on cloud type. Typically, an order of magnitude increase in the aerosol number concentration (in the diameter range 0.1 - 3.0 μm) results in a factor of 2 to 5 increase in droplet

Aerosols

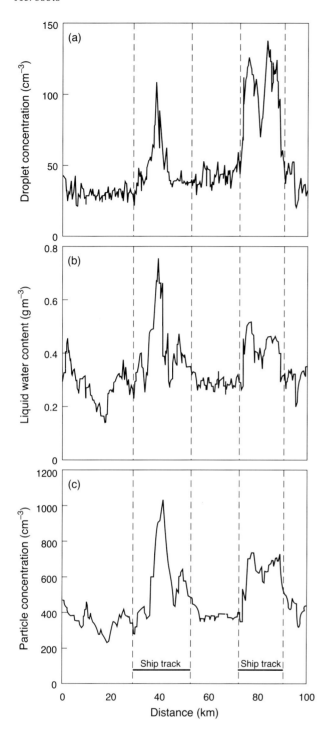

Figure 3.9: Observations of the changes in microphysical properties of stratocumulus clouds on a level flight through cloud in which ship tracks are present (from Radke *et al.*, 1989). The ship tracks contain high aerosol particle and cloud droplet concentrations leading to increased cloud water content.

concentration; it implies a decrease in the droplet radius at cloud top of 1.0 - 2.0 μm, depending on cloud thickness.

3.9.2 Cloud Concentration Nuclei (CCN) Population and Aerosol Sources

Aerosol particles which have the potential to act as CCN are produced by a variety of processes, some of which are subject to anthropogenic influence. The relation between the CCN activity spectrum and the aerosol mass loading is complex since, as aerosol ages, more of the mass resides in particles which can act as CCN at low supersaturations.

3.9.2.1 Sulphates and Nitrates

Although the oxidation of precursor gases is a minor source of total aerosol mass, the resulting particles can be very effective as CCN because they are hygroscopic and their small size enables them to be transported up to cloud height in large numbers. This is particularly the case for aerosols derived from SO_2 and NO_x emissions or from natural DMS emissions. Positive correlations between CCN and sulphate have been observed: the scatter in the data however shows that other factors influence this relationship (Nguyen *et al.*, 1983).

It may be inferred from the measurements of Hudson (1991) that growth of CCN is rapid in the case of anthropogenic CCN, but slower for natural CCN. Model calculations show that the time needed to transform DMS into CCN active at 0.2-0.3% supersaturation can be 1-7 days, depending on the oxidation mechanism assumed (Lin and Chameides, 1993; Raes, 1993). This difference is important since the time would be either smaller or larger than the time-scales of synoptic meteorological processes. Charlson *et al.* (1987) postulated a potential feedback mechanism since the DMS source might be temperature or radiation sensitive. Certain links of this mechanism have received observational support (Ayers and Gras, 1991; Hegg *et al.*, 1991) and support from calculations (Raes, 1993) showing a positive correlation between CCN and DMS emissions. Satellite observations by Durkee *et al.* (1991) show plumes of aerosol extending downwind from urban pollution sources and also from regions which would be expected to act as source regions for DMS. The analysis highlighted aerosol smaller than about 0.5 μm; if such aerosol is composed of sulphate particles, it would be expected that these plumes would be regions of high CCN concentration. The multi-year observational programme at Cape Grim has yielded a data set showing clear seasonal correlations between DMS, sub-micrometer sulphate aerosol, CCN and cloud optical depth (Ayers *et al.*, 1991, Ayers and Gras, 1991, Boers *et al.*, 1994). Nitrates produced by gaseous reactions are highly soluble in water and hence would be expected also to act as good CCN.

CCN may also be enhanced by the presence of gas phase nitric acid in heavily polluted air (Kulmala et al., 1993).

3.9.2.2 Organics

The type of organic aerosols produced by combustion are related to the composition of the fuel consumed. Hydrophilic properties of the combustion aerosol depend on the relative oxygen content of the fuel, which may be hydrogen rich fossil fuel or oxygen enriched terrestrial biofuels including fossil sources. Smoke particles produced from combustion of woody materials contain polar organic compounds (Simoneit et al., 1993) and are effective CCN. Hudson et al. (1991) show that while most of the particles in smoke produced by burning vegetation act as CCN, only 1-2% of the particles produced by burning jet fuel were active under identical conditions. This probably is a result of the former containing soluble material which improves the effectiveness of the particle to act as a CCN. Novakov and Penner (1993) have shown that organic condensates, some from biomass burning, can also play a role in the CCN budget, and may be at least as important as sulphate aerosols in some regions; Rogers et al. (1991) also suggest that biomass burning is a major source of CCN.

3.9.2.3 Soot

Soot produced by biomass burning will contain hydrophobic and hydrophilic components and its ability to form CCN will depend on the fuel consumed. However, the incorporation of soot into cloud droplets may reduce the albedo (Kaufman and Nakajima, 1993) owing to its strong absorbing properties although the small number of measurements made to date do not indicate a significant contribution by soot to short-wave absorption by clouds (Heintzenberg, 1988; Twohy et al., 1989; Charlson and Heintzenberg, 1995).

3.9.2.4 Mineral Dust

Mineral dust aerosol is usually formed in the coarse mode and such particles are generally insoluble. They are therefore less effective as CCN than many of the soluble particles. However, mixed particles formed from mineral dust with a soluble coating resulting from gas-particle conversion processes may be very effective as CCN. At the present time the importance of such particles as CCN is not known.

3.9.2.5 Sea Salt

Aerosol particles, composed largely of sodium chloride, are produced by the evaporation of droplets formed at the surface by bubble bursting or during the formation of spray droplets under strong wind conditions. Latham and Smith (1990) suggest that, due to the dependence of the activity spectrum of the sea salt aerosol on the wind speed, a potential climate feedback exists, although the relative importance of sea salt aerosol and sulphate from DMS in the marine atmosphere has not been fully quantified.

3.9.3 Cloud Radiative Properties

The effect of clouds on radiative transfer is dependent on the distribution of water between liquid and ice as well as on the cloud particle size distribution and shape. It is through its effect on these parameters that aerosol may indirectly influence radiative transfer.

3.9.3.1 Water Clouds

For visible wavelengths, scattering dominates the radiative transfer and, for the same water content, an increase in droplet concentration and reduction in the mean droplet size leads to an increased reflectivity, as shown in Figure 3.10. For a cloud layer with a typical thickness of 100-200 m, the change from a maritime to a continental droplet population will increase the reflectivity from less than 60% to more than 70%. Charlson et al. (1987) suggested that a 30% increase in the droplet concentration in marine stratocumulus could increase the planetary solar albedo by around 0.02. For such clouds having optical thicknesses in the range 5 to 50, characteristic of much low level stratus and stratocumulus cloud, the changes in albedo are only weakly dependent on cloud optical thickness and the solar zenith angle (Fouquart et al., 1990).

At infrared wavelengths, the main effect of water cloud is to absorb and emit radiation, scattering is generally unimportant. Since absorption and emission are dependent on the liquid water content and only weakly on the droplet size distribution, at least for cloud droplet radii less than about 10 μm, the effects of changes in the spectrum due to changes in CCN are small, except where the changes lead to an increase in cloud water content or lifetime. However, Kaufman and Nakajima (1993) have demonstrated that the interaction of aerosols from biomass burning with clouds may result in a large increase in absorption due to the presence of particles containing carbon.

3.9.3.2 Ice Clouds

As the formation of water clouds depends on the presence of CCN, so the formation of ice clouds often depends on the presence of particles which act as ice nuclei. In the absence of such nuclei, clouds would exist in the form of supercooled water at temperatures warmer than about −40°C. The presence of ice significantly modifies the

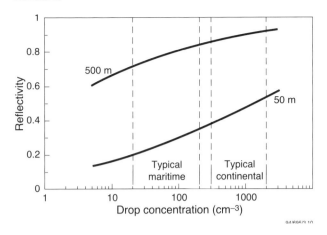

Figure 3.10: Calculated reflectance of horizontally uniform cloud layers of thickness 500 and 50 m, each with a water content of 1 g m^{-3}, as a function of assumed droplet concentration (from Bohren, 1987).

precipitation growth processes, often leading to more rapid growth of large particles.

Atmospheric aerosol particles may also be important in determining the radiative properties of cirrus. Unlike low level water clouds, the temperature difference between ice clouds and the surface is large, so they have a significant greenhouse effect which is sensitive to ice crystal concentration. The presence of small crystals, observed by Wielicki et al. (1990), and others, will also be sensitive to ice nucleus concentration; such particles have a significant effect on the albedo of cirrus. Quantification of these effects awaits further research.

Sassen et al. (1989) have reported the occurrence of very thin "sub-visible" cirrus with optical thicknesses less than 0.03 which have a measurable effect on the long-wave radiation budget. Other observations have also indicated that such clouds are composed of small ice particles. The occurrence and properties of such clouds will be particularly sensitive to aerosol in the upper troposphere resulting from volcanic eruptions as well as tropospheric sources such as aircraft emissions.

3.9.4 The Indirect Effect of Anthropogenic CCN on Cloud Albedo

The global radiative forcing due to the effect of anthropogenic aerosols on cloud albedo has not yet been calculated in a rigorous way. Estimates are based on the notion that anthropogenic emissions lead to an increase in the aerosol number concentration, and eventually to an increase in cloud droplet number concentrations. For a constant liquid water content, an increase in cloud droplets reduces their mean radius and results in an increase of the cloud albedo (Twomey, 1977). Grassl (1988) suggested that this could be more important than the direct aerosol effect. This phenomenon is readily demonstrated by the impact of ship emissions on the reflectivity of overlying clouds (see Figure 3.9 and Coakley et al., 1987). Applying this to marine stratiform clouds, which cover 25% of the Earth, Charlson et al. (1987) calculated that an increase of 30% in the droplet population would yield a global forcing of about -1.7 Wm^{-2}. Similar estimates have been obtained from sensitivity studies with general circulation models (Slingo, 1990). A study by Han et al. (1994), based on satellite data, showed that mean cloud droplet radii in the Northern Hemisphere are 0.8 μm smaller over land and 0.4 μm smaller over the ocean than in the Southern Hemisphere. If these differences were due only to excess anthropogenic aerosols, and applying this again to marine stratiform clouds, it would correspond to an increase in the number concentration of around 15% and a forcing of -0.8 Wm^{-2} in the Northern Hemisphere. Calculations by Jones et al. (1994) using a general circulation model suggest that the effect of anthropogenic aerosol is to reduce the effective droplet radius in low level clouds in the Northern Hemisphere by more than 2 μm compared with the Southern Hemisphere, and over continental regions by more than 1 μm compared with the Northern Hemisphere ocean regions.

The estimates above suggest that the climate system could react significantly to changes in global CCN or cloud droplet number concentrations of several tens of percent. There is abundant evidence that industrial emissions and biomass burning increase CCN and cloud droplet concentrations by up to several orders of magnitude (Leaitch et al., 1986, Pueschel et al., 1986, Kaufmann and Nakajima, 1993), suggesting that the indirect effect might indeed be very important. However, questions remain concerning the geographical extent of these aerosol increases and how they might eventually contribute to the global burden. Although it is possible to calculate the change in aerosol mass during long-range transport, it is still difficult to do the same for the aerosol and CCN number concentration; coagulation and coalescence in clouds will reduce the aerosol number concentration but growth within cloud increases the number of CCN at lower supersaturations for subsequent clouds. The lifetime of CCN number concentration is therefore expected to be shorter than that of the associated aerosol mass, which would limit the geographical extent of the anthropogenic indirect effect.

The different dynamics of aerosol mass and aerosol number concentration is the main reason why both the direct and indirect effect cannot be related in a linear way to the source strength of pollutants. For a constant aerosol mass, changes in the size distribution of the aerosol can still result in changes in the optical and activation properties of the aerosol. This dependence on the size distribution is critical when determining the number of

CCN, since a single aerosol particle will either be activated or not. The dependence is somewhat less critical when determining the scattering coefficient since a single particle always scatters a range of wavelengths. Dealing with aerosol dynamics in global circulation models is one of the major difficulties which prevents an *ab initio* calculation of the radiative forcing by aerosols and involves the use of incomplete and unverified physical models.

The measurements reveal that the CCN-sulphate mass relationship is sub-linear, so that a doubling in the sulphate mass does not double the population of CCN. Two factors that may reduce the indirect effect as the aerosol concentration is increased are:

i) the availability of water vapour which, at high CCN concentrations, limits the fraction of CCN that is activated to cloud droplets,

ii) the saturation in the cloud reflectivity for increasing cloud droplet concentrations (see Figure 3.10).

The indirect effect is expected to be largest for emissions into unpolluted air (Twomey, 1991) while the direct effect is expected to keep increasing with increased emissions.

3.9.5 Precipitation Formation and Cloud Lifetime

The growth of cloud particles to form precipitation-sized particles is a complex and poorly quantified process. The speed of growth depends on the size spectrum of the cloud drops and is slower if the cloud is composed of large numbers of relatively small droplets. Consistent with this, precipitation is often observed to fall from low level maritime clouds with a water content of less than 0.1 g m^{-3}, whereas precipitation is seldom observed from continental clouds with water contents below 0.3 g m^{-3} (e.g., Hobbs and Rangno, 1985).

The role of ice nuclei in the precipitation process is even less clear than that of CCN. Although there is evidence that the initiation of the ice phase depends on the presence of such particles (Vali, 1985), secondary processes are also important. These secondary processes are sensitive to the drop size distribution (Mossop, 1978) and hence potentially sensitive to CCN concentrations. Albrecht (1989) and Liou and Ou (1989) pointed out that since increasing aerosol concentration would lead to reduced droplet sizes, it would also lead, if the liquid water content remained constant, to reduced precipitation efficiency and increased cloud lifetime. It is of note that the ship track observations of Radke *et al.* (1989) show drizzle formation is inhibited by the presence of high concentrations of aerosol. Fravalo *et al.* (1981) and Turton and Nicholls (1987) showed that the diurnal variation of marine stratocumulus was dependent on the liquid water content and that changes in water content could have a significant effect on the diurnal variation of the solar radiation reaching the surface.

Another possibility is that surface-active materials (e.g., polar organic molecules) might change the evaporation rates of droplets, thereby changing the longevity of droplets themselves. This presumably could lead to changes in the lifetime of clouds and thus the cloud cover.

The effects of increases in aerosol concentration on cloud lifetime are not confined to water clouds. Mitchell and Carter (1993) have calculated the effect of changes in the CCN population on the microphysical development of mixed phase clouds. It was shown that, in a non-steady state situation, an increase in the concentration of CCN typical of those observed in polluted areas could reduce the snowfall rate by almost 25%. Under steady state conditions, the precipitation rate must balance the condensation rate, but an increase in CCN resulted in a 50% increase in the cloud liquid water content. The climatic impact of increased particle concentration and increased lifetime of high level cirrus clouds could be of opposite sign to similar changes in low-level clouds.

Existing measurements of global cloud properties do not have sufficient accuracy to determine long-term change; thus the global significance of aerosol effects on clouds has been conjectural. It has been shown however that widespread observations of a reduction in the diurnal cycle of surface air temperature over land (Karl *et al.*, 1993) implies an increase in low level cloud cover since 1950, largely confined to land areas (Hansen *et al.*, 1994) which agrees with observations (Henderson-Sellars, 1992). Although the observations and their analysis do not identify a specific cause for the change, the restriction of cloud cover changes to land areas suggests that it may be related to increased anthropogenic aerosol concentrations.

3.10 Future Trends
3.10.1 Industrialisation

Table 3.1 shows that approximately 10% of the mass of the aerosol entering the atmosphere is of anthropogenic origin and that much of this is due to industrial sources, although changes in agricultural practice, including extensive biomass burning, are also significant sources. However, in some regions, the mass of anthropogenic aerosol dominates over natural aerosol. Increasing industrialisation could therefore lead to increases in aerosol loadings, although improved technology (electrostatic precipitators to remove solid particles and flue gas scrubbers to remove the gases which create aerosol through gas-particle conversion) has the potential to reduce the strength of some industrial sources. Due to the relatively short lifetime of aerosol particles in the atmosphere (even of those injected into the stratosphere) the effects of changes in source strengths will be seen fairly quickly. In this respect, the effects of aerosol are very unlike those of most anthropogenic greenhouse gases.

Future changes in the radiative forcing by aerosol particles are most likely to result from changes in the industrial sulphur dioxide emissions and, to a smaller degree, in the pattern of biomass burning. Changes in the natural emissions of aerosols and aerosol precursors caused, for example, by changes in climate may also occur (see e.g., IPCC, 1990). The current trends in anthropogenic sulphur dioxide emissions are downwards in Europe and North America: between 1980 and 1990 the emissions in these regions decreased by 28% and 15% respectively (Schang et al., 1994; NAPAP, 1990). On the other hand, emissions in China, India and several other Asian countries are increasing rapidly (Foell and Green, 1990).

World Energy Council Scenarios (WEC, 1993) (Table 3.8) suggest that over the period 1990-2020 sulphur emissions will decrease by at least 50% over North America and Western Europe while increasing by more than 200% over Africa and the East Asia region. Schematic projections into the future of anthropogenic sulphur dioxide emissions have also been made by Galloway (1989).

In various scenarios formulated in the 1992 IPCC report (IPCC, 1992) the global sulphur dioxide emissions associated with fossil fuel combustion range from 76 Tg S/yr in 1990 to 82-141 TgS/yr in 2025 and 55-232 TgS/yr in the year 2100. More recent estimates of sulphur emissions (e.g., Muller, 1992; Spiro et al., 1992) suggest a slightly lower figure for 1990 emissions with a best estimate of about 70 TgS/yr. The projected values for 2025 in the IPCC(1992) scenarios are quite high compared with Table 3.8. A factor influencing the emissions is likely to be concern over local environmental effects such as acid deposition. Although these figures relate to emissions, the short lifetime of aerosols suggests that the changes in emissions will give proportionate changes in atmospheric aerosol burdens.

3.10.2 Feedbacks

There are several processes by which aerosol concentrations are related to climate, hence providing the potential for climate feedback effects, although no strong feedback mechanisms have been conclusively demonstrated. Of the natural sources, those of marine aerosol and mineral dust would be expected to depend on local meteorological conditions and on climate-sensitive agricultural practices. The strengths of natural sources of several of the precursor gases are also dependent on meteorological conditions (e.g., stability, radiative flux and soil moisture) as well as on the local agricultural practice, while the major sink, precipitation scavenging, is dependent on rainfall patterns. These processes provide a further potential feedback mechanism, although over land the aerosol formed by gas-particle conversion comprises a small fraction of the total loading. The importance of these feedback processes has not yet been quantified.

Some feedback processes are also possible with anthropogenic aerosol sources. Significant climate change could result in a redistribution of power and transport requirements as well as the location of industrial sources of aerosol.

3.11 Likely Developments

Although the general principles of the processes by which

Table 3.8: Present and projected annual emissions of sulphur (Mt) from World Energy Council (WEC, 1993). Present figures relate to 1990 while maximum and minimum projections are shown for 2020, according to different control scenarios.

Region	1990	2020 (min)	2020 (max)
North America	12.1	3.0	6.2
Latin America	3.2	5.7	13.5
Western Europe	10.4	2.2	4.6
Central and Eastern Europe	3.9	0.9	2.6
CIS	12.4	2.3	9.5
Middle East and North Africa	2.2	4.4	9.6
Sub-Saharan Africa	1.9	3.6	10.8
Pacific	15.1	15.1	26.2
South Asia	3.4	5.7	17.8
World	64.6	42.9	100.8

aerosols affect climate are generally well established, there are many uncertainties involved in quantifying the effects, and many difficulties in the development of parametrizations for use by numerical models of climate. The problems of characterising the anthropogenic aerosol content by size and chemical composition, and the spatial and temporal variability of aerosol properties, are crucial to the development of improved estimates of direct radiative forcing. Further work is also needed to relate the physical properties of different types of aerosol particle to their radiative properties. The major problems in the estimation of the indirect radiative forcing concern the relationship between aerosol particle concentrations and cloud droplet or ice crystal populations and the impact of microphysical changes on the optical properties of clouds which often have considerable spatial and temporal variability. Developments over the next few years will help to resolve some of the outstanding problems through the combined use of in-situ observations, satellite measurements and numerical models.

3.11.1 Local Measurements

It is presently possible to make local measurements, in clear air, of the aerosol size distribution in the size range 10^{-3} to 10 μm, together with the total number of particles, although there are problems measuring the concentration of interstitial aerosol particles in clouds. It is also possible to measure the mean particle composition within several particle size ranges and the total mass of aerosol. The development of improved techniques for size-resolved chemical analysis will clarify the relation between aerosol size distribution and source strengths. Ground based and satellite remote sensing is essential to relate the detailed local measurements to aerosol forcing on a regional or global scale and many developments are likely in the availability and scope of such measurements over the next few years.

Detailed information on the height dependence of aerosol backscatter, absorption coefficient and optical depth can only be obtained using aircraft, although most of the necessary instrumentation is available. Aircraft observations are also necessary to clarify the way in which aerosol optical properties vary with humidity as the hygroscopic elements of the particles grow. Studies are already planned to make detailed measurements at a variety of locations where the aerosols are believed to have different origins and characteristics. One experiment is currently under way over the ocean near Tasmania, where anthropogenic influences are small, and another will be undertaken in the North Atlantic, where, depending on the air trajectory, both polluted and relatively clean conditions may be encountered. Clearly, such detailed observations are only feasible at a limited number of sites and it will be necessary to support these with observations which may be more limited in scope but global in coverage, if the global impact of aerosols is to be better quantified.

Chemical characterisation of the climatically important aerosol types (sulphates, organics, anthropogenic soil dust and soot) is necessary both to allow the development of quantitative connections to aerosol or aerosol-precursor sources and to provide critical information on which to base estimates of particle refractive index. It is necessary to utilise appropriate sampling methods that allow the size dependence of composition and physical properties to be determined and to provide standardisation and intercalibration of measurement methods. The development of specialised techniques for molecular speciation of organics should occur over the next few years.

3.11.2 Ground-based and Satellite Monitoring

Long-term monitoring of aerosol profiles, for example the balloon-borne measurement programme at Laramie, Wyoming (Hofmann, 1993) and the CN/CCN monitoring programme at Cape Grim, Tasmania (Gras, 1994) have helped to quantify the local trends in aerosol. However it is necessary to measure the global and regional trends in the aerosol characteristics if their changing impact on the Earth's radiation budget is to be determined.

Long-term monitoring of the global distribution of aerosols in the boundary layer is being developed by the WMO Global Atmosphere Watch and this network should provide geographical and seasonal distributions of aerosols. Instrumentation at some of these sites will also provide information on aerosol optical properties and on CCN as well as on the chemical composition of the aerosols. These sites will also provide calibration and verification of remote sensing techniques.

Satellite remote sensing which enables aerosol size and loading to be estimated, is essential if the global impact of aerosols is to be properly assessed. Only satellite monitoring can achieve the global coverage and the necessary spatial resolution to measure such inhomogeneous fields as aerosols. However, remote sensing of aerosol properties poses many problems. Penner et al. (1994) identify a number of satellite instruments which will be flown within the next ten years and offer the possibility of improved measurements of aerosol physical properties on the global scale. The development of instruments which provide polarisation information will enable considerable advances to be made. However, it is essential that such measurements be combined with aerosol chemical measurements for which remote sensing is not presently feasible.

3.11.3 Modelling

The integration of local measurements with global fields of aerosol properties requires the use of aerosol models which can then be used to estimate the contribution of different aerosols to the net aerosol forcing. Such models should include the variety of aerosol sources and removal processes. Present models rely on many empirical relationships and pay little attention to the variety of aerosol types. However, the developments outlined above should, over the next few years, provide the observational support necessary for the development of improved models.

References

Albrecht, B.A., 1989: Aerosols, cloud microphysics and fractional cloudiness. *Science,* **245**, 1227-1230.

Allard, P., J. Carbonelle, D. Dajlevic, J. LeBronec, P. Morel, M.C. Robe, J.M. Maurenas, R. Faivre-perret, D. Martin, J.C. Saubroux and P. Zettwoog, 1991: Eruptive and diffusive emissions of CO_2 from Mount Etna. *Nature,* **351**, 387-391.

Andreae, M.O., 1983: Soot carbon and excess fine potassium: Long-range transport of combustion-derived aerosols. *Science,* **220**, 1148-1151.

Andreae, M.O., 1986: The ocean as a source of atmospheric sulphur compounds. In: *The Role of Ar-Sea Exchange In Geochemical Cycling* . P. Buat-Ménard (ed.), Reidel, Dordrecht, pp331-362.

Andreae, M.O., 1991: Biomass burning : its history, use and distribution and its impact on environmental quality and global climate. In: *Global Biomass Burning: Atmospheric, Climatic and Biospheric Implications*. J.S. Levine (ed.), MIT Press, pp3-21.

Andreae, M.O., 1993: Global distribution of fires from space. *Eos Trans.* **AGU 74**, 129-135.

Andreae, M.O., 1995: Climate effects of changing atmospheric aerosol levels. In: *World Survey of Climatology, Vol. XVI Future Climate of the World*. A. Henderson-Sellers (ed), Elsevier, Amsterdam. (In press.)

Andreae, M.O., T.W. Andreae, R.J. Ferek and H. Raemdonck, 1984: Long-range transport of soot carbon in the marine atmosphere. *Science of the Total Environment,* **36,** 73-80.

Andreae, M.O., H. Berresheim, T.W. Andreae, M.A. Kritz, T.S. Bates and J.T. Merrill, 1988: Vertical distribution of dimethylsulphide, sulphur dioxide, aerosol ions, and radon over the north-east Pacific Ocean. *J. Atmos. Chem.,* **6,** 149-173.

Austin, P.H., M.B. Baker, A.M. Blyth and J.B. Jensen, 1985: Small scale variability in warm continental cumulus clouds. *J. Atmos. Sci.,* **42,** 1233-1138

Ayers, G.P. and J.L. Gras, 1991: Seasonal relationship between cloud condensation nuclei and aerosol methane sulphonate in marine air. *Nature,* **353,** 834-835.

Ayers, G.P., J.P. Ivey and R.W. Gillett, 1991: Coherence between aerosol cycles of dimethylsulphide, methane sulphonate and sulphur in marine air. *Nature,* **349,** 404-406.

Balkanski, Y.J., D.J. Jacob, G.M. Gardner, W.C. Graustein and K.K. Turekian, 1993: Transport and residence times of continental aerosols inferred from a global three-dimensional simulation of ^{210}Pb. *J. Geophys. Res.,* **98,** 20,573-20,586.

Boers, R., G.P. Ayers and J.L. Gras, 1994: Coherence between seasonal variation in satellite-derived cloud optical depth and boundary layer CCN concentrations at a mid-latitude Northern Hemisphere station. *Tellus,* **46B,** 123-131.

Ball, R.J. and G.D. Robinson, 1982: The origin of haze in the Central United States and its effect on solar irradiation. *J. Appl. Met.,* **21,** 171-188

Blanchard, D.C., 1983. The production, distribution, and bacterial enrichment of the seasalt aerosol in air-sea exchange of gases and particles. In: *Air-Sea Exchange Of Gases And Particles,.* P.S. Liss and W.G.N. Slinn (eds), Reidel, Boston, USA, pp 407-454.

Blanchet, J-P., 1989: Towards estimation of climatic effects due to arctic aerosols. *Atmos. Env.,* **11,** 2609-2625.

Bohren, C.F., 1987: Multiple scattering of light and some of its observable consequences. *Amer. J. Phys.,* **55,** 524-533.

Boutron, C.F., U. Gorlach, J.-P. Candelone, M.A. Bolshov and R.J. Delmas, 1991: Decrease in anthropogenic lead, cadmium and zinc in Greenland snows since the late 1960s. *Nature,* **353,** 153-156.

Braham, R.R. Jr., 1974: Cloud physics of urban weather modification - a preliminary report. *Bull. Am. Met. Soc.,* **55,** 100-106.

Brenguier, J.L., 1990: Parametrization of the condensation process in small-scale non-precipitating cumuli. *J. Atmos. Sci.,* **47,** 1127-1148.

Buat-Ménard, P., U. Ezat and A. Gaudichet, 1983: Size distribution and mineralogy of aluminosilicate dust particles in tropical Pacific air and rain. In: H.R. Pruppacher, R.G. Semonin and W.G.N. Slinn (eds), *Precipitation Scavenging, Dry Deposition And Resuspension*; pp1259-1269, Elsevier.

Cachier, H., M.P. Bremond and P. Buat-Ménard, 1989: Carbonaceous aerosols from different tropical biomass burning sources. *Nature,* **340,** 371-373.

Cambray, R.S., P.A. Cawse, J.A. Garland, J.A. B. Gibson, P. Johnson, G.N.J. Lewis, D. Newton, L. Salmon and B.O. Wade, 1987: Observations on radioactivity from the Chernobyl accident. *Nuclear Energy,* **26,** 77-110.

Chamberlain, A.C., 1991: *Radioactive Aerosols*. Cambridge University Press, 255pp.

Charlson, R.J., 1988: Have the concentrations of tropospheric aerosol particles changed? In: F.S. Rowland and I.S.A. Isaksen (eds)*The Changing Atmosphere*, J Wiley & Sons Ltd, Chichester, England pp79-90..

Charlson, R.J. and J. Heintzenberg (eds), 1995: Aerosol forcing of climate (Ch. 3) *Proc. Dahlem Workshop, 24-29 April, 1994.* J. Wiley & sons. Berlin. (In press.)

Charlson, R.J., J.E. Lovelock, M.O. Andreae and S.G. Warren, 1987: Oceanic phytoplankton, atmospheric sulphur, cloud albedo and climate. *Nature,* **326,** 655-661.

Charlson, R.J., J. Langner and H. Rodhe, 1990: Sulphate aerosol and climate. *Nature,* **348,** 22.

Charlson, R.J., J. Langner, H. Rodhe, C.B. Leovy and S.G. Warren, 1991: Perturbation of the Northern Hemisphere radiative balance by backscattering from anthropogenic sulphate aerosols. *Tellus,* **43A**-13, 152-163.

Charlson, R.J., S.E. Schwartz, J.M. Hales, R.D. Cess, J.A. Coakley, J.E. Hansen and D.J. Hofmann, 1992: Climate forcing by anthropogenic aerosols. *Science*, **255**, 422-430.

Church, T.M., J.M. Taramontano, D.M. Whelpdale, M.O. Andreae, J.N. Galloway, W.C. Keene, A.H. Knap and J. Tokos, 1991: Atmospheric and precipitation chemistry over the North Atlantic Ocean; shipboard results from April-May 1984. *J. Geophys. Res.*, **96**, 18,705-18,725.

Clarke, A.D., 1989: Aerosol light absorption by soot in remote environments. *Aerosol Sci. Tech.*, **10**, 161-171.

Clarke, A.D. and R.J. Charlson, 1985: Radiative properties of the background aerosol: absorption component of extinction. *Science*, **229**, 263-265.

Coakley, J.A. Jr., R.D. Cess and F.B. Yurevich, 1983: The effect of tropospheric aerosols on the Earth's radiation budget: a parametrization for climate models. *J. Atmos. Sci.*, **40**, 116-138.

Coakley, J.A., R.L. Bernstein and P.A. Durkee, 1987: Effect of ship-track effluents on cloud reflectivity. *Science*, **237**, 1020-1022.

Cooper, W.A., 1989: Effects of variable growth histories on droplet size distributions, Part 1, Theory, *J. Atmos. Sci.*, **46**, 1301-1311

Covert, D.S., A.P. Waggoner, R.E. Weiss, N.C. Ahlquist and R.J. Charlson, 1980: Atmospheric aerosols, humidity and visibility. In *The Character And Origins Of Smog Aerosols. A Digest Of Results From The California Aerosol Characterisation Experiment (ACHEX)*, G.M.Hidy, P.K. Mueller, D. Grosjean, B.R. Appel and J.J. Wesolowski (eds), Wiley and Sons, pp559-581.

D'Almeida, G.A., P. Koepke and E.P. Schettle, 1991: *Atmospheric Aerosols: Global Climatology And Radiation Characteristics.* A Deepak Publ., Hampton, VA, USA.

Duce, R.A., 1978: Speculations on the budget of particulate and water phase non-methane organic carbon in the global troposphere. *Proc.Appl.Geophys.*, **116**, 244-273.

Duce, R.A., P.S. Liss, J.T. Merrill, E.L. Atlas, P. Buat-Ménard, B.B. Hicks, J.M. Miller, J.M. Prospero, R. Arimoto, T.M. Church, W. Ellis, J.N. Galloway, L. Hansen, T.D. Jickells, A.H. Knap and K.H. Reinhardt, 1991: The atmospheric input of trace species to the world ocean. *Global Biogeochemical Cycles,* **5**, 193-259.

Dulac, F., P. Buat-Ménard, U. Ezat, S. Melki and G. Bergametti, 1989: Atmospheric imput of trace metals to the western Mediterranean : uncertainties in modelling dry deposition from cascade impactor data. *Tellus*, **41B**, 262-378.

Dulac, F., D. Tanre, G. Bergametti, P. Buat-Ménard, M. Desbois and D. Sutton, 1992: Assessment of the African dust mass over the western Mediterranean Sea using Meteosat data. *J. Geophys. Res.*, **97**, 2489-2506.

Durkee, P.A., F. Pfeil, E. Frost and R. Shema, 1991: Global analysis of aerosol particle characteristics. *J. Atmos. Env.*, **25A**, 2457-2471.

Dutton, E.G. and J.R. Christy, 1992: Solar radiative forcing at selected locations and evidence for global lower tropospheric cooling following the eruptions of El Chichon and Pinatubo. *Geophys. Res. Lett.*, **19**, 2313-2316.

Ellis, H.T. and R.F. Pueschel, 1971: Solar radiation: absence of air pollution trends at Mauna Loa. *Science*, **172**, 845-846.

Engardt, M.and H. Rodhe, 1992: A comparison between patterns of temperature trends and sulphur aerosol pollution. *Geophys. Res. Lett.*, **20**, 117-120

Erickson, D.I., J.J. Walton, S.J. Ghan and J.E. Penner, 1991: Three dimensional modelling of the global sulphur cycle, a first step. *Atmos. Env.*, **25A**, 2513-2520.

Flowers, E.C., R.A. McCormick and K.R. Kurfis, 1969: Atmospheric turbidity over the United States, 1961-1966. *J.Appl.Met.*, **8**, 955–962.

Foell,W.K. and C.W. Green, 1990: *Acid Rain In Asia: An Economic, Energy And Emission Overview.* Paper presented at the Second Workshop on Acid Rain in Asia, Asian Inst. of Technology, Bangkok, 19-22 Nov 1990.

Fouquart, Y. and H. Isaka, 1992: Sulphur emission, CCN, clouds and climate: a review. *Annales Geophysicae,* **10**, 462-471.

Fouquart, Y., J.C. Buriez, M. Herman and R.S. Kandel, 1990: The influence of clouds on radiation: a climate-modelling perspective. *Rev. Geophys.*, **28**, 145-166.

Fravalo, C., Y. Fouquart and R. Rosset, 1981: The sensitivity of a model of low stratiform clouds to radiation. *J. Atmos. Sci.*, **38**, 1049-1062.

Fraser, R.S., Y.J. Kaufman and R.L. Mahoney, 1984: Satellite measurements of aerosol mass and transport. *Atmos. Env.*, **18**, 2577-2584.

Galbally, I.E. and C.R. Roy, 1978: Exhalation of nitric oxide from soils. *Nature,* **256**, 217-221.

Galloway, J.N., 1989: Atmospheric acidification: predictions for the future. *Ambio*, **18**, 161-166.

Galloway, J.N. and H. Rodhe, 1991: Regional atmospheric budgets of S and N: how well can they be quantified? *Proc. Roy. Soc. Edinburgh*, **97B**, 61-80.

Graedel, T.E., 1979: Turpenoids in the atmosphere. *Rev. Geophys. Space Phys.*, **17**, 937-947.

Gras, J.L., 1994: CN, CCN and particle size in Southern Ocean air at Cape Grim. *Atmos.Res.* (In press.)

Grassl, H., 1988: What are the radiative and climatic consequences of the changing concentration of atmospheric aerosol particles? In: *The Changing Atmosphere* , F.S. Rowland and I.S.A. Isaksen (eds), J Wiley & Sons Ltd, Chichester, England, pp79-90.

Han, Q.W., W.B. Rossow and A.A. Lacis, 1994: Near-global survey of effective droplet radii in liquid water clouds using ISCCP data. *J. Climate*, **7**, 465-497.

*Hansen, A.D.A., R.S. Artz, A.A.P. Pszenny and R.E. Larson, 1990: Aerosol black carbon and radon as tracers for air mass origin over the North Atlantic Ocean. *Global Biogeochemical Cycles*, **4**, 189-199.

+Hansen, J.E., W. Rossow and J. Fung, 1990: The missing data on global climate change. *Issues in Science and Technology*, 62-69.

Hansen, J.E., M. Sato and R. Ruedy, 1994: Long-term changes of the diurnal temperature cycle: implications about mechanisms of global climate change. In: *US Dept of Energy Report.,* G. Kukla and T. Karl (eds).

Hegg, D.A., L.R. Radke and P.V. Hobbs, 1991: Measurements of Aitken nuclei and cloud condensation nuclei in the marine atmosphere and their relation to the DMS-cloud-climate hypothesis. *J.Geophys.Res.*, **96**, 18727-18773.

Hegg, D.A., R.J. Ferek and P.V. Hobbs, 1993: Light scattering and cloud condensation nucleus activity of sulphate aerosol measured over the Northeast Atlantic Ocean. *J. Geophys. Res.*, **98**, 14887-14894.

Heintzenberg, J., 1982: Size-segregated measurements of particulate elemental carbon and aerosol light absorption at remote locations. *Atmos. Env.*, **16**, 2461-2469.

Heintzenberg, J., 1988: A processor-controlled multi-sample sort photometer. *Aerosol Sci. Technol.*, **8**, 227-233.

Heintzenberg, J., 1989: Fine particles in the global troposphere: a review. *Tellus*, **41B**, 149-160.

Helmes, L. and R. Jaenicke 1988: Long-term series of atmospheric turbidity estimated from records of sunshine duration and cloud cover. In: *Aerosols and Climate*, P.V.Hobbs and M.P.McCormick (eds), Deepak Publ. pp139-146.

Henderson-Sellars, A., 1992: Continental cloudiness changes this century. *Geo. Journal*, **27.3**, 255-262.

Hildemann, L.M., G.R. Markowski and G.R. Cass, 1991: Chemical composition of emissions from urban sources of fine organic aerosol. *Envir. Sci. Tech.*, **25**, 744-759.

Hobbs, P.V. and A.L. Rangno, 1985: Ice particle concentrations in clouds. *J. Atmos. Sci.*, **42**, 2523-2549.

Hobbs, P.V., L.F. Radke and S.E. Shumway, 1970: Cloud condensation nuclei from industrial sources and their apparent influence on precipitation in Washington. *J. Atmos. Sci.*, **27**, 81-89.

Hobbs, P.V., J.R. Tuell, D.A. Hegg, L.F. Radke and M.W. Elgrowth, 1982: Particles and gases in the emissions from the 1980-81 volcanic eruption of Mt. St. Helens. *J. Geophys. Res.*, **87**, 11062-11086.

Hofmann, D.J., 1991: Aircraft sulphur emissions. *Nature*, **349**, 659.

Hofmann, D.J., 1993: Twenty years of balloon borne tropospheric aerosol measurements at Laramie, Wyoming. *J. Geophys. Res.*, **98**, 12753-12766.

Hoppel, W.A., 1979: Measurements of the size distribution and CCN supersaturation spectrum of sub-micron aerosols over the ocean. *J. Atmos. Sci.*, **36**, 2006-2015.

Hudson, J.G., 1991: Observations of anthropogenic cloud condensation nuclei. *Atmos. Env.*, **25A**, 2449-2455.

Hudson, J.G. and C.F. Rogers, 1986: Relationship between critical supersaturation and cloud droplet size; implications for cloud mixing processes. *J. Atmos. Sci.*, **43**, 2341-2359.

Hudson, J.G., J. Hallett, and C.F. Rogers, 1991: Field and laboratory measurements of cloud-forming properties of combustion aerosols. *J. Geophys. Res.*, **96**, 10847-10859.

Huebert, B.J., G. Lee and W.L. Warren, 1990: Airborne aerosol inlet passing efficiency measurements. *J. Geophys. Res.*, **95**, 16369-16381.

Hunter, D.E., S.E. Schwartz, R. Waggoner and C.M. Benkovits, 1993: Seasonal, latitudinal and secular variations in temperature trend: evidence for influence of anthropogenic sulphate. *Geophys. Res. Lett.*, **20**, 2455-2458.

Husar, R.B. and D.E. Patterson, 1987: Haze climate of the United States. EPA/600/S3-86/071.

Husar, R.B., D.E. Patterson, J.M. Holloway, W.E. Wilson and T.G. Ellestad, 1979: Trends of Eastern US haziness since 1948. In *Proc. Fourth Symp. on Atmos. Turbulence, Diffusion and Air Pollution, Reno, NV*, Amer. Met.Soc. pp249-256.

Husar, R.B., J.M. Holloway and D.E. Patterson, 1981: Spatial and temporal pattern of eastern US haziness: a summary. *Atmos. Env.* **15**, 1919-1928.

IPCC (Intergovernmental Panel on Climate Change), 1990: *Climate Change: the IPCC Scientific Assessment*, J.T. Houghton, G.J. Jenkins and J.J. Ephraums (eds). Cambridge University Press, Cambridge, UK.

IPCC, 1992: *Climate Change 1992: The Supplementary Report to the IPCC Scientific Assessment*, J.T. Houghton, B.A. Callander and S.K. Varney (eds). Cambridge University Press, Cambridge, UK.

Jankowiak, I. and D. Tanre, 1992: Satellite climatology of Saharan dust outbreaks: method and preliminary results. *J.Climate*, **5**, 646-656.

Jones, A., D.L. Roberts and A. Slingo, 1994: A climate model study of the indirect radiative forcing by anthropogenic sulphate aerosols. *Nature*, **370**, 450-463.

Junge, C.E., 1963: *Air Chemistry and Radioactivity*. Academic Press, London.

Kai, K., Y. Okada, O. Uchino, I. Tabata, H. Nakamura and T. Takasugi, 1988: Lidar observation and numerical simulation of a Kosa (Asian dust) over Tsukuba, Japan, during the spring of 1986. *J. Met. Soc. Japan*, **66**, 812-828.

Kaminski, U., and P. Winkler, 1988: Increasing submicron particle mass concentration at Hamburg. II, Source discussion. *Atmos. Env.*, **22**, 2879-2883.

Karl, T.R., G. Kukla, V. Razuvayev, M.G. Changery, R.G. Quayle, R.R. Heim, D.R. Easterling and C.B. Fu, 1991: Global warming; evidence for asymmetric diurnal temperature change. *Geophys. Res. Lett.* **18**, 2253-2256.

Karl, T.R., P.D. Jones, R.W. Knight, G. Kukla, N. Plummer, V. Razuvayev, K.P. Gallo, J. Lindseay, R.J. Charlson and T.D. Peterson, 1993: A new perspective on recent global warming. *Bull. Am. Met. Soc.*, **74**, 1007-1023.

Kaufman, Y.J., 1993: Aerosol optical thickness and atmospheric path radiance. *J.Geophys.Res.*, **98**, 2677-2692.

Kaufman, Y.J. and T. Nakajima, 1993: Effect of Amazon smoke on cloud microphysics and albedo: analysis from satellite imagery. *J. Appl. Met.*, **32**, 729-744.

Kaufman, Y.J., C.J. Tucker and I. Fung, 1990: Remote sensing of biomass burning in the tropics. *J.Geophys.Res.*, **95**, 9927-9939.

Kent, G.S., M.P. McCormick and S.K. Schaffner, 1991: Global optical climatology of the free tropospheric aerosol from 1.0 μm satellite occultation measurements. *J. Geophys. Res.*, **96**, 5249-5268.

Kiehl, J.T. and B.P. Briegleb, 1993: The relative roles of sulphate aerosols and greenhouse gases in climate forcing. *Science*, **260**, 311-314.

King, M.D., L.F. Radke and P.V. Hobbs, 1993: Optical properties of marine stratocumulus clouds modified by ships. *J. Geophys. Res.*, **98**, 2729-2739.

Kulmala, M., A. Laaksonen, P. Korhonen, T. Vesala, T. Ahonen and J.C. Barrett, 1993: The effect of atmospheric nitric acid vapour on cloud condensation nucleus activation. *J. Geophys. Res.*, **98**, 22949-22958.

Langner, J. and H. Rodhe, 1991: A global three-dimensional model of the tropospheric sulphur cycle. *J. Atmos. Chem.*, **13**, 255-263.

Langner, J., H. Rodhe, P.J. Crutzen and P. Zimmermann, 1992: Anthropogenic influence on the distribution of tropospheric sulphate aerosol. *Nature,* **359,** 712-716.

Latham, J. and M.H. Smith, 1990: Effect on global warming of wind-dependent aerosol generation at the ocean surface. *Nature,* **347,** 372-373.

Lawson, D.R. and W. Winchester, 1979: Atmospheric sulphur aerosol concentrations and characteristics from the South American continent. *Science,* **205,** 1267-.

Leaitch, W. R. and G.A. Isaac, 1994: On the relationship between sulphate and cloud droplet number concentration. *J.Climate,* **7,** 206-212.

Leaitch, W.R., J.W. Strapp, G.A. Isaac, and J.G. Hudson, 1986: Cloud droplet nucleation and cloud scavenging of aerosol sulphate in polluted atmospheres. *Tellus,* **38B,** 328-344.

Lelieveld, J. and J. Heintzenberg, 1992: Sulphate cooling effect on climate through in-cloud oxidation of anthropogenic SO_2. *Science,* **258,** 117-120.

Legrand, M. and R.J. Delmas, 1987: A 220 yr continuous record of volcanic H_2SO_4 in the Antarctic Ice Sheet. *Nature,* **327,** 671-676.

Levine, J.S.(ed) 1991: Global Biomass Burning; Atmospheric, Climatic And Biospheric Implications. MIT Press, pp.xxv-xxx and other chapters.

Lin, X. and W.L. Chameides, 1993 : CCN formation from DMS oxidation without SO_2 acting as an intermediate. *Geophys. Res. Lett.,* **20,**579-582

Liou, K.-N. and S.-C. Ou, 1989: The role of cloud microphysical processes in climate: an assessment from a one-dimensional perspective. *J.Geophys. Res.,* **15,** 8599-8607.

Losno, R., G. Bergametti and P. Carlier, 1992: Origins of atmospheric particulate matter over the North Sea and the Atlantic Ocean. *J. Atmos. Chem.,* **15,** 333-352

Luecken, D.J., C.M. Berkowitz and R.C. Easter, 1991: Use of a three dimensional cloud chemistry model to study trans-Atlantic transport of soluble sulphur species. *J. Geophys. Res.,* **96,** 22477-22490.

Marley, N.A., J.S.Gaffney and M.M. Cunningham, 1993: Aqueous greenhouse species in clouds, fogs and aerosols. *Envir. Sci. Technol.,* **27,** 2864-2869.

Martin, G.M., D.W. Johnson and A. Spice, 1994: The measurement and parametrization of effective radius of droplets in warm stratocumulus clouds. *J. Atmos. Sci.,* **50,** 1823-1842.

Mayewski, P.A., W.B. Lyons, M.J. Spencer, M. Twickler, W. Dansgaard, B. Koci, C.I. Davidson and R.E. Honrath, 1986: Sulphate and nitrate concentrations from a South Greenland ice core. *Science* **232,** 975-977.

Mayewski, P.A., W.B. Lyons, M.J. Spencer, M. Twickler, C.F. Buck and S. Whitlow, 1990: An ice core record of atmospheric response to anthropogenic sulphate and nitrate. *Nature,* **346,** 554-556.

McCormick, M.P., P. Harnill, T.J. Pepin, W.P. Chu, T. Swissler and L.R. McMaster, 1979: Satellite studies of the stratospheric aerosol. *Bull. Am. Met. Soc.,* **60,** 1038-1046.

Mitchell, D.A., and E.J. Carter, 1993: The dependence of cloud LWC and precipitation rate on CCN and the resulting effect on chemical wet deposition fluxes. *IAMAP/IAHS Joint Symposium, Yokohama, Japan,* 11-23 June, 1993.

Monahan, E.C., 1986: The ocean as a source for atmospheric particles. In: *The Role of Air-Sea Exchange in Geochemical Cycling.* NATO-ASI series(C) vol. 185, P. Buat-Ménard (ed.), Reidel. pp129-163.

Mossop, S.C., 1978: The influence of drop size distribution on the production of secondary ice particles during graupel growth. *Quart. J. R. Met.Soc.,* **104,** 1-13.

Muller, J.-F., 1992: Geographical distribution and seasonal variation of surface emissions and deposition velocities of atmospheric trace gases. *J. Geophys. Res.,* **97,** 3787-3804.

NAPAP, 1990: Acidic deposition: *State of Science and Technology Report 24.* US Government Printing Office, Washington, DC, 129 pp.

Novakov, T. and J.E. Penner, 1993: Large contribution of organic aerosols to closed condensation nuclei concentrations. *Nature,* **365,** 823-826.

Nguyen, B.C., B. Bonsang and A. Gaudry, 1983: The role of the ocean in the global atmospheric sulphur cycle. *J. Geophys. Res.,* **88,** 10903-10914.

Nyeki, S.A.P., I. Colbeck, R.M. Harrison and T. Hall, 1994: A multi-channel integrating nephelometer to measure real-time atmospheric aerosol scattering co-efficients. *Meas. Sci. Technol.,* **5,** 593-599.

Ogren, J.A. and R.J. Charlson, 1983: Elemental carbon in the atmosphere: cycle and lifetime. *Tellus* **35B,** 241-254.

Penner, J.E., R.E. Dickinson and C.A. O'Neill, 1992: Effects of aerosol from biomass burning on the global radiation budget. *Science* **256,** 1432-1434.

Penner, J.E., R.J. Charlson, J.M. Hales, N. Laulainen, R. Leifer, T. Novakov, J. Ogren, L.F. Radke, S.E. Schwartz and L. Travis, 1994 : Quantifying and Minimising Uncertainty of Climate Forcing by Anthropogenic Aerosols. *Bull. Am. Met. Soc.,* **75,** 375-400.

Petrenchuk, O.P., 1980: On the budget of sea salts and sulphur in the atmosphere. *J. Geophys. Res.,* **85,** 7439-7444.

Prospero, J.M., 1990: Mineral aerosol transport to the North Atlantic and North Pacific: the impact of African and Asian sources. In: *The Long-range Transport of Natural and Contaminated Substances,* A.H. Knap (ed), NATO ASE series, **297,** 59-88.

Pueschel, R.F., C.C. van Valin, R.C. Castillo, J.A. Kadlecek and E. Ganor, 1986: Aerosols in polluted versus nonpolluted air masses: long-range transport and effects on clouds. *J.Clim. Appl. Met.,* **25,** 1908-1917.

Pye, K., 1987: *Aeolian Dust And Dust Deposition.* Academic Press, London.

Radke, L.F., J.A. Coakley, and M.D. King, 1989: Direct and remote sensing observations of the effects of ships on clouds. *Science,* **246,** 1146-1148.

Raes, F, 1993: Entrainment of free tropospheric aerosols as a regulating mechanism for cloud condensation nuclei in the remote marine boundary layer. *J.Geophys. Res.* (In press.)

Raga, G.B. and P.R. Jonas, 1993: On the link between cloud-top radiative properties and sub-cloud aerosol concentrations. *Quart. J. R. Met. Soc.,* **119,** 1419-1425.

Rehkopf, J., 1984: Ein zweidimensionales globales Ausbreitungsmodell für Aerosole und Schwefelkomponenten. PhD Thesis, Hamburg, Germany, 1984.

Rodhe, H.,1978: Budgets and turn-over times of atmospheric

sulphur compounds. *Atmos. Env.*, **12**, 671-678.

Rodhe, H. and I. Isaksen, 1980: Global distribution of sulphur compounds in the troposphere estimated in a height/latitude transport model. *J. Geophys. Res.*, **85**, 7408-7409.

Rogers, C.F., J.G. Hudson, B. Zielinska, R.L. Tanner, J. Hallett and J.G. Watson, 1991: Cloud condensation nuclei from biomass burning. *Proc. Chapman Conf. on Global Biomass Burning*, MIT Press.

Roosen, G.R., R.J. Angionne and C.H. Klemcke, 1973: Worldwide variations in atmospheric transmission observations. *Bull. Am. Met. Soc.*, **54**, 307-316.

Samson, J.A., S.C. Barbard, J.S. Obremski, D.C. Riley, J.J. Black and A.W. Hogan, 1990: On the systematic variation in surface aerosol concentration at the South Pole. *Atmos. Res.*, **25**, 385-396.

Sassen, K., M.K. Griffin and G.C. Dodd, 1989: Optical scattering and microphysical properties of subvisual cirrus clouds, and climatic implications. *J. Appl. Met.*, **28**, 91-98.

Savoie, D.L. and J.M. Prospero, 1989: Comparison of oceanic and continental sources of non-seasalt sulphate over the Pacific Ocean. *Nature*, **339**, 685-687.

Schang, J., U. Pederson, J.E. Skjelmoen, K. Arnesen and A. Bartoniova, 1994: Data Reports - Annual summaries, *EMEP-CCC Report 4/90/4* (and previous reports). Norwegian Institute for Air Research.

Schmidt, M., 1974: Verlauf der Kondensationskernkonzentration an zwei süddeutschen Stationen. *Meteorol. Rundsch.*, **27**, 151-153.

Simkin, T. and L. Sibert, 1984: Explosive eruptions in space and time: durations, intervals and a comparison of the worlds active volcanic belts. In: *Explosive Volcanism: Inception, Evolution and Hazards, Studies in Geophysics*, Natl. Acad. Press, Washington DC, USA, 110-121.

Simkin, T., L. Siebert, L. Mcclelland, D. Bridge, C. Newhall and J.H. Latter, 1981: *Volcanoes Of The World: A Regional Directory, Gazetteer And Chronology Of Volcanism During The Last 1000 Years*. Hutchinson Ross Publishing Company, Stroudsburg, Pennsylvania, USA.

Simoneit, B.R.T. and M.A. Mazurek, 1981: Air pollution; the organic constituents. *Crit. Rev. Envir. Control*, **11**, CRC Press, 219-276.

Simoneit, B.R.T., W.F. Rogge, M.A. Mazurek, L.J. Standley and G.R. Cass, 1993: Lignan pyrolysis products, lignans and resin acids as specific tracers of plant classes in emissions from biomass combustion. *Envir. Sci. Tech*, **27**, 2533-2541.

Slingo, A., 1990: Sensitivity of the Earth's radiation budget to changes in low clouds. *Nature*, **343**, 49-51.

Slinn, W.G.N., 1983: Air-to-sea transfer of particles. In: *Air-Sea Exchange of Gases and Particles*, P.S. Liss and W.G.N. Slinn (eds), 209-405, Reidel.

SMIC, 1971. *Study Of Man's Impact On Climate*. MIT Press, Cambridge, MA., USA.

Spiro, P.A., D.J. Jacob and J.A. Logan, 1992: Global inventory of sulphur emissions with 1° × 1° resolution. *J.Geophys.Res.*, **97**, 6023-6036.

Stoiber, R.E., S.N. Williams and B. Huebert, 1987: Annual contribution of sulphur dioxide to the atmosphere by volcanoes. *J. Volcanol. Geotherm. Res.*, **33**, 1-8.

Stonehouse, B., 1986: *Arctic Air Pollution*, Cambridge University Press.

Tarrason, L. and T. Iversen,1992: The influence of North African anthropogenic sulphur emissions over western Europe. *Tellus*, **44B**, 114-132.

Taylor, K.E. and J.E. Penner, 1994: Response of the climate system to atmospheric aerosols and greenhouse gases. *Nature*, **369**, 734-737.

Toon, O.B. and J.B. Pollack, 1976: A global average model of atmospheric aerosol for radiative transfer calculations. *J. Appl. Met.*, **15**, 225-245.

Turco R.F., C.B. Toon, R.C. Whitten, J.B. Pollack and P. Hamill, 1983: The global cycle of particulate elemental carbon: a theoretical assessment, In: *Proc. 4th International Conference on Precipitation Scavenging, Dry Deposition and Resuspension*, H.R. Pruppacher, R.G. Semonin and W.G.N. Slinn (eds), Elsevier, NY, pp1337-1351.

Turton, J. D. and S. Nicholls, 1987: A study of the diurnal variation of stratocumulus using a multiple mixed layer model. *Quart. J. R. Met. Soc.*, **113**, 969-1009.

Twohy, C.H., A.D. Clarke, S.G. Warren, L.F. Radke and R.J. Charlson, 1989: Light absorbing material extracted from cloud droplets and its effect on cloud albedo. *J.Geophys. Res.*, **94**, 8623-8631.

Twomey, S.A., 1977: The influence of pollution on the shortwave albedo of clouds. *J. Atmos. Sci.*, **34**, 1149-1152.

Twomey, S.A. 1991: Aerosols, clouds and radiation. *Atmos. Env.*, **25A**, 2435-2442.

Twomey, S.A., K.A. Davidson and K.J. Seton, 1978: Results of 5 years' observations of cloud nucleus concentrations at Robertson, New South Wales. *J. Atmos. Sci.*, **35**, 650-656.

Twomey, S.A., M. Piepgrass and T.L. Wolfe, 1984: An assessment of the impact of pollution on global cloud albedo. *Tellus*, **36B**, 356-366.

Vali, G., 1985: Atmospheric ice nucleation - a review. *J. Rech. Atmos.*, **19**, 105-115.

Warneck, P. 1988: *Chemistry of the Natural Atmosphere*. Academic Press, Inc., San Diego, 757pp.

Warner, J. and S. Twomey, 1967: The production of cloud nuclei by cane fires and the effect on cloud drop concentrations. *J. Atmos. Sci.*, **24**, 704-706.

Warren, S.G. and A.D. Clarke, 1990: Soot in the atmosphere and snow surface of Antarctica. *J. Geophys. Res.*, **95**, 1811-1816.

WEC, 1993: *Energy For Tomorrow's World - The Realities, The Real Options And The Agenda For Achievement*. World Energy Council, St. Martin's Press, New York, USA, 320pp.

White, W., 1990: The contribution of fine particle scattering to total extinction. Sections 4.1-4.4. In: *Visibility: Existing and Historical Conditions - Causes and Effects*. Acidic Deposition State of Science and Technology Report **24**, National Acid Precipitation Assessment Program, US Government Printing Office, Washington, DC.

Wielicki, B.A., J.T. Suttles, A.J. Heymsfield, R.M. Welch, J.D. Spinhirne, M.L.C. Wu, D.O. Starr, L. Parker and R.F. Arduini, 1990: The 27-28 October 1986 FIRE IFO cirrus case study - Comparison of radiative transfer theory with observations by satellite and aircraft. *Mon. Wea. Rev.*, **118**, 2356-2376.

Yamamoto, G.M., M. Tanaka and K. Arao, 1971: Secular variation of atmospheric turbidity over Japan. *Met. Soc.*, **49**, 859-865.

Yue, G K, M.P. McCormick, W.P. Chu, P. Wang and M.T. Osborn, 1989: Comparative studies of aerosol extinction measurements made by SAMII and SAGEII satellite experiments. *J.Geophys. Res.*, **94**, 8412-8424.

Zimmermann, P.H., 1987: MOGUNTIA: A handy global tracer model. In *Proceedings of the 16th NATO/CCMS International Technical Meeting on Air Pollution Modelling and Its Application, Lindau, Germany,* Reidel, Dordrecht, pp.593-608.

4

Radiative Forcing

K.P. SHINE, Y. FOUQUART, V. RAMASWAMY, S. SOLOMON, J. SRINIVASAN

Contributors:
M.O. Andreae, J. Angell, G. Brasseur, C. Brühl, R.J. Charlson, M.D. Chou, J.R. Christy, T. Dunkerton, E. Dutton, B.A. Fomin, C. Granier, H. Grassl, J. Hansen, Harshvardhan, D. Hauglustaine, P. Hobbs, D.J. Hoffman, L. Hood, N. Husson, I. Karol, Y.J. Kaufman, J. Kiehl, S. Kinne, M.K.W. Ko, K. Labitzke, H. Le Treut, A. McCulloch, A.J. Miller, M. Molina, E. Nesme-Ribes, A.H. Oort, J.E. Penner, S. Pinnock, V. Ramanathan, A. Robock, E. Roeckner, M.E. Schlesinger, K. Sassen, G.-Y. Shi, A.N. Trotsenko, W.-C. Wang.

CONTENTS

Summary	167
4.1 Introduction	169
4.1.1 Definitions of Radiative Forcing	169
4.2 Greenhouse Gases	171
4.2.1 Spectroscopy	171
4.2.2 Calculating the Radiative Forcing	171
4.3 Radiative Forcing due to Changes in Ozone	173
4.3.1 Introduction	173
4.3.2 Radiative Forcing due to Changes in Stratospheric Ozone	173
4.3.3 Significance of the Forcing due to Stratospheric Ozone Loss	177
4.3.4 Temperature Changes in the Lower Stratosphere	177
4.3.5 Sensitivity of Surface-Troposphere Climate	179
4.3.6 Radiative Forcing due to Changes in Tropospheric Ozone	179
4.3.7 Radiative Forcing due to Total Atmospheric Ozone Change	180
4.3.8 Outstanding Issues	180
4.4 Effects of Tropospheric and Stratospheric Aerosols	181
4.4.1 Tropospheric Aerosols	181
4.4.1.1 The direct aerosol effect	181
4.4.1.2 Indirect effect of anthropogenic aerosols on cloud albedo	183
4.4.1.3 Summary of anthropogenic aerosol effects	185
4.4.2 Radiative Forcing due to Stratospheric Aerosol	186
4.4.2.1 Introduction	186
4.4.2.2 Stratospheric aerosol radiative effects	186
4.4.2.3 Observations and simulations of climatic effects	187
4.4.2.4 Summary	189
4.5 Solar Variability	189
4.5.1 Observations of Variability in Solar Irradiance since 1978	189
4.5.2 Inferences of Variability in Solar Irradiance on Longer Time-scales	190
4.5.3 Correlations between Climate and Solar Variability	191
4.5.4 Summary	192
4.6 Other Forcings	192
4.7 Estimates of Total Forcing	192
4.7.1 Relative Confidence in Estimates of Radiative Forcing	192
4.7.2 Best Estimates of Forcings since Pre-industrial Times	193
4.7.3 Estimates of Forcing from Observed Temperature Records	195
4.7.4 Future Forcing	196
4.8 Forcing-Response Relationships	197
4.8.1 Background	197
4.8.2 Forcing-Response Relationships for Well-mixed Greenhouse Gases	197
4.8.3 Forcing-response Relationship for Other Forcings	198
4.8.4 Summary	199
References	199

SUMMARY

The concept of radiative forcing

- Global-mean radiative forcing is a valuable concept for giving at least a first-order estimate of the potential climatic importance of various forcing mechanisms. However, there could be limits to its utility because the relationship between global-mean radiative forcing and global-mean surface temperature change may not be as simple as previously thought. For example, its general applicability has recently been questioned in the cases of forcing due to ozone and tropospheric aerosols, where changes in concentration are highly variable horizontally and/or vertically.

Greenhouse gases

- The direct global-mean radiative forcing due to changes in concentrations of the greenhouse gases (carbon dioxide (CO_2), methane (CH_4), nitrous oxide (N_2O) and the halocarbons) since pre-industrial times is 2.4 Wm^{-2} and is believed to be accurate to within 15%. This value is essentially unchanged from previous IPCC assessments.
- The global-mean radiative forcing due to increases in tropospheric ozone since pre-industrial times is positive and is estimated to be a few tenths of a Wm^{-2}. Since the late 1970s, decreases in stratospheric ozone have led to a global-mean radiative forcing estimated to be about -0.1 Wm^{-2}. The accuracy of these estimates is presently limited by incomplete knowledge of the temporal, horizontal and vertical changes in ozone.

Tropospheric aerosols

- The direct global-mean radiative forcing due to increases in sulphate aerosol amounts since pre-industrial times is negative and may lie in the range -0.25 to -0.9 Wm^{-2}, but there are substantial uncertainties due to the lack of detailed knowledge of the amount of sulphate aerosols and their radiative properties. The effect of aerosols emitted as a result of biomass burning has received much less attention and is even more uncertain; the direct global-mean radiative forcing since pre-industrial times may lie in the range -0.05 to -0.6 Wm^{-2}. Soot carbon in tropospheric aerosols has the potential to reduce the magnitude of the aerosol forcing; however global estimates are not yet available.
- Estimates exist for the indirect effect of aerosols on cloud droplet sizes. The radiative forcing is believed to be negative and may be of a similar magnitude to the direct radiative forcing due to aerosol changes but the uncertainty is much larger. The possible effects of aerosols on cloud liquid water content and cloud cover are just beginning to be investigated.

Volcanic aerosols

- The volcanic eruption of Mt. Pinatubo in June 1991 resulted in a large enhancement to the stratospheric aerosol layer; this caused a transient but large global-mean negative forcing that peaked at about 4 Wm^{-2} and exceeded 2 Wm^{-2} for about one year. Calculated radiative forcings are broadly consistent with values deduced from satellite observations.
- General Circulation Model (GCM) simulations of the effect of the Mt. Pinatubo eruption on surface and lower stratospheric temperatures show encouraging agreement with observations. This suggests that the GCM's temperature response to a transient large-scale radiative forcing of a large magnitude is reasonable.

Solar variability

- Extension of the current understanding of the relationship between observed changes in solar output and other indicators of solar variability suggests that long-term increases in solar irradiance since the 17th century Maunder Minimum might have been climatically significant. A global-mean radiative forcing of a few tenths of a Wm^{-2} since 1850 has been suggested, but uncertainties are large.

Problem in combining global-mean radiative forcing from different mechanisms

- An estimate of the net global mean radiative forcing due to all human activity is not presented because the usefulness of combining estimates of global-mean radiative forcing of differing signs, and resulting from different spatial patterns, is not currently understood. For example, if, by coincidence, the global-mean radiative forcing were to be zero, due to the cancellation of positive and negative forcings from different mechanisms, this cannot be taken to imply the absence of regional-scale or possibly even global climate change.

Radiative Forcing

4.1 Introduction

This chapter considers recent advances in our knowledge of the factors which, by perturbing the planetary radiation budget, might result in climate change. Such a perturbation is referred to here as "radiative forcing" – some authors use the term "climate forcing" for the same concept.

The concept of radiative forcing (e.g., IPCC 1990 and 1992; and see Sections 4.1.1 and 4.8) is based on earlier climate model calculations which show that there is an approximately linear relationship between the global-mean radiative forcing at the tropopause and the equilibrium global mean surface temperature change. Importantly, model calculations have shown that, for a number of forcing mechanisms, the relationship is relatively unaffected by the nature of the forcing (e.g., whether it be due to a change in greenhouse gas concentration or solar output), provided it is appropriately defined (see Section 4.1.1). If the global mean radiative forcing is given by ΔF (Wm^{-2}) and the global mean surface temperature response is ΔT_s (in K) then

$$\Delta T_s = \lambda \Delta F \tag{4.1}$$

where λ is a climate sensitivity parameter. λ is determined by a number of processes such as water vapour feedback, cloud feedbacks and ice-albedo feedback and its computed value is found to vary greatly amongst different GCMs (over a range of at least 0.3 to 1.4 K/(Wm^{-2})), mostly because of uncertainties in the calculation of cloud feedbacks (see e.g., IPCC 1990 and 1992). While knowledge of the climate feedbacks is of crucial importance for climate prediction, consideration of recent work in these areas is beyond the scope of this chapter, but will be discussed in more detail in the next full IPCC assessment in 1995.

Section 4.2 to 4.6 will concentrate on forcing due to changes in:

(a) greenhouse gas concentrations, including ozone;
(b) aerosols, both in the troposphere (mainly as a result of human activity) and in the stratosphere (mainly as a result of volcanic activity);
(c) solar output.

In the cases of forcing due to changes in lower stratospheric ozone and volcanic aerosols, changes in surface and lower stratospheric temperatures add useful supplementary information; since these responses are not considered elsewhere in this report, they will be briefly considered in this chapter. Section 4.7 provides an overall summary of our knowledge of radiative forcing.

The radiative forcing concept encapsulated in Equation (4.1) has served the climate community well over the past two decades. Recently its general applicability has been questioned for forcing due to changes in both ozone and sulphate aerosols where changes in concentration are highly variable horizontally and/or vertically. Section 4.8 reviews studies regarding this topic which are, as yet, too preliminary to allow firm conclusions to be drawn. However, they do suggest that the climate sensitivity (λ in Equation (4.1)) may be significantly different for the forcing due to spatially inhomogenous changes in tropospheric aerosols and in ozone than it is for the well-mixed greenhouse gases. In particular, the work suggests that the degree of cancellation, in the global mean, between the positive greenhouse forcing and the negative tropospheric aerosol forcing may be a poor guide to eventual climate response. For example, if, by coincidence, the global-mean radiative forcing were to be zero, due to an exact cancellation of positive and negative forcings with different spatial characteristics, this could not be taken to imply the absence of regional-scale or possibly even global climate change. We continue to compare global mean radiative forcings in this chapter, but we do so with caution. The resolution of this issue will be an important topic for future IPCC reports.

It should also be noted that if a climate change mechanism leads to altered conditions in the stratosphere, the dynamics of the troposphere can also be affected (e.g., Rind *et al.*, 1990, 1992); thus there is the potential for surface/tropospheric climate change to be induced by mechanisms other than by directly perturbing the radiation budget.

Despite these possible problems, global-mean radiative forcing remains a useful concept for comparing the potential climatic effects of changes in different greenhouse gases, with the possible exception of ozone. It is also a useful single number for comparing different estimates of tropospheric aerosol forcing or ozone forcing.

4.1.1 Definitions of Radiative Forcing

In IPCC (1990) and IPCC (1992) a specific definition of radiative forcing was adopted:

> The radiative forcing of the surface-troposphere system (due to a change, for example, in greenhouse gas concentration) is the change in net irradiance (in Wm^{-2}) at the tropopause AFTER allowing for stratospheric temperatures to re-adjust to radiative equilibrium, but with surface and tropospheric temperatures held fixed at their unperturbed values.

This follows earlier work (e.g., Ramanathan *et al.*, 1985, Hansen *et al.*, 1981 and references therein). The tropopause is chosen because, in simple models at least, it is considered that in a global and annual mean sense the surface and troposphere are so closely coupled that they behave as a single thermodynamic system. The adjustment of stratospheric temperatures could be considered as a feedback; however, it is counted as part of the forcing because the timescale of the adjustment is a few months, compared with the decadal time-scale required for the surface-atmosphere system to adjust to the forcing, due to the large thermal inertia of the oceans. One additional feature of allowing for stratospheric adjustment is that the change in the net irradiance is then the same at the tropopause as at the top of the atmosphere; this is not the case when stratospheric temperatures are unadjusted (see Hansen et al., 1981).

The utility of radiative forcing will be the subject of Section 4.8, but it is useful here to illustrate the potential importance of the adjustment process using 1-dimensional radiative-convective model results reported by Rind and Lacis (1993) (see Table 4.1).

The surface temperature response to an imposed radiative forcing, ΔT_o, is the so-called no-feedback case, in which clouds, water vapour and surface albedo are held fixed. For this case, the equivalent form of Equation (4.1) is:

$$\Delta T_o = \lambda_o \Delta F$$

where λ_o is the climate sensitivity in the absence of feedbacks and ΔF is the adjusted forcing and is the same as that in Equation (4.1). This form will now be shown to be more useful than an alternative approach using the instantaneous forcing such that:

$$\Delta T_o = \lambda_i \Delta F_i$$

where the subscript "i" refers to the instantaneous forcing.

Table 4.1 shows values of ΔF_i, ΔF, λ_i and λ_o for a number of different forcing mechanisms. The difference between λ_i and λ_o shows the strength and sign of the adjustment process. The sign of the adjustment process depends on whether the change in forcing leads to a heating or a cooling of the stratosphere with a consequent increase or decrease in the thermal infrared emission from the lower stratosphere to the troposphere. For a doubling of CO_2, the stratosphere cools so that the adjusted forcing is about 6% less than the instantaneous. For an increase in CFC-11 from 0 to 1 ppbv the lower stratosphere warms and this makes the adjusted forcing 7% higher than the instantaneous. For changes in stratospheric ozone, the sign of λ_o reverses between using instantaneous and adjusted forcing. For other forcing mechanisms, such as changes in methane, or solar output, the adjustment process is less important.

The most important conclusion from Table 4.1 is that λ calculated using adjusted forcing is much less dependent on the forcing mechanism than that computed using the instantaneous forcing.

Cess et al. (1985) also reported similar experiments using a 1-dimensional radiative-convective model, which had no stratosphere; these showed that forcing due to changes in solar output, CO_2 and tropospheric aerosols result in a similar climate sensitivity, although heavily-absorbing soot aerosols were found to have a markedly different sensitivity.

In preparing this review some difficulty has been experienced in intercomparing work performed by different authors because some have applied the term "radiative forcing" to the instantaneous change in tropopause irradiance, not allowing for any change in stratospheric temperature. In other publications it is not clear which definition of radiative forcing has been adopted. The forcing should be calculated as a global mean using appropriate temperature, humidity and cloud

Table 4.1: Surface temperature changes using a 1-D radiative-convective model (assuming no feedbacks) for a range of forcing mechanisms, together with the radiative forcing at the tropopause (ΔF) and the climate sensitivity parameter (λ_o). The subscript "o" refers to the no-feedback case. The subscript "i" refers to the instantaneous forcing with no feedbacks. These results are from Rind and Lacis (1993) and Lacis (pers. comm.). (Note that surface temperature changes would be about 1 to 4 times larger if climate feedbacks were included – see IPCC (1992)). The changes in radiative properties used here are meant to be illustrative and do not necessarily represent actual or projected changes.

Forcing Mechanism	ΔT_o (K)	ΔF_i (Wm^{-2})	ΔF (Wm^{-2})	λ_i K/(Wm^{-2})	λ_o K/(Wm^{-2})
CO_2 300-600ppmv	1.31	4.63	4.35	0.28	0.30
CH_4 0.28-0.56ppmv	0.16	0.53	0.52	0.30	0.30
N_2O 0.16-0.32ppmv	0.27	0.96	0.92	0.29	0.30
CFC-11 0-1 ppbv	0.07	0.21	0.22	0.31	0.30
CFC-12 0-1 ppbv	0.08	0.26	0.28	0.32	0.30
O_3 50% reduction at all altitudes	-0.38	0.26	-1.23	-1.47	0.31
Solar Const +2% at all wavelengths	1.35	4.48	4.58	0.30	0.30
Stratospheric Aerosol τ = 0.15	-0.99	-3.42	-3.3	0.29	0.30

conditions – again, it is not always clear, in published estimates, what conditions are being used for calculations.

Since the adjustment process is generally most important for changes in greenhouse gases, we suggest that, in future, the greenhouse gas radiative forcing be referred to as either:

(i) **instantaneous radiative forcing** if no change in stratospheric temperature is accounted for, or
(ii) **adjusted radiative forcing** if the stratospheric temperature has been allowed to re-adjust to the instantaneous forcing.

Assumptions concerning the vertical profiles of temperature and all radiatively active constituents (including cloudiness) should be stated.

4.2 Greenhouse Gases

Our understanding of the enhanced greenhouse effect is dependent on two broad areas. First we need to understand the fundamental radiative properties ("spectroscopy") of the gases involved. Next, these spectroscopic data need to be included in radiative transfer models to calculate the radiative forcing due to changes in gas concentration, for a given atmospheric profile of temperature, water vapour and other trace gases, and cloudiness.

In this section the discussion will concentrate on the radiative forcing of greenhouse gases as a result of direct emissions of that gas. This is referred to as the *direct* greenhouse forcing. Greenhouse gas concentrations can change not only as a result of emissions of that gas, but when emissions of other gases lead to chemical reactions which alter the concentration of that gas; this is termed the *indirect* greenhouse forcing. Changes in ozone appear to be the most important indirect effect; these will be considered in Section 4.3.

4.2.1 Spectroscopy

In the thermal infrared (approximately 4-500 μm) molecules absorb and emit radiation by changing the energy with which they vibrate and/or rotate; the wavelengths of the vibration/rotation transitions occur over narrow spectral intervals. Laboratory and theoretical studies are required to determine the wavelengths, strengths and widths of these transitions; in the case of heavy molecules, such as the halocarbons, the individual transitions are so closely-packed, that they are generally not resolved in laboratory observations.

The main databases of spectral parameters of atmospheric gases are subject to periodic update; the two main catalogues, HITRAN (Rothman *et al.*, 1992) and GEISA (Husson *et al.*, 1992) have both been substantially revised since IPCC (1992). These revisions are based on improvements to both laboratory measurements and theoretical techniques. Of the gases of most direct importance to radiative forcing, ozone and methane have undergone the most substantial revision, and many previously neglected weak spectral lines have been added. A detailed assessment of the effect of these revisions has not yet been reported; however, Fomin *et al.* (1993) report that the net irradiance at the tropopause, evaluated for a mid-latitude clear-sky atmosphere, changes by less than 1% between using the 1986 and 1992 versions of HITRAN. It seems unlikely that the effect of these revisions on radiative forcing will be greater than 5% but there is a need to assess both the effect of the changes, and the potential impact of remaining uncertainties.

Continuum absorption, especially by water vapour, is also of importance in calculating radiative forcing. In spite of recent progress in describing the effect, further theoretical and laboratory investigations are required to resolve remaining uncertainties.

Further work on the infrared absorption cross-sections of halocarbons has been reported; this is particularly important for some of the HCFCs (hydrochlorofluorocarbons) and HFCs (hydrofluorocarbons) as some of the data used in IPCC (1990) were from a single source. Comparisons of strengths of many CFCs (chlorofluorocarbons) and HCFC-22 are presented in McDaniel *et al.* (1991), Cappellani and Restelli (1992) and Clerbaux *et al.* (1993). For newer HCFCs, HFCs and perfluorocarbons, measurements are more limited; these are tabulated in WMO (1994).

As examples, for HCFC-123, HCFC-141b, and HFC-142b, the spread of results is more than 25% of the mean cross-section; for HFC-134a the spread is about 10%. Detailed descriptions, including temperatures and pressures of the measurements, are not available for all the datasets, so it is difficult to comment on the discrepancies; only Cappellani and Restelli (1992) and Clerbaux *et al.* (1993) have published cross-sections for a number of HFC/HCFCs over a range of temperatures.

4.2.2 Calculating the Radiative Forcing

In Section 4.7.2 updated calculations of radiative forcing due to changes in greenhouse gas concentration will be presented; in this section the basis on which these calculations are made will be assessed.

A whole hierarchy of different radiative transfer schemes are available to compute the radiative forcing, ranging from line-by-line models through to so-called wide-band models (Ellingson *et al.*, 1991). In addition, the results from such models can be represented by relatively simple empirical formulae, such as those presented in IPCC (1990) and Shi (1992), to allow rapid, and reasonably accurate, computation of forcing.

Many details of the radiation schemes (such as methods of handling clouds, spectral overlap of gases, treatment of water vapour continuum) affect the radiative forcing and are not handled in the same way by all schemes (see Ellingson et al., 1991).

The ultimate test of such models is their ability to reproduce observed irradiances given the observed state of the atmosphere. Until recently, the quality of observations was generally inadequate to assess the models; now high quality experimental data (see e.g., Ellingson et al., 1992) are becoming available and should provide valuable checks on the realism of radiative models.

The radiative forcing due to changes in concentrations of CO_2, CH_4, N_2O, CFC-11, CFC-12 presented in IPCC (1990) have been re-assessed by Kratz et al. (1993) and Shi and Fan (1992), although neither set of authors accounted for stratospheric adjustment. The historical forcing computed for the period 1765-1990 is found to agree with the expressions in IPCC (1990) to within 10% for all gases except for the direct greenhouse forcing due to methane (Table 4.1 indicates that adjustment is not an issue for methane). Kratz et al. (1993), Shi and Fan (1992) and Lelieveld et al. (1993) report forcing values of up to 20% greater than that given in IPCC (1990). The recommended value for the radiative forcing due to methane may need to be revised in the near future, but additional work will be necessary before we can make a precise recommendation. There appears to be no pressing reason to alter the expressions for the other gases. Table 4.2 lists the radiative forcing strength of these gases, relative to CO_2, for small perturbations from present-day conditions; the empirical formulae in Table 2.2 of IPCC (1990) should be used for larger perturbations. It must be emphasised that the relative weakness of the radiative forcing due to changes in CO_2, on a per molecule or per unit mass basis, is more than compensated for by the large increase in its atmospheric concentration, since pre-industrial times, relative to that of other gases. As discussed in Section 4.7.2, CO_2 makes by far the largest individual contribution to the greenhouse gas radiative forcing over that period.

Despite the general agreement between the earlier IPCC reports and more recent calculations, a similar degree of agreement is not found for radiative transfer calculations performed in GCMs used for climate prediction. Cess et al. (1993) report an intercomparison of instantaneous clear-sky radiative forcing due to a doubling of CO_2 from 15 different GCMs. The results deviated by as much as 20% from reference line-by-line calculations; some of the difference was due to the neglect of some minor absorption bands, and in particular those near 10 μm. The size of the deviation should not be taken to indicate the uncertainty in calculating the CO_2 forcing; it is more indicative of weakness in the radiative transfer schemes used in some GCMs.

New estimates of radiative forcing for a whole range of other minor species have been reported recently. These include the HFCs, HCFCs and perfluorinated molecules. Molecules which are created by the destruction of halocarbons have the potential to cause a radiative forcing, but their lifetimes are believed to be too short for them to be of importance (see Section 5.2.5 and WMO, 1994).

HCFCs and HFCs

The radiative forcing due to these gases was reported in IPCC (1990) and taken from Fisher et al. (1990); more recent calculations include those of Shi (1992), Brühl (pers. comm.) and Clerbaux et al. (1993). In general, the results from these sources differ because different radiation schemes and spectroscopic data were used, and different assumptions were made about vertical profiles of temperatures, clouds, overlapping species and the inclusion of stratospheric adjustment. When results are presented as ratios of forcing to other gases (e.g., to CFC-11), the absolute forcing of the reference gas should be, but has not always been, stated.

Table 4.2: Adjusted radiative forcing per unit mass and per unit molecule increase in atmospheric concentration, relative to CO_2. The table shows direct forcing only, for small perturbations around present day conditions. For larger changes use expressions in Table 2.2 of IPCC (1990).

Gas	ΔF for per unit mass relative to CO_2	ΔF for per unit molecule relative to CO_2
CO_2	1	1
CH_4	58	21
N_2O	206	206
CFC-11	3,970	12,400
CFC-12	5,750	15,800

The IPCC (1990) values for this class of gases were instantaneous forcing. The adjusted forcing can be up to 10% greater than the instantaneous forcing (see WMO, 1994).

A more detailed comparison of the radiative forcing due to HFCs and HCFCs is presented in WMO (1994). The variation of the relative forcings from different sources show little consistency. The same spectral data in different radiation schemes do not always give the same relative forcings amongst the HFCs and HCFCs; and schemes using different spectral data do not always show the differences that would be anticipated from the values used for cross-sections. The results from Brühl (pers. comm.) and Clerbaux et al. (1993) generally show no systematic difference compared with IPCC (1990). There are only two gases, HFC-125 and HFC-152a, for which there is a consistent and large deviation from IPCC (1990) and new values are recommended here. For all other gases IPCC (1990) values are retained, and these are estimated to be accurate to within about 25% although more work is necessary to establish this.

Table 4.3 presents values for a wide variety of gases and from a variety of sources. There is a considerable reliance on unpublished material which is regrettable but unavoidable at present.

Since the list of such molecules is increasing, it is useful to divide it into groups on the basis of current use and anticipated production. The categories used here (A.McCulloch, pers. comm.) are:

(i)   in production now, and likely to be widely used;
(ii) in production now, for specialised end use;
(iii) under consideration for specialised end use.

Perfluorocarbons

The radiative effects of CF_4 and C_2F_6 have undergone renewed interest, although values have been reported by earlier studies (e.g., Wang et al., 1980; Ramanathan et al., 1985; Hansen et al., 1989). These gases possess such a high degree of symmetry that most of their absorption occurs in a very narrow spectral region and detailed assessment probably requires use of the cross-sections in a line-by-line code (Isaksen et al., 1992). For C_2F_6 accurate measurements of the cross-section have only recently become available The best estimate for both gases (Isaksen et al. (1992)) is given in Table 4.3. The value for C_6F_{14} is an average from Roehl et al. (1994) and Ko (pers. comm.). The average value for SF_6 from Ko et al. (1993) and Stordal et al. (1993) is also given in the table; Ko et al. (1993) indicate that there is a 10-20% spread in available cross-section data for this molecule.

Stratospheric water vapour

The radiative effects of changes in tropospheric water vapour concentrations are classed as a climate feedback and were discussed in IPCC (1992), and will be further discussed in the next full IPCC assessment in 1995. Increases in stratospheric water vapour might also result from a climate feedback, but the interest here is that they are expected to increase as a result of an increased rate of methane oxidation as methane concentrations rise. In IPCC (1990), the direct effect of methane was increased by 30% to account for the stratospheric water vapour increases whilst WMO (1991) reports a range of 22-38%. However, these numbers have been disputed. Lelieveld et al. (1993) and Wuebbles (pers. comm.) compute the indirect forcing to be about 5% of the direct methane effect, whilst Hauglustaine et al. (1994) find it to be about 1%. Rind and Lacis (1993), on the other hand, calculate the effect to reach 25% by the year 2100 (there is a typographical error in the paper indicating the value is 50%) whilst the effect is much less for earlier years. There appears to be a growing consensus around a smaller value and we propose that the radiative forcing of stratospheric water vapour changes due to methane changes be taken to be of order of 5% of the methane radiative forcing. There are no available observations with which to constrain model estimates of decadal time-scale changes in stratospheric water vapour; the accuracy of models depends on their ability to describe methane and water vapour in the lower stratosphere.

4.3 Radiative Forcing due to Changes in Ozone

4.3.1 Introduction

Chapter 2 discussed the evidence for changes in both tropospheric and stratospheric ozone as a result of human activity. In this section the consequences of these changes on radiative forcing are discussed. The possible links between changes in solar output and changes in ozone will be discussed in Section 4.5.

4.3.2 Radiative Forcing due to Changes in Stratospheric Ozone

The principal features outlined in IPCC (1992) and WMO (1991) concerning the net radiative forcing of the surface-troposphere system due to ozone depletion in the lower stratosphere were:

(i) the distinction between the solar component which acts to heat the surface-troposphere system and the long-wave component which tends to cool it;
(ii) the difference between the instantaneous forcing (referred to as "Mode A" in WMO, 1991) and the forcing calculated using adjusted stratospheric temperatures (referred to as "Mode B" in WMO, 1991); the consequent cooling of the lower stratosphere enhances the long-wave radiative effects to give a net negative forcing.

THE GREENHOUSE EFFECT OF INCREASED CONCENTRATIONS OF CARBON DIOXIDE

It is sometimes stated that, because there is so much CO_2 already in the atmosphere, it is "saturated" and extra CO_2 can have no additional greenhouse effect. The purpose of this box is to present a simplified explanation as to why this is a misconception.

When gases are present in small concentrations (the halocarbons are examples) the radiative effect of a gas is almost linear in concentration; doubling the concentration of, for example, CFC-11 will approximately double its greenhouse effect. This is not the case for greenhouse gases present in larger concentrations, for well-understood reasons (see e.g., Goody and Yung, 1989). Doubling the concentration of CO_2 from its present day concentrations leads to a 10-20% increase in the total greenhouse effect due to CO_2 – this factor is well-understood and has been included in climate models for several decades.

Figure 4.1 illustrates aspects of the increased greenhouse effect of CO_2 using a detailed radiative transfer model (see footnote below for details). Figure 4.1a shows the spectral variation in the net infrared irradiance ("flux") at the tropopause in Wm^{-2}/cm^{-1}. The shape of the curve is dictated by two factors. First, the Planck function determines the maximum amount of energy that can be emitted at a given wavelength and temperature. At typical atmospheric temperatures, the maximum lies between 10 and 15 μm; at wavelengths shorter than 5 μm little can be emitted. The second factor is the absorbing properties of the atmosphere, which is dictated by the presence of greenhouse gases (of which water vapour is the most important, followed by CO_2) and clouds.

If the troposphere were transparent to infrared radiation, then the irradiance reaching the tropopause would be the same as that leaving the surface. However greenhouse gases and clouds absorb the radiation emitted by the surface over a range of wavelengths; they emit energy in all directions, but since the temperature generally decreases with altitude in the troposphere, less on average is emitted upwards than is absorbed from below, so less reaches the tropopause. The downward radiation from the stratosphere into the troposphere is also an important factor.

CO_2, like many other gases, absorbs and emits radiation by changing the energy at which it vibrates and rotates. The wavelengths of CO_2 absorption are grouped into bands (see e.g., Goody and Yung, 1989) with a strong absorption band centred near 15 μm (Figure 4.1b). The centre of the 15 μm absorption band is so strong that the radiation reaching the tropopause comes from very close to the tropopause and, more importantly, the CO_2 in the stratosphere emits as much downwards as the troposphere emits upwards – the net irradiance is thus very close to zero (see Figure 4.1a – note: the feature at about 10 μm is mainly due to ozone).

Figure 4.1c shows the modelled effect of an instantaneous change in CO_2 on the net irradiance at the tropopause. (The change in concentration between 1980 and 1990 is chosen for the purposes of illustration.) All other factors, such as cloudiness and temperature are held fixed. This plot indicates significant changes in irradiance. At the centre of the 15 μm band, the increase in CO_2 concentration has almost no effect – the CO_2 absorption is indeed saturated at these wavelengths. Away from the band centre CO_2 is less strongly absorbing so that an increase in CO_2 concentration does have an effect. The net irradiance at the tropopause decreases – this corresponds to positive radiative forcing that would tend to warm the climate system. As more and more CO_2 is added to the atmosphere, more of its spectrum will become saturated – *but there will always be regions of the spectrum which remain unsaturated and capable of enhancing the greenhouse effect if CO_2 concentrations are increased*. An example is the 10 μm band system. As shown in Figure 4.1b, it is about 1 million times weaker than the peak of the 15 μm band but its contribution to the irradiance change in the lower frame is much higher than might be anticipated; as CO_2 concentrations increase, the 10 μm band would increase in its importance relative to the 15 μm band.

The saturation effect is partly responsible for the fact that CFC molecules are about 10,000 times more effective at enhancing the greenhouse effect than molecules of CO_2. However, for every extra CFC molecule in the atmosphere since pre-industrial times, there are around 70,000 more CO_2 molecules – *thus, the relative weakness of CO_2 per molecule is more than compensated by the large absolute increase in the number of CO_2 molecules in the atmosphere.*

Note:

The radiative transfer calculations were performed using the standard narrow band code of Shine (1991) which has a spectral resolution of 10 cm^{-1}. The atmosphere used is a Northern Hemisphere mean for January, including the effects of clouds, water vapour, ozone, carbon dioxide and a number of other gases, although the above argument is not sensitive to the details. The spectrum in the middle frame is the sum of the linestrengths in each of the 10 cm^{-1} intervals at a temperature of 250 K, using the HITRAN database (Rothman *et al.*, 1992); the units are $cm^{-1}/(kgm^{-2})$. In the lower frame, the irradiance change is the instantaneous radiative forcing.

Radiative Forcing

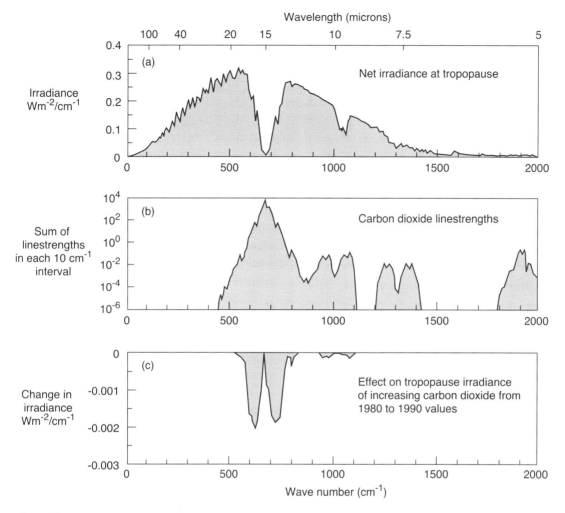

Figure 4.1: An illustration that additional amounts of CO_2 in the atmosphere do enhance the greenhouse effect – the details of the calculations are given in the footnote to the box. (a) Net infrared irradiance (Wm^{-2}/cm^{-1}) at the tropopause from a standard radiative transfer code using typical atmospheric conditions; (b) Representation of the strength of the spectral lines of CO_2 in the thermal infrared; note the logarithmic scale. (c) Change in net irradiance at the tropopause (in Wm^{-2}/cm^{-1}) on increasing the CO_2 concentration from its 1980 to 1990 levels, whilst holding all other parameters fixed. Note that the change in irradiance at the wavelength of maximum absorption, as shown in (b), is essentially zero, while the most marked effects on the irradiance are at wavelengths at which CO_2 is less strongly absorbing.

These features have been supported by several model studies (Ramaswamy *et al.*, 1992; Wang *et al.*, 1993; Shine, 1993; Karol and Frolkis, 1992), as well as by an intercomparison exercise (Shine *et al.*, 1995; WMO, 1994). An expanded discussion of recent developments in this area can be found in WMO (1994).

Accurate knowledge of the magnitude of the ozone loss in the lower stratosphere in different regions is important in evaluating the global radiative forcing. While IPCC (1992) reported the forcings due to ozone loss in the mid-to-high latitudes, the SAGE observations (McCormick *et al.*, 1992) suggest a depletion in the tropical regions as well. Thus, the low-latitude regions could also experience a negative radiative forcing due to the local stratospheric ozone loss (Schwarzkopf and Ramaswamy, 1993), but there is little supporting evidence for such changes from other measurements (WMO, 1994).

The radiative forcing is strongly governed by the shape of the vertical profile of the ozone change, particularly in the vicinity of the tropopause (Wang *et al.*, 1980; Lacis *et al.*, 1990). While it is unambiguously clear that a loss of ozone in the lower stratosphere will lead to a negative radiative forcing of the surface-troposphere system, the precise value is dependent on the assumed shape of the ozone change with altitude. In the absence of reliable data near the tropopause, the different assumptions made in model computations of the forcing can lead to a significant difference in the estimates (Wang *et al.*, 1993; Schwarzkopf and Ramaswamy, 1993).

Wang *et al.* (1993) report instances where the loss of stratospheric ozone results in a warming rather than a cooling of the surface-troposphere system; this can be explained by the fact that the position of the tropopause in these calculations is such that some of the ozone-sonde

Table 4.3: *Radiative forcing due to halocarbons per unit mass and per molecule increase in atmospheric concentration, relative to CFC-11. The table shows direct forcings only. The absolute radiative forcing due to CFC-11 is taken from IPCC (1990) and is 0.22 dX Wm^{-2} where dX is the perturbation to the volume mixing ratio of CFC-11 in ppbv.*

Gas		ΔF per unit mass relative to CFC-11	ΔF per unit molecule relative to CFC-11	Source
CFCs and other chlorinated species				
CFC-11	$CFCl_3$	1.00	1.00	IPCC (1990)
CFC-12	CF_2Cl_2	1.45	1.27	IPCC (1990)
CFC-113	$CF_2ClCFCl_2$	0.93	1.27	IPCC (1990)
CFC-114	CF_2ClCF_2Cl	1.18	1.47	IPCC (1990)
CFC-115	CF_3CF_2Cl	1.04	1.17	IPCC (1990)
*Carbon tetrachloride	CCl_4	0.41	0.46	IPCC (1990)
Methyl chloroform	CH_3CCl_3	0.23	0.22	IPCC (1990)
HFC/HCFCs in production now, and likely to be widely used				
HCFC-22	CHF_2Cl	1.37	0.86	IPCC (1990)
HCFC-141b	CH_3CFCl_2	0.73	0.62	IPCC (1990)
HCFC-142b	CH_3CF_2Cl	1.12	0.82	IPCC (1990)
HFC-134a	CF_3CH_2F	1.04	0.77	IPCC (1990)
*HFC-32	CH_2F_2	1.06	0.40	Fisher (pers.comm.)
HFC/HCFCs in production now, for specialised end use				
HCFC-123	CF_3CHCl_2	0.72	0.80	IPCC (1990)
HCFC-124	CF_3CHFCl	0.88	0.87	IPCC (1990)
*HFC-125	CF_3CHF_2	1.03	0.90	Granier (pers. comm.)
HFC-143a	CH_3CF_3	1.03	0.63	IPCC (1990)
*HFC-152a	CHF_2CH_3	1.02	0.49	Granier (pers. comm.)
*HCFC-225ca	$CF_3CF_2CHCl_2$	0.72	1.07	Granier (pers. comm.)
*HCFC-225cb	$CClF_2CF_2CHClF$	0.87	1.29	Granier (pers. comm.)
HFC/HCFCs under consideration, for specialised end use				
*HFC-23+	CHF_3	1.59	0.81	Fisher (pers. comm.)
*HFC-134	CHF_2CHF_2	1.08	0.80	Fisher (pers. comm.)
*HFC-143	CH_2FCHF_2	0.85	0.52	Clerbaux and Colin (1994)
*HFC-227	CF_3CHFCF_3	0.95	1.17	Mean of Brühl and Fisher (pers. comm.)
*HFC-236	$CF_3CH_2CF_3$	1.06	1.17	Fisher (pers. comm.)
*HFC-245	$CHF_2CF_2CFH_2$	0.95	0.93	Fisher (pers..comm.)
*HFC43-10mee	$C_5H_2F_{10}$	0.86	1.58	Fisher (pers. comm.)
Fully fluorinated substances				
* CF_4		0.69	0.44	Isaksen *et al.* (1992)
* C_2F_6		1.36	1.37	Isaksen *et al.* (1992)
* C_3F_8		0.77	1.05	Roehl *et al.* (1994)
*cC_4F_8	perfluorocyclobutane	1.00	1.46	Fisher (pers.comm.)
*C_6F_{14}		0.75	1.84	Mean of Roehl *et al.* 1994 and Ko (pers comm.)
*SF_6		2.75	2.92	Mean of Ko *et al.* (1993) and Stordal *et al.* (1993)
Other species				
CFC-13	$CClF_3$	1.37	1.04	Mean of Brühl and Fisher (pers comm.)
$CHCl_3$		0.090	0.078	Fisher (pers. comm.)
CH_2Cl_2		0.23	0.14	Fisher (pers. comm.)
CF_3Br (Halon 1301)		1.19	1.29	IPCC (1990)
CF_3I		1.20	1.71	Pinnock (pers. comm.)

\+   Also emitted as a by-product of other halocarbon production.
*   Indicates value amended from IPCC (1990), or gas not previously listed: the mass factor for CCl_4 has altered due to a typographical error in IPCC (1990).

observed increases in ozone are attributed to the lower stratosphere. Hauglustaine *et al.* (1994) also find that the decrease in stratospheric ozone causes a warming of the surface-troposphere system. They used a 2-D chemical-dynamical-radiative model to simulate changes in the concentration of a number of gases including ozone since pre-industrial times. The sign of their stratospheric ozone effect appears to be because, in the Northern Hemisphere at least, their model simulates less ozone depletion in the lower stratosphere than is indicated by recent observations; if ozone loss occurs in the upper stratosphere it can cause a positive radiative forcing (see Lacis *et al.*, 1990).

The overall effect of observed stratospheric ozone depletion on radiative forcing has not been significantly updated since WMO (1991) which reported an adjusted radiative forcing of about -0.1 Wm^{-2} between 1980 and 1990. Hansen *et al.* (1993a) compute a global mean adjusted forcing of -0.2±0.1 Wm^{-2} between 1970 and 1990. Such values represent a small but not negligible offset to the total greenhouse forcing from changes in CO_2, CH_4, N_2O and the CFCs which resulted in a forcing of about 0.45 Wm^{-2} between 1980 and 1990.

Model-dependent factors may also be significant for the accuracy of the computed forcing. An intercomparison exercise (Shine *et al.*, 1995; WMO, 1994), with tightly specified conditions, found that, although all the participating models produced a negative forcing, there was a substantial spread (more than a factor of two). Most of this spread results from differences in calculations at solar wavelengths; the spread seems due more to the approximations used by the different groups rather than an inherent uncertainty in modelling solar radiation in the atmosphere.

4.3.3 Significance of the Forcing due to Stratospheric Ozone Loss

The weight of evidence suggests that heterogeneous chemical reactions involving the decay products of halocarbons are the main cause of the observed global ozone depletion (WMO, 1994). Since several of these compounds, particularly the CFCs, exert a (direct) positive effect, the (indirect) negative radiative forcing due to the chemically-induced ozone losses has the potential to substantially reduce the overall contribution of the halocarbons, particularly over the past decade.

Daniel *et al.* (1994) have partitioned the total direct and the total indirect forcing among the various halocarbons. The indirect effect is strongly dependent upon the effectiveness of each halocarbon for ozone destruction. Each bromine atom released in the stratosphere is estimated to contribute more to the indirect effect than each chlorine atom. About 50% of the 1980-1990 indirect forcing of about -0.1 Wm^{-2} is attributed to the CFCs; since the CFCs dominate the direct positive forcing due to all halocarbons (estimated to be about 0.1 Wm^{-2}), their net effect is about 50% of the direct effect because of ozone loss. Carbon tetrachloride and methyl chloroform each contribute 20% to the indirect forcing, while bromocarbons contribute 10%; since these species contribute little to the direct effect, their net effect is likely to be negative. This analysis suggests that the net halocarbon forcing was probably quite strong and positive in the 1960s and 1970s (i.e., before the period of marked ozone losses), but that the rate of increase in forcing since then has probably decreased owing to the indirect component attributable to ozone depletion.

As with all radiative forcings, there is a significant question concerning the degree to which positive and negative forcings of a different nature (such as the CFC and ozone forcings) can be directly intercompared.

4.3.4 Temperature Changes in the Lower Stratosphere Models

Fixed dynamical heating (FDH) and radiative-convective models compute significant temperature decreases in the lower stratosphere as a result of the stratospheric ozone changes of the past decade (WMO, 1991; Miller *et al.*, 1992; Shine, 1993; Karol and Frolkis, 1992; Ramaswamy and Bowen, 1994). It is this temperature change that determines, to a substantial extent, the negative forcing due to the ozone losses (IPCC, 1992; WMO, 1991).

McCormack and Hood (1994) calculate the temperature decreases using an FDH model employing the ozone changes deduced from the the Solar Backscattered Ultra Violet instrument on the Nimbus-7 satellite for the period 1979-1991; the temperature changes are comparable to or slightly less than the decadal change inferred from satellite and radiosonde data in regions where the observed trends are statistically significant.

GCM studies with imposed ozone losses in the lower stratosphere also obtain a temperature decrease in this region. Hansen *et al.* (1993a) obtain a cooling in the lower stratosphere that is consistent with the global mean decadal trend (-0.4°C) inferred from radiosonde observations. Another GCM study (Ramaswamy, pers. comm.) finds a similar cooling of the lower stratosphere and shows that the FDH temperature changes exhibit a qualitatively similar zonal pattern as the GCM results. A three-dimensional chemical-radiative-dynamical investigation of the climatic effects due to the Antarctic ozone losses (Mahlman *et al.*, 1994), in which the transport of ozone and the ozone losses are handled explicitly, reveals a decrease of temperature in the Southern Hemisphere lower stratosphere that is consistent with the observed trends. An important aspect of such GCM experiments is that the dynamical response results in temperature changes at

Observations

There is a considerable literature reporting observational evidence of a cooling of the lower stratosphere which is discussed in more detail in WMO, 1994. The detection of long-term trends in lower stratospheric temperatures is made difficult because of the episodic and frequent volcanic eruptions which cause a major perturbation to those temperatures (see Section 4.4.2.4) as well as other sources of natural variability. An additional problem concerns the quality of available radiosonde data (Gaffen, 1994; Parker and Cox, 1994). Changes in instrumentation, ascent times and reporting practices introduce a number of time varying biases which have not yet been properly characterised; they indicate the need for some caution when using data primarily intended for weather forecasting in climate trend analysis.

Based on all available radiosonde reports for the period December, 1963 to November, 1988, Oort and Liu (1993) infer a trend in the global lower stratospheric (100-50 mbar) temperature of -0.4±0.12°C/decade; the cooling trend is apparent during all seasons. Latitudinal profiles of the estimated trends show that the cooling of the lower stratosphere has occurred everywhere, but that the strongest temperature decreases (-1°C/decade) have occurred in the Southern Hemisphere high latitudes, strongly suggesting an association with the Antarctic "ozone hole". These results are in generally good agreement with earlier estimates by Angell (1988) using a subset of 63 sonde stations and corroborate the findings in Miller *et al.* (1992). Labitzke's (pers. comm.) analysis of Northern Hemisphere sonde data indicates an annual mean trend of -0.2 to -0.4°C/decade between 1965 and 1992 at most latitudes between 30 and 80 mbar, although the trend varies greatly from month to month. In addition there is an indication that the trend during springtime is more negative over the period 1979 to 1993 than over the period 1965 to 1992.

Channel 4 of the Microwave Sounder Unit (MSU) on the NOAA polar-orbiting satellites has been used to monitor trends in lower stratospheric temperature (between about 120 and 40 mbar) since 1979. Christy and Drouilhet (1994) report a trend of -0.26°C/decade for the period January, 1979 to November, 1992, but comment that, because of the effects of the volcanoes, its significance is hard to assess. For the period 1979 to 1991, Randel and Cobb (1994), using MSU data, infer a significant cooling of the lower stratosphere over the Northern Hemisphere mid-latitudes in winter and spring (with a peak exceeding -1.5°C/decade) and over Antarctica in the Southern Hemisphere spring (peak exceeding -2.5°C/decade) (Figure 4.2); the overall space-time patterns are similar to those determined for ozone trends except in southern mid- to high latitudes in winter. The Northern Hemisphere trends derived from MSU data are in good agreement with the sonde analysis from the Free University of Berlin (McCormack and Hood, 1994; Labitzke, pers. comm.)

In summary, the available analyses indicate that the lower stratosphere has, on a global-mean basis, cooled by about 0.25-0.4°C/decade in recent decades although more work on the quality of the archived data sets is clearly warranted.

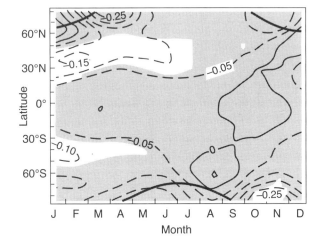

Figure 4.2: Latitude-time sections of zonal mean total ozone (in Dobson Units (DU)/year) and lower stratospheric zonal mean temperature trends (in °C/yr) for the period from 1979 to 1991 from Randel and Cobb (1994). The ozone data are from the Total Ozone Mapping Spectrometer (TOMS). The temperature data are from Channel 4 of the Microwave Sounding Unit (see text for details). Stippling denotes regions where the statistical fits are not different from zero at the 2 sigma level. The hatched areas on the upper plot indicate polar night, where TOMS does not measure. The bold lines on the lower plot also indicate the edge of the polar night.

Inferences

The cooling of the lower stratosphere, either from the long-term records or those over the past decade, is too strong to be attributable to increases in the well-mixed greenhouse gases (mainly CO_2) alone (Miller et al., 1992; Hansen et al., 1993a; Shine, 1993; Ramaswamy and Bowen, 1994). In contrast, models employing the observed ozone losses yield a global temperature decrease that is broadly consistent with observations. This strongly suggests that, among the trace gases, stratospheric ozone change is the dominant contributor to the observed cooling trends. However, the potential competing effects due to as yet unknown changes in other radiative constituents (e.g., ice clouds, water vapour, tropospheric aerosols and tropospheric ozone: Hansen et al., 1993a; Ramaswamy and Bowen, 1994) makes it difficult to rigorously quantify the precise contribution by ozone to the temperature trends.

It is encouraging that both the FDH models and the GCMs yield a lower stratospheric cooling that is consistent with the magnitude inferred from observations. Precise agreement should not be expected as, in all the model studies, the lower stratospheric temperature changes are subject to uncertainties related to the assumed vertical and horizontal distribution of the ozone change, and there is uncertainty in the observed temperature trends.

4.3.5 Sensitivity of Surface-Troposphere Climate

Hansen et al. (1993c) have used a GCM to evaluate the relative effects due to the well-mixed greenhouse gases and ozone upon the surface temperature. A sequence of model runs with the 1970-1990 increases in the well-mixed greenhouse gases only is compared with a sequence including observed ozone changes; the members of each sequence differ only in their initial conditions. It is estimated that the 1970-1990 modelled surface warming (0.35°C) is reduced by 15% due to the ozone changes. Although there is a considerable spread among the different GCM realisations in the sequence of experiments performed, the results indicate that the cooling induced by ozone loss has the potential to reduce the warming effect obtained due to the halocarbon increases over the time period considered. Despite the fact that only one GCM study has been performed to study the surface temperature response, the results are consistent with expectations from the radiative forcing and the simple relation usually assumed for the forcing-response relationship. However, as discussed in Section 4.8, the forcing-response relationship (i.e., the climate sensitivity) may be sensitive to the altitude at which the ozone perturbation is introduced in the model.

4.3.6 Radiative Forcing due to Changes in Tropospheric Ozone

As discussed in Chapter 2 there is observational and model evidence for an increase in tropospheric ozone since pre-industrial times at least in some areas; this is believed to be due to anthropogenic emissions of hydrocarbons, oxides of nitrogen and carbon monoxide. An increase in tropospheric ozone leads to a positive forcing that would tend to warm the surface-troposphere system and has been the subject of several recent studies (e.g., Lacis et al., 1990; WMO, 1991; Wang et al., 1993; Schwarzkopf and Ramaswamy, 1993; Hauglustaine et al., 1994). Wang et al.'s (1993) study using ozonesonde observations indicates that tropospheric ozone changes have the potential to yield a substantial local forcing comparable to the effect of all other greenhouse gases. Hauglustaine et al. (1994) use a 2–D radiative-dynamical-chemical model to estimate that changes in tropospheric ozone since pre-industrial times have contributed a global mean forcing of 0.55 Wm^{-2} over that period, with the dominant contribution in the mid-latitudes of the Northern Hemisphere. Their computed tropospheric ozone change is in reasonable agreement with other similar studies (WMO, 1994) and is not inconsistent with the few available long-term surface observations; global upper tropospheric ozone changes, which are an important component of the forcing, are however not well quantified by observations. Fishman (1991), using observed trends, estimates that between 1965 and 1985 a 1% per year trend in tropospheric ozone applied over the entire Northern Hemisphere implies an approximate global mean forcing of 0.15 Wm^{-2}, or about 20% of the effect of well-mixed greenhouse gases; however since this study uses evidence from a few northern mid-latitude sites, the extrapolation of ozone changes to the whole hemisphere may not be valid.

Marenco et al. (1994) have used observations from France in the late 19th century together with recent observations of the meridional distribution of tropospheric ozone to estimate a global-mean forcing since pre-industrial times of 0.6 Wm^{-2}. They used the simple expression for tropospheric ozone forcing (0.02 Wm^{-2}/ppbv change in tropospheric ozone) used in IPCC (1990); this expression is reasonable for first-order estimates when the ozone change is distributed throughout the troposphere (WMO, 1991; IPCC, 1992).

Given that there have been so few detailed studies of the global or hemispheric radiative forcing due to tropospheric ozone, and there is considerable reliance on chemical models in inferring the change in ozone on large scales, it is difficult to characterise a range for the forcing. Nevertheless, the available modelling results and observational information indicate a positive forcing, since pre-industrial times, of the order of a few tenths of a Wm^{-2}; this would make tropospheric ozone changes of comparable significance to changes of methane over the same period. The increases in tropospheric ozone will be highly regional in nature and so will the positive radiative forcing associated with them. Indeed, since the sources of the precursors of ozone are similar to those of sulphate aerosols in the troposphere, the regional pattern of forcing may be similar in size, but of opposite sign, to the radiative forcing associated with the aerosols (see Section 4.4.1).

In an intercomparison exercise (Shine *et al.*, 1995), similar to that for the stratospheric ozone loss, a specific tropospheric ozone increase profile was also prescribed. All the participating models obtain a positive forcing due to increasing tropospheric ozone; again, however, there is a spread in the forcing – in this case, by about 50%, indicating a non-negligible difference in the radiative treatments in the various models.

4.3.7 Radiative Forcing due to Total Atmospheric Ozone Change

While it is clear that a tropospheric ozone increase alone would lead to a positive radiative forcing and that this would be opposite to the effect due to the lower stratospheric losses, the sign of the net effect is uncertain (WMO, 1985; Lacis *et al.*, 1990; Karol and Frolkis, 1992; Schwarzkopf and Ramaswamy, 1993; Wang *et al.*, 1993). Based on the global-mean estimates above, it seems likely that radiative forcing due to changes in tropospheric ozone has dominated since pre-industrial times. Over the past decade or so, stratospheric ozone loss is likely to have dominated the forcing as the stratospheric losses are more marked than the tropospheric changes over the same period.

4.3.8 Outstanding Issues

Radiative forcing due to a specified ozone change, as a function of the ozone altitude, is qualitatively well understood. Nevertheless, as revealed by the intercomparison exercise, approximate radiative methods appear to differ in their estimates; it is important that such differences be understood. However, the principal limitation inhibiting an accurate estimate of the global ozone forcing is the lack of knowledge of the precise vertical profile of change with latitude and season.

In contrast to the well-mixed greenhouse gases, ozone change causes a more complicated radiative forcing – neither the ozone profile nor its change are uniform in the horizontal and vertical. The spatial patterns of the radiative forcing differ for changes in ozone and the well-mixed gases. In addition, even the apportionment between the surface and the troposphere is different for ozone relative to the well-mixed greenhouse gases (WMO, 1985; 1991). Ozone forcings have not been used for systematic climate studies analogous to those carried out for the well-mixed greenhouse gases. It remains to be determined the degree to which the irradiance change at the tropopause is a reasonable indicator of the surface temperature response in the case of ozone changes.

As discussed in WMO (1991) and WMO (1994), there is evidence that heterogeneous chemistry on sulphate aerosol leads to enhanced ozone loss. The observations of unusually low ozone in the northern mid-latitudes during the winter of 1992-93 and spring of 1993 (see Chapter 2) suggest a possible link with aerosols produced as a result of the Mt. Pinatubo eruption. Such a volcanic eruption-ozone link would imply an enhancement of the transient negative radiative forcing owing to the presence of unusually large volcanic sulphate aerosol concentrations (see Section 4.4.2).

Additional complications in the determination of the ozone forcing are uncertainties in the feedbacks related to chemical processes. One example of this is the stratospheric ozone loss-OH-methane lifetime connection. A decrease of stratospheric ozone could lead to an increase of UV radiation in the troposphere, thereby affecting tropospheric photochemical processes and providing a potential feedback via changes in tropospheric ozone and OH (e.g., Madronich and Granier, 1992, 1994). For example, methane lifetimes could decrease leading to lower concentrations than would be expected in the absence of stratospheric ozone depletion. A lower stratospheric cooling due to the ozone loss can affect, for example, the water vapour mixing ratios there, with the potential to alter heterogeneous chemical reactions. Also, consequent changes in methane in the troposphere could be accompanied by changes in stratospheric water vapour which, in turn, would affect the radiation balance. The radiative forcing calculations so far have mostly employed observed ozone changes. It is extremely desirable to have interactive climate-chemistry-transport model investigations that simulate correctly the observed ozone changes and account for the chemical feedbacks in computing the forcing.

Aircraft emissions have the potential to alter upper tropospheric/lower stratospheric ozone (see Chapter 2). The tentative conclusion of WMO (1994) is that the radiative forcing associated with the ozone change is similar in size to that resulting from the emissions of CO_2 from aircraft: aircraft are responsible for about 3% of the fossil fuel emissions of CO_2. The increase in high altitude cloudiness, via contrails, and changes in water vapour at the same levels, might have a climate consequence but we know of no evidence to support a substantive effect.

4.4 Effects of Tropospheric and Stratospheric Aerosols

4.4.1 Tropospheric Aerosols

Aerosols in the troposphere produced by anthropogenic activities have the potential to change the globally averaged temperature of the Earth. Aerosols may increase the albedo both directly, by back-scattering incoming solar radiation, and indirectly, by increasing the concentrations of cloud condensation nuclei and therefore cloud droplets, which would also result in more back-scattered solar radiation. The latter is also suspected to induce further effects through the decrease of precipitation, thus affecting cloud lifetime and, in turn, cloud cover. On the other hand, carbonaceous particles are strong absorbers at solar wavelengths; over regions of high surface albedos and in cloudy areas they can substantially decrease the amount of radiation reflected to space. Globally, this latter effect is expected to be small and aerosols should increase the planetary albedo; however the influence of aerosol absorption is only beginning to receive attention and further research is needed.

Desert dust is also an important component of the atmospheric aerosol load; the degree to which its distribution has been affected by human activity is not known.

Three important distinctions must be made between tropospheric aerosol forcing and greenhouse forcing. Firstly, estimates of the forcing due to tropospheric aerosols are based on modelled, rather than observed, distributions of the aerosols and their optical properties; for most of the greenhouse gases, the concentration changes are from observations. Secondly, the lifetime of tropospheric aerosols is a few weeks, while for most greenhouse gases the lifetimes range from decades to centuries. Hence most of the aerosols in the troposphere at the present time are the result of emissions during the past few weeks; the enhanced greenhouse gas concentrations are the accumulated effect of releases over the past 10 to 100 years. As pointed out in IPCC (1992), this means that the concentration of tropospheric aerosols will respond far more rapidly to changes in emissions than will the concentrations of greenhouse gases. Thirdly, the aerosol forcing is more heterogeneous in space and time. Thus, global-mean forcing might not reflect its climatic importance as will be discussed in Section 4.8. For tropospheric aerosols there is little difference between the instantaneous and adjusted radiative forcing; most values in the literature are instantaneous forcings.

Although the bulk of sulphate emissions occur in the Northern Hemisphere this does not mean that the Southern Hemisphere might not be significantly affected. Firstly, as discussed in Section 4.4.1.2, the generally cleaner conditions in the Southern Hemisphere make cloud properties more susceptible to increased sulphate concentrations; secondly, as discussed in Section 4.8.3, climate model studies indicate that a radiative forcing applied mainly in the Northern Hemisphere can have a significant effect in the Southern Hemisphere.

4.4.1.1 The direct aerosol effect

Since IPCC (1992) there has been increasing interest in the role of tropospheric aerosols resulting from industrial activity and biomass burning. Chapter 3 considers the various aspects of characterising aerosols for estimating their impact on radiative forcing; in this section we concentrate on the most recent estimates of the aerosol radiative forcing. In the discussion of radiative forcing, the approach in Chapter 3 was a zero-dimensional box model which highlighted some of the causes of uncertainty. Here we consider a more comprehensive approach.

An accurate assessment of the effect of tropospheric aerosols on the radiation balance requires knowledge of:
(i) their spatial and temporal distribution and
(ii) their radiative properties.

Sulphate aerosols

The first estimate of the spatial distribution of the radiative forcing due to sulphate aerosols was discussed in IPCC (1992). It was calculated with a simple aerosol radiative transfer model (Charlson *et al.*, 1991, 1992) applied to the spatial and temporal distribution of sulphate aerosols calculated over the globe obtained from an idealised 3–D chemical-dynamical model (Langner and Rodhe, 1991). Large negative forcings of -2 to -4 Wm^{-2} were found over the eastern USA, central Europe and China, with a Northern Hemispheric averaged sulphate forcing of -1.1 Wm^{-2} (-0.6 Wm^{-2} for the global-mean forcing) since pre-industrial times. The major sources of uncertainty in these calculations are related to the two factors listed above, i.e., sulphate distribution and optical properties.

A more sophisticated radiative transfer calculation has been performed by Kiehl and Briegleb (1993) on the same sulphate distribution (Figure 4.3). They also calculated the spatial and temporal distribution of the instantaneous greenhouse forcing (Figure 4.4) using the pre-industrial to 1990 trace gas changes from IPCC (1990). They then compared the aerosol direct forcing to the greenhouse forcing to assess the relative role of the aerosol's negative forcing to that of the positive greenhouse forcing.

Kiehl and Briegleb (1993) employed temperature and moisture distributions from ECMWF analyses, cloud data from a GCM which were adjusted using observational data, climatological ozone data and a realistic distribution of surface albedo. The sulphate optical properties were based on assuming a log-normal size distribution and employing H_2SO_4 refractive index data (calculations were also carried out for ammonium sulphate indices of refraction). Although the spatial distribution (Figure 4.3) of the anthropogenic aerosol forcing was found to be similar to that obtained by Charlson *et al.* (1991), the magnitude of the forcing is roughly a factor of two smaller: the Northern Hemisphere averaged forcing is -0.43 Wm^{-2} and the global-mean forcing is -0.3 Wm^{-2} since pre-industrial times. The source of the difference between these two studies was mostly attributed to differences in the optical properties of the aerosols that were used in the simulations (asymmetry parameters and spectral dependence of the aerosol extinction: the asymmetry parameter is a measure of the amount of radiation scattered in the forward direction relative to the backward direction).

The radiative forcing due to sulphate aerosols plus greenhouse gases, derived by Kiehl and Briegleb (1993) indicated that over the eastern USA, central Europe and eastern China (see Figure 4.5) the negative forcing by the aerosols offsets a significant amount of the positive greenhouse forcing. Note that for the global average, the instantaneous greenhouse forcing calculated by Kiehl and Briegleb is +2.1 Wm^{-2} compared to a -0.3 Wm^{-2} for aerosol forcing. It must be emphasised that the geographical pattern of forcing shown in Figure 4.5 must not be taken to indicate the geographical pattern of climate response because the atmospheric circulation leads to a non-local response to a localised forcing (see Section 4.8).

Hansen *et al.* (1993a), using the same sulphate distribution, but allowing for some aerosol absorption, obtained a global mean forcing of -0.25 Wm^{-2}.

A global-mean forcing by anthropogenic sulphate aerosols of -0.9 Wm^{-2} has been obtained by Taylor and Penner (1994) with a coupled sulphur chemistry and climate model. The chemical model used was similar to, but independent of, the model used by Langner and Rodhe. It is important to note that the global-mean forcing could range as low as -0.6 Wm^{-2} if the dependence of aerosol extinction on relative humidity was accounted for in the manner of Kiehl and Briegleb. The remainder of the difference was partly attributed to larger sulphate concentrations in the Taylor and Penner study, leading to larger aerosol optical thickness, and to a larger seasonal variation of sulphate amounts.

The very large sensitivity of the results of these simulations to the prescription of the aerosol's optical properties clearly indicates that improvements in calculating the direct sulphate aerosol forcing require more observational data on the chemical and physical properties of the aerosol as well as more refined sulphate distribution calculations.

Biomass burning

As noted in Chapter 3 the aerosols produced by the combustion of biomass account for a significant part of the most radiatively active aerosols of the accumulation mode (0.1 to 1 μm). The global-mean direct aerosol forcing due to biomass burning has been estimated to be -0.8 Wm^{-2} (Penner *et al.*, 1992). This estimate is based upon simple box model calculations and, according to Penner (pers.

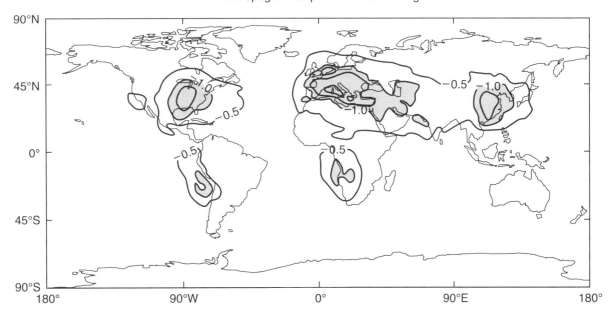

Figure 4.3: Geographic distribution of annual mean direct radiative forcing (Wm^{-2}) from anthropogenic sulphate aerosols (after Kiehl and Briegleb, 1993). The calculations use the sulphate distribution calculated by Langner and Rodhe (1991). The contour interval is 0.5 Wm^{-2} and values below -1Wm^{-2} are shaded.

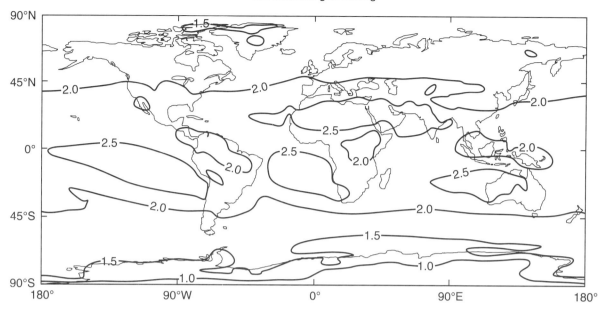

Figure 4.4: Geographic distribution of annual-mean instantaneous greenhouse forcing, pre-industrial to present, (Wm^{-2}) from CO_2, CH_4, N_2O, CFC-11 and CFC-12 (after Kiehl and Briegleb, 1993). The gases are uniformly mixed throughout the atmosphere; the spatial variability in the forcing results from a combination of variations in temperature, humidity and cloudiness. The effects of stratospheric and tropospheric ozone changes, and a number of CFCs, are not included. The contour interval is 0.5 Wm^{-2}.

comm.), it would be reduced by roughly a factor of 2 if it were made with a 3–D model. A further reduction of a factor of 2 applies if the period of reference is fixed to 1850 since significant changes in biomass burning are thought to have occurred prior to 1850 (Andreae, 1994). This gives a global and annual mean forcing of -0.2 Wm^{-2} since pre-industrial times. The calculations of Hansen et al. (1993a) give about half this value (-0.12 Wm^{-2}) for the same period of reference. According to Chapter 3, these values could be in error by nearly a factor of 3, giving a possible range of -0.05 to -0.6 Wm^{-2} since pre-industrial times, and is in addition to the negative forcing due to sulphate aerosol.

The spatial distribution of aerosols from biomass origin is also heterogeneous and mostly localised in the continental tropical areas; in addition, it is strongly seasonal. As for the sulphate aerosols, their climatic influence might not be assessed correctly by using a single global-mean radiative forcing.

The contribution of elemental carbon in aerosols produced from biomass burning can significantly modify their optical characteristics by increasing the absorption efficiency; however, at present, there are no global estimates of the forcing due to elemental carbon or organic based aerosols produced by biomass burning.

The question of the global climatic response to large localised soot injections has been considered with the burning oil wells in Kuwait during the Gulf War in early 1991. The response seems to have been small outside the neighbouring region. A series of model simulations by Bakan et al. (1991) concluded that the reason for that was the relatively low initial height of the soot injection; however, a simulation performed with a fourfold increase of the soot emission led to a large global response.

4.4.1.2 Indirect effect of anthropogenic aerosols on cloud albedo

The global energy balance is very sensitive to cloud albedo, particularly for marine stratus clouds which cover about 25% of the Earth. Cloud albedo is itself calculated to be sensitive to changes in cloud droplet number concentration. This droplet number depends, in a complicated manner, on the concentration of cloud condensation nuclei (CCN) which, in turn, depends on aerosol concentration. Through this indirect effect, the radiative forcing caused by anthropogenic aerosol might be further increased. This concerns both sulphate aerosols and aerosols from biomass burning. The problem, however, is extremely complex and many related key issues such as the influence of cloud entrainment and mixing processes remain mostly unknown. This matter is made more challenging because the ability of an aerosol particle to act as a CCN under the low supersaturations found in clouds depends both on its size and its chemical composition. Two features of the latter are important: the water-soluble fraction and the presence of substances (e.g., organic species) that influence surface tension. Thus substantial chemical understanding is also needed before the CCN/cloud albedo/climate problem can be considered as well posed.

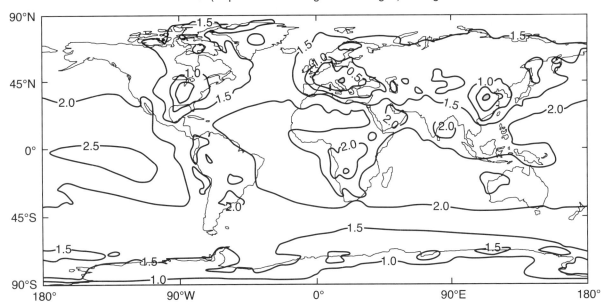

Figure 4.5: Geographic distribution of the annual-mean instantaneous radiative forcing (Wm^{-2}) since pre-industrial times due to changes in both greenhouse gases and sulphate aerosols (direct effects only) – the field is the sum of those shown in Figures 4.3 and 4.4 (after Kiehl and Briegleb, 1993). Note that the global-mean forcing is positive (1.8 Wm^{-2}) but there are small regions where the negative forcing due to sulphate aerosols is greater than the positive greenhouse forcing, so that the local sum is negative. The contour interval is 0.5 Wm^{-2} and values below zero are shaded.

Although the link between CCN concentration and cloud droplet size distributions is not firmly established, direct observations of the impact of CCN on cloud albedos have already been reported in Coakley *et al.* (1987) which showed the impact of aerosols concentrations on cloud reflectances in ship stack exhausts. Kim and Cess (1993) used monthly averaged Earth Radiation Budget Experiment (ERBE) observations to compare the variations with longitude of the cloud albedos of two strongly polluted ocean areas of the Northern Hemisphere (North Atlantic and North Pacific) to those of remote clean areas of the Southern Hemisphere (Indian Ocean and South Pacific). They found some evidence of a decrease of the cloud albedos with distance from the continent in the Northern Hemisphere whereas observations in the Southern Hemisphere showed no similar variation. However, a direct link with aerosols was not established by that paper, so the results are, at most, only suggestive of an aerosol-cloud albedo link.

Satellite observations have also been used to estimate the effect of smoke aerosols on the properties of clouds over the Brazilian Amazon Basin (Kaufman and Nakajima, 1993). This analysis was based on several tens of thousands of bright clouds. It concluded that a large reduction in drop size (from 15 to 9 μm) occurred when smoke was present but it also showed that the reflectances of those clouds indeed decreased, probably due to the increased absorption by graphitic carbon. For those particular clouds, the indirect effect of aerosols would be a positive forcing. These observations exemplify the important influence of the chemical composition of aerosols, in particular that of carbonaceous components. Indeed, the multiple scattering that occurs in dense clouds considerably amplifies the effect of even a small portion of absorbing particles. The influence of aerosols on cloud absorption is a complicating factor that needs to be considered (Stephens and Tsay, 1990; Chylek and Hallett, 1992) before a realistic estimate of aerosol forcing can be made.

Model Simulations

Jones *et al.* (1994), using fields from the United Kingdom Meteorological Office Unified Forecast/Climate Model, calculated the indirect radiative forcing of sulphate aerosols due to a modification of cloud droplet size. The cloud radiative properties were calculated from the cloud liquid water content and effective radius. In that calculation, the effective radius was related to aerosol concentrations using an empirical relationship based on observational data collected for a variety of meteorological conditions (Martin *et al.*, 1994) and the distribution of column sulphate aerosol loading was derived by Langner and Rodhe (1991). The indirect effect at the top of the atmosphere was found to be about -1.3 Wm^{-2} in the global annual mean since pre-industrial times. The Northern

Hemisphere forcing is found to be 1.6 times greater than that in the Southern Hemisphere; this is small compared to the factor of 3 difference in the direct aerosol forcing calculated by Kiehl and Briegleb (1993) using the same sulphate distribution. The reason is that the lower the pre-industrial concentration of CCN, the more susceptible the cloud albedo, and hence forcing, is to an increase in sulphate concentration; this acts to amplify the effect of aerosol increases on the forcing in regions such as the southern oceans. For a similar reason, Jones et al. (1994) point out that the forcing per unit mass of sulphate will have been greater earlier in the century when cloud albedo was more susceptible to changes in CCN.

The estimates of Jones et al. (1994) can be compared with previous estimates based on simpler models. Kaufman and Chou (1993) found a forcing of -0.45 Wm^{-2} and Charlson et al. (1992) estimated -1 Wm^{-2}.

These preliminary estimates of the radiative forcing due to the indirect effect of aerosols were based on calculations of the variations of cloud reflectances associated with variations in droplet size, the cloud liquid water content remaining constant. Observations (Radke et al., 1989) suggest, however, that cloud liquid water may vary in relation to droplet number. Clouds with smaller droplets tend to retain their liquid water through inhibition of the formation of drizzle. As a consequence, for constant external conditions (turbulent surface fluxes, entrainment, sea surface temperature, incident solar energy, etc.), clouds with smaller droplets would tend to have liquid water concentrations that are larger than those of similar clouds with larger droplets. The resulting effect of these mechanisms should be to enhance the space and time averaged albedo of the areas of strong sulphate aerosol concentration.

In recent model simulations, attempts have been made to account for these complex interactions between radiation, cloud microphysics (e.g., cloud droplet size) and precipitation. GCM simulations appear to exhibit a large sensitivity to these microphysical processes: a simulation performed with the LMD GCM using fixed sea surface temperatures (Boucher et al., 1994a, 1994b) investigated the effect that an increase in cloud droplet concentration would have on cloud properties by imposing a fourfold increase, for illustrative purposes. The consequent changes in cloud liquid water content and cloud cover led to a decrease in absorbed solar radiation which was approximately the same size as the decrease due to changes in cloud albedos resulting from variations in droplet size. Hence changes in cloud amount and liquid water content could lead to a significant enhancement of the indirect effect of aerosols.

These model calculations confirm that through various mechanisms the aerosol indirect effect might make a very important contribution to radiative forcing. However, many questions remain open, such as that of the role of absorbing particles; in addition, the observations on which these simulations are based are still scarce, the evidence is tenuous and further observational confirmations are clearly needed. In addition, the representation of clouds in GCMs is still one of the major sources of uncertainty in climate modelling, the indirect aerosol effect adding to that uncertainty. As a consequence we can have little confidence in the reliability of available estimates.

4.4.1.3 Summary of anthropogenic aerosol effects

Although it might appear very likely that anthropogenic aerosols (including those resulting from biomass burning) do influence the radiative balance of the planet by back-scattering incoming solar radiation, the exact value of this effect cannot be calculated with confidence at this time due to inadequate knowledge of their sources, properties and geographical distribution. This uncertainty will persist until new observational data can be collected on a global scale with the required accuracy (Hansen et al., 1993b).

The advisability and utility of comparing global-mean forcing of anthropogenic aerosols with that due to greenhouse forcing is questionable because the two types of forcing are fundamentally different in nature. The inhomogeneity of the geographical and temporal distribution of aerosol forcing might lead to a climatic response different from that resulting from a more geographically uniform forcing, even with the same global-mean forcing (see Section 4.8).

Current estimates place the value of the global-mean forcing since pre-industrial times due to direct effects of anthropogenic sulphate aerosols in the range -0.25 to -0.9 Wm^{-2}. Inclusion of other aerosols from biomass burning could cause an additional forcing of -0.05 to -0.6 Wm^{-2}.

Although some observations have confirmed that, at least in some cases, changes in CCN concentration resulting from aerosol concentration enhancements do have a significant impact on cloud albedos, many questions arise that remain essentially unanswered. Under these conditions any estimate is very uncertain. The presently available estimates have shown that the indirect effect of aerosols might be of similar size to the direct effect, but observations have also shown that the influence of absorbing particles could lead to a decrease in cloud albedos in some circumstances. To permit even initial assessment of the magnitude of the various mechanisms involved in this effect, innovative and exploratory work is required.

4.4.2 Radiative Forcing due to Stratospheric Aerosol

4.4.2.1 Introduction

The stratospheric aerosol layer is of importance for the radiation budget of the stratosphere and for the surface-troposphere system. An enhancement of sulphate particle concentrations is observed following volcanic eruptions which inject sulphur-containing gases into the stratosphere that are then converted to aerosols (see Chapter 3). Large concentrations of these aerosols following volcanic eruptions perturb the climate by:

(a) warming the lower stratosphere through the enhanced absorption of solar and long-wave radiation;

(b) reducing the net radiation available to the surface-troposphere system (i.e., a negative surface-troposphere radiative forcing).

However, owing to the short residence times of these aerosols in the stratosphere (with an e-folding time of 1-2 years), the radiative forcing produced by a single volcanic eruption lasts for only a few years. The spatial and the temporal distribution of the forcing is governed by aerosol microphysics and large-scale dynamical processes. The 1991 eruption of Mt. Pinatubo in the Philippines stands out as probably the major volcanic perturbation of this century (Sato *et al.*, 1993).

There is a possibility that the background stratospheric aerosol amount has increased as a result of human activity (Hofmann, 1990). However, even if all the aerosol in the stratosphere just prior to the Mt. Pinatubo eruption was due to human activity, which is very unlikely, the radiative forcing would be less than -0.2 Wm^{-2}, if the 1990 value of the optical depth given by Sato *et al.* (1993) is used.

4.4.2.2 Stratospheric aerosol radiative effects

Model Computations

The radiative forcing due to stratospheric aerosols is mainly determined by the optical depth and the area-weighted mean (or effective) radius (Lacis *et al.*, 1992). The surface-troposphere forcing is less sensitive to other characteristics such as the aerosol composition and the altitude. For particle effective radii below 2 μm, which is the regime that is normally observed, there is a globally averaged cooling; above this size, the infrared radiative effects would dominate and there would be a globally averaged warming of the surface-troposphere system. The sensitivity of the radiative forcing is estimated to be about -2.5 to -3 Wm^{-2} for a midvisible optical depth of 0.1 (Lacis *et al.*, 1992; Russell *et al.*, 1993). The forcing is spatially inhomogeneous and evolves temporally as well, consistent with the microphysical mechanisms governing the aerosol formation and sinks, and the global scale transport of volcanic gases and aerosols.

In a model investigation of the aerosol produced by the eruption of Mt. Pinatubo, Hansen *et al.* (1992) considered a global mean midvisible optical depth that has a value of about 0.15 ten months after the eruption, decaying exponentially after that with a one-year time constant. The corresponding global-mean forcing is estimated to have had a maximum value of about -4 Wm^{-2}, and a value of about -1 Wm^{-2} until about two years following the eruption. The volcanic aerosol forcing over the first two years after the eruption is comparable with the greenhouse gas forcing over the past century and is substantially greater than greenhouse forcing over the past decade (~0.4 Wm^{-2}). However, the effect of the aerosols from a single eruption is felt for only a short duration unlike the greenhouse gases which have longer lifetimes. For such a transient forcing only a small fraction of the possible equilibrium temperature change (Equation (4.1)) can be realised because of the thermal inertia of the climate system.

Kinne *et al.* (1992) estimate a heating of the tropical lower stratosphere of about 0.3 K/day due to the Mt. Pinatubo aerosols. This heating anomaly resulted in:

a) an increase in the local temperature;

b) an additional mean upward motion for the aerosol cloud leading to adiabatic cooling;

c) reductions in ozone concentrations as a consequence of the enhanced upward motions.

Each of these processes operated on a different time-scale; maximum lower stratospheric temperatures were observed after ~90 days; maximum ozone losses of about 1.5 ppmv occurred after 140 days.

Observations

Numerous observations and/or inferred information concerning the Mt. Pinatubo aerosols have been documented (e.g., GRL (1992) and see Chapter 3). The optical depth estimates during the first year ranged from ~0.1-0.3 and varied with location. The most optically thick portions of the cloud resided between 20-25 km, and were confined to the 10°S-30°N zone during the early period (McCormick *et al.*, 1992). Within 2-3 months, high stratospheric optical depths were observed to at least 70°N; the aerosol loading in the Southern Hemisphere was also enhanced. The aerosol loading has now decayed to near pre-eruption levels.

Directly observed narrow and broad band total solar irradiance effects attributable to the eruption of Mt. Pinatubo have been reported by both surface-based and satellite observations (Valero and Pilewskie, 1992; Dutton and Christy, 1992; Dutton *et al.*, 1994; Minnis *et al.*, 1993; Saunders, 1993). The observations during the first two years generally agree closely with the theoretically expected reductions in the solar radiation at the surface, as estimated from optical depth measurements. The reductions range from <5 to 20 Wm^{-2} in the diurnal mean depending on the associated aerosol optical depth. Figure 4.6 shows the anomaly in low latitude net irradiance at the top of the atmosphere, as measured by the ERBE (Minnis et al., 1993; updated by Minnis, 1994). The observations indicate a decrease of about 5 Wm^{-2} in the absorbed solar radiation in the period following the eruption.

The satellite observations of Minnis et al. suggest the possibility of significant cloud modification and hence the presence of an indirect aerosol radiative forcing due to the volcanic aerosols. Polarisation lidar observations by Sassen (1992) in the Northern mid-latitudes following the Mt. Pinatubo eruption indicate that the aerosols could have affected the microphysical and optical characteristics of cirrus clouds in the upper troposphere. A model study (Jensen and Toon, 1992) also suggests that the aerosols have a potential for the formation of cirrus clouds, with the extent of the effect depending on the aerosol composition and size distribution. While cirrus clouds so affected could make an important contribution to radiative forcing, global estimates of such indirect forcing are not possible at present.

4.4.2.3 Observations and simulations of climatic effects
Surface-troposphere
The time-series of anomalies from satellite observations reveals that there was a nearly global, but non-uniform, tropospheric cooling following the eruption of Mt. Pinatubo (Dutton and Christy, 1992). This has been shown to be coincident with the observed reduction in solar radiation reaching the troposphere. The summer 1992 cooling was greatest in the continental interiors (Hansen et al., 1993a).

Hansen et al. (1992) have conducted a GCM investigation of the climatic impact due to the Mt. Pinatubo aerosols, using their calculations of forcing discussed in Section 4.4.2.2. The model predicts a global cooling that represents a dramatic but temporary break in the warm conditions of recent years. The cooling maximised about fifteen months after the eruption time. There is good agreement between the model predictions and the observations (see Figure 4.7b and c). The model-computed surface cooling ranges from 0.4 to 0.6°C while the observed cooling estimates in 1992 (~a year after the eruption) range from 0.3 to 0.5°C (Hansen et al., 1993a).

Graf et al. (1993) have used a global aerosol microphysics-transport model to obtain the aerosol

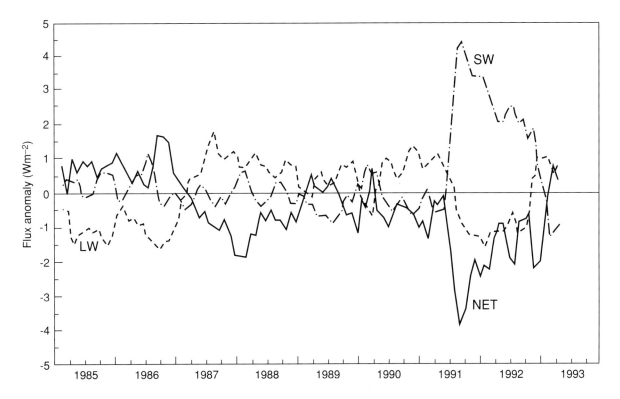

Figure 4.6: Time series of smoothed wide field of view Earth Radiation Budget Experiment long-wave (LW), short-wave (SW) and net (LW-SW) irradiance anomalies (in Wm^{-2}) between 40°N and 40°S relative to the 5 year (1985-1989) monthly mean (after Minnis et al., 1993, updated by Minnis, 1994). The deviation starting in mid-1991 is mainly due to the Mt. Pinatubo eruption – the net anomaly in August (about -4 Wm^{-2}) is almost three times higher than the standard deviation computed between 1985 and 1989.

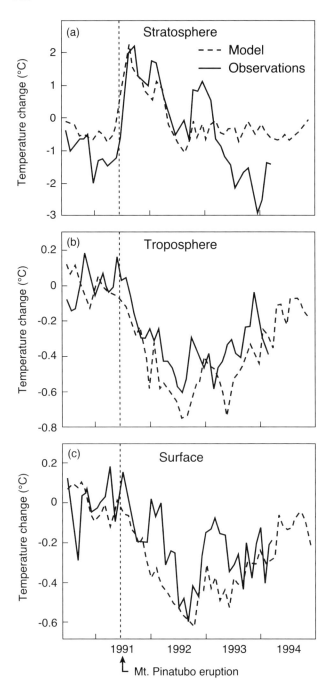

Figure 4.7: Observed and modelled (from the GISS GCM) monthly mean temperature changes over the period of the Mt. Pinatubo eruption (updated from Hansen et al., 1993a). The eruption is indicated by the vertical dashed line. (a) Stratospheric temperatures are from satellite observations and show the 30 mb zonal mean temperature at 10°S and are supplied by M.Gelman, NOAA; the model results are the 10-70 mb layer at 8 to 16°S. The zero is the mean for 1978 to1992. (b) Tropospheric temperatures are from satellite observations and are supplied by J. Christy, Univ of Alabama; the observations and model results are essentially global. The zero is given by the mean for the 12 months preceding the eruption. (c) Surface temperatures are derived from meteorological stations; the observations and model results are essentially global. The zero is given by the mean for the 12 months preceding the eruption. Note that the model results use a simple prediction of the way the optical thickness of the volcanic cloud varied with time, made soon after the eruption, rather than detailed observations of the evolution of the cloud.

distribution under volcanically perturbed conditions; this is used to compute the radiative forcing. In turn, the forcing is employed in a GCM to study the response of the climate system. There is a significant warming of the aerosol-containing layers in the tropics and the mid-latitudes, consistent with observations. The warming increases the equator-to-pole temperature gradient in the upper troposphere and lower stratosphere. There is also an increase in the polar night jet in the Northern Hemisphere, with strong zonal winds that extend down into the troposphere.

By removing the influence of El Niño – Southern Oscillation events, Robock and Mao (1994) have identified the patterns of climate response to large volcanoes on a regional and seasonal basis. The cooling lasts approximately 2 years, with the Northern Hemisphere response being larger, and with an amplitude of approximately 0.1 to 0.2°C averaged over the 6 largest eruptions of the past century. A particularly interesting pattern is the warming over the Northern Hemisphere continents in the winter following tropical eruptions and in the second winter following high latitude eruptions. This pattern is found both in observations (Robock and Mao, 1992) and in GCM simulations (Graf et al., 1993). The radiative forcing from major eruptions (including Mt. Pinatubo) is believed to alter atmospheric circulation patterns, so as to increase the transfer of heat from the oceans to Northern Hemisphere continents during the winter following the eruption.

Stratosphere

A number of studies have reported warming of the lower stratosphere following the eruption of Mt. Pinatubo (Labitzke and McCormick, 1992; Angell, 1993; Hansen et al., 1993a; Spencer and Christy, 1993; Christy and Drouilhet, 1994). Recent observations are discussed in more detail in WMO (1994).

Angell (1993) finds that the warming of the lower stratosphere following the eruption of both Agung (in 1963) and El Chichon (in 1982) was greatest in the equatorial zone and least in the polar zones. The warming following the El Chichon eruption was slightly greater than that following the Agung eruption everywhere except the south polar zone. Preliminary analysis indicated that, in the northern extratropics and the tropics, the warming following the Mt. Pinatubo eruption was comparable to the warming that followed Agung and El Chichon which is consistent with an analysis of radiosonde data by Labitzke (see WMO, 1994). However, in southern temperate and polar zones, the warming following Mt. Pinatubo is considerably greater, perhaps due to a contribution from the eruption of Cerro Hudson in Chile. Globally, the warming following the Mt. Pinatubo eruption was greater than that after the El Chichon and Agung eruptions.

Lower stratospheric temperature data deduced from Channel 4 of MSU (see Section 4.3.4) show that the eruption of Mt. Pinatubo gave a slightly greater global mean warming (about 1.1°C) than El Chichon (about 0.7°C) compared to the immediate pre-eruption temperatures (Christy and Drouilhet, 1994).

Both MSU and radiosondes are in general agreement that the post-eruption warming is similar in the Northern Hemisphere for both El Chichon and Mt. Pinatubo. The greater warming in the Southern Hemisphere following Mt. Pinatubo is consistent in both the MSU and radiosonde analyses.

Hansen *et al.* (1993a) show that the tropical warming in the lower stratosphere associated with Mt. Pinatubo is well simulated by the GISS GCM with an imposed idealised volcanic aerosol cloud (see Figure 4.7a).

4.4.2.4 Summary

By far the most encouraging development in the study of volcanic aerosol impacts has been the opportunity to compare model computations of radiative forcings with observations taken following the eruptions of El Chichon and Mt. Pinatubo, particularly the latter. The computed forcings appear to be in reasonable agreement with inferences from observations. Further, GCM simulations, especially based on the Mt. Pinatubo eruption, appear to yield transient global mean temperature responses that are reasonably consistent with observations, both for the lower stratosphere and the surface. In this regard, the Mt. Pinatubo eruption, being of a large magnitude, has afforded a test of the ability of current GCMs to simulate the climate system's response to a large transient external radiative perturbation. The preliminary work so far provides fair optimism for the credibility of the models on short time-scales. However, considerably more work in comparing model simulations with observations remains to be done, especially evaluating the degree of agreement in the spatial and the seasonal responses, and in extending these findings to forcing with longer time-scales.

Since the stratospheric aerosol forcing is a transient one, its climatic effect on the decadal scale requires a knowledge of the duration of the forcing. A key issue in this regard is the necessity of knowing the spatial and temporal evolution in the aerosol optical depth and size characteristics; this requires a continuous monitoring of the stratosphere. The aerosol microphysical processes and the potential for indirect effects through cirrus cloud modifications also need to be studied.

The radiative forcing from Mt. Pinatubo-type volcanic eruptions is large but transitory. The magnitude of the negative forcing during the first two years (> 1 Wm^{-2}) following the eruption easily overwhelms the decadal increments in the greenhouse gas forcing (~0.4 Wm^{-2}).

The transient cooling tendency at the surface induced by such a negative forcing has the potential to dominate the global surface temperature record for a few years.

4.5 Solar Variability

Since the Sun provides the energy which drives the climate system, variations in solar output are obviously a potential mechanism for driving climate change. Quantification of that role is made difficult because reliable observations of the solar irradiance have only been made from satellites since the late 1970s. These observations can be used to estimate changes in solar output on longer time-scales by making use of relationships with other indicators of solar variability which are more easily observed from the surface. Records for some of these indicators extend back to the 17th century. However, these relationships are based on an incomplete understanding of solar physics, so there are substantial uncertainties in the estimated change in solar output.

In addition, there are a number of studies which have shown significant correlations between indicators of solar variability and changes in climate. The reality of any connection is often controversial, especially when there is no obvious physical mechanism that can provide a quantitative explanation.

4.5.1 Observations of Variability in Solar Irradiance since 1978

Since the late 1970s, the variations of both the integrated and spectrally resolved solar irradiance have been precisely measured from a number of different space-borne instruments. These have demonstrated that the total solar irradiance (sometimes less accurately known as the solar constant) is certainly not strictly constant, having varied by about 0.1% during solar cycle 21 (1979 to 1990) (Hickey *et al.*, 1988; Willson and Hudson, 1988; Lee *et al.*, 1987).

The radiative forcing due to changing solar output is obtained by multiplying the change in total solar irradiance by a factor of (1 – planetary albedo)/4 or 0.175; the factor of 4 is associated with the fact that the Earth's surface area is four times the cross-sectional area of the Earth as seen from the Sun whilst the factor (1 – planetary albedo) ensures that only solar radiation actually absorbed by the Earth-atmosphere system is accounted for. With this scaling a 0.1% change in total solar irradiance would be equivalent to a radiative forcing of 0.24 Wm^{-2} if all the change at the top of the atmosphere was experienced at the tropopause. In this section all changes quoted in Wm^{-2} are instantaneous radiative forcings rather than changes in total solar irradiance itself.

While the radiative forcing associated with a change of 0.1% in total solar irradiance is a significant fraction of that believed to be induced by changes in greenhouse gas concentration over about the period of a decade, there are several factors that are believed to make this comparison an overestimate of the true role of the Sun upon recent climate. First, the thermal inertia of the ocean-atmosphere system leads to lag times between radiative forcing and climate response that are much longer than a decade. Therefore the steady accumulation of greenhouse gases in the atmosphere over many decades can ultimately cause a climatic impact far greater than the cyclic solar changes (which average to a net of near zero over 11 years or so). The realised surface temperature change associated with the most recent 11-year solar cycle forcing is estimated to be in the order of 0.02-0.06°C (Hoffert et al., 1988; Foukal and Lean, 1990; Lee et al., 1993), much smaller than the corresponding equilibrium response of about 0.2°C to a sustained forcing due to a 0.1% change in solar output.

A second factor that affects the interpretation of observed changes in solar irradiance is the fact that the spectral variation of the change plays a key role in determining where the incoming energy is absorbed. Ultraviolet radiation at wavelengths shorter than about 0.3 μm is absorbed nearly entirely above the tropopause, and is estimated to represent about 20% of the change in the Sun's total output over an 11-year cycle; the region between 0.3 and 0.4 μm makes up an additional 13% whose penetration to the troposphere is more limited than that at longer wavelengths (Lean, 1991; Lean et al., 1992a). Thus the direct forcing of the troposphere by solar irradiance changes at the wavelengths that effectively penetrate there is less than 80% of the total spectrally integrated change in the solar irradiance (Lean, 1991).

However, the possibility of feedbacks that could modify the propagation of UV radiation changes to the tropopause should also be considered. Haigh (1994) made the important point that enhancements in the irradiance at wavelengths less than 0.25 μm could lead to ozone increases in the lower stratosphere that would act to shield the troposphere from extraterrestrial irradiance changes. At winter high latitudes less solar radiation could reach the troposphere during periods of high solar activity.

4.5.2 Inferences of Variability in Solar Irradiance on Longer Time-scales

It is possible that solar changes over longer time-scales may have been larger than those since 1978, and therefore more important for understanding climate forcing on time-scales of centuries or more.

One approach to inferring historical changes in solar output assumes that the Sun's brightening can be attributed to the irradiance effect of photospheric structures called faculae, which, like dark sunspots, are regions of enhanced magnetic activity (Lean, 1991). Foukal and Lean (1990) estimated solar cycle irradiance variations since 1874, based on an empirical analysis of the relationship between sunspot number (as a surrogate for faculae) and directly observed total irradiance (corrected for sunspot blocking) since 1979, coupled with historical measurements of sunspot number. Their results suggest that the mean total irradiance has been rising steadily since about 1945, with the largest peaks so far in about 1980 and 1990. The long-term change since 1874 was of the order 0.05% (about 0.1 Wm^{-2}). The method of Lean and co-workers suggests a solar output decrease of about 0.24% (about 0.6 Wm^{-2}) during the Maunder Minimum (mainly in the 17th century) as compared to contemporary values (Lean et al., 1992b). The larger variability was speculated to arise from changes in background solar emission in addition to changes associated with the 11-year cycle alone as determined by Foukal and Lean (1990).

A different approach to inferring variability in solar output was presented by Nesme-Ribes and Mangeney (1992) and Nesme-Ribes et al. (1993), who consider evidence for changes in the solar convection pattern and in the solar diameter within the framework of the solar dynamo model. They include detailed historical observations of the motion of sunspots, which allow inference of the solar surface rotation rate. This in turn can be related to the solar meridional circulation pattern and to a corresponding change in kinetic energy and total solar irradiance. They inferred a variation of the solar irradiance during the Maunder Minimum of about 0.5% (about 1 Wm^{-2}). If this interpretation is correct, it very likely implies a smaller role for changes in solar irradiance during the past century, when the Sun has apparently been much more active (e.g., Foukal and Lean, 1990).

Hoyt and Schatten (1993) estimated the changes in solar irradiance over the past several centuries using five different surrogates for solar activity. They derive a possible forcing of around 0.4 Wm^{-2} between 1850 and the present day. This is considerably smaller than the estimated greenhouse forcing of more than 2 Wm^{-2}, but such solar changes may have made a non-negligible contribution to climate change earlier this century.

Lockwood et al. (1992) compared the brightness fluctuations of the Sun with eight years of observations of 33 Sun-like stars. They concluded that the Sun is in an unusually steady phase compared to similar stars, and suggest that the extrapolation of current measurements of solar brightness to previous epochs based on contemporary solar indicators could be risky. Lockwood et al. (1992) emphasise the fact that future solar-induced climate fluctuations could exceed those of the historical past. However, Foukal (1994) points out that our understanding of the Sun's magnetic activity is inconsistent with any change in solar output greater than those that have been inferred for the past few centuries.

4.5.3. Correlations between Climate and Solar Variability

Suggestive correlative evidence for an enhanced role of the Sun in forcing climate has been presented by many authors, including Labitzke and Van Loon (1993, and references therein), Reid (1991), Friis-Christensen and Lassen (1991) and Tinsley and Heelis (1993). In these studies, the correlations between solar indices such as sunspots and solar cycle length, and observed characteristics of the atmosphere (e.g., temperature at particular locations, global average sea surface temperature, etc.) are examined. Some authors have questioned the usefulness of solar-cycle correlation studies, noting that undersampling and/or aliasing of other periodic atmospheric phenomena could lead to spurious results (e.g., Teitlebaum and Bauer, 1990; Salby and Shea, 1991; Dunkerton and Baldwin, 1992). A combination of, for example, biannual and quasi-biennial oscillations could induce 10- to 12-year periodicities and hence lead to correlations similar to those observed, but unrelated to solar forcing (Dunkerton and Baldwin, 1992). In addition, internal variability of the climate system, or other mechanisms such as changes in ocean circulation, could lead to 10- to 12-year oscillations over short records which could be misinterpreted as solar signals.

Labitzke and Van Loon (1993) noted remarkably high correlations between stratospheric temperatures and solar indices (such as solar emissions at a wavelength of 10.7 cm), as well as an apparent propagation of such correlations in the form of planetary-scale temperature patterns throughout the troposphere (with amplitudes exceeding 1°C in some cases). Kodera (1993) showed, by examining running correlations, that undersampling is not the origin of the Labitzke/Van Loon oscillation. However, it is clear that the large changes in temperature noted by Labitzke and Van Loon (1993) and others are inconsistent with observed changes in the radiative forcing associated with the well-documented total solar irradiance fluctuations of at least the past decade (and very probably the past century). Observations also show that the ultraviolet region of the spectrum is subject to much larger fluctuations over the 11-year solar cycle, i.e., 5-10% at 0.2 μm, and 2-6% at 0.21-0.25 μm, decreasing to less than 1% by 0.29 μm (Rottman, 1988; Cebula et al., 1992; Lean et al., 1992a). These wavelengths are all absorbed well above the tropopause. If this forcing resulted only in local stratospheric temperature changes, then there would be little direct impact upon surface climate. Nevertheless, the likely stratospheric dynamical responses to this forcing and the physical mechanisms that could transmit and possibly amplify their effects in the troposphere need also to be considered.

Recent model assessments of the effects of observed ultraviolet flux variations over the 11-year solar cycle predict relatively minor perturbations of the order of 1% for total ozone, 1°C for stratospheric temperatures, and less than 1 ms^{-1} for stratospheric zonal wind (Huang and Brasseur, 1993; Wuebbles et al., 1991). Observational evidence for the solar forcing of the middle atmosphere is not straightforward because of the short record (from 1 to 3 cycles), so that no relationship can be considered well-established statistically. Recent analysis of stratospheric ozone, wind and temperature data by Hood et al. (1993) and Randel and Cobb (1994) suggest solar-related changes in all of these parameters that are qualitatively consistent with model predictions but quantitatively much larger than expected (e.g., the 23±9 ms^{-1} zonal wind oscillation noted by Hood et al., 1993). Particularly noteworthy is the work of Kodera and co-workers (e.g., Kodera et al., 1990; Kodera, 1991) suggesting a strong solar-cycle dependence of the strength of the upper stratospheric polar night jet. Kodera et al. (1990) noted that changes in the position and strength of the zonal winds near the stratopause in December appeared to influence the tropospheric westerlies observed in the following February, perhaps via wave-mean flow interactions. Such changes are qualitatively consistent with the findings of Labitzke and Van Loon (1993), which require a mechanism capable of changing the climatology of the stratosphere and troposphere throughout the winter season dependent upon solar activity.

Kodera et al. (1991) presented a GCM study of the modulation of the stratospheric circulation during the Northern Hemisphere winter by the quasi-biennial oscillation (QBO) and by solar activity. By imposing a 40% modulation of the ultraviolet heating, they obtained an oscillation quite similar to that observed by Labitzke and Van Loon (1993), but noted that this forcing far exceeds that justifiable by solar observations. Further, Hamilton (1990) examined historical records of surface temperature and pressure dating back to the late 1800s. Combining these data with a wide range of plausible time-series for QBO phase, he showed that no reproduction of Van Loon and Labitzke's (1988) solar correlation results of these quantities was possible during the period from 1875 to 1936, provided that the QBO phases then followed the same pattern as in modern observations. Thus it appears that these oscillations, if real, remain inconsistent with contemporary observations of solar output and theoretical models of atmospheric dynamics.

Friis-Christensen and Lassen (1991) found a high correlation between solar cycle length and Northern Hemisphere land temperatures. Hoyt and Schatten (1993) used solar cycle length as one of their parameters in deducing a quantitative variation in solar output over recent centuries. They also find a good correlation between solar output and Northern Hemisphere surface temperatures over the past century. However their results imply that a 0.14% increase in solar output (equivalent to a forcing of 0.34 Wm^{-2}) causes a surface warming of 0.5°C; this is a high climate sensitivity which, if applied to the 4 Wm^{-2} forcing associated with doubling the concentration of CO_2, would result in a warming of about 6°C. Thus the hypothesis that variability in solar irradiance explains the observed surface temperature variations over the past century is inconsistent with our current understanding of climate sensitivity and would require a dramatically different forcing-response relationship for solar forcing than for other forcing mechanisms; there is no known physical mechanism and no modelling evidence to support such a difference.

4.5.4 Summary

The characterisation of variability in solar irradiance is one element in efforts to understand climate change. In roughly descending order of scientific confidence:

(i) Observations since the late 1970s indicate that the change in radiative forcing between the maximum and minimum of the 11-year solar cycle is about 0.24 Wm^{-2}. Although this is a substantial fraction of, for example, the decadal increment of radiative forcing due to greenhouse gases, the fact that it is cyclic means that such changes in solar irradiance are likely to play only a negligible role in surface-troposphere climate variations.

(ii) Extension of current understanding of the relationship between observed solar irradiance change and other indicators of solar variability indicates that long-term increase in solar irradiance since the 17th century might have been climatically significant. A radiative forcing of about 0.3 Wm^{-2} since 1850 has been suggested, but uncertainties are large.

(iii) Observations of Sun-like stars suggest that solar variability could be larger in the future, but current understanding of the Sun indicates that this variability might not exceed the changes inferred for the Sun during the past few centuries.

(iv) Studies using limited records indicate correlations of winds and temperatures with the solar cycle. However, their interpretation remains controversial on statistical grounds. No physical mechanism has been proposed that is quantitatively consistent with the relationships implied by the correlations.

4.6 Other Forcings

IPCC (1990) briefly discussed two other possible radiative forcing mechanisms – due to changes in surface albedo and "Milankovitch" variations in the Earth's orbital parameters. It was concluded that, for the global-mean, the likely forcings were negligible compared to the effects of other mechanisms on the century time-scale; there appears to be no need to change this conclusion.

If, in future, the emphasis shifts from global and annual mean forcings to the need to know the forcing at more regional levels (see Section 4.8) then there may be a need for re-assessment.

The annual and hemispheric mean radiative forcing due to orbital changes are negligible on the century time-scale. However, Smits *et al.* (1993) show that between 1765 and 1990 the Northern Hemispheric mean incident solar radiation has increased by 1 Wm^{-2} in April and decreased by 1 Wm^{-2} in September; to convert to a forcing change these values should be multiplied by about 0.7 to account for the planetary albedo.

An exploratory calculation of the effect of changes in surface albedo might assume that 20% of the tropical rain forest has been lost over a 30 year period (Henderson-Sellers and Gornitz, 1984) with the surface albedo changing from 0.13 to 0.16. Assuming the changes in planetary albedo to be about one-third of the surface albedo change (due to effects of clouds and atmospheric gases) then for a diurnal-mean incident solar radiation of 400 Wm^{-2}, the 30 year radiative forcing actually over the tropical forests would be about -0.8 Wm^{-2}. Locally this could be a significant fraction of, but the opposite sign to, the greenhouse forcing over the same period (+1.2 to 1.5 Wm^{-2}) but it must be emphasised that this surface albedo effect would occur only over 7% of the globe. However, the assumed change in surface albedo has not yet been confirmed by recent global satellite surveys (e.g., Whitlock *et al.*, 1994); the surface albedo fields from such surveys do not yet agree in either absolute value or spatial gradient. A complete reassessment would need to consider other possible changes in surface albedo.

4.7 Estimates of Total Forcing

Two general approaches can be used to estimate past forcing. One is simply to compare the best estimates of each forcing mechanism, an approach used in IPCC (1990) and updated below. A second method attempts to combine forcing estimates with observed temperature changes and simple climate models; the degree of agreement between modelled and observed temperatures is then taken as some constraint on the sizes of forcings. Recent work in this area is reviewed in Section 4.7.3.

4.7.1 Relative Confidence in Estimates of Radiative Forcing

The method by which the radiative forcing is currently estimated varies from mechanism to mechanism; there is an associated variation in our level of confidence in the estimates. The basis of the forcing estimates has been discussed in the appropriate sub-sections but it is useful to summarise the overall picture.

Direct greenhouse forcing: Radiative forcing due to changes in concentrations of gases such as CO_2, CH_4, N_2O and the CFCs is calculated using radiative transfer models and accurate direct observations of changes in concentrations over the past few decades; changes in concentration since the pre-industrial period are also believed to be reasonably well characterised. Some work remains to be done on the spectroscopic properties of these molecules and on the observational verification of the radiative transfer methods used to calculate the forcing. Although it is difficult to quantify rigorously, the uncertainty in the total anthropogenic greenhouse forcing is probably about ±15%. Confidence in the estimates is relatively high.

Stratospheric ozone and water vapour: Many of the estimates of the forcing due to changes in stratospheric ozone are based on observed trends over the past decade. Outside of the Antarctic springtime, the natural variability of ozone is so large that confidence in the cause of the trends is less than that associated with the "direct" greenhouse gases. The vertical distribution of the trend is not well established. For stratospheric water vapour, observations to date are inadequate for trend detection, and forcing estimates rely entirely on models. Confidence in the estimates is low.

Tropospheric aerosols and ozone: For both these potential forcing mechanisms, calculations rely mainly on model estimated changes in concentrations with some constraints provided by limited observations; the models require knowledge of emissions of precursors and their distribution and the chemical transformations that generate and destroy the aerosols and ozone, as well as the advective processes; confidence in our current ability to model these processes is not high (see Chapters 2 and 3). There are additional significant uncertainties in the optical properties of the aerosols. Confidence in forcing calculations is low. The forcings due to aerosols and tropospheric ozone changes are of opposite sign, but might be of a similar magnitude and may have similar spatial patterns. Two points of importance are that firstly, our confidence in the estimates of both tropospheric ozone forcing and aerosol forcing is low, and secondly, the two processes are related, as the combustion processes leading to the production of aerosols or their precursors also generally lead to the production of ozone precursors. It may therefore be desirable to consider their combined effect.

Confidence in the radiative forcing due to the impact of aerosols on clouds is very low. This is mainly due:

(i) to the difficulties in relating changes in cloud properties to changes in aerosol concentrations and the lack of observational evidence of trends in cloud properties;

(ii) the added impact of carbonaceous material on the radiative properties of clouds;

(iii) the difficulties in representing clouds in atmospheric models.

Volcanic aerosols in the stratosphere: The eruptions of Mt. Pinatubo, and to a lesser extent El Chichon, caused large but transient impacts on surface and top-of-the-atmosphere radiation; the changes in irradiance have been amenable to direct observation. The radiative forcing resulting from Mt. Pinatubo is arguably the best characterised radiative forcing. Confidence in the forcing due to earlier eruptions decreases as there is progressively less knowledge of the strength and nature of individual eruptions. Confidence for eruptions 100 years ago is quite low. Confidence in the effect of volcanic aerosols on, for example, forcing due to changes in cirrus cloud properties or stratospheric ozone is very low.

Solar variations: Direct satellite-based observations of the variation in solar output only extend back until 1978. Although the measurements are of high precision and stability, the radiative forcing at the tropopause is much less clear because of factors such as wavelength variations in the change in incoming solar energy and solar-driven changes in stratospheric ozone. Variations in forcing prior to 1978 have a very low confidence as they rely on indirect techniques which depend on an incomplete understanding of solar physics and/or the assertion that variations since 1978 are representative of variations on much longer time-scales.

4.7.2 Best Estimates of Forcings since Pre-industrial Times

Global-mean forcings are discussed here, but an important caveat is necessary. Even if the global-mean forcing/response relationship discussed in Sections 4.1.1 and 4.8 is correct, the degree of cancellation of global-mean temperature change due to opposing forcing mechanisms may not be a good guide to the cancellation on more regional scales. This applies to both the tropospheric aerosol forcing and the forcing due to changes in ozone. While forcings over smaller geographical areas could be quoted, this is not felt to be helpful: local forcing may bear little resemblance to local response because heat is transported by the atmosphere and ocean.

Greenhouse Forcing: Recent estimates of the direct forcing due to greenhouse gas changes since about 1850 are reasonably consistent. Although slightly different periods are chosen and different trace gas concentration changes may be used, estimates are 2.3 Wm^{-2} (IPCC, 1990), 2.1 Wm^{-2} (Hansen et al., 1993a), 2.6 Wm^{-2} (Hauglustaine et al., 1994), 2.1 Wm^{-2} (Kiehl and Briegleb, 1993) and 2.3 Wm^{-2} (Shi and Fan, 1992). The first three of these are adjusted forcings, the latter two are instantaneous forcings.

Using the revised pre-industrial to 1992 changes in trace gas concentrations listed in Table 2.1 and the forcing-concentration relationships from IPCC (1990) (Table 2.2) and from Table 4.3, an adjusted radiative forcing of 2.45 Wm^{-2} is calculated and is estimated to be accurate to within ±15%. By gas this is: 1.56 Wm^{-2} for CO_2; 0.47 Wm^{-2} for CH_4; 0.14 Wm^{-2} for N_2O; 0.06 Wm^{-2} for CFC-11 and 0.14 Wm^{-2} for CFC-12; other gases give about 0.08 Wm^{-2}, with CCl_4, HCFC-22 and CFC-113 contributing most of this.

For the indirect adjusted forcing, uncertainties are larger. Hansen et al. (1993a) estimate -0.2 Wm^{-2} for stratospheric ozone change from 1970 to 1990; this is somewhat higher than would be inferred from Ramaswamy et al. (1992) who found a forcing of about -0.1 Wm^{-2} for 1980 to 1990, which was a period of more marked ozone depletion. Hauglustaine et al. (1994) calculate an increase in tropospheric ozone since pre-industrial times which resulted in a forcing of 0.55 Wm^{-2}; there are insufficient similar calculations to characterise an uncertainty in this estimate, but, as discussed earlier, a value of order of a few tenths of a Wm^{-2} is supported by other work. A range of 0.2 to 0.6 Wm^{-2} is tentatively suggested; the upper limit corresponds to the simple estimate of Marenco et al. (1994) in which a pre-industrial ozone mixing ratio of 10 ppbv is assumed. For the effect of methane-induced changes in stratospheric water vapour, assuming a forcing of 5% of the direct methane forcing yields a value of about 0.02 Wm^{-2}.

Tropospheric Aerosol Forcing: The discussion in Section 4.4 led to estimates of the direct effect of sulphate aerosols of between -0.25 and -0.9 Wm^{-2} since about 1850. The corresponding value for aerosols from biomass burning is -0.05 to -0.6 Wm^{-2}. However, we cannot rule out the possibility that the effects may lie outside these ranges.

The indirect effect of aerosols, due to cloud albedo changes, is **much** more uncertain. Kaufman and Chou (1993) report a value of -0.45 Wm^{-2} since 1850 due to sulphate aerosols, although they believe that it may be more negative, while Jones et al. (1994) report a value of -1.3 Wm^{-2}. Penner et al. (1992) conclude that the indirect effect of biomass burning must be comparable with the sulphate effect. Hansen et al. (1993a) estimate a combined biomass/sulphate indirect effect of -0.5 Wm^{-2}, with at least a factor of two uncertainty. However, Kaufman and Nakajima's (1993) observational study indicates that the effect of biomass burning may be to *decrease* the albedo of some clouds. This may act to reduce the effect of aerosol changes on cloud radiative properties. A tentative range of 0 to -1.5 Wm^{-2} is used here for the total indirect forcing due to tropospheric aerosols; the upper value is simply the same value as the combined sulphate/biomass direct forcing. However, we emphasise that there can be *little* confidence in such a range on the basis of present understanding.

An overall estimate of the effect of tropospheric aerosol is impossible to make with any confidence. On the basis of a simple addition of the above estimates, a range of -0.3 Wm^{-2} to -3 Wm^{-2} is plausible but it is possible that the actual forcing lies outside this range. These estimates may be subject to *substantial* revision in the future; this is in contrast to the greenhouse forcing which we believe to be far better characterised. Hence, on a global mean basis at least, the effect of tropospheric aerosols could range from a partial to a substantial offset of the greenhouse forcing; however, we emphasise again that cancellation of forcings on a global mean may not be a good guide to the regional climatic influence of individual forcings.

Volcanic Aerosols: Peak global-mean forcings of up to -4 Wm^{-2} have been estimated as a result of the Mt. Pinatubo eruption (Hansen et al., 1992; Minnis et al., 1993) although this was an unusually large eruption. Averaged over several years the forcing due to individual eruptions is -0.5 Wm^{-2} or less but such values are still of significance compared to other forcing mechanisms. Over the past century it is unlikely that there have been large trends in forcing due to stratospheric aerosols, although observational support for such a statement is lacking. Nevertheless, within the century, volcanic eruptions have been distributed unevenly in time, such that the period from about 1925 to 1960 was notably lacking in eruptions of potential climatic importance compared with the periods before and after (see e.g., Robock, 1991). Sato et al.'s (1993) chronology of stratospheric aerosol optical depth since 1850 also suggests important inter-decadal changes, although they acknowledge the significant uncertainties associated with their estimates. Hence, over shorter periods, the variation in stratospheric aerosol loading may have caused a significant variation in forcing.

Solar Variability: A number of recent studies suggest changes since the Maunder Minimum larger than those given in IPCC (1990). In terms of top of the atmosphere forcing since the Maunder Minimum, estimates include +0.6 Wm^{-2} (Lean et al., 1992b), +0.9 Wm^{-2} (Hoyt and Schatten, 1993) and +1 Wm^{-2} (Nesme-Ribes et al., 1993). Since about 1850, the forcing may have been about half these values; a mid-range value of 0.3 Wm^{-2} is taken, with a lower limit of 0.1 Wm^{-2} on the basis of Foukal and Lean's (1990) study and an upper limit of 0.5 Wm^{-2} based on Nesme-Ribes et al. (1993). Superimposed on this increase, a shorter time-scale variation has been proposed. Such changes could be of significance for the earlier part of the century when greenhouse/aerosol forcings were much smaller; their significance since then depends to a large extent on the degree of compensation between the

Radiative Forcing

greenhouse and aerosol forcing, although confidence in the estimates of solar forcing is very low.

Summary: Figure 4.8 shows our estimate of the radiative forcing due to different climate change mechanisms from about 1850 to 1990 based on the above discussion and subject to the caveats about the utility of global-mean forcing discussed in this chapter. An estimate of the net radiative forcing due to all human activity is not presented as the usefulness of combining estimates of global-mean radiative forcings of different signs, and resulting from different spatial patterns, is not currently understood. The different levels of confidence in such estimates also complicates any attempt to combine them.

4.7.3 Estimates of Forcing from Observed Temperature Records

An alternative method for assessing past forcings is to use statistically based comparisons with observed changes in surface temperature, either directly, or else by using a climate model with some representation of oceanic thermal inertia to calculate the evolution of the temperature. This modelled temperature series can then be compared with the observed series and some measure of explained variance can be computed.

Three recent examples of this technique using simple global or hemispheric energy balance models are Kelly and Wigley (1992), Schlesinger and Ramankutty (1992) and Shi *et al.* (1993).

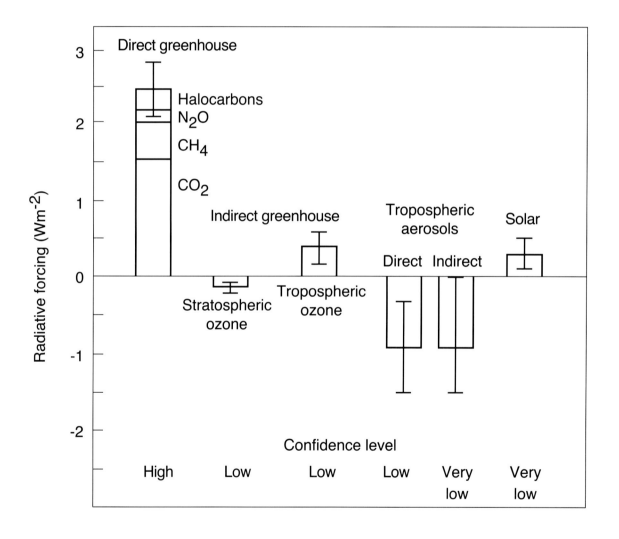

Figure 4.8: Estimates of the global and annual-mean radiative forcing (Wm^{-2}) from 1850 to 1990 for a number of potential climate change mechanisms. The height of the column indicates a mid-range estimate of the forcing whilst the error bars show an estimate of the uncertainty range. Our confidence that the actual forcing lies within this range varies greatly from mechanism to mechanism; a subjective confidence level is indicated for each mechanism as either high, low or very low. The contributions of individual greenhouse gases are indicated in the first column. The episodic and transient nature of the effects of volcanic eruptions means that their relative importance depends greatly on the precise period for which forcings are quoted; their effects are not shown in the figure, but they can be of importance on decadal time-scales.

Kelly and Wigley (1992) use their best estimate of greenhouse gas and sulphate aerosol and impose a speculative representation of the effect of solar cycle length variations on solar constant with an undetermined amplitude B. They find the climate sensitivity and B which yield the largest value of explained variance. The results show that the explained variance is highest (61.6%) when the solar cycle effects are included on their own; however, it is only slightly lower (60.7%) when greenhouse gases and tropospheric aerosols are included. The results demonstrate that the explained variance is a very poor guide to the contribution of different forcing mechanisms.

Schlesinger and Ramankutty (1992) adopt a similar approach and, further, explore the sensitivity to the size of the sulphate aerosol forcing. They obtain the highest explained variance when including greenhouse gases, solar cycle effects and tropospheric aerosols, for a pre-industrial to 1978 contribution of aerosols of -1.5 Wm^{-2}. The proposed solar cycle effect does improve their explained variance and reduces the climate sensitivity (by a factor of two) but again a wide range of different forcings are possible which result in only a small change in explained variance.

Neither of these two studies included the effects of volcanic aerosols. Shi *et al.* (1993) include an estimate of the effect of volcanoes in addition to greenhouse gases and, neglecting changes in tropospheric aerosols, achieve an explained variance exceeding 75%. They do not vary solar constant with solar cycle; instead, using the estimates of solar constant variation from Foukal and Lean (1990), which includes only a small long-term forcing trend, they find that the impact of solar variations is negligible. Thus a quite different mix of forcing mechanisms is found to produce a good agreement with observations.

Further problems in this energy balance model approach are:

(a) they ignore the effect of unforced natural variability; this variability might require that an appreciable fraction of the inter-decadal variance does not need to be explained in terms of forcing mechanisms (Stouffer *et al.*, 1994);

(b) they assume that the surface temperature record is of uniform reliability since the mid-19th century so that the modelled temperatures are required to fit the observations equally well for all periods;

(c) they either assume values for climate sensitivity and effective thermal inertia, or use them as floating variables in the maximisation of explained variance;

(d) they assume the same climate sensitivity for all forcings.

Whilst these studies produce interesting and suggestive results, the ambiguity in the results and the neglect of unforced variability makes it unclear whether they provide useful information in addition to that provided by the more direct estimates discussed in Section 4.7.2.

Hansen *et al.* (1993a) use a simplified version of the GISS GCM with a simple ocean to explore the dependence of the modelled temperature evolution when a variety of forcings are imposed. Although they do not quantify the degree of agreement between observed and modelled temperature changes, they do conclude that the inclusion of solar forcing and tropospheric aerosol forcing does achieve a greater consistency with observations than greenhouse gases alone. They also argue that the net tropospheric aerosol forcing cannot be much greater than their 1850-1990 estimate of -0.9 Wm^{-2}.

4.7.4 Future Forcing

New scenarios of possible future concentrations of greenhouse gases and aerosols are not presented in this report, so earlier estimates from IPCC (1992) are not updated (see also Wigley and Raper (1992) and Wigley (1994)). Between 1990 and 2100 the positive forcing due to the well-mixed greenhouse gases (i.e. excluding ozone) ranges from 3.4 to 8.5 Wm^{-2} depending on assumptions about population growth, economic growth, resource availability and policy. For all assessments CO_2 is the dominant greenhouse gas, contributing between 76 and 84% of the total.

Wigley and Raper (1992) also include estimates of the impact of sulphate aerosol forcing (which includes direct and indirect effects) which are in the range +0.2 to -1.4 Wm^{-2} for the period 1990-2100. The positive forcing occurs for certain scenarios where sulphur emissions decline over the century; because the aerosols are so short-lived, they respond quickly to emission changes and the aerosol concentration decreases. Given the large uncertainties in estimating the contemporary aerosol radiative forcing and in estimating future aerosol concentrations (see Table 3.8), we can have little confidence in such estimates.

Kaufman and Chou (1993) extend their estimates of sulphate aerosol indirect forcing to the year 2060. They use the IPCC (1990) Scenarios with future SO_2 emissions taken to be a fixed fraction of fossil fuel consumption. In their model the effectiveness of sulphur emissions at changing cloud albedo decreases as the aerosol concentration increases. Thus while they estimate that the indirect sulphate effect might offset 50% of the CO_2 forcing in 1980, the offset might reduce to 20% in 2060. In common with other authors, Kaufman and Chou draw attention to the way sulphate aerosol concentrations will respond much more rapidly to emission changes than CO_2 concentrations. It must again be pointed out that such estimates are subject to very large uncertainties given our present understanding, although the study does highlight some important issues.

Daniel *et al.* (1994) have produced projections of future forcing due to halocarbons including their effect on stratospheric ozone concentrations. They consider the relative effectiveness of each molecule in releasing chlorine and bromine in the lower stratosphere and highlight the role of bromocarbons in generating the negative radiative forcing due to ozone depletion. Their future projections are based on assumptions concerning the phase out of CFCs and the growth rate in emissions of HFCs. Daniel *et al.*'s projections have the negative forcing due to ozone decreasing to zero by about 2050 after which the forcing due to the HFCs dominates; the forcing between 1990 and 2100 is about 0.15 Wm^{-2}. Wigley (1994), using a simpler method for the ozone effect, and using IS92 Scenarios (IPCC, 1992), obtains a higher forcing change for this period, with a range 0.2 to 0.4 Wm^{-2}. Such estimates will be sensitive to assumptions about the precise mix of halocarbons used and, more particularly, to the assumed growth rates in emissions.

Other factors influencing radiative forcing, such as possible changes in tropospheric ozone, are also difficult to characterise. Future variations in natural forcings due to volcanic aerosols and solar variability cannot be characterised.

4.8 Forcing-Response Relationships

In Section 4.1.1, the rationale for using adjusted radiative forcing to intercompare different forcing mechanisms was presented. In a 1-D radiative convective model, the climate sensitivity parameter was found to be remarkably independent of the forcing mechanism. In this section the relationship between radiative forcing and climate response in higher dimensional models is briefly discussed. There are two distinct issues: firstly, the extent to which the climate sensitivity parameter (λ in Equation (4.1)) is independent of forcing mechanism and secondly, the extent to which global-mean radiative forcing can indicate changes on smaller scales; this has become important with the recognition of more regional scale forcings resulting from changes in tropospheric ozone and aerosols. In particular, even if λ was independent of forcing mechanism, the global-mean surface temperature change implied by summing forcings with opposing signs and differing spatial patterns may give a misleading impression of the size of regional-scale climate changes.

4.8.1 Background

Manabe and Wetherald (1980), Hansen *et al.* (1984) and Nesme-Ribes *et al.* (1993) have found the climate sensitivity of a GCM to perturbations of CO_2 amount and solar output to be very similar as were the geographical variations of the surface temperature change. This is despite the fact that the surface-atmosphere partitioning of these forcings is quite different with much more of the modelled solar forcing directly affecting the surface; the latitudinal distribution of the forcings is also different. Potter and Cess (1984), using a zonally averaged model, reached a similar conclusion when comparing forcing due to the imposition of background tropospheric aerosols and due to changes in solar output. Global-mean surface temperature responses were within 5% and at any one latitude the response was within 10% for the same magnitude of forcing from the two mechanisms.

These studies led to the conclusion that global-mean temperature change, at the surface and within the troposphere (but not the stratosphere), is primarily governed by the global-mean radiative forcing, and that latitudinal variations in temperature are reasonably independent of forcing mechanism.

These studies have supported the use of radiative forcing as an appropriate index with which to assess both the relative and absolute climatic impacts of changes in forcing. However, there are clearly limits to the applicability of the concept. Palaeoclimatic simulations (e.g., Kutzbach, 1992) which adopt different orbital parameters to those of the present day are able to generate marked regional climate responses, despite the fact that the global mean incident solar radiation is unaltered (even the changes on time-scales of 10^5 years, involving variation in orbital eccentricity, are small).

A related aspect is the dependence of the climate sensitivity to the sign of the forcing. GCM experiments with mixed-layer oceans can show greater sensitivity to negative forcings than positive forcings due mainly to the fact that the snow/sea-ice albedo effect becomes more important in a colder climate (e.g., Wetherald and Manabe, 1975; Manabe *et al.*, 1991 and Roeckner *et al.*, 1994). However, this asymmetry in response was very much smaller in the coupled ocean-atmosphere GCM study of Manabe *et al.* (1991) when they compared the transient response to a halving and doubling of CO_2 concentration. This indicates that caution is necessary in interpreting forcing-response relationships in models with simplified oceans.

4.8.2 Forcing-Response Relationships for Well-mixed Greenhouse Gases

Wang *et al.* (1991, 1992) used a GCM with a mixed-layer ocean to investigate whether "effective CO_2"[1] amount can be used as a proxy for the effect of an increase in a mixture of greenhouse gases. Wang *et al.* (1992) found that the

---

[1] Effective CO_2 is the amount of CO_2 that would give the same radiative forcing since pre-industrial times as that resulting from changes in all greenhouse gases.

global-mean surface temperature response due to the mixture of greenhouse gases was within 4% of that due to an effective CO_2 amount giving the same instantaneous forcing. Deviations in regional temperature response could exceed 20%, for example, over parts of North America. However, in agreement with earlier studies, the zonal mean surface temperature changes due to the two different forcing mechanisms were within 10%.

Wang et al. (1992) compared equal instantaneous forcing (of 3.1 Wm^{-2}) due to effective CO_2 and for the mixture of gases. Allowing for stratospheric adjustment would lead to a decrease in the forcing which would be more marked for effective CO_2 than it would be for the mixture of gases. Application of the ratios of adjusted/unadjusted forcings from Table 4.1 gives an adjusted CO_2 forcing of 2.9 Wm^{-2} and for the mixture of gases it would be 3.0 Wm^{-2}; the ratio of these two forcings is about the same as the ratio of the global-mean warmings calculated by Wang et al. Taking account of adjustment would also lead to a slight decrease in the regional differences in temperature change for the two forcings. Hence the tropospheric climate sensitivity to different mixtures of greenhouse gases appears very similar, particularly when stratospheric adjustment is included.

Despite the fact that the changes in concentration of the greenhouse gases such as CO_2, CH_4, N_2O and the CFCs are globally relatively uniform, Figure 4.4 illustrates that the resulting radiative forcing is anything but uniform, owing to the influence of regional variations in temperature, cloudiness and humidity. Thus the well-behaved nature of the forcing-response relationship for greenhouse gas changes is *not* a simple result of spatial uniformity in its forcing.

4.8.3 Forcing-response Relationship for Other Forcings

Several recent sets of experiments, summarised below, have examined the response to more localised forcing both in the vertical and horizontal.

Hansen et al. (1993b) used an idealised version of the GISS GCM to examine the effect of a hypothetical forcing of 4 Wm^{-2} applied either entirely at the surface, or entirely in individual model layers. The forcing applied at the surface is included in three different ways – in one experiment it is applied equally over the globe, in another, it is applied only polewards of 30°, and in a third it is only applied equatorwards of 30°; in all three cases, the global mean is the same. A forcing applied at latitudes greater than 30° is found to cause a factor of 1.9 more warming than the same forcing at latitudes less than 30°. After accounting for stratospheric adjustment, the sensitivity applied in individual layers is reasonably independent of altitude above the lower troposphere; however, when applied at the surface, or in the near-surface layer, there is a different surface temperature response, most markedly when cloud feedbacks are allowed.

Hansen et al. (1993b) also investigated the dependence of the forcing-response relationships to very idealised changes in ozone, again applied sequentially in each model layer. The climate sensitivity is markedly dependent on the height of the ozone perturbation in cases where cloud feedbacks are allowed.

Taylor and Penner (1994) used a version of the NCAR Community Climate Model with a mixed-layer ocean and computed the pre-industrial to present day forcing due to the direct effect of sulphate aerosols and observed changes in CO_2 concentration. They performed integrations with CO_2 changes alone, sulphate changes alone and with both forcings combined. The climate sensitivity (calculated from the instantaneous forcing) varied markedly between experiments ranging from 1K/Wm^{-2} for sulphate alone to 2.6 K/Wm^{-2} for sulphate and CO_2. Thus in this model the global mean forcings from different mechanisms are a poor guide to climate response. The pattern of climatic response is found to be markedly different to the pattern of forcing; as an example, in the sulphate-only case the Southern Hemisphere forcing is only 20% of the Northern Hemisphere forcing, and yet the Southern Hemisphere surface temperature response is about 70% of the Northern Hemisphere value. An outstanding issue is the high climate sensitivity of the model (which implies a warming of 6.4°C for a doubling of CO_2) compared to other GCMs.

Roeckner et al. (1994) report experiments with the ECHAM GCM using a mixed-layer ocean. Sulphate forcing is mimicked by changing the surface albedo using the pattern of anthropogenic sulphur loading from Langner and Rodhe (1991). Using this pattern, but reversing the sign to give a positive forcing, it is found that the climate sensitivity is identical to that derived from an increased CO_2 experiment despite the large differences in the vertical and horizontal patterns of the forcings. However, it is markedly different when applied as a negative forcing, due to the asymmetry of the snow/sea-ice albedo feedback. As with Taylor and Penner's (1994) study, there was a considerable response remote from the areas of forcing, in the sulphate-only case.

Chen (1994) used the GFDL GCM with a mixed-layer ocean and fixed cloud amounts to examine the dependence of the climate sensitivity; the forcing was imposed in the model by either doubling CO_2 or by perturbing the cloud radiative properties (via the liquid water path and effective radius) either globally or in specific regions in six different ways. The model's global-mean climate sensitivity was found to be virtually independent of the sign, magnitude or location of the forcing, although the latitudinal variation of the climate response differed amongst the experiments.

4.8.4 Summary

The relationship between radiative forcing and climate response is crucially important to the development of radiative forcing (and hence any global warming potential) as an appropriate index. For a wide range of forcing mechanisms the forcing/response relationship appears similar. However, some studies indicate that climate sensitivity may be markedly different for forcings which have a strong vertical or horizontal structure, although a clear consensus has yet to emerge. However, in model calculations where two opposing radiative forcings with differing spatial patterns are applied, the global-mean surface temperature change does not give an adequate indication of the changes on smaller spatial scales. The need for continued careful experimentation with both idealised and realistic forcings in a range of different models is clearly apparent before firm conclusions can be drawn.

References

Andreae, M.O., 1994: Climatic effects of changing atmospheric aerosol levels. In: *World Survey of Climatology, Vol XVI: Future Climates of the World,* A. Henderson-Sellers (ed).

Angell, J.K., 1988: Variations and trends in tropospheric and stratospheric global temperatures, 1958-1987. *J. Climate*, **1**, 1296-1313.

Angell, J.K., 1993: Comparison of stratospheric warming following Agung, El Chichon and Pinatubo volcanic eruptions. *Geophys. Res. Lett.*, **20**, 715-718.

Bakan, S., A. Chlond, U. Cubasch, J. Feichter, H. Graf, H. Grassl, K. Hasselmann, I. Kirchner, M. Latif, E. Roeckner, R. Sausen, U. Schlese, D. Schriever, I. Schult, U. Schumann, F. Sielmann, and W. Welke, 1991: Climate response to smoke from the burning oil wells in Kuwait. *Nature* **351**, 367-371.

Boucher, O., H.Le Treut and M.B. Baker, 1994a: Sensitivity of a GCM to changes in cloud droplet concentration. In: *Preprints of 8th Conference on Atmospheric Radiation,* American Meteorological Society, Boston, pp.558-560.

Boucher, O., H.Le Treut and M. Baker, 1994b: Precipitation and Radiation Modelling in a GCM: Introduction of Cloud Microphysical Processes, submitted to *J. Geophys. Res.*

Cappellani, F. and G. Restelli, 1992: Infrared band strengths and their temperature dependence of the hydrohalocarbons HFC-134a, HFC-152a, HCFC-22, HCFC-123 and HCFC-142b. *Spectrochimica Acta*, **48A**, 1127-1131.

Cebula, R.P., M.T. Deland and B.M. Schlesinger, 1992: Estimates of solar variability using the Mg II index from the NOAA 9 satellite. *J. Geophys. Res.*, **97**, 11613-11620.

Cess, R.D., G.L. Potter, S.J. Ghan and W.L. Gates, 1985: The climatic effects of large injections of atmospheric smoke and dust: a study of climate feedback mechanisms with one and three dimensional climate models. *J.Geophys. Res.*, **90**, 12937-12950.

Cess, R.D., M-H. Zhang, G.L. Potter, H.W. Barker, R.A. Colman, D.A. Dazlich, A.D. Del Genio, M. Esch, J.R. Fraser, V. Galin, W.L. Gates, J.J. Hack, W.J. Ingram, J.T. Kiehl, A.A. Lacis, H. Le Treut, Z.-X. Li, X.-Z. Liang, J.-F. Mahfouf, B.J. McAveney, V.P. Meleshko, J-J. Morcrette, D.A. Randall, E. Roeckner, J.-F. Royer, A.P. Sokolov, P.V. Sporyshev, K.E. Taylor, W.-C. Wang and R.T. Wetherald, 1993: Uncertainties in CO_2 radiative forcing in atmospheric general circulation models. *Science* **262**, 1252-1255.

Charlson, R.J., J. Langner, H. Rodhe, C.B. Leovy and S.G. Warren, 1991: Perturbation of the Northern Hemisphere radiative balance by backscattering from anthropogenic sulphate aerosols. *Tellus,* **43A-B**, 152-163.

Charlson, R.J., S.E. Schwartz, J.M. Hales, R.D. Cess, J.A. Coakley, Jr., J.E. Hansen and D.J. Hofmann, 1992: Climate forcing by anthropogenic aerosols. *Science* **255**, 422-430.

Chen, C-T., 1994: Sensitivity of the simulated global climate to perturbations in low cloud microphysical properties. Ph.D. thesis, Princeton.

Christy, J.R. and S.J. Drouilhet, 1994: Variability in daily, zonal mean lower stratospheric temperatures. *J. Climate*, **7**, 106-120.

Chylek, P. and J. Hallett, 1992: Enhanced absorption of solar radiation by cloud droplets containing soot particles in their surface. *Quarterly J. Roy. Meteor. Soc.* **118**, 167-172.

Clerbaux, C., and R. Colin, 1994: Determination of the infrared cross-section and global warming potential of 1, 1, 2-trifluoroethane (HFC-143). *Geophys. Res. Lett.*, **21**, 2377-2380.

Clerbaux, C., R. Colin, P.C. Simon and C. Granier, 1993: Infrared cross-sections and global warming potentials of 10 alternative hydrohalocarbons. *J. Geophys. Res.*, **98**, 10491-10497.

Coakley, J.A.Jr., R.L. Bernstein and P.A. Durkee, 1987: Effects of ship-stack effluents on cloud reflectivities. *Science,* **237**, 1020-1022.

Daniel, J.S., S. Solomon and D.L. Albritton, 1994: On the evaluation of halocarbon radiative forcing and global warming potentials. *J. Geophys. Res.* (in press)

Dunkerton, T.J., and M.P. Baldwin, 1992: Modes of interannual variability in the stratosphere. *Geophys. Res. Lett.*, **19**, 49-52.

Dutton, E.G. and J.R. Christy, 1992: Solar radiative forcing at selected locations and evidence for global lower tropospheric cooling following the eruptions of El Chichon and Pinatubo. *Geophys. Res. Lett.*, **19**, 2313-2316.

Dutton, E.G., P. Reddy, S. Ryan, J. De Luisi, 1994: Features and effects of aerosol optical depth observed at Mauna Loa, Hawaii, 1982-1992. *J. Geophys. Res.*, **99**, 8295-8306.

Ellingson, R.G., J. Ellis and S.B. Fels , 1991: The intercomparison of radiation codes used in climate models: Long wave results. *J. Geophys. Res*, **96**, 8929-8953

Ellingson, R.G., W.J. Wiscombe, J. DeLuisi, V. Kunde, H. Melfi, D. Murcray and W. Smith, 1992: The Spectral Radiation Experiment (SPECTRE): Clear Sky Observations and their use in ICRCCM and ITRA. In: *IRS '92: Current Problems in Atmospheric Radiation*, S. Keevalik and O. Karner (eds), Deepak Publishing, pp 451-453.

Fisher, D.A., C.H. Hales, W.-C. Wang, M.K. W.Ko and N.-D. Sze, 1990: Model calculations of the relative effects of CFCs and their replacements on global warming. *Nature,* **344**, 513-516.

Fishman, J., 1991: The global consequences of increasing tropospheric ozone concentrations. *Chemosphere*, **22**, 685-695.

Fomin, B.A., A.N. Rublev and A.N. Trotsenko, 1993: Using line parameter databases in accurate modelling of radiative heat exchange in a scattering and absorbing atmosphere. In: *Atmospheric Spectroscopy Applications-ASA Reims 93:Workshop Proceedings,* A. Barbe and L. Rothman (eds), pp290-293.

Foukal, P., 1994: Stellar luminosity variations and global warming. *Science,* **264,** 238-239.

Foukal, P. and J. Lean, 1990: An empirical model of total solar irradiance variation between 1874 and 1988. *Science,* **247,** 556-558.

Friis-Christensen, E. and K. Lassen, 1991: Length of the solar cycle: an indicator of solar activity closely associated with climate. *Science,* **254,** 698-700.

Gaffen, D.J. 1994: Temporal inhomogeneities in radiosonde temperature records. *J Geophys. Res.,* **99,** 3667-3676.

Goody, R.M. and Y.L. Yung, 1989: Atmospheric Radiation. Oxford University Press.

Graf, H.-F., I. Kirchner, A. Robock and I. Schult, 1993: Pinatubo eruption winter climate effects: models versus observations. *Climate Dynamics,* **9,** 81-93.

GRL, 1992. Mount Pinatubo Special Section, *Geophys. Res. Lett.,* 19, 149-218.

Haigh, J.D., 1994: The role of stratospheric ozone in modulating the solar radiative forcing of climate. *Nature,* **370,** 544-546.

Hamilton, K., 1990: A look at the recently proposed solar QBO-weather relationship. *J. Climate,* **3,** 497-503.

Hansen, J.E., D. Johnson, A. Lacis, S. Lebedeff, P. Lee, D. Rind and G. Russell, 1981: Climate impacts of increasing carbon dioxide. *Science,* **213,** 957-966.

Hansen, J.E., A. Lacis, D. Rind, G. Russell, P. Stone, I. Fung, R. Ruedy and J. Lerner 1984: Climate sensitivity: analysis of feedback mechanisms. *Geophysical Monographs,* **29,** 130-163.

Hansen, J.E., A. Lacis and M. Prather, 1989: Greenhouse effect of chlorofluorocarbons and other trace gases. *J.Geophys. Res.,* **94,** 16417-16421

Hansen, J.E., A. Lacis, R. Ruedy and M. Sato, 1992: Potential climate impact of Mount Pinatubo eruption. *Geophys. Res. Lett.,* **19,** 215-218.

Hansen, J.E., A. Lacis, R. Ruedy, M. Sato and H. Wilson, 1993a: How sensitive is the world's climate? *National Geographic Research and Exploration,* **9,** 142-158.

Hansen, J.E., M. Sato, A. Lacis and R. Ruedy, 1993b: Climatic impact of ozone change. In: *The Impact on Climate of Ozone Change and Aerosols, Background Material from the Joint Workshop of IPCC Working Group I and the International Ozone Assessment Panel, Hamburg , May, 1993.*

Hansen, J. E., W. Rossow and I. Fung, 1993c: Long-term monitoring of global climate forcings and feedbacks. *NASA CP3234, Proceedings of workshop held at Goddard Institute for Space Studies, New York, February 3-4, 1992,* 91pp.

Hauglustaine, D.A., C. Granier, G.P. Brasseur and G. Megie, 1994: The importance of atmospheric chemistry in the calculation of radiative forcing on the climate system. *J. Geophys. Res.,* **99,** 1173-1186.

Henderson-Sellers, A. and V. Gornitz, 1984: Possible climatic impacts of land cover transformations, with particular emphasis on tropical deforestation. *Climatic Change,* **6,** 231-257.

Hickey, J. R., B.M. Alton, H.L. Kyle and D. Hoyt, 1988: Total solar irradiance measurements by ERB/Nimbus-7: A review of nine years. *Space Sci. Rev.,* **48,** 321-342 .

Hoffert, M.I., A. Frei and V.K. Narayanan, 1988: Application of Solar Max ACRIM data to analysis of solar-driven climatic variability on Earth, *Clim. Change,* **13,** 267-285.

Hofmann, D.J., 1990: Increase in stratospheric background sulphuric acid aerosol mass in the past 10 years. *Science,* **248,** 996-1000.

Hood, L.L., J.L. Jirikowic and J.P. McCormack, 1993: Quasi-decadal variability of the stratosphere: Influence of long-term solar ultraviolet variations, *J. Atmos. Sci.,* **50,** 3941-3958.

Hoyt, D.V. and K.H. Schatten, 1993: A discussion of plausible solar irradiance variations, 1700-1992. *J. Geophys. Res.,* **98**(A11), 18895-18906.

Huang, T.Y.W. and G.P. Brasseur, 1993: Effect of long-term solar variability in a two-dimensional interactive model of the middle atmosphere. *J. Geophys. Res.,* **98,** 20413-20427.

Husson, N., B. Bonnet, N.A. Scott and A. Chedin,1992: Management and study of spectroscopic information: The GEISA program. *J. Quant. Spectrosc. Radiat. Transf.,* **48,** 509-518

IPCC (Intergovernmental Panel on Climate Change), 1990: *Climate Change: the IPCC Scientific Assessment,* J.T. Houghton, G.J. Jenkins and J.J. Ephraums (eds) Cambridge University Press, Cambridge, UK. 365 pp.

IPCC, 1992: *Climate Change 1992: The Supplementary Report to the IPCC Scientific Assessment,* J.T. Houghton, B.A. Callander and S.K. Varney (eds), WMO/UNEP, Cambridge University Press, Cambridge, UK. 200pp.

Isaksen I.S.A., C. Brühl, M. Molina, H. Schiff, K.P. Shine and F. Stordal, 1992:An assessment of the role of CF_4 and C_2F_6 as greenhouse gases. *Centre for International Climate and Energy Research, University of Oslo,* Policy Note 1992:6.

Jensen, E. and O.B. Toon, 1992: The potential effects of volcanic aerosols on cirrus-cloud microphysics. *Geophys. Res. Lett.,* **19,** 1759-1762.

Jones, A., D.L. Roberts and A. Slingo, 1994 : A climate model study of the indirect radiative forcing by anthropogenic sulphate aerosols. *Nature,* **370,** 450-453.

Karol, I.and V.A. Frolkis, 1992: Model evaluation of the radiative and temperature effects of the ozone content changes in the global atmosphere of 1980s. In: *Ozone in the Troposphere and Stratosphere,* Proceedings of the Quadrennial Ozone Symposium, 1992, Charlottesville, VA. NASA Conference Publication 3266, pp409-412

Kaufman, Y.J. and M.D. Chou, 1993: Model simulations of the competing climatic effects of SO_2 and CO_2. *J. Climate.,* **6,** 1241-1252.

Kaufman, Y.J. and T. Nakajima, 1993: Effect of Amazon smoke on cloud microphysics and albedo – Analysis from satellite imagery. *J. Appl. Meteor.,* **32,** 729-744.

Kelly, P.M., and T.M.L. Wigley, 1992: Solar cycle length, greenhouse forcing, and global climate. *Nature,* **360,** 328-330.

Kiehl, J.T. and B.P. Briegleb, 1993: The relative role of sulphate aerosols and greenhouse gases in climate forcing. *Science,* **260,** 311-314.

Kim, Y. and R.D. Cess, 1993: Effect of anthropogenic sulphate aerosols on low level cloud albedo over oceans. *J. Geophys. Res.,* **98,** 14883-14885.

Kinne, S., O.B. Toon and M. Prather, 1992: Buffering of stratospheric circulations by changing amounts of tropical ozone: A Pinatubo case study. *Geophys. Res. Lett.*, **19**, 1927-1930.

Ko, M.K.W., N.D. Sze, W.-C. Wang, G. Shia, A. Goldman, F.J. Murcray, D.G. Murcray and C.P. Rinsland, 1993: Atmospheric sulphur hexafluoride: sources, sinks and greenhouse warming. *J. Geophys. Res.*, **98**, 10499-10507.

Kodera, K., 1991: The solar and equatorial QBO influences on the stratospheric circulation during the early northern-hemisphere winter. *Geophys. Res. Lett.*, **18**, 1023-1026.

Kodera, K., 1993: The quasi-decadal modulation of the influence of the equatorial QBO on the north-polar stratospheric temperatures. *J. Geophys. Res.*, **98**, 7245-7250.

Kodera, K., K. Yamazaki, M. Chiba and K. Shibata, 1990: Downward propagation of upper stratospheric mean zonal wind perturbation to the troposphere. *Geophys. Res. Lett.*, **17**, 1263-1266.

Kodera, K., M. Chiba and K. Shibata, 1991: A General Circulation Model study of the solar and QBO modulation of the stratospheric circulation during the Northern Hemisphere winter. *Geophys. Res. Lett.*, **18**, 1209-1212.

Kratz, D.P., M.-D. Chou and M.M.-H. Yan, 1993: Infrared radiation parameterizations for the minor CO_2 bands and for several CFC bands in the window region. *J. Climate*, **6**, 1269-1281.

Kutzbach, J.E., 1992: Modelling large climatic changes of the past. In:*Climate System Modelling*, K.E.Trenberth (ed), Cambridge University Press, pp 669-688.

Labitzke, K. and M.P. McCormick, 1992: Stratospheric temperature increases due to Pinatubo aerosols. *Geophys. Res. Lett.*, **19**, 207-210.

Labitzke, K., and H. van Loon, 1993: Some recent studies of probable connections between solar and atmospheric variability. *Annales Geophysicae*. (In press)

Lacis, A.A., D.J. Wuebbles and J.A. Logan, 1990: Radiative forcing of climate by changes in the vertical distribution of ozone. *J. Geophys. Res.*, **95**, 9971-9981.

Lacis, A.A., J.E. Hansen and M. Sato, 1992: Climate forcing by stratospheric aerosols. *Geophys. Res. Lett.*, **19**, 1607-1610.

Langner, J. and H. Rodhe, 1991: A global three-dimensional model of the tropospheric sulphur cycle. *J. Atmos. Chem.*, **13**, 225–263.

Lean, J., 1991: Variations in the sun's radiative output. *Rev. Geophys.*, **29**, 505-535.

Lean, J., M. Van Hoosier, G. Brueckner and D. Prinz, 1992a: SUSIM/UARS observations of the 120 to 300 nm flux variations during the maximum of the solar cycle: inferences for the 11-year cycle. *Geophys. Res. Lett.*, **19**, 2203-2206.

Lean, J., A. Skumanich and O. White, 1992b: Estimating the Sun's radiative output during the Maunder minimum. *Geophys. Res. Lett.*, **19**, 1591-1594.

Lee, R.B., B.R. Barkstrom and R.D. Cess, 1987: Characteristics of the ERBE solar monitors. *Appl. Opt.*, **26**, 3090-3096.

Lee, R.B., W.C. Bolden, M.A. Gibson, J. Paden, D.K. Pandey, S. Thomas and R.S. Wilson, 1993: Correlation of solar irradiance and atmospheric temperature variations derived from spacecraft radiometry. *Adv. Space. Res.* (In press)

Lelieveld, J., P.J. Crutzen and C. Brühl, 1993: Climate effects of atmospheric methane. *Chemosphere*, **26**, 739-768.

Lockwood, G.W., B.A. Skiff, S.L. Baliunas and R.R. Radick, 1992: Long-term solar brightness changes estimated from a survey of Sun-like stars. *Nature*, **360**, 653-655.

Madronich, S. and C. Granier 1992: Impact of recent total ozone changes on tropospheric ozone photodissociation, hydroxyl radicals and methane trends. *Geophys.Res.Lett.*, **19**, 465-467.

Madronich, S. and C. Granier 1994: Tropospheric chemistry changes due to increased UV-B radiation. In: *Stratospheric Ozone Depletion – UV-B Radiation in the Biosphere*, R.H. Biggs and M.E.B. Joyner (eds), Springer, pp.3-10.

Mahlman, J.D., J.P. Pinto and L.J. Umscheid, 1994: Transport, radiative and dynamical effects of the Antarctic ozone hole: A GFDL "SKYHI" model experiment. *J. Atmos. Sci.*, **51**, 489-508.

Manabe S. and R.T.Wetherald, 1980: On the distribution of climate change resulting from an increase of CO_2 content of the atmosphere. *J. Atmos. Sci.*, **37**, 99-118.

Manabe S., R.J. Stouffer, M.J. Spelman and K. Bryan, 1991: Transient response of a coupled ocean-atmosphere model to gradual changes of atmospheric CO_2. Part 1: Annual mean response. *J. Climate*, **4**, 785-818.

Marenco, A., H. Gouget, P. Nédelec, J-P. Pagés and F. Karcher, 1994: Evidence of a long-term increase in tropospheric ozone from Pic du Midi data series – consequences: positive radiative forcing. *J.Geophys.Res.* **99**, 16617-16632.

Martin, G.M., D.W. Johnson and A. Spice, 1994: The measurement and parameterization of effective radius of droplets in warm stratocumulus clouds. *J.Atmos.Sci.*, **51**, 1823-1842.

McCormack, J. P. and L.L. Hood, 1994: Relationship between ozone and temperature trends in the lower stratosphere: Latitudinal and seasonal dependences. *Geophys. Res. Lett.* **21**, 1615-1618.

McCormick, M.P., R.E. Veiga, and W.P. Chu, 1992: Stratospheric ozone profile and total ozone trends derived from the SAGE I and II data. *Geophys. Res. Lett.*, **19**, 269-272.

McDaniel, A.H., C.A. Cantrell, J.A. Davidson, R.E. Shetter and J.G. Calvert, 1991: The temperature dependent infrared absorption cross-sections for chlorofluorocarbons: CFC-11, CFC-12, CFC-13, CFC-14, CFC-22, CFC-113, CFC-114 and CFC-115. *J. Atmos. Chem*, **12**, 211-227.

Miller, A.J., R.M. Nagatani, G.C. Tiao, X.F. Niu, G.C. Reinsel, D. Wuebbles and K. Grant, 1992: Comparisons of observed ozone and temperature trends in the lower stratosphere. *Geophys. Res. Lett.*, **19**, 929-932.

Minnis, P., 1994: Radiative forcing by the 1991 Mt. Pinatubo Eruption. *Preprints of 8th Conference on Atmospheric Radiation* American Meteorological Society, Boston, pp J 9-J11.

Minnis, P., E.F. Harrison, G.G. Gibson, F.M. Denn, D.R. Doelling and W.L. Smith, Jr., 1993: Radiative forcing by the eruption of Mt. Pinatubo deduced from NASA's Earth Radiation Budget Experiment data. *Science*, **259**, 1411-1415.

Nesme-Ribes, E. and A. Mangeney, 1992: On a plausible physical mechanism linking the Maunder minimum to the little ice age. *Radiocarbon,* **34**, 263-270.

Nesme-Ribes, E., E.N. Ferreira, R. Sadourny, H. Le Treut and Z.X. Li, 1993: Solar dynamics and its impact on solar irradiance and the terrestrial climate. *J. Geophys. Res.*, **98**(A11), 18923-18935.

Oort, A.H. and H. Liu, 1993: Upper air temperature trends over the globe, 1958-1989. *J. Climate*, **6**, 292-307.

Parker, D.E. and D.I. Cox, 1994: Towards a consistent global climatological rawinsonde database. *Int. J. Climatology*. (In press)

Penner, J.E., R.E. Dickinson and C.A. O'Neill, 1992: Effects of aerosol from biomass burning on the global radiation budget. *Science*, **256**, 1432-1434.

Potter, G.L. and R.D. Cess, 1984: Background tropospheric aerosols: implications within a statistical-dynamical climate model. *J. Geophys. Res.*, **89**, 9521-9526.

Radke, L.F., J.A. Coakley, Jr. and M.D. King, 1989: Direct and remote sensing observations of the effects of ships on clouds. *Science*, **246**, 1146-1149.

Ramanathan, V., R.J. Cicerone, H.B. Singh and J.T. Kiehl, 1985: Trace gas trends and their potential role in climate change. *J. Geophys. Res.*, **90**, 5547-5566.

Ramaswamy, V. and M.M. Bowen, 1994: Effect of changes in radiatively active species upon the lower stratospheric temperatures. *J. Geophys. Res.*, **99**, 18909-18921.

Ramaswamy, V., M.D. Schwarzkopf and K.P. Shine, 1992: Radiative forcing of climate from halocarbon-induced global stratospheric ozone loss. *Nature*, **355**, 810-812.

Randel, W.J. and J.B. Cobb, 1994 : Coherent variations of monthly mean total ozone and lower stratospheric temperature. *J. Geophys. Res.*. **99**, 5433-5447.

Reid, G.C., 1991: Solar total irradiance variations and the global sea surface temperature record. *J. Geophys. Res.*, **96**, 2835-2844.

Rind D. and A. Lacis 1993: The role of the stratosphere in climate change. *Surveys in Geophysics*, **14**, 133-165.

Rind D., R. Suozzo, N.K. Balachandran and M.J. Prather, 1990: Climate change and the middle atmosphere, Part I: The doubled CO_2 climate. *J. Atmos. Sci.*, **47**, 475-494.

Rind D., N.K. Balachandran and R. Suozzo, 1992: Climate change and the middle atmosphere, Part II: The Impact of volcanic aerosols. *J. Climate*, **5**, 189-208.

Robock, A. 1991: The volcanic contribution to climate change of the past century. In: *Greenhouse-Gas-Induced Climatic Change: A critical appraisal of simulations and observations*, M.E. Schlesinger (ed.), Elsevier, pp 429-443.

Robock, A. and J. Mao, 1992: Winter warming from large volcanic eruptions. *Geophys. Res. Lett.*, **19**, 2405-2408.

Robock, A. and J. Mao, 1994: The volcanic signal in surface temperature observations. Submitted to *J. Climate*.

Roeckner, E., T. Siebert and J. Feichter, 1994: Climatic response to anthropogenic sulphate forcing simulated with a general circulation model. To appear in *Proceedings of Dahlem Workshop "Aerosol Forcing of Climate"*. Wiley.

Roehl, C., D. Boglu, C. Brühl and G. Moortgat, 1994: Infrared band intensities and global warming potentials of CF_4, C_2F_6, C_3F_8, C_4F_{10}, C_5F_{12} and C_6F_{14}. (In preparation.)

Rothman, L.S., R.R. Gamache, R.H. Tipping, C.P. Rinsland, M.A.H. Smith, D.C. Benner, V.Malathy Devi, J-M. Flaud, C. Camy-Peyret, A. Perrin, A. Goldman, S.T. Massie, L.R. Brown and R.A. Toth, 1992: The HITRAN molecular database: editions of 1991 and 1992. *J. Quant. Spectrosc. Radiat. Transfer*, **48**, 469-507.

Rottman, G.J., 1988: Observations of solar UV and EUV variability. *Adv. Space Res.*, **8**(7), 53-66.

Russell, P.B., J.M. Livingston, E.G. Dutton, R.F. Pueschel, J.A. Reagan, T.E. De Foor, M.A. Box, D. Allen, P. Pilewskie, B.M. Herman, S.A. Kinne and D.J. Hoffman 1993: Pinatubo and pre-Pinatubo Optical Depth Spectra: Mauna Loa measurements, comparisons, inferred particle size distributions, radiative effects and relationships to Lidar data. *J.Geophys.Res.*, **99**, 22699-22985.

Salby, M.L.and D.J. Shea, 1991: Correlations between solar activity and the atmosphere: An unphysical explanation. *J. Geophys. Res.*, **96**, 22579-22595.

Sassen, K., 1992: Evidence for liquid-phase cirrus cloud formation from volcanic aerosols: Climatic implications. *Science*, **257**, 516-519.

Sato, M., J.E. Hansen, M.P. McCormick and J.B. Pollack, 1993: Stratospheric aerosol optical depths, 1850-1990. *J. Geophys. Res.*, **98**, 22987-22994.

Saunders, R., 1993: Radiative properties of Mt.. Pinatubo volcanic aerosols over the tropical Atlantic. *Geophys. Res. Lett.*, **20**, 137-140.

Schlesinger, M.E. and N. Ramankutty, 1992: Implications for global warming of intercycle solar-irradiance variations. *Nature*, **360**, 330-333.

Schwarzkopf, M.D. and V. Ramaswamy, 1993: Radiative forcing due to ozone in the 1980s: Dependence on altitude of ozone change. *Geophys. Res. Lett.*, **20**, 205-208.

Shi, G., 1992: Radiative forcing and greenhouse effect due to the atmospheric trace gases. *Science in China (Series B)*, **35**, 217-229.

Shi, G. and X. Fan, 1992: Past, present and future climatic forcing due to greenhouse gases. *Adv. in Atmos. Sci.*, **9**, 279-286.

Shi, G.Y., J.D. Guo, X.B. Fan, B. Wang and L.X. Wang, 1993: Climate change and its causes. *Proceedings of the International Conference on Regional Environment and Climate Changes in East Asia*. Taipei, Taiwan.

Shine, K.P., 1991: On the relative greenhouse strength of gases such as the halocarbons. *J. Atmos. Sci.*, **48**, 1513-1518.

Shine, K.P., 1993: The greenhouse effect and stratospheric change. In: *The role of the stratosphere in global change*, M-L. Chanin (ed.), NATO ASI series, Springer.

Shine, K.P., B.P. Briegleb, A.S. Grossman, D. Hauglustaine, H. Mao, V. Ramaswamy, M.D. Schwarzkopf, R. van Dorland and W.-C. Wang, 1995: Radiative forcing due to changes in ozone: a comparison of different codes. In: *Atmospheric ozone as a climate gas* W-C. Wang and I.S.A. Isaksen(eds), NATO ASI series, Springer. (In press)

Smits, I., T. Fichefet, C. Tricot and J.-P. van Ypersele, 1993: A model study of the time evolution of climate at the secular time-scale. *Atmósfera*, **6**, 255-272.

Spencer, R.W. and J.R. Christy, 1993: Precision lower stratospheric temperature monitoring with the MSU: Validation and results 1979-1991. *J. Climate,* **6**, 1194-1204.

Stephens, G.L. and S-C. Tsay, 1990: On the cloud absorption anomaly. *Quarterly J. Roy. Meteor. Soc.*, **116**, 671-704.

Stordal, F., B. Innset, A.S. Grossman and G. Myhre, 1993: SF_6 as a greenhouse gas; an assessment of Norwegian and global sources and the global warming potential. *Norwegian Institute for Air Research, Report,* OR 15/93.

Stouffer, R.J., S. Manabe and K.Y. Vinnikov, 1994: Model assessment of the role of natural variability in recent global warming. *Nature.,* **367**, 634-636.

Taylor, K. and J.E. Penner, 1994: Response of the climate system to atmospheric aerosols and greenhouse gases. *Nature,* **369**, 734-737.

Teitlebaum, H. and P. Bauer, 1990: Stratospheric temperature eleven year variations: Solar cycle influences or stroboscopic effect? *Ann. Geophys.,* **8**, 239-242.

Tinsley, B.A. and R.A. Heelis, 1993: Correlations of atmospheric dynamics with solar activity: evidence for a connection via the solar wind, atmospheric electricity, and cloud microphysics. *J. Geophys. Res.,* **98**, 10375-10384.

Valero, F.P.J. and P. Pilewskie, 1992: Latitudinal survey of spectral optical depths of the Pinatubo volcanic cloud: Derived particle sizes, columnar mass loadings, and effects on planetary albedo. *Geophys. Res. Lett.,* **19**, 163-166.

Van Loon, H. and K. Labitzke, 1988: Association between the 11-year solar cycle, QBO, and the atmosphere, part II: surface and 700 mb in the Northern Hemisphere winter. *J.Climate,* **1**, 905-920.

Wang, W.-C., J.P. Pinto and Y.L. Yung, 1980: Climatic effects due to halogenated compounds in the Earth's atmosphere. *J. Atmos. Sci.,* **37**, 333-338.

Wang, W.-C., M.P. Dudek, X.-Z. Liang and J.T. Kiehl, 1991: Inadequacy of effective CO_2 as a proxy in simulating the greenhouse effect of other radiatively active gases. *Nature,* **350**, 573-577

Wang, W.-C., M.P. Dudek and X-Z. Liang, 1992: Inadequacy of effective CO_2 as a proxy in assessing the regional climate change due to other radiatively active gases. *Geophys. Res.Lett.,* **19**, 1375-1378

Wang, W.-C., Y-C. Zhuang and R.D. Bojkov, 1993: Climate implications of observed changes in ozone vertical distributions at middle and high latitudes of the Northern Hemisphere. *Geophys. Res. Lett.,* **20**, 1567-1570.

Wetherald, R.T. and S. Manabe, 1975: The effect of changing the solar constant on the climate of a general circulation model. *J. Atmos.Sci.,* **32**, 2044-2059.

Whitlock, C., T. Charlock, W. Staylor, R. Pinker, I. Laszlo, A. Ohmura, H. Gilgen, R. DiPasquale, S. LeCroy and N. Ritchey, 1994: The first global WCRP surface radiation budget data set. Submitted to *Bull. Amer. Meteorol. Soc.*

Wigley, T.M.L., 1994 : The contribution from emissions of different gases to the enhanced greenhouse effect. In: *The Rio Convention on Climate Change: The New Regime and the Agenda for Research,* T. Hanisch (ed), Westview Press.

Wigley, T.M.L. and S.C.B. Raper, 1992: Implications for climate and sea level of revised IPCC emissions scenarios. *Nature,* **347**, 293-300.

Willson, R.C. and H.S. Hudson, 1988: Solar luminosity variations in solar cycle 21. *Nature,* **332**, 810-812.

WMO, 1985: Atmospheric Ozone 1985. *Global Ozone Research and Monitoring Project Report No. 16.* World Meteorological Organisation, Geneva.

WMO, 1991: Scientific Assessment of Ozone Depletion; 1991. *Global Ozone Research and Monitoring Project Report No 25.* World Meteorological Organisation, Geneva.

WMO, 1994: Scientific Assessment of Ozone Depletion, 1994: *Global Ozone Research and Monitoring Project Report No. 37.* World Meteorological Organisation, Geneva.

Wuebbles, D.J., D.E. Kinnison, K.E. Grant and J. Lean, 1991: The effect of solar flux variations and trace gas emissions on recent trends in stratospheric ozone and temperature. *J. Geomagn. Geoelect.,* **43**, Suppl., Part 2, 709-718.

5

Trace Gas Radiative Forcing Indices

D.L. ALBRITTON, R.G. DERWENT, I.S.A. ISAKSEN, M. LAL, D.J. WUEBBLES

Contributors:
C. Brühl, J.S. Daniel, D. Fisher, C. Granier, S.C. Liu, K. Patten, V. Ramaswamy, T.M.L. Wigley

CONTENTS

Summary   209

5.1 Introduction   211
  5.1.1 Utility of Trace-Gas Radiative Forcing Indices   211
    5.1.1.1 Changes in radiative forcing as the basis for measures of induced climate change   211
    5.1.1.2 Uses of radiative forcing indices   211
  5.1.2 Definition of Relative Radiative Forcing Indices   212
    5.1.2.1 General characteristics   212
    5.1.2.2 Primary factors   212
    5.1.2.3 Additional factors   214
    5.1.2.4 Formulations of radiative forcing indices   215
  5.1.3 General Scientific Limitations of a Simple Index   216

5.2 Calculation of Global Warming Potentials   217
  5.2.1 Reference Molecule   217
  5.2.2 Spectral Properties of the Present and Future Atmosphere   218
    5.2.2.1 Trace gas composition   218
    5.2.2.2 Water vapour and clouds   219
  5.2.3 Lifetimes   220
  5.2.4 Direct GWPs   221
  5.2.5 Indirect Effects   223
    5.2.5.1 General characteristics   223
    5.2.5.2 Indirect effects upon the GWP of CH_4 225
  5.2.6 The Product of GWP and Estimated Current Emissions   226

5.3 A Perspective on GWPs   227
  5.3.1 The Insights Gained From Ozone Depleting Potentials (ODPs)   227
  5.3.2 Characteristics Relevant to Uses of GWPs in Policy Formulation   229

References   230

SUMMARY

Radiative forcing indices have been developed as a relative measure of the potential globally averaged warming effect on the surface–troposphere system arising from emission of a set amount (e.g., 1 kilogram) of a variety of trace gases. These gases can exert a radiative forcing of the climate system both directly and indirectly. Direct forcing occurs when the emitted gas is itself a greenhouse gas. Indirect forcing occurs when chemical transformation of the original gas produces or destroys a gas or gases that themselves are greenhouse gases. The indices reflect the cumulative radiative forcing over some chosen time period of interest. The choice of time horizon is a user choice that depends upon issues such as the time-scales of the climate system and the emphasis on long-term versus short-term potential climate changes. Future changes in the numerical indices are likely as research in related areas yields improved input to the calculations, implying that whatever framework is adopted for the use of these indices, it must have flexibility to incorporate what could be substantial changes in the specified numerical values of the indices.

The set of Global Warming Potentials (GWPs) for greenhouse gases that is presented in this chapter is an updating and expansion of that presented in the 1992 assessment of the Intergovernmental Panel on Climate Change (IPCC, 1992). The time horizons of the GWPs are 20, 100, and 500 years, as was the case in IPCC (1992). GWPs have a number of important limitations. The GWP concept is difficult to apply to gases that are very unevenly distributed and to aerosols. For example, relatively short-lived pollutants such as the nitrogen oxides and the volatile organic compounds (precursors of ozone, which is a greenhouse gas) vary markedly from region to region within a hemisphere and their chemical impacts are highly variable and non-linearly dependent upon concentrations. Further, the indices and the estimated uncertainties are intended to reflect global averages only, and do not account for regional effects. They do not include climatic or biospheric feedbacks, nor do they consider any environmental impacts other than those related to climate. The major changes since IPCC (1992) are the following:

- *GWPs for 16 new chemical species have been calculated, bringing the number now available to 38.*

The new species are largely hydrofluorocarbons (HFCs), which are being manufactured as substitutes for the chlorofluorocarbons (CFCs), and the very long-lived fully fluorinated compounds: SF_6 and the perfluorocarbons. SF_6 is manufactured mainly for insulation of electrical equipment. The main source of CF_4 and C_2F_6 is production as accidental by-products of aluminium manufacture, while other perfluorocarbons have been suggested as potential CFC and halon substitutes.

- *The decay response of a pulse of CO_2 in the atmosphere, which is the reference for the GWPs, was derived from a balanced carbon model in which sinks are constrained to match sources that was deemed (see Chapter 1) representative of the current understanding of the global carbon cycle.* The reference for the GWPs of IPCC (1990, 1992) was a three-parameter fit involving an unbalanced carbon model. The present GWPs are 10-20% larger than they would otherwise have been because of the different reference.

- *For those species addressed in IPCC (1992), a majority of the GWP values are larger, typically by 10-30%.* These changes are largely due to (i) changes in the carbon dioxide (CO_2) reference noted above and (ii) improved values for atmospheric lifetimes, particularly those species that are removed by chemical reactions in the troposphere.

- *Both the direct and indirect components of the GWP of methane (CH_4) have been estimated.* The indirect component is now deemed to be better understood than at the time of the previous IPCC reports. The effects that were incorporated include methane's influence on lengthening its own atmospheric decay response time and on production of tropospheric ozone and stratospheric water vapour. The sum of these effects is estimated to increase substantially the total GWP for CH_4 compared to its direct GWP. Uncertainties in evaluating these effects arise from incomplete understanding of dynamical and chemical atmospheric processes (especially the use of the decay response time described in Chapter 2). The product of the methane GWP and its estimated annual anthropogenic emission is comparable to that of CO_2 over a 20-year time horizon and is a significant fraction for longer time horizons.

- *The sensitivity of the GWPs to a changing future atmosphere has been explored.* As was done in IPCC (1990, 1992), the composition of the background atmosphere used in the GWP calculations presented herein was the present-day abundances of CO_2, CH_4, and nitrous oxide (N_2O), which were assumed constant into the future. Since this is not a realistic assumption, the sensitivity of the GWPs to a changing future atmosphere was also explored. For example, a choice of an increasing CO_2 abundance (from 360 ppmv currently to 650 ppmv by the end of the 22nd century) in the background atmosphere would produce 20% larger GWPs for the longer time horizons. The sensitivity of the direct GWPs to some of the changes likely to be associated with an altered climate, namely altered water vapour and clouds, was also explored. The results suggest that the changes in GWPs would not be substantial.

- *A typical uncertainty, relative to the CO_2 reference, in the direct GWPs presented herein is estimated to be ±35%.* The range stems from uncertainties in the relative radiative forcing per molecule and lifetimes in the atmosphere. Larger uncertainties apply to the estimation of indirect effects, but these are difficult to quantify.

Reliable *indirect* GWPs are difficult to estimate for the ozone-depleting gases (e.g., CFCs and the bromine-containing halons) and hence are not included in this report. Better estimates of the negative, indirect component of the GWP arising from the reduced radiative forcing from ozone losses in the lower stratosphere require a detailed understanding of several factors including the vertical profile of the ozone loss (see Chapter 4), the role of bromine and chlorine in destroying stratospheric ozone, and the photochemical effectiveness of each ozone-depleting gas.

Reliable radiative forcing indices for gases that form atmospheric aerosols (e.g., sulphur dioxide, SO_2) cannot currently be formulated meaningfully, chiefly because of the lack of understanding of many of the processes involved (e.g., composition of the aerosols, radiative properties) and because of uncertainties regarding the climate response to the inhomogeneous spatial distributions of the aerosols (Chapter 3).

5.1 Introduction

This chapter addresses the *numerical indices* that can be used to provide a simple representation of the contributions of emissions of different atmospheric trace gases to greenhouse warming. In general terms, such indices are used to estimate the relative impact of emission of a fixed amount of one greenhouse gas compared to another for globally averaged radiative forcing of the climate system over a specified time scale. This focus leads to several key questions. What types of indices are there? How representative are they of the relative contributions of the greenhouse gases to the radiative forcing of the climate system? How well can all of the contributions of a trace gas be embodied in a single index? What are the most policy-relevant aspects of the uncertainties associated with radiative forcing indices?

In addressing these questions, this chapter draws heavily on the information in the preceding chapters of this report (cross-referenced herein), the earlier climate-system reports of the Intergovernmental Panel on Climate Change (IPCC) (IPCC, 1990; 1992), the ozone-depletion reports of the United Nations Environment Programme (UNEP) and the World Meteorological Organisation (WMO) (WMO 1990, 1992, 1995), and recent journal publications. While the major objective of the text that follows is to update the information on radiative forcing indices, another goal is to note the aspects of such indices that involve user choices, with the aim of stimulating comments that could enhance the usefulness of the corresponding section on this topic in the forthcoming IPCC full scientific assessment of the understanding of climate change (IPCC, 1995).

In this introduction we summarise the utility of such radiative forcing indices, describe the types of indices that have been developed, and outline the physical quantities and choices upon which they depend. In the subsequent sections of this chapter, we describe the calculations of the indices considered, outline the input data used and assumptions made in the calculations, show the sensitivity of the results to some of the specifications and assumptions, present the resulting numerical indices, describe their uncertainties, and discuss key features of the radiative forcing indices that are likely to be of prime interest to their users.

5.1.1 Utility of Trace-Gas Radiative Forcing Indices

5.1.1.1 Changes in radiative forcing as the basis for measures of induced climate change

As described in Chapter 4 (see in particular Table 4.1), calculations made with climate models indicate that, for well-mixed greenhouse gases at least, the relationship between changes in the globally integrated adjusted *radiative forcing at the tropopause and global-mean surface temperature* changes is independent of the gas causing the forcing. Furthermore, similar studies indicate that, to first order, this *"climate sensitivity"* is relatively insensitive to the type of forcing agent (e.g., changes in the atmospheric concentration of a well-mixed greenhouse gas such as CO_2 or changes in the solar radiation reaching the atmosphere). Spatial variations of radiative forcing can influence regional climate responses (e.g., Wang *et al.*, 1991, 1993). Such variations are not currently considered in the formulation of relative indices for climate change, but could be important (e.g., see Section 4.4.1.1).

The first-order forcing-response relationship is the scientific basis for using changes in radiative forcing as the common mode for (i) comparing some of the natural causes of climate change (e.g., solar forcing) to some of the human-caused interferences in the climate system (e.g., well-mixed greenhouse gases) or (ii) intercomparing the relative radiative-forcing roles of the numerous greenhouse gases. This practice is further strengthened by two facts. First, the indices represent radiative forcing, not climate response. There are fewer uncertainties associated with calculating changes in radiative forcing than with calculating climate-system responses, e.g., global mean surface temperature, sea level, or precipitation. Second, the indices are relative quantities, not absolute. Calculating the alteration in radiative forcing due to addition of a fixed amount of greenhouse gas A relative to the same addition of greenhouse gas B can be done more accurately than having to know the absolute alteration in radiative forcing due a change in a single greenhouse gas alone. The GWP may be usefully thought of as a proxy for assessment of the relative roles of different greenhouse gases, assuming that the climate system responds linearly to perturbations in radiative forcing.

5.1.1.2 Uses of radiative forcing indices

The availability of a simple and representative index that quantifies the relative roles of additions of the various greenhouse gases in causing climate change would have considerable practical value (i) in the assessment of the relative or aggregate contributions of the many human activities that can cause such changes and (ii) in the potential decisions as to how best to minimise the magnitude of the impact of such activities on the climate system. Indeed, as nations approach the discussion of policies that could reduce the likelihood of human perturbation of the Earth's climate system, some countries are incorporating comparative trace-gas indices in their formulation of approaches to reducing their trace-gas emissions. Furthermore, the Parties of the UN Framework Convention on Climate Change will seek enhanced scientific input regarding such indices as one of the tools for meeting the Convention's commitment to "prevent dangerous anthropogenic interference with the climate system".

Provided international agreement can be reached on a set of indices that would place the various greenhouse gases on an equivalent scale, this could allow countries to choose a course of action that is the most appropriate for them (economically, technically, socially, etc.) in meeting their commitment to help reduce radiative forcing. Examples of the utility of these indices would be the following:

- *They can be used to rank the emissions of the various countries*, i.e., a radiative-forcing-scaled net emission for each country could include all greenhouse gases on a common scale and hence could represent that country's current or projected net contribution to greenhouse warming.

- *They can be the quantitative basis to a collective "basket" approach to greenhouse-gas emission reductions within a given country.* For example, the product of a compound's radiative-forcing index and the amount by which the emissions of that compound are reduced (or the removal of that compound is increased) could yield the equivalent reduction in global radiative forcing achieved. Therefore, this index could also be the quantitative basis of the "trading" of one type of emission reduction for another within a national goal or quota.

- *They can contribute to economic assessments of trade-offs between alternative technologies*, e.g., weighting by radiative forcing indices in comparing the merits of one fuel against another. Such comparisons could also be modified to include efficiency and other technological factors.

- *They can function as a quantitative signal to governmental and industrial policy makers*, thereby encouraging some activities and discouraging others. For example, the flagging of possible candidate substitutes for the chlorofluorocarbons (CFCs) that have high radiative forcing indices could identify them for possible avoidance.

- *Together with other information, they can be used as a part of the basis for total environmental impact assessments where several different factors or actions are considered.* For example, environmental labelling of products or environmental audits of industrial processes.

5.1.2 Definition of Relative Radiative Forcing Indices

5.1.2.1 General characteristics

The relative radiative forcing indices that have been formulated to date have two noteworthy general features:

A relative radiative forcing index is not purely a geophysical quantity, such as is a change of temperature. Rather, these indices are user-oriented constructs whose calculation involves not only an understanding of a few relevant Earth-system processes (e.g., radiation transfer and chemical removal), but also some policy-oriented choices (e.g., a selection of the time span of interest). Hence, such indices per se are not subject to observation and testing in the sense of many climate-system predictions, but are best judged by (i) their representativeness of the overall radiative forcing role of the specified trace gas and (ii) their overall usefulness to those who formulate and establish policies regarding the greenhouse gases. We differentiate below the characteristics that are based on the understanding of geophysical processes and those that stem from policy-oriented choices.

The focus in the formulation of radiative forcing indices is on the effect of anthropogenic emissions of greenhouse gases. Chapters 1 and 2 describe how emissions control atmospheric concentrations, while Chapter 4 describes how the atmospheric concentration of a chemical species determines its contribution to the total radiative forcing. Chapter 4 also illustrates how the changes in greenhouse gas concentrations that have occurred in the past have altered the total radiative forcing (also see Figures 2.2 to 2.4 of IPCC, 1990). However, from the viewpoint of formulating policies for reductions in greenhouse gases, indices that denote the *relative* radiative roles of anthropogenic emissions, rather than *absolute* concentration changes caused by human activities, have received more attention (see Section 2.2.1 of IPCC, 1990). That emission-based emphasis continues here.

5.1.2.2 Primary factors

Common to all greenhouse gases are three major factors – two technical and one user-oriented– that determine the relative contribution of addition of 1 kilogram of a greenhouse gas to radiative forcing and hence are the primary input in the formulation, calculation, and use of relative radiative forcing indices. Assessment of the absolute, rather than relative, radiative forcing involves the additional factor of total quantity emitted and is addressed both in Chapter 4 and in Figure 5.7.

Technical factors

Factor 1: *The strength with which a given species absorbs long wave radiation and the spectral location of its absorbing wavelengths.* Chemical species differ markedly in their abilities to absorb long-wave radiation. Overlaps of the absorption spectra of various chemical species with one another (especially H_2O, CO_2, and, to a lesser extent, O_3) are important factors. In addition, while the absorption of infrared radiation by many greenhouse gases varies linearly with their concentration, a few important ones display non-linear behaviour (e.g., CO_2, CH_4, and N_2O). For those gases, the relative radiative forcing will depend upon concentration and hence upon the scenario adopted for the future trace-gas atmospheric abundances. A key factor in the greenhouse role of a given species is the location of its absorption spectrum relative to the region in the absorption of atmospheric water vapour through which most outgoing planetary thermal radiation escapes to space. Consequently, *other things being equal*, chemical species that have strong absorption band strengths in the relatively weak water vapour "window" are more important greenhouse gases than those that do not. This is illustrated in Figure 5.1, which shows how the instantaneous radiative forcings due to the pulse emission of 1 kilogram of various long-lived gases change as the concentrations decay away in time after they have become well mixed (e.g., about a year after injection into the atmosphere). The relevant point here is on the left-hand scale at $t = 1$, namely, that the radiative forcing of an equal emission of the various gases can differ by as much as four orders of magnitude. Laboratory studies of molecular radiative properties are a key source of the basic information needed in the calculation of radiative forcing indices. The status of such spectroscopic data of greenhouse gases is discussed in detail in Section 4.2.

Factor 2: *The lifetime or response time of the given species in the atmosphere.* Greenhouse gases differ markedly in how long they reside in the atmosphere once emitted. Greenhouse gases that persist in the atmosphere for a long time are more important, other things being equal, in radiative forcing than those that are shorter lived. This point is also illustrated in Figure 5.1. The initial dominance of the band strength in determining the instantaneous radiative forcing at early times can be overwhelmed by the lifetime factor at later times. As described in Chapters 1 and 2, the atmospheric lifetime of a gas is determined largely by the Earth-system process(es) that remove the gas from the atmosphere, e.g., the stratospheric photolytic decomposition of N_2O and the chemical destruction of CH_4 in the troposphere by the hydroxyl radical. Knowledge of these varied and often complex processes stems from laboratory investigations, field studies, and theoretical modelling. The status of this understanding of the atmospheric lifetime of greenhouse gases is further assessed in Chapters 1 and 2. After emission into the current or projected atmosphere, the time-scale for removal of most greenhouse gases is equivalent to the lifetime (see definitions in Chapter 2). In some cases, the time scale for removal of a gas from the atmosphere cannot be simply characterised or is dependent upon the perturbation and/or the background atmosphere and other sources; in these cases (chiefly CO_2 and CH_4) we refer to removal of a pulse as the response time or decay response.

User-oriented factor

Factor 3: *The time period over which the radiative effects of the species are to be considered.* Since many of the responses of the Earth's climate to changes in radiative forcing are long (e.g., the centennial-scale warming of the oceans), it is the *cumulative* radiative forcing of a greenhouse gas, rather than its instantaneous value, that is of primary importance to crafting a relevant radiative forcing index. As a consequence, such indices involve an integral over time. Rodhe (1990) has noted that the choice of time interval can be compared to cumulative-dosage effects in radiology. IPCC (1990, 1992) used integration *time horizons* of 20, 100, and 500 years in calculating the indices. Figure 5.2 shows the integrals of the decay functions in Figure 5.1 for a wide range of time horizons. It illustrates the need for the user of the radiative forcing indices to select the time period of consideration. A strongly absorbing, but short-lived, gas like HCFC-225ca will contribute more radiative forcing in the short term than a weaker-absorbing, but longer-lived, gas like N_2O; however, in the longer term, the reverse is true. Methane is a key greenhouse gas discussed extensively in Section 5.2.5; its integrated radiative forcing would lie below that of N_2O and reach a plateau more quickly because of its shorter lifetime.

The spread of numerical values of the radiative forcing indices reported in Section 5.2.4 below largely reflects the influence of these three major factors.

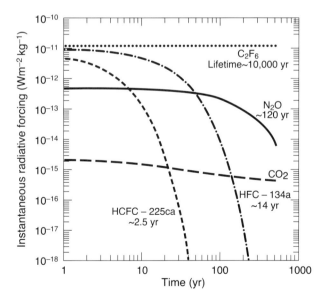

Figure 5.1: Instantaneous radiative forcing ($Wm^{-2}\ kg^{-1}$) versus time after a pulse release for several different greenhouse gases. The CO_2 decay response function and other aspects of the calculations are described in Section 5.2.1.

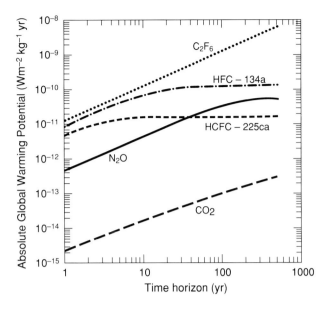

Figure 5.2: Integrated radiative forcing (Wm^{-2} kg^{-1} yr) for a range of greenhouse gases (from Figure 5.1).

5.1.2.3 Additional factors
There are other factors, some of them substantial, that influence the degree to which a given gas contributes to radiative forcing.

Indirect effects

In addition to their direct radiative effects described above, many chemical species also have indirect effects on radiative forcing that arise largely from atmospheric chemical processes. For example, a radiatively important product of the oxidative removal of CH_4 is stratospheric water vapour, a greenhouse gas. Hence, the total contribution of CH_4 emissions to radiative forcing should ideally be the sum of the direct and indirect effects. In the case of CH_4, the indirect contribution may be roughly comparable and additive to the direct effect.

Indirect effects may also be opposite in sign to the direct effects. An example is the set of chemicals that deplete the stratospheric ozone layer. In addition to the well-known direct (positive) contributions to radiative forcing of ozone-depleting species, the resulting ozone loss in the lower stratosphere reduces the radiative forcing at the tropopause, thereby introducing a negative component to the indices of these compounds. Observations of ozone loss suggest that this forcing varies substantially with season and location and is also significant in the global average (Section 4.3).

Another class of chemical species that is relevant to radiative forcing is the set that are not significant greenhouse gases themselves, but whose emissions affect the concentrations of greenhouse gases and aerosols via atmospheric chemical processes. The nitrogen oxides and volatile organic compounds from surface and/or high altitude pollution sources are examples of this class, since they are capable of forming ozone and changing the oxidising capacity of the atmosphere.

As explained in Chapter 2, the atmospheric chemical processes that drive many of these indirect effects are difficult to characterise adequately, thereby limiting the ability of current models to calculate the potentially important indirect contributions of many gases. Particularly for species that are chemically removed from the atmosphere on short time-scales (days to months, such as volatile organic compounds), chemical and radiative impacts are believed to be highly non-linear and regional in nature, thus complicating the derivation of global radiative forcing. While these effects will not be quantified here, they could be important.

Role of a changing atmosphere

As noted above, the radiative properties of the atmosphere, which are determined by atmospheric composition, influence the degree to which a given greenhouse gas contributes to greenhouse forcing. Since the atmospheric concentrations of absorbing gases continually change, this aspect of a radiative forcing index is a time-varying quantity. Indeed, a somewhat "circular" situation cannot be avoided: an index is calculated for an assumed set of future greenhouse gas concentrations and then the subsequent use of that number in policy decisions can cause the future concentrations to be different from those originally assumed! In past IPCC assessments, the then-current atmospheric abundances of greenhouse gases were assumed, for simplicity, to be fixed throughout the integration over the time horizon. In this chapter, we explore the sensitivity of radiative forcing indices to future scenarios of atmospheric composition. Similarly, since ambient water vapour and clouds are also major factors in atmospheric absorption, the sensitivity of calculated indices to likely climate-induced changes in these parameters is also examined. The sensitivity of GWPs to atmospheric aerosols is likely to be similar to or less than the sensitivity to clouds because of similarities in radiative properties. These sensitivity tests provide some insight into how the infrared transmission and hence the indices would be different in a potentially perturbed future climate or altered aerosol distribution. However, this analysis does not consider possible future changes in atmospheric chemistry, circulation, biospheric processes, or other climate feedbacks that could result in significant changes in the lifetimes of many trace gases and hence in their GWPs.

The consideration of a changing future atmosphere here does not imply that the definition of the GWP should be

Trace Gas Radiative Forcing Indices

altered to include such changes. Rather, the sensitivity tests described below suggest that the relative integrated radiative forcing of different greenhouse gases is not radically altered when several important aspects of likely future atmospheric change are taken into account, thereby strengthening the use of GWPs under a range of conditions.

5.1.2.4 Formulations of radiative forcing indices

Global Warming Potential

Based on the major factors summarised above, the relative potential of a specified emission of a greenhouse gas to contribute to a change in future radiative forcing, i.e., its Global Warming Potential (GWP), has been expressed as the time-integrated radiative forcing from the instantaneous release of 1 kg of a trace gas expressed relative to that of 1 kg of a reference gas (IPCC, 1990):

$$GWP(x) = \frac{\int_0^{TH} a_x \bullet [x(t)] dt}{\int_0^{TH} a_r \bullet [x(t)] dt} \quad (5.1)$$

where TH is the time horizon over which the calculation is considered; a_x is the climate-related, radiative forcing due to a unit increase in atmospheric concentration of the gas in question; $[x(t)]$ is the time-decaying abundance of a pulse of injected gas; and the corresponding quantities for the reference gas are in the denominator. The adjusted radiative forcings per kilogram, a, are derived from infrared radiative transfer models. As noted above, a_r is a function of time when future changes in CO_2 are considered. Time-dependent changes in a_x or lifetimes are not explicitly considered here. The trace gas amounts, $[x(t)]$ and $[r(t)]$, remaining after time t are based upon the atmospheric lifetime or response time of the gas in question and the reference gas, respectively (see Chapters 1 and 2).

The reference gas has been taken generally to be CO_2, since this allows a comparison of the radiative forcing role of the emission of the gas in question to that of the dominant greenhouse gas that is emitted as a result of human activities, hence of the broadest interest to policy considerations. Figure 5.3 shows the GWPs for C_2F_6, HFC-134a, HCFC-225ca, and N_2O for different time horizons with CO_2 as the reference gas. This figure illustrates the way in which the lifetime of a gas dictates the variation of its GWP over different time horizons. When the lifetime of the gas in question differs substantially from the response time of the reference gas, the GWP is sensitive to the choice of time horizon, i.e., for HFC-134a and HCFC-225ca (shorter) and for C_2F_6 (longer). When the lifetime of the gas in question is comparable to the response time of CO_2 (nominally on the order of 150 yr, although it is clear that the removal of CO_2 cannot be adequately described by a single, simple exponential lifetime; see Section 5.2.1 and Chapter 1), the GWP is relatively insensitive to choice of time horizon, i.e., for N_2O. The initial GWPs (for time horizons much shorter than the lifetime of the gas or the reference) simply reflect the relative radiative forcing per molecule compared to the reference gas (see Figure 5.1). For longer time horizons, those species that decay more rapidly than the reference gas display sharply decreasing GWPs, with the slope of the decay being dependent mainly on the lifetime of the gas in question. Gases with lifetimes much longer than that of the reference gas (e.g., C_2F_6) display steeply increasing GWPs over their lifetimes.

A key aspect of GWP calculations is thus the choice of the reference gas. Unfortunately from the perspective of its selection as the reference gas, the atmospheric response time of CO_2 has the largest scientific uncertainty of the major greenhouse gases. As described in Chapter 1, the uptake of CO_2 is a complex process involving the biosphere, ocean, ocean-atmosphere exchange rates, deep ocean sediments, etc. Furthermore, CO_2 is also re-circulated among these reservoirs at an exchange rate that is poorly known at present, and it appears that the budget of CO_2 is difficult to balance with current information. As a result, when CO_2 is used as the reference, the numerical values of the GWPs of all greenhouse gases are likely to change, perhaps substantially, in the future simply because research will improve the understanding of the removal processes of CO_2.

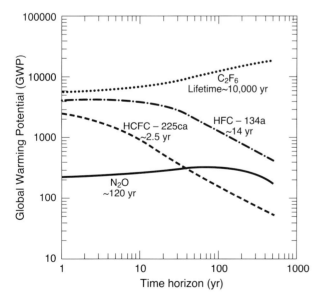

Figure 5.3: Global Warming Potentials (GWPs) for a range of greenhouse gases with differing lifetimes, as in Figures 5.1 and 5.2, using CO_2 as the reference gas.

Absolute Global Warming Potential

A variant formulation (e.g., Wigley, 1994a,b) is to consider simply the integrated radiative forcing of the gas in question:

$$AGWP(x) = \int_0^{TH} a_x \bullet [x(t)] dt \quad (5.2)$$

[Units: Wm^{-2} kg^{-1}yr]

The advantage of this formulation is that the index is independent of the calculated Absolute Global Warming Potential (AGWP) of CO_2. An important disadvantage is that the absolute value of radiative forcing depends upon many factors that are poorly known, such as the distributions and radiative properties of clouds (e.g., Cess et al., 1993). Figure 5.2 shows the AGWPs for several gases as a function of time horizon.

Other formulations

Fisher et al. (1990) and WMO (1990) presented GWP calculations for CFCs and HCFCs integrated over the entire lifetimes of the species rather than to fixed time horizons. It can be shown that if lifetimes do not vary with time and if radiative forcing varies linearly with concentration, this definition is equivalent to the IPCC (1990) formulation over infinite time horizons (see Lashof and Ahuja, 1990; Rodhe, 1990). Fisher et al. (1990) also used CFC-11 as their reference gas, since they were primarily interested in the relative radiative forcing among CFCs and their substitutes. In contrast, Hammond et al. (1990) suggested use of the instantaneous relative radiative forcings (Figure 5.1) rather than their integrated values over chosen time horizons (Figures 5.2 and 5.3). Section 5.1.2.2 discussed the user-oriented factors relating to the need for and choices of time horizon.

Lashof and Ahuja (1990) used a similar definition to IPCC (1990) and argued for use of infinite time horizons together with "discount rates" designed to reflect increasing uncertainty with time (e.g., to account for the possibility that new technologies will emerge to solve problems). Other authors such as Reilly and Richards (1993) have argued that GWPs should include damage factors, which could tend to offset future discounts. Harvey (1993) proposed an alternative GWP index that accounts for the duration of capital investments in the energy sector. The possibility of coupling such factors to the GWP definition requires detailed study of economics and policy implications, together with the requirement for scientific accuracy, and is beyond the scope of the present review.

Finally, it may be argued that the GWP should be based on the *actual temperature response* (or "realised warming") rather than radiative forcing. Clearly, the temperature responses (and associated changes such as sea level) are not only the ultimate "end result" associated with climate studies, but are also more readily understood by non-scientists. However, as noted as Section 4.8, evaluation of the temperature response requires understanding of the responses of climate to changes in radiative forcing, which would introduce additional uncertainties in a trace-gas index beyond those associated with radiative forcing changes.

5.1.3 General Scientific Limitations of a Simple Index

The conceptual framework outlined above captures the major facets of the relative roles of greenhouse gases in altering radiative forcing of the climate system. Inaccuracies in a calculated GWP stem from uncertainties in the input data and processes noted above, e.g., uncertain molecular band strengths and inadequately characterised chemical processes, notably those that influence indirect effects. Estimates of the effects of those uncertainties have been given in Chapters 1 – 4, and many are explicitly noted in association with the results presented below. Other more user-oriented inaccuracies and limitations have been discussed elsewhere in this chapter.

The need to specify a time horizon is a user choice, not a GWP uncertainty. Similarly, the choice of a reference future atmosphere for the calculation of a GWP is not, in a strict sense, an uncertainty, but rather is largely an agreed-upon selection of a future scenario for atmospheric composition. Further, the GWP as defined here is only a measure of relative radiative forcing, not a measure of potential damage resulting from possible climate change that includes economic or other variables. These limitations should be kept in mind in economic or policy analysis.

However, there are some conceptual problems associated with the scientific formulation of a single index for all gases (IPCC, 1990, 1992). A few are worth emphasising:

- *Insofar as it is intended to be globally representative, the GWP concept is difficult to apply to gases that are very unevenly distributed and to aerosols.* As noted in Chapter 2, the shorter-lived gases frequently have very inhomogeneous spatial distributions. For example, carbon monoxide abundances are larger in the Northern Hemisphere than in the Southern, reflecting the relative magnitudes of human-influenced emissions. Furthermore, relatively short-lived pollutants such as the nitrogen oxides and the volatile organic compounds (important precursors of ozone) vary markedly from region to region within a hemisphere and their chemical impacts are highly variable and non-linearly dependent upon concentrations. The accuracy of calculation and meaning of a globally averaged forcing, particularly relative to a more-evenly distributed gas like CO_2, is hence weakened. Aerosols also have spatially inhomogeneous sources and very short atmospheric lifetimes; hence, their spatial distributions can be quite variable. As a result, the comparison of the radiative forcing by aerosols and ozone precursors compared to that of CO_2 is problematic (Section 4.4), but potentially quite important.

- *GWPs for greenhouse gases with vertical profiles or latitudinal variations of radiative forcing that are markedly different from the reference gas have larger inaccuracies.* The vertical and latitudinal distribution of the radiative perturbation of a greenhouse gas can influence the partitioning of the long-wave radiative-flux perturbation in the troposphere-surface climate system. Gases with markedly different distributions of their radiative flux perturbations may influence the climate system (particularly regionally) in different ways that are not reflected simply in the GWP ratio (Wang *et al.*, 1991). Thus, the use of a common scale afforded by GWPs may be weakened (see Section 4.8).

- *Warming and cooling components of the GWPs for each of the ozone depleting gases may be difficult to characterise.* Ozone losses in the lower stratosphere are larger at higher latitudes, hence, the associated cooling has a latitudinal variation that differs from that of the direct warming due to the ozone depleters. Further, the globally averaged cooling is sensitive to the vertical profile of the ozone change (see Section 4.3), and the ozone depletion caused by each gas depends upon its chemical properties. Incorporating such different geographic forcings and photochemical effects from direct and indirect components into a single radiative forcing index likely requires detailed study of chemical ozone-loss processes and associated radiative forcing that is beyond the scope of the present assessment.

5.2. Calculation of Global Warming Potentials

5.2.1 Reference Molecule

Carbon dioxide has generally been adopted as the reference molecule for GWP calculations and is a principal greenhouse gas for many policy concerns. However, IPCC (1990) underlined the implications of such a choice for uncertainties in GWPs. Wigley (1994a,b) has emphasised the uncertainty in accurately defining the denominator for GWP calculations if CO_2 is used as the reference molecule, and suggested the use of "Absolute" or AGWPs based on Equation 5.2. Wuebbles *et al.* (1994b) have also noted the importance of uncertainties in the carbon cycle for calculations of GWPs when CO_2 is used as the reference. Current carbon cycle models do not include possible climate feedbacks. It is important to note that the budget of CO_2 must be carefully balanced in some way, with detailed accounting of trends, sources, and sinks (see Chapter 1). While recognising these issues, Caldeira and Kasting (1993) discuss feedback mechanisms that tend to offset some of these uncertainties for GWP calculations.

Carbon dioxide added to the atmosphere decays in a highly complex fashion, showing an initial fast decay over the first 10 years or so, followed by a more gradual decay over the next 100 years or so, and a very slow decline over the thousand year time scale, mainly reflecting transfer processes in the biosphere, ocean, and deep ocean sediments, respectively. Because of the different time constants for CO_2 removal processes, the uptake of CO_2 is quite different from that of other trace gases and is not well described by a single lifetime (Moore and Braswell, 1994). Figure 5.4 shows the decay over time of a pulse of CO_2 from the range of carbon cycle models discussed in Chapter 1. Even after several centuries, a portion of the pulse remains in the atmosphere because of slow transfer to the deep ocean sinks. Figure 5.4 shows that the range in decay response among the various models considered in this assessment is rather small; the "Bern" model as used in Chapter 1 of this document was chosen as representative and used for the decay response of CO_2 for the GWP calculations presented in Section 5.2.4 below.

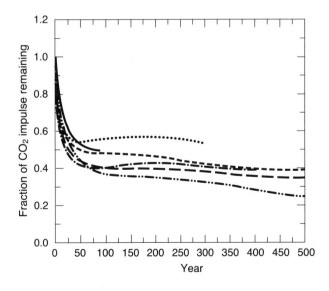

Figure 5.4: Impulse response for an injection of CO_2 versus time from several different carbon cycle models (see Chapter 1) assuming future CO_2 concentrations follow the S650 profile shown in Figure 1.12 (see Section 5.2.2.1).

5.2.2 Spectral Properties of the Present and Future Atmosphere

The spectral absorption characteristics of most of the major greenhouse gases are now rather well known as a result of laboratory studies (see Chapter 4 and references therein). Weak spectral lines for ozone and methane have been included in databases such as HITRAN, but are unlikely to be important for GWP analyses (see Chapter 4). The radiative band strengths of many HFCs and HCFCs have been measured in the laboratory (e.g., Clerbaux *et al.*, 1993). While significant uncertainties do remain in band strengths for some of these molecules, as described in Chapter 4 and in more detail in WMO (1994), it is important to note that their concentrations and emission rates in the present atmosphere are quite small, so that their contribution to current radiative forcing is not presently significant. As noted in Chapter 4, the radiative forcings per molecule used in this report are very close to those recommended in IPCC (1990) and are believed to be accurate to within about 25%. The estimated radiative forcings per molecule of HFC-125 and -152a have changed relative to IPCC (1992), by +15% and +8%, respectively.

To provide realistic evaluations of GWPs for specified time horizons and estimate their uncertainties, future changes in the radiative properties of the atmosphere must be considered. Some of these changes to the present state can be estimated based upon scenarios (e.g., CO_2 concentrations), while others are dependent upon the evolution of the entire climate system and are poorly known (e.g., clouds and water vapour).

5.2.2.1 Trace gas composition

In IPCC (1990), an unchanging atmosphere was assumed with the trace gas concentrations, hence the radiative properties of the atmosphere, unchanging in time. However, likely (but unforecastable) changes in CO_2, CH_4, or N_2O concentrations will lead to future changes in the radiative forcing per molecule of those gases (and perhaps others whose spectral bands overlap with them), as noted previously. The radiative properties of CO_2 are particularly sensitive to changes in concentration, since the large optical depth of CO_2 in the current atmosphere makes its radiative emission depend logarithmically on concentration. Thus, the forcing for a particular incremental change of CO_2 will become smaller in the future, when the atmosphere is expected to contain a larger concentration of the gas. In the case of CH_4 and N_2O, there is a square-root dependence of the forcing on their respective concentrations (IPCC, 1990); hence, just as for CO_2, the forcings due to a specified increment in either gas are expected to become smaller for future scenarios. For the other trace gases considered here, the present and likely future values are such that the direct radiative forcing is linear with respect to their concentrations and hence is independent of the scenario.

We explore here the impact of future changes in CO_2 concentrations upon GWP calculations and compare these results to those obtained for an unchanging atmosphere. For example, Figure 5.5 illustrates the decay response of a pulse of CO_2 from the Bern carbon cycle model for constant atmospheric CO_2 and for the S650 CO_2 scenario discussed in detail in Chapter 1. The S650 Scenario includes CO_2 concentrations that increase from the 1993 level of about 356 ppmv to a stabilised value of about 650 ppmv near the end of the 22nd century. The differences occur largely at the later times (>75 yr).

For reference to earlier IPCC reports, we also show in Figure 5.5 the CO_2 response function that was used in IPCC (1990, 1992). The fractional CO_2 decay was parametrized as a sum of three exponential functions crudely characterising the CO_2 removal processes:

$$F[CO_2(t)] = 0.30036\exp(-t/6.993)+0.34278\exp(-t/71.109)$$
$$+0.35686\exp(-t/815.727). \qquad (5.3)$$

The derivation was based upon an unbalanced ocean-atmosphere model (Oeschger *et al.*, 1975). Its favourable comparison to the results of the balanced ocean models in Chapter 1 could be largely fortuitous, since a cancellation of effects could have occurred in its derivation. The faster initial (first several decades) decay of added CO_2 calculated in the current models reflects rapid uptake by

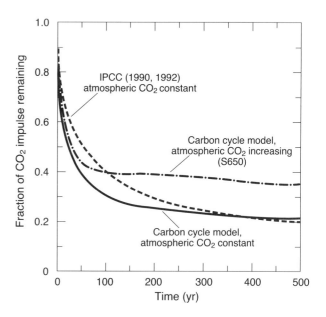

Figure 5.5: Impulse response for an injection of CO_2 versus time for the adopted carbon model for two future CO_2 scenarios: (i) constant current CO_2 concentrations and (ii) the increasing CO_2 concentrations of the S650 profile shown in Figure 1.12. Also shown for reference is the three-exponential function used to represent the CO_2 decay response in IPCC (1990, 1992).

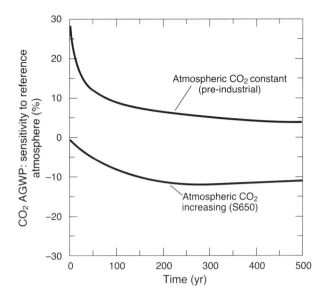

Figure 5.6: The sensitivity of the AGWP of CO_2 to the choices of future CO_2 scenarios other than constant current concentrations: (i) constant pre-industrial concentrations (280 ppmv) and (ii) the increasing concentrations of the S650 profile shown in Figure 1.12.

the biosphere and is believed to be an important improvement.

Figure 5.5 illustrates the fact that the dependence of the AGWP of CO_2 upon choice of future atmospheric CO_2 concentrations is not a highly sensitive one. This critical point is further quantified in Figure 5.6, which shows, for the adopted CO_2 decay response, how the AGWP would differ for choices of future atmospheric CO_2 scenarios other than constant current concentrations. A constant atmosphere at pre-industrial values (280 ppmv) would yield AGWPs for CO_2 different by less than 20% for all time horizons. Similarly, the increasing CO_2 concentrations of the S650 Scenario would yield AGWPs that are smaller by 15% or less as compared to constant current concentrations. Note that values of -15% do not imply a change in sign of the AGWP or GWP, only that the AGWP becomes as much as 15% smaller when possible increases in future CO_2 abundances up to 650 ppmv are considered rather than constant current values. The effects are small because decreases in the radiative forcing per molecule due to the increasing CO_2 atmospheric abundance tend to be opposite in sign to those due to the changed CO_2 decay response of balanced CO_2 models (see Caldeira and Kasting, 1993 and Wigley, 1994a). This similarity in AGWPs suggests that the relative radiative effect of each added kilogram of gas is not strongly dependent upon future CO_2 abundances, but the total absolute radiative forcing will, of course, be strongly dependent upon future abundances (e.g., the total number of kilograms of added gas).

We also considered the possible evolution of the radiative forcing of CH_4 and N_2O and the interplay between the spectral overlap of these two gases using the IS92a Scenario published in the Annex of IPCC (1992). If the calculations were made with the IS92a CH_4 and N_2O scenarios rather than with the constant current values, the direct GWPs of CH_4 would decrease by 2 to 3%, and the 20-, 100-, and 500-yr GWPs of N_2O would decrease by 5, 10, and 15%, respectively. The impact of the adopted future scenarios for CO_2, CH_4, and N_2O on the radiative forcing of other trace species was not considered here.

5.2.2.2 Water vapour and clouds
Water vapour

While it is likely that water vapour will change in a future climate state, the effect of such changes upon the direct GWPs of the great majority of molecules of interest here is expected to be quite small. For example, the model of Clerbaux *et al.*, (1993) was used to test the sensitivity of the direct GWP for CH_4 to changes in water vapour. Even for changes as large as 30% in water vapour concentration, the calculated GWP of CH4 changed by only a small percentage (C. Granier, personal communication, 1993). For many other gases whose radiative impact occurs largely in the region where water vapour's absorption is relatively weak, similar or smaller effects are likely.

Clouds

Clouds composed of water drops or ice crystals possess absorption bands in virtually the entire terrestrial infrared spectrum. By virtue of this property, they modulate considerably the infrared radiation escaping to space from the Earth's surface and atmosphere. Since cloud tops generally have lower temperatures than the Earth's surface and the lower part of the atmosphere, they reduce the outgoing infrared radiation. This reduction depends mainly on cloud height and optical depth. The higher the cloud, the lower is its temperature and the greater its reduction in infrared emission. On the other hand, higher clouds (in particular high ice clouds) tend to have low water content and limited optical depth. Such clouds are partially transparent, which reduces the infrared trapping effect.

The absorption bands of several trace gases overlap significantly with the spectral features of water drops and ice crystals, particularly in the window region. Owing to the relatively strong absorption properties of clouds, the greenhouse effect of many trace molecules is diminished in the presence of clouds. The spectral overlap between a molecule and cloud acts to reduce the radiative forcing associated with the increase in the concentration of the molecule considered. For example, calculations with a high-spectral resolution algorithm (V. Ramaswamy, personal communication, 1993) show that the radiative forcing due to a 0-1 ppbv increase (uniform with height) in CFC-11 in an overcast mid-latitude summer atmosphere with cloudtop at 3 km is 37% less than the corresponding clear-sky value (0.35 Wm^{-2}). The reduction from the clear-sky value is as high as 83% if an overcast sky with cloudtop at 10 km is considered. The difference between clear-sky and the overcast sky forcings depends on the absorption features of the trace gas considered. In the case of CF_4, the corresponding reductions from the same model in the overcast sky conditions with respect to the clear-sky value (0.097 Wm^{-2}), are 22% (cloudtop at 3 km) and 81% (cloudtop at 10 km), respectively. However, it is important to note that the impact of changes in clouds upon GWPs depends upon the ratio between the change in radiative forcing of the gas considered and that of the reference gas, not the absolute change in radiative forcing of the gas alone (i.e., the AGWP, which is strongly affected).

Possible future changes in cloud amounts and characteristics are a key element in predicting the absolute climate responses to greenhouse forcing. While it is not possible at present to fully predict future changes in cloud properties, Table 5.1 illustrates the range of sensitivity of GWPs to these effects by presenting calculations for clear-sky conditions and calculations including a three-layer cloud model. The table shows that the presence or absence of clouds results in estimated changes of the adjusted relative radiative forcings of the molecules considered here of typically less than 12%. While these estimates are derived only from a single model and only for a particular three-layer cloud model, the consideration of a complete absence of clouds is an extreme test. Thus, uncertainties in future cloud cover due to climate change are unlikely to substantially impact GWP calculations.

5.2.3 Lifetimes

The steady-state lifetimes of atmospheric gases are largely dictated by their photolysis and/or oxidation by reaction with OH, O(1D), and, in a few cases, Cl. The estimated lifetimes of trace gases considered in this report were updated to account for improved understanding. Updated reaction rate constants and absorption cross-sections for the relevant chemical destruction processes are based upon laboratory measurements summarised in JPL (1992) and the additional sources indicated in Table 5.2 (see also WMO, 1995).

Methylchloroform is a relatively short-lived gas destroyed mainly by reaction with the reactive OH radical in the troposphere (although its oceanic uptake must also be considered, see Butler et al., 1991). Observations of the trend in methylchloroform concentration, together with information regarding emissions and other sinks (e.g., oceanic uptake), can be used to deduce its lifetime and hence the globally averaged atmospheric OH concentration (Prinn et al., 1992). The lifetimes of other key gases destroyed by OH (i.e., CH_4, HCFCs, HFCs) can then be inferred relative to that of methylchloroform (see, e.g., Prather and Spivakovsky, 1990) with greater accuracy than would be possible from *a priori* calculations of the complete tropospheric OH distribution, so long as the gases are long-lived (hence relatively well-mixed). It is likely that methane is also destroyed in part by uptake to soil (IPCC, 1992), but this process is believed to be relatively slow and makes a small contribution to the removal of this gas from the atmosphere (see Chapter 2). Possible soil or ocean sinks are not considered for any other species.

Photochemical destruction mechanisms for

Table 5.1: Change in adjusted relative radiative forcings with and without clouds for selected molecules, from the model of Clerbaux et al. (1993).

Species	Chemical formula	Percentage change in adjusted relative radiative forcing per molecule compared to CO_2, due to incorporation of cloud feedbacks
CFC-11	$CFCl_3$	-11
CFC-12	CF_2Cl_2	-12
HCFC-22	CF_2HCl	-11
HCFC-123	$C_2F_3HCl_2$	-11
HCFC-141b	$C_2FH_3Cl_2$	-10
HCFC-142b	$C_2F_2H_3Cl$	-11
HFC-134a	CH_2FCF_3	-12
HFC-125	C_2HF_5	-11
HFC-152a	$C_2H_4F_2$	-12
Methane	CH_4	-8
Nitrous oxide	N_2O	-1

perfluorocarbons such as CF_4, C_2F_6, and C_6F_{14} are extremely slow. For CF_4 and C_2F_6, the dominant loss process is likely to be decomposition in the airflow through combustors, resulting in lifetimes of the order of 50,000 and 10,000 years, respectively (Ravishankara et al., 1993), due to the strong chemical bonds of these species. Larger perfluorocarbons such as C_6F_{14} have lifetimes exceeding a thousand years.

5.2.4 Direct GWPs

New direct GWPs of many gases were calculated for this report with the radiative transfer models developed at the National Center for Atmospheric Research – NCAR (Breigleb, 1992; Clerbaux et al., 1993), Lawrence Livermore National Laboratory – LLNL (Wuebbles et al., 1994a,b), the Max Planck Institüt für Chemie – Mainz (Brühl, 1993), the Indian Institute of Technology (Lal and Holt, 1991, updated in 1993), and the University of Oslo (Fuglesvedt et al., 1994). These were compared to the values given in IPCC (1990). The radiative forcing a-factors (see Equation 5.1) used in Chapters 4 and 5 of this report are summarised in Tables 4.2 and 4.3[1]; they were largely unchanged from IPCC (1990), with additional halocarbons added. In addition, the studies of Ko et al. (1993) and of Solomon et al. (1994) were used in deriving GWPs for SF_6 and CF_3I, respectively. Table 5.2 presents a composite summary of the recommended GWP results. With the exception of CF_3I, all of the molecules considered have lifetimes in excess of several months and thus can be considered reasonably well-mixed; only an upper limit rather than a value is presented for CF_3I.

There have been some significant changes in the values used for the lifetimes of trace gases in this study as compared to the IPCC (1992) assessment. The estimates for the lifetimes of many of the gases destroyed primarily by reaction with tropospheric OH (e.g., HCFC-22, HCFC-141b, HCFC-142b, etc.) are about 15% shorter than in IPCC (1992), due mainly to recent studies suggesting a shorter lifetime for CH_3CCl_3 based upon improved calibration methods and upon an oceanic sink (Butler et al., 1991). Similarly, the estimates for the lifetimes of gases destroyed mainly by photolysis in the stratosphere (e.g., CFC-12, CFC-113, H-1301) are about 10% shorter than in IPCC (1992) due to a shorter estimated lifetime for CFC-11 and related species. Lifetime estimates of a few other gases have also changed due to improvements in the understanding of their specific photochemistry (e.g., note that the lifetime for CFC-115 is now estimated to be about 1700 years, as compared to about 500 years in earlier assessments).

Several new gases proposed as CFC and halon substitutes are considered here for the first time, such as HCFC-225ca, HCFC-225cb, HFC-227ea, and CF_3I. Table 5.2 also includes for the first time a full evaluation of the GWPs of several fully fluorinated species, namely SF_6, CF_4, C_2F_6, and C_6F_{14}. SF_6 is used mainly for insulation of electrical equipment (Ko et al., 1993), while CF_4 and C_2F_6 are believed to be produced mainly as accidental by-products of aluminium manufacture. C_6F_{14} and other perfluoroalkanes have been proposed as potential CFC substitutes. The very long lifetimes of these gases (Ravishankara et al., 1993) lead to large GWPs over long time-scales.

The reference (see Equation 5.1) for the GWPs in Table 5.2 is the adopted CO_2 decay response from the carbon cycle model (see Section 5.2.1 and Chapter 1). The GWP calculations were carried out with background atmospheric trace gas concentrations held fixed at 354 ppmv (see Section 5.2.2.1 and the sensitivity tests discussed therein).

The uncertainty in the GWP of any trace gas other than CO_2 depends upon the uncertainties in the AGWP of CO_2 and the AGWP of the gas itself. The uncertainties in the relative values of AGWPs for various gases depends upon the uncertainty in relative radiative forcing per molecule (estimated to be about 25% for most gases, as shown in Chapter 4) and on the uncertainty in the lifetimes of the trace gas considered (which are likely to be accurate to about 10% for CFC-11 and CH_3CCl_3 as shown in Chapter 2, and perhaps 20-30% for other gases derived from them). Combining these dominant uncertainties (in quadrature) suggests uncertainties of less than ±35% in the direct AGWPs for nearly all of the trace gases considered in Table 5.2 . We note that a few of the newest CFC substitutes (namely HFC-236fa, -245 ca, and -43-10 mee) have larger uncertainties in lifetimes since fewer kinetic studies of their chemistry have been reported to date. Uncertainties in the AGWPs for CO_2 depend upon uncertainties in the carbon cycle (see Chapter 1) and on the future scenario for CO_2. The effect of the latter uncertainty is likely to be relatively small, and a range is illustrated in Figure 5.6 for two very different illustrative cases.

The direct GWPs given in Table 5.2 can be readily converted to other frameworks such as AGWPs, GWPs for a changing atmosphere, and GWPs using as reference either a specific carbon cycle model or the three-parameter fit employed in IPCC (1990, 1992). Note in particular that Table 5.3 presents the relevant factors to carry out conversions to scenarios in which much different future CO_2 abundances are assumed.

- To convert to AGWP units, the numbers in Table 5.2 should be multiplied by the AGWP for the adopted Bern CO_2 carbon cycle model, with fixed CO_2 (354 ppmv) scenario (i.e., Line 1 in Table 5.3) and

---

[1] a factors are equivalent to the ΔF per unit mass values listed in Tables 4.2 and 4.3.

Table 5.2: *Global Warming Potentials (mass basis), referenced to the AGWP for the adopted carbon cycle model CO_2 decay response and future CO_2 atmospheric concentrations held constant at current levels. Only direct effects are considered, except for methane.*

Species	Chemical Formula	Lifetime and Reference years		Global Warming Potential (time horizon)		
				20 years	100 years	500 years
CFCs						
CFC-11	$CFCl_3$	50±5	(b)	5000	4000	1400
CFC-12	CF_2Cl_2	102	(c)	7900	8500	4200
CFC-13	$CClF_3$	640	(a)	8100	11700	13600
CFC-113	$C_2F_3Cl_3$	85	(c)	5000	5000	2300
CFC-114	$C_2F_4Cl_2$	300	(a)	6900	9300	8300
CFC-115	C_2F_5Cl	1700	(a)	6200	9300	13000
HCFCs, etc.						
Carbon tetrachloride	CCl_4	42	(c)	2000	1400	500
Methylchloroform	CH_3CCl_3	5.4±0.6	(b)	360	110	35
HCFC-22 (†††)	CF_2HCl	13.3	(d)	4300	1700	520
HCFC-141b (†††)	$C_2FH_3Cl_2$	9.4	(d)	1800	630	200
HCFC-142b (†††)	$C_2F_2H_3Cl$	19.5	(d)	4200	2000	630
HCFC-123 (††)	$C_2F_3HCl_2$	1.4	(d)	300	93	29
HCFC-124 (††)	C_2F_4HCl	5.9	(d)	1500	480	150
HCFC-225ca (††)	$C_3F_5HCl_2$	2.5	(d)	550	170	52
HCFC-225cb (††)	$C_3F_5HCl_2$	6.6	(d)	1700	530	170
Bromocarbons						
H-1301	CF_3Br	65	(c)	6200	5600	2200
Other						
HFC-23 (†)	CHF_3	250	(j)	9200	12100	9900
HFC-32 (†††)	CH_2F_2	6	(d)	1800	580	180
HFC-43-10mee (†)	$C_5H_2F_{10}$	20.8	(h)	3300	1600	520
HFC-125 (††)	C_2HF_5	36.0	(d)	4800	3200	1100
HFC-134 (†)	CHF_2CHF_2	11.9	(e)	3100	1200	370
HFC-134a (†††)	CH_2FCF_3	14	(d)	3300	1300	420
HFC-152a (††)	$C_2H_4F_2$	1.5	(d)	460	140	44
HFC-143 (†)	CHF_2CH_2F	3.5	(k)	950	290	90
HFC-143a (††)	CF_3CH_3	55	(d)	5200	4400	1600
HFC-227ea (†)	C_3HF_7	41	(g)	4500	3300	1100
HFC-236fa (†)	C_3H2F_6	250	(h)	6100	8000	6600
HFC-245ca (†)	C_3H3F_5	7	(d)	1900	610	190
Chloroform (††)	$CHCl_3$	0.55	(d)	15	5	1
Methylene chloride (††)	CH_2Cl_2	0.41	(d)	28	9	3
Sulphur hexafluoride	SF_6	3200	(a)	16500	24900	36500
Perfluoromethane	CF_4	50000	(a)	4100	6300	9800
Perfluoroethane	C_2F_6	10000	(a)	8200	12500	19100
Perfluorocyclobutane	$c\text{-}C_4F_8$	3200	(a)	6000	9100	13300
Perfluorohexane	C_6F_{14}	3200	(a)	4500	6800	9900
Methane*	CH_4	14.5±2.5	(i)	62±20	24.5±7.5	7.5±2.5
Nitrous oxide	N_2O	120	(c)	290	320	180
Trifluoroiodomethane	CF_3I	<0.005	(f)	<5	<<1	<<<1

* Includes direct and indirect components (see Section 5.2.5)
(†) Indicates HFC/HCFCs under consideration for specialised end use (see Chapter 4)
(††) Indicates HFC/HCFCs in production now for specialised end use (see Chapter 4)
(†††) Indicates HFC/HCFCs in production now and likely to be widely used (see Chapter 4)
a   Ravishankara, *et al.* (1993).
b   Prather, private communication 1993, based on the forthcoming NASA CFC report and other considerations.
c   Average of reporting models in NASA CFC report. Scaled to CFC-11 lifetime.
d   Average of JPL (1992) and IUPAC (1992) with 277 K rate constants for OH+halocarbon scaled against $OH+CH_3CCl_3$ and lifetime of tropospheric CH_3CCl_3 of 6.6 yr. Stratospheric lifetime from WMO (1992)
e   DeMore (1993). Used 277 K OH rate constant ratios with respect to CH_3CCl_3, scaled to tropospheric lifetime of 6.6 yr for CH_3CCl_3.
f   Solomon *et al.* (1994).
g   Zhang *et al.* (1994) and Nelson et al. (1993) with 277 K rate constants for OH+halocarbon scaled against $OH+CH_3CCl_3$ and lifetime of tropospheric CH_3CCl_3 of 6.6 yr.
h   W. DeMore (personal communication, 1994) with 277 K rate constants for OH+halocarbon scaled against $OH+CH_3CCl_3$ and lifetime of tropospheric CH_3CCl_3 of 6.6 yr.
i   Includes the dependence of the response time on CH_4 abundance (see Chapter 2)
j   Schmoltner *et al.* (1993) with 277 K rate constants for OH+halocarbon scaled against $OH+CH_3CCl_3$ and lifetime of tropospheric CH_3CCl_3 of 6.6 yr.
k   Barry *et al.* (1994) with 277 K rate constants for OH+halocarbon scaled against $OH+CH_3CCl_3$ and lifetime of tropospheric CH_3CCl_3 of 6.6 yr.

multiplied by 1.291×10^{-13} to convert the AGWP of CO_2 from *per* ppmv to *per* kg on a global basis.

- To convert to GWP units using one of the other indicated carbon cycle models and/or trace-gas future scenarios, the numbers in Table 5.2 should be multiplied by the AGWP for the adopted Bern CO_2 carbon cycle model, fixed CO_2 (354 ppmv) scenario (Line 1 in Table 5.3) and divided by the AGWP value in Table 5.3 for the carbon cycle model and/or scenario in question.

- To convert to GWPs that are based on the same reference as was used in IPCC (1990,1992), the numbers in Table 5.2 should be multiplied by the AGWP for the adopted Bern CO_2 carbon cycle model, fixed CO_2 (354 ppmv) scenario (Line 1 in Table 5.3) and divided by the AGWP value in Table 5.3 for the CO_2-like gas, IPCC (1990) decay function, fixed CO_2 (354 ppmv) (i.e., last line in Table 5.3).

5.2.5 Indirect Effects
5.2.5.1 General characteristics

In addition to the direct forcing caused by injection of infrared absorbing gases to the atmosphere, some compounds can also modify the radiative balance through indirect effects relating to chemical transformations. When the full interactive chemistry of the atmosphere is considered, a very large number of possible indirect effects can be identified, ranging from the production of stratospheric water vapour as an indirect effect of H_2 trends (see, e.g., Khalil and Rasmussen, 1990) to changes in the HCl/ClO ratio and hence in ozone depletion resulting from CH_4 injections.

The effects arising from such processes are difficult to quantify in detail (see Chapter 2), but many are highly likely to represent only small perturbations to the direct GWP and to global radiative forcing. For example, there is no chemical evidence at present to suggest formation of breakdown products of HCFCs and HFCs that are sufficiently long-lived to significantly affect their GWPs (see WMO, 1995). Recent work has shown that the production of products such as fluoro- and chlorophosgene and organic nitrates from the breakdown of HCFCs, HFCs, and CFCs is unlikely to represent a significant indirect effect on the GWPs of those species, due to the removal of these water-soluble products in clouds and rain (Kindler *et al.*, 1994; WMO, 1995). Similarly, the addition of HCFCs and HFCs to the atmosphere can, in principal, affect the oxidising capacity of the lower atmosphere and hence their lifetimes, but the effect is completely negligible for reasonable assumed abundances of these trace gases.

Table 5.4 summarises some key stratospheric and tropospheric chemical processes that do represent important indirect effects for GWP estimates. We emphasise that the processes indicated here are not intended to be inclusive of all possible effects, but rather to emphasise those most likely to be important. The processes and the current state of understanding of them are examined in Chapter 2. We summarise here the GWP-relevant aspects.

CH_4: change in CH_4 lifetime

Small changes in CH_4 concentrations can significantly affect the atmospheric OH concentration, rendering the response time for the decay of the added gas substantially longer than that of the ensemble (i.e., longer than the nominal 10-yr lifetime for CH_4). This is due to the non-linear chemistry associated with relaxation of the coupled OH-CO-CH_4 system (see Lelieveld *et al.*, 1993; Prather,

Table 5.3: Absolute GWPs (AGWPs) (Wm^{-2} yr $ppmv^{-1}$)†.

	Time horizon		
Case	20 year	100 year	500 year
CO_2, Bern carbon cycle model, fixed CO_2 (354 ppmv)	0.235	0.768	2.459
CO_2, Bern carbon cycle model, S650 Scenario	0.225	0.702	2.179
CO_2, Wigley carbon cycle model, S650 Scenario	0.248	0.722	1.957
CO_2, Enting carbon cycle model, S650 Scenario	0.228	0.693	2.288
CO_2, LLNL carbon cycle model, S450 Scenario	0.247	0.821	2.823
CO_2, LLNL carbon cycle model, S650 Scenario	0.246	0.790	2.477
CO_2, LLNL carbon cycle model, S750 Scenario	0.247	0.784	2.472
CO_2-like gas, IPCC (1990) decay function, fixed CO_2 (354 ppmv)	0.267	0.964	2.848

† Multiply these numbers by 1.291×10^{-13} to convert from per ppmv to per kg.

Table 5.4: Important indirect effects on GWPs

Species	Indirect effect	Sign of effect on GWP
CH_4	Changes in response times due to changes in tropospheric OH	+
	Production of tropospheric O_3	+
	Production of stratospheric H_2O	+
	Production of CO_2 (for certain sources)	+
CFCs, HCFCs,	Depletion of stratospheric O_3	−
Bromocarbons	Increase in tropospheric OH due to enhanced UV	−
CO	Production of tropospheric O_3	+
	Changes in response times due to changes in tropospheric OH	+
	Production of tropospheric CO_2	+
NO_x	Production of tropospheric O_3	+
NMHCs	Production of tropospheric O_3	+
	Production of tropospheric CO_2	+

1994; and Chapter 2 for further details). This effect was discussed in IPCC (1990) and IPCC (1992) as an indirect effect on OH concentrations, and thus is not new. It arises through the fact that small changes in OH due to addition of a small pulse of CH_4 slightly affect the rate of decay of the much larger amount of CH_4 in the background atmosphere, thereby influencing the net removal of the added pulse. It is critical to note that the exact value of the CH_4 pulse response time depends upon a number of key factors, including the absolute amount of CH_4, size of the pulse, etc., making its interpretation complex and case-dependent. The detailed explanation of the effect is presented in Prather (1994) and in Chapter 2, which forms the basis for the present assessment.

CH_4: production of stratospheric H_2O

The production of stratospheric water vapour by methane oxidation involves relatively simple chemistry and represents a potentially significant indirect effect (Brühl, 1993; Lelieveld *et al.*, 1993). It is clear from theory and observations (Jones *et al.*, 1986) that nearly two water vapour molecules are produced per methane molecule destroyed. Calculations of the impact of this process on the CH_4 GWP presented below employ multi-dimensional models for methane and water vapour transport, together with a detailed treatment of stratospheric chemistry.

CH_4, CO, NO_x, NMHCs: production of tropospheric ozone

Tropospheric O_3 is a strong greenhouse gas (see Chapter 4), but one whose abundance is strongly coupled to the distributions of reactive OH and NO_x (NO + NO_2), making accurate calculation of the tropospheric O_3 distribution a notoriously difficult problem in atmospheric chemistry.

Tropospheric ozone production is initiated by oxidation of reduced carbon compounds, including CO, CH_4, and non-methane hydrocarbons (NMHCs). These compounds and NO_x are emitted from a variety of human-influenced sources. Hence, production of tropospheric O_3 represents an indirect effect of these species (see Chapter 2).

There are substantial inhomogeneities and other complexities to contend with regarding the indirect GWPs of CO, NO_x, and the NMHCs. The lifetime of NO_x in the troposphere is only a few days, while those of CO and NMHCs are of the order of months to days. The short lifetime of NO_x coupled with scattered sources (especially from anthropogenic emissions) leads to highly variable surface concentrations ranging from a few tenths to tens of ppbv in the polluted boundary layer. For this range of NO_x abundances, the photochemistry of O_3 production in the troposphere is a complex and highly non-linear function of added NO_x (e.g., Lin *et al.*, 1988; Thompson, 1992). This non-linearity, together with the difficulty of accurately calculating the transport and removal of NO_x species, imply that the indirect effect of ozone production from surface NO_x emissions will be subject to large uncertainties. While NMHCs and CO have somewhat longer lifetimes ranging from days to months, these gases are also strongly influenced by local sources, transport processes (particularly venting from the boundary layer), and non-linear chemistry. These factors render calculation of the indirect GWPs associated with their surface release extremely uncertain and hence their GWPs will not be calculated in this report. This does not imply that they are not significant for radiative forcing.

The indirect effect of ozone production through addition of CH_4 also depends upon the abundance and distribution of NO_x in the troposphere. However, the long lifetime of CH_4 (order of 10 years) implies that injections of this gas will be distributed throughout the troposphere, and thus its distribution is not dependent upon the details of transport phenomena, such as boundary layer venting and convection. It is also important to recognise that ozone production in the mid- and upper troposphere is more effective for radiative forcing than that near the surface (see Chapter 4), emphasising the radiative role of chemical ozone production processes that can take place in the free troposphere. Much of the indirect effect of ozone production through CH_4 chemistry likely takes place in the mid- to upper troposphere, where NO_x abundances generally exceed the approximate 10-30 pptv threshold at which the chemistry switches from ozone depletion to ozone production (see Ehhalt and Drummond, 1988; Fehsenfeld and Liu, 1993). These considerations suggest that the indirect effect of ozone production for CH_4 GWPs may be calculated with greater confidence than those of CO, NO_x, or NMHCs, although it is not completely independent of emissions of these gases.

In summary, we present here the indirect GWP effect of tropospheric ozone production only for CH_4. Additional GWP quantification must await further study of the model intercomparisons described in Chapter 2, additional field, laboratory, and theoretical characterisation of the processes involved in tropospheric ozone production, and further work on the radiative forcing impacts of regional and global ozone changes described in Chapter 4.

CFCs, HCFCs and bromocarbons: depletion of stratospheric ozone

Stratospheric ozone depletion by halocarbons causes significant cooling near the tropopause and hence is believed to induce a negative radiative forcing of the surface-troposphere system (Ramaswamy *et al.*, 1992; IPCC, 1992, WMO, 1992;). A recent General Circulation Model (GCM) calculation suggests that the globally averaged radiative flux change at the tropopause due to ozone depletion can be related to surface temperature responses (see Chapter 4). However, quantitative evaluation of the magnitude of this effect requires detailed knowledge of the vertical profile of ozone loss (Schwarzkopf and Ramaswamy, 1993). Further, a full evaluation of the radiative impact of each halocarbon requires detailed consideration of the photochemical effectiveness of each contributing gas (particularly bromocarbons), and future scenarios for halocarbon and ozone change (see Daniel *et al.*, 1994). Hence, the indirect effects of halocarbons will not be included here, but will be considered in WMO (1995), where the photochemistry of ozone depletion is described in detail. A point worth noting is the possibility that, for a given halocarbon, the balance of the direct (positive) and indirect (negative) radiative forcing components may lead to a "negative GWP". This clearly would introduce a new factor that would have to be dealt with in the use of such indices in policy decisions, underlining the difficulty of considering gases with multiple, and very different, environmental impacts using a single simple index. Multiple impacts could require more sophisticated policy tools.

Changes in N_2O and CH_4 can also impact stratospheric ozone, but these effects are believed to be considerably smaller than those related to increases in halogenated gases.

5.2.5.2 Indirect effects upon the GWP of CH_4

Recent research studies of the indirect effects on the GWP of methane include those of Lelieveld and Crutzen (1992), Brühl (1993), Lelieveld *et al.* (1993) and Hauglustaine *et al.*, (1994a,b). In this report, we consider those results together with inputs from Chapters 2 and 4 of this document. The relative radiative forcing for methane itself compared to CO_2 on a per molecule basis is given in Table 4.2 and is used here. Eight multi-dimensional models were used to study the chemical response of the atmosphere to a 20% increase in methane as discussed in Section 2.9. The calculated range of ozone increases from the full set of tropospheric models considered in Section 2.9 provides insight regarding the likely range in ozone production. Uncertainties in these calculations include those related to the NO_x distributions employed in the various models, formulation of transport processes, and other factors discussed in detail in Chapter 2. The estimated uncertainty in the indirect GWP for CH_4 from tropospheric ozone production given below is based upon the calculated mid- to upper tropospheric ozone response of the models to the prescribed methane perturbation at northern mid-latitudes (see Figure 2.16) and consideration of the current inadequacies in the understanding of many relevant atmospheric processes. The calculated ozone changes from the model simulations of Chapter 2 for a 20% increase in methane implies an indirect effect that is about 25±15% of the direct effect of methane (or 19±12% of the total), using the infrared radiative code of the NCAR model. A similar number is estimated in Chapter 4. The upper end of this range is close to that presented in IPCC (1990).

Release of CH_4 leads to increased stratospheric water vapour through photochemical oxidation; estimates of this indirect effect are of the order of 5% or less of the direct effect of methane (4% of the total) based on the discussion in Chapter 4; current results from the LLNL, NCAR, and Mainz radiative/photochemical two-dimensional models; and the published literature (e.g., Lelieveld and Crutzen, 1992; Brühl, 1993; Lelieveld *et al.* 1993; Hauglustaine *et al.*, 1994a,b). We adopt 5% of the direct effect in the table below, which is smaller than the value quoted in IPCC (1990).

Each injected molecule of CH_4 ultimately forms CO_2,

representing an additional indirect effect which would increase the GWPs by approximately 3 for all time horizons (see IPCC, 1990). However, as noted by Lelieveld and Crutzen (1992), this indirect effect is unlikely to apply to biogenic production of CH_4 from most sources (e.g., from rice paddies), since the ultimate source of the carbon emitted as CH_4 in this case is CO_2, implying no net gain of carbon dioxide. While non-biogenic methane sources such as mining operations do lead indirectly to a net production of CO_2, this methane is often included in national carbon production inventories. In this case, consideration of CO_2 production in the GWP could lead to "double-counting", depending upon how the GWPs and inventories are combined. As shown in Table 2.3, most human sources of methane are biogenic, with another large fraction being due to coal mines and natural gas. Thus, the indirect effect of CO_2 production does not apply to much of the CH_4 inventory, and is not included in Table 5.5 (in contrast to IPCC (1990), where this effect was included).

As in Table 5.2, the GWPs were calculated relative to the CO_2 decay response of the Bern carbon cycle model with a constant current CO_2 and CH_4 atmosphere. Table 5.5 summarises the composite result for methane GWPs and its uncertainty, and considers the breakdown of the effects among various contributing factors. The ranges in CH_4 GWPs shown in Table 5.5 reflect the uncertainties in response time, lifetime, and indirect effects as discussed below. We assume a lifetime of methane in the background atmosphere of 10±2 years (which is consistent with Table 2.3). However, the response time of an added pulse is assumed to be much longer (12-17 years based upon Chapter 2). The total GWPs reported in IPCC (1990) including indirect effects are within the ranges shown in Table 5.5. The longer response time adopted here for methane perturbations is responsible for a large part of the change in methane GWP values compared to the nominal values including direct effects only in the IPCC (1992) report (although the fact that indirect effects were likely to be comparable to the direct effect was noted). This change is based entirely on the analysis presented in Chapter 2 used to define the methane response time for this report (see Prather, 1994). The decay response has been thoroughly tested only for small perturbations around a background state and continuing input flux approximately representative of today's atmosphere. It would be different if, for example, large changes in methane emissions were to occur in the near future. It is also believed to be sensitive to other chemical factors such as the sources of carbon monoxide. The GWP determined in this manner is similarly valid for relatively small perturbations, e.g., those that would be required to stabilise concentrations at current levels rather than continuing the small trend (order 1%/yr) observed in the past decade (see Chapter 2). However, the GWP shown in Table 5.5 cannot be used to estimate the radiative forcing that occurred since pre-industrial times, when methane concentrations more than doubled.

5.2.6 The Product of GWP and Estimated Current Emissions

Based upon the relative radiative forcing indices that are the focus of this chapter and an estimate of the total anthropogenic emissions of these compounds, the contributions of current emissions of each gas or group of gases to the total radiative forcing expected for *future* time horizons can be estimated. It is useful to compare these quantities with the assessments of total forcing in the *current* atmosphere presented in Section 4.7, where aerosol, volcanic, and solar absolute forcings are also estimated.

It is clear that the anthropogenic emissions of many compounds are known only approximately, and we present Figure 5.7 only as an approximate guide to consideration of relative effects. For this purpose, we assumed approximate average emissions of each gas over the decade of the 1980s. We assumed annual anthropogenic emissions for CO_2 of about 6.6±1.3 GtC/yr (about 19,400-29,000 Tg/yr) for industrial and land use sources of CO_2 based on Table 1.3 (taking into account fossil fuel combustion, cement manufacture and net global land-use emissions). The annual anthropogenic emissions and their uncertainties for CH_4 and N_2O were obtained from Tables 2.3 and 2.4, respectively. We estimate annual anthropogenic emissions of about 1.5 Tg/yr for the sum of

Table 5.5: *Total GWP for CH_4, including indirect effects, referenced to the AGWP computed for the CO_2 decay response of the Bern carbon cycle model and future CO_2 atmospheric concentrations held constant at current levels.*

	Time horizon		
	20 year	100 year	500 year
Total CH_4 GWP, including indirect effects and 12-17 year response time	42-82	17-32	5-10
Fraction of total GWP due to tropospheric O_3 change	19±12%	19±12%	19±12%
Fraction of total GWP due to stratospheric H_2O change	4%	4%	4%

Trace Gas Radiative Forcing Indices 227

the major CFCs (CFC-11, CFC-12, CFC-113, CFC-114, and CFC-115), methylchloroform, and halons, 0.005 Tg/yr for SF_6 and 0.03 Tg/yr for CF_4. Combining these emission estimates with the GWPs for 20- 100- and 500-year time horizons (see Tables 5.2 and 5.5) yields the approximate future integrated forcing from anthropogenic emissions shown in Figure 5.7. The uncertainty ranges shown reflect estimated uncertainties in emissions for CO_2, CH_4, and N_2O as indicated above, as well as uncertainties in GWPs for CH_4 and the ozone-depleting gases. While they are clearly not major greenhouse gases on a global basis at current emission rates for time horizons considered in Figure 5.7, SF_6 and CF_4 are included to illustrate the impact of small releases of gases with high relative GWP and because of their long lifetimes (see, e.g., the values for the 500-year time horizon). In addition, they are significant fractions of the total product of GWP times emissions of a few specific countries. For CH_4, the range includes the uncertainties in GWPs discussed above and presented in Table 5.5. For the sum of CFCs and halons, it is possible that the net GWP is small or even negative when the indirect effects of ozone depletion are considered, and the range depicted in the figure includes the upper limit represented by the direct effects of these gases only. The figure illustrates that future anthropogenic radiative forcings due to current emissions are likely to be induced primarily by CO_2 and methane (but this depends upon the future state of the atmosphere through its effect on the methane response time as noted above and based upon Chapter 2; Figure 5.7 refers to a future concentration of methane close to today's value). Further quantifying the relative magnitudes of the two, as well as the effects of other gases, requires better understanding of emissions, response times, and indirect effects.

5.3 A Perspective on GWPs

5.3.1 The Insights Gained From Ozone Depleting Potentials (ODPs)

The international negotiations that led to the 1987 Montreal Protocol on Substances that Deplete the Ozone Layer provide an example of the utility of trace-gas indices. The Ozone Depletion Potential (ODP) of a chlorine- or bromine-containing compound is a measure of its relative ability to destroy stratospheric ozone. The Protocol tabulated such values in its "controlled substances" list. Although not used widely, the provisions of the Protocol allow a country to "trade" between substances, using the ODP as the scaling factor. Furthermore, the ODPs have been used to set limits, relative to the CFCs, on the HCFCs that are to replace the CFCs.

The use of ODPs in the Montreal Protocol had several characteristics that are relevant to the consideration of

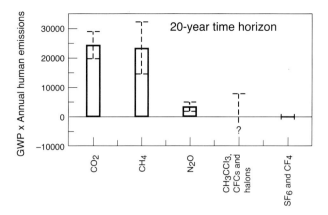

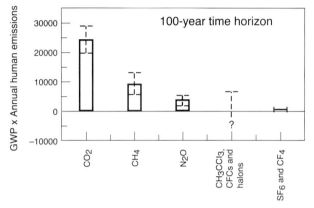

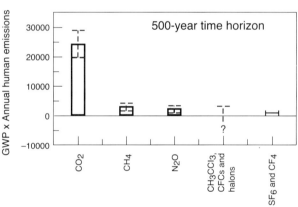

Figure 5.7: The product of GWP with estimated annual anthropogenic emissions typical of the 1980s for various gases, for time horizons of 20, 100 and 500 years. Ranges indicate the range of possible uncertainties considered here in estimating GWPs and total anthropogenic emissions. The estimated emissions are approximate. Note that uncertainties in the indirect radiative forcing of CFCs and halons due to ozone depletion imply that the net GWPs for these gases could be negative, but this is presently uncertain.

GWPs or AGWPs playing a similar role in the Framework Convention on Climate Change:

- *Single Values for the ODPs.* The Legal Drafting Groups that wrote the final text of the Protocol required that single values be stated for the ODPs, despite the fact that a scientific estimate always has some level of uncertainty and hence a numerical range within which the ODP is likely to lie. The median value of the estimated range of possible values was the one provided to the Protocol. The same single-value issue is likely to emerge in discussions of the use of GWPs in implementing the Climate Convention with, for example, a "basket" approach within which greenhouse gases can be "traded" on a scaled basis.

- *Periodic Updates of ODPs From Science Assessments.* The Montreal Protocol requires that periodic scientific assessments be made of the state of the science of the ozone layer. Updated ODPs have been part of these assessments, e.g., WMO (1992). Thus, in principle, the "certified" values can be altered if the advance in scientific understanding warrants it. However, the rapidity with which policy has moved subsequently to phase out all of the halocarbons that have high ODPs (e.g., the 1990 Amendments) has meant that the "basket" feature of the Protocol and the mechanisms for updating ODPs is unlikely to be exercised further. Nevertheless, the updating provision of the Montreal Protocol suggests that policy formulation under the Climate Convention could accommodate, via a periodic scientific assessment process, the inevitable improvements that will occur in the specification of GWPs. Indeed, the sequence of IPCC reports are illustrative examples of such updates to policy-relevant scientific information.

- *Model-Dependent ODPs Acceptable.* ODPs are the product of a calculation based on a model of how halogens interact with stratospheric ozone. Since models are approximations of reality and since model formulations differ, so do the ODPs that they produce. The scatter among model results was not considered by policy makers as a debilitating complexity. It was simply taken as one indication of uncertainty. Since GWP calculations are an analogous situation, the spread of calculations from different models could be handled in the same fashion in formulating a recommended set of current values.

- *ODPs' Time Horizons.* Since the calculation of an ODP involves integrating its contribution to ozone depletion into the future, a choice of time horizon must be made in the calculation. The original values supplied to policy makers in the scientific assessments were based only on an infinite time horizon, which was accepted as the policy basis when anticipated ozone changes were far in the future. However, with the recognition of large observed ozone depletions in the current atmosphere and with the development of short-lived substitutes for the long-lived chlorofluorocarbons, ODPs with short time horizons were developed. These have been used for some policy considerations, particularly those associated with selecting HCFCs that would limit chlorine input from these substitutes over the next 5 to 15 years, during which atmospheric halogen concentrations and hence ozone losses are expected to attain their peak values. This experience suggests that scientific considerations relating to time-scales of likely or observed atmospheric responses can help clarify choices relating to time horizons.

- *Complexity Introduced by the Bromine/Chlorine Interplay.* The ODP of a bromine-containing species depends on the level of atmospheric chlorine assumed in the calculation. The scientific assessment panel described this added complexity to the decision makers and explained that the calculated ODPs were based on the current chlorine levels. These ODP values were accepted, demonstrating that such a level of complexity can be handled by the Protocol process and that an agreed-upon value of a variable, whose future value cannot be fully predicted, can be assimilated by the policy process. This experience bodes well for the assimilation of similar GWP complexities by the policy process, e.g., the dependence of GWP for CH_4 on the future CH_4 and N_2O concentration scenarios.

However, it is important to note that GWPs are more complex than ODPs. Hence, the use of GWPs in the future negotiations of the Framework Convention on Climate Change could be more difficult than those associated with the use of ODPs in the Montreal Protocol. The reasons are severalfold:

- *The sinks (hence atmospheric lifetimes) of most of the major ozone-depleters – chlorofluorocarbons and bromocarbons – are well known.* There is a greater degree of uncertainty for the major human-influenced greenhouse gas – CO_2. Hence, the reference (denominator) for GWPs, and their numerical values for all species, are likely to change more than was the case for ODPs, which used CFC-11 as the reference species.

- *Stratospheric ozone depletion involved largely one family of chemical compounds – the halocarbons.* In contrast, there are several different types of greenhouse gases, thereby invoking the need to understand a much broader set of processes in the greenhouse gas issue, e.g., stratospheric photochemical removal *vis-a-vis* biospheric uptake. Hence, broad quantification of GWPs will be more difficult.

- *The response of the stratosphere to injection of halocarbons is relatively rapid (time scale of the order of five years), while many of the climate responses to radiative forcing are much longer (order of decades to centuries).* Thus, the role of time-scales in greenhouse gas decisions and time-horizon choices in GWPs could be fundamental. For example, a relatively short response time of an affected system allows earlier observations of the response and, in principle, more opportunity to use such observations in policy decisions. The opposite is true for long response times.

- *The producers of the halocarbons are the industrialised countries.* In contrast, every country produces and/or removes (e.g., forest sink for CO_2) greenhouse gases. As a result, the spectrum of countries that will be involved in using GWPs and in negotiating emission reductions will be much broader, implying more complex negotiations.

- *The halocarbons are linked to a limited part of the industrial sector.* Greenhouse gases permeate many aspects of the economies of most nations, since the energy, transportation, and agricultural sectors are all involved. As above, the spectrum of the industrial sector involved in technically defining the ways in which GWPs are used and in which emission reductions can be accomplished will be broader, with similar implications.

5.3.2 Characteristics Relevant to Uses of GWPs in Policy Formulation

The choice of time horizon for GWPs

While the selection of a time horizon of a radiative forcing index is largely a "user" choice (i.e., a policy decision), some scientific points are relevant to that selection:

- Policy-relevant climate-change phenomena exist at both ends of the climate-change time spectrum:

    (i) If the policy emphasis is to help guard against the possible occurrence of potentially abrupt, non-linear climate responses in the relatively near future, then a choice of a 20-year time horizon would yield an index that is relevant to making such decisions regarding appropriate greenhouse gas abatement strategies. In addition, if the speed of potential climate change is of greatest interest (rather than the eventual magnitude), then a focus on shorter time horizons can be used.

    (ii) Similarly, if the policy emphasis is to help guard against long-term, quasi-irreversible climate or climate-related changes (e.g., the very slow build up of and recovery from sea level changes that are controlled by slow processes such as warming of the ocean), then a choice of a 100-year or 500-year time horizon would yield an index that is relevant to making such decisions regarding appropriate greenhouse gas abatement strategies.

With this awareness, policies could choose to be a mix of emphases. GWPs with differing time horizons can aid in establishing such a mix. Indeed, that was the case in the Montreal Protocol deliberations, in which the long-lived, high-ODP gases were the initial focus and the shorter-lived, lower-ODP gases were a subsequent focus.

- The scientific uncertainties are very substantial for Earth-system processes that occur on long time-scales. This fact implies larger inaccuracies for GWPs with longer time horizons compared to those with shorter time horizons.

GWPs and natural sources

The *nature* of the emission source of a compound – i.e., whether it is natural, anthropogenic, or both – does not figure directly into the calculation of a GWP (or an ODP). It is important, of course, for understanding budgets and trends. Other things being equal, the fact that a compound has large natural sources does not imply a less-accurate or a less-meaningful GWP.

Awareness of improving science

Users of GWPs need to be aware that future changes in numerical indices are likely as research in related areas yields improved input to the calculations. This implies that whatever framework is adopted for the use of these indices, it must have flexibility to incorporate what could be substantial changes in the specified numerical values of the indices. The use of ODPs in the Montreal Protocol indicates that the science/policy greenhouse dialogue can deal fruitfully with similar climate-related complexities as well.

References

Barry, J., H.W. Sidebottom, J. Treacy and J. Franklin, 1994: Kinetics and mechanism for the atmospheric oxidation of 1,1,2-trifluoroethane. *Int. J. Chem. Kin.* (In press).

Briegleb, B., 1992: Longwave band model for thermal radiation in climate studies. *J. Geophys. Res.*, **97**, 11475-11486.

Brühl, C., 1993: The impact of the future scenarios for methane and other chemically active gases on the GWP of methane. *Chemosphere*, **26**, 731-738.

Butler, J. H., J.W. Elkins, T.M. Thompson, B.D. Hall, T.H. Swanson and V. Koropalov, 1991: Oceanic consumption of CH_3CCl_3: Implications for tropospheric OH. *J. Geophys. Res.*, **96**, 22347-22355.

Caldeira, K. and J.F. Kasting, 1993: Global warming on the margin. *Nature.*, **366**, 251-253.

Cess, R.D., M-H. Zhang, G.L. Potter, H.W. Barker, R.A. Colman, D.A. Dazlich, A.D. Del Genio, M.Esch, J.R. Fraser, V. Galin, W.L. Gates, J.J. Hack, W.J. Ingram, J.T. Kiehl, A.A. Lacis, H. Le Treut, Z-X. Li, X-Z. Liang, J-F. Mahfouf, B.J. McAveney, V.P. Meleshko, J-J. Morcrette, D.A. Randall, E. Roeckner, J-F. Royer, A.P. Sokolov, P.V. Sporyshev, K.E. Taylor, W-C. Wang and R.T. Wetherald, 1993: Uncertainties in CO_2 radiative forcing in atmospheric general circulation models. *Science*, **262**, 1252-1255.

Clerbaux, C., R. Colin, P.C. Simon and C. Granier, 1993: Infrared cross sections and global warming potentials of 10 alternative hydrohalocarbons. *J. Geophys. Res.*, **98**, 10491-10497.

Daniel, J.S., S. Solomon and D.L. Albritton, 1994: On the evaluation of halocarbon radiative forcing and global warming potentials. *J. Geophys. Res.* (Accepted)

DeMore, W.B, 1993: Rate constants for the reactions of OH with HFC-134a and HFC-134. *Geophys. Res. Lett.*, **20**, 1359-1362.

Ehhalt, D.H., and J. Drummond, 1988: NO_x sources and the tropospheric distribution of NO_x during STRATOZ III. In: *Tropospheric Ozone: Regional and Global Scale Interactions*, I. S. A. Isaksen (ed.), D. Reidel, Boston, pp217-237.

Fehsenfeld, F.C. and S.C. Liu, 1993: Tropospheric ozone: Distribution and sources. In: *Global Atmospheric Chemical Change*, C.N. Wewitt and W.T. Sturges (eds.), Elsevier, London, pp169-231.

Fisher, D.A., C.H. Hales, W-C. Wang, M.K.W. Ko and N.D. Sze, 1990: Model calculations on the relative effects of CFCs and their replacements on global warming. *Nature*, **344**, 513-516.

Fuglestvedt, J.S., J.E. Jonson and I.S.A. Isaksen, 1994: Effects of reduction in stratospheric ozone on tropospheric chemistry through changes in photolysis rates. *Tellus.* (Accepted)

Hammond, A. L., E. Rodenburg and W. Moomaw, 1990: Commentary in *Nature*, **347**, 705-706.

Harvey, L.D., 1993: A guide to global warming potentials (GWPs). *Energy Policy*, pp24-34, January, 1993.

Hauglustaine, D.A., C. Granier, G.P. Brasseur and G. Mégie, 1994a: The importance of atmospheric chemistry in the calculation of radiative forcing on the climate system. *J. Geophys. Res.*, **99**, 1173-1186.

Hauglustaine, D.A., C. Granier, and G.P. Brasseur, 1994b: Impact of increased methane emissions on the atmospheric composition and related radiative forcing on the climate system. *Environmental Monitoring Assessment*. (Accepted)

IPCC, 1990: *Climate Change: The Scientific Assessment.* J.T. Houghton, G.J. Jenkins and J.J. Ephraums (eds.), Cambridge University Press, Cambridge, UK.

IPCC, 1992: *Climate Change 1992: The Supplementary Report to the IPCC Scientific Assessment.* J.T. Houghton, B.A. Callander and S.K. Varney (eds.), Cambridge University Press, Cambridge, UK.

IPCC, 1995: *Climate Change: The Scientific Assessment.* (In preparation)

IUPAC, 1992: Evaluated kinetic and photochemical data for atmospheric chemistry. *J. Phys. Chem. Ref. Data*, **21**, no. 6.

JPL, 1992: Chemical kinetics and photochemical data for use in stratospheric modelling,. *JPL report,* Number **92-20.**

Jones, R.L., J.A. Pyle, J.E. Harries, A.M. Zavody, J.M. Russell and J.C. Gille, 1986: The water vapour budget of the stratosphere studied using LIMS and SAMS satellite data. *Quart. J. Roy. Met. Soc.*, **112**, 1127-1143.

Khalil, M.A. and R.A. Rasmussen, 1990: Global increase of atmospheric molecular hydrogen. *Nature*, **347**, 743-744.

Kindler, T.P., W.L. Chameides, P.H. Wine, D.M. Cunnold, F.N. Alyea and J.A. Franklin, 1994: The fate of atmospheric phosgene and the ozone depletion potentials of its parent compounds: CCl_4, C_2Cl_4, C_2HCl_3, CH_3CCl_3, $CHCl_3$. *J. Geophys. Res.* (Submitted)

Ko, M.K.W., N.D. Sze, W-C. Wang, G. Shia, A. Goldman, F.J. Murcray, D.G. Murcray and C.P. Rinsland, 1993: Atmospheric sulphur hexafluoride: Sources, sinks, and greenhouse warming. *J. Geophys. Res.*, **98**, 10499-10507.

Lal, M. and T. Holt, 1991: Ozone depletion due to increasing anthropogenic trace gas emissions: role of stratospheric chemistry and implications for future climate. *Clim. Res.*, **1**, 85-95.

Lashof, D.A. and D.R. Ahuja, 1990: Relative contributions of greenhouse gas emissions to global warming. *Nature*, **344**, 529-531.

Lelieveld, J. and P.J. Crutzen, 1992: Indirect chemical effects of methane on climate warming. *Nature*, **355**, 339-342.

Lelieveld, J., P.J. Crutzen and C. Brühl, 1993: Climate effects of atmospheric methane. *Chemosphere*, **26**, 739-768.

Lin, X., M. Trainer and S.C. Liu, 1988: On the nonlinearity of the tropospheric ozone production. *J. Geophys. Res.*, **93**, 15879-15888.

Moore, B. III, and B.H. Braswell, Jr., 1994: The lifetime of excess atmospheric carbon dioxide. *Global Biochem. Cycles*, **8**, 23-28.

Nelson, D.D., M.S. Zahinsen and C.E. Kolb, 1993: OH reaction kinetics and atmospheric lifetimes of $CF_3CFCHCF_3$ and CF_3CH_2Br,. *Geophys. Res. Lett.*, **20**, 197-200.

Oeschger, H., U. Siegenthaler, U. Schotterer and A. Gegelmann, 1975: A box diffusion model to study the carbon dioxide exchange in nature. *Tellus*, **27**, 168-192.

Prather, M.J., 1994: Lifetimes and Eigenstates in atmospheric chemistry. *Geophys. Res. Lett.*, **21**, 801-804.

Prather, M.J. and C.M. Spivakovsky, 1990: Tropospheric OH and the lifetimes of hydrochlorofluorocarbons (HCFCs). *J. Geophys. Res.*, **95**, 18723-18729.

Prinn, R., D. Cunnold, P. Simmonds, F. Alyea, R. Boldi, A.

Crawford, P. Fraser, D. Gutzler, D. Hartley, R. Rosen, and R. Rasmussen, 1992: Global averaged concentration and trend for hydroxyl radicals deduced from ALE/GAGE trichloroethane (methylchloroform) data for 1978-1990. *J. Geophys. Res.,* **97**, 2445-2462.

Ramaswamy, V., M.D. Schwarzkopf and K.P. Shine, 1992: Radiative forcing of climate from halocarbon-induced global stratospheric ozone loss. *Nature,* **355**, 810-812.

Ravishankara, A.R., S. Solomon, A.A. Turnipseed and R.F. Warren, 1993: Atmospheric lifetimes of long-lived species. *Science,* **259**, 194-199.

Reilly, J. M. and K.R. Richards, 1993: Climate change damage and the trace gas index issue. *Env. Res. Econ.,* **3**, 41-61.

Rodhe, H., 1990: A comparison of the contributions of various gases to the greenhouse effect. *Science,* **248**, 1217-1219.

Schmoltner, A.M., R.K. Talukdar, R.F. Warren, A.Mellouki, L. Goldfarb, T. Gierczak, S.A. Mckeen and A.R. Ravishankara, 1993: Rate coefficients of several hydrofluorocarbons with OH and $O(^1D)$ and their atmospheric lifetime. *J. Phys. Chem.,* **97**, 8976-8982.

Schwarzkopf, M.D. and V. Ramaswamy, 1993: Radiative forcing due to ozone in the 1980's: dependence on altitude of ozone change. *Geophys. Res. Lett.,* **20**, 205-208.

Solomon, S., J.B. Burkholder, A.R. Ravishankara and R.R. Garcia, 1994: On the ozone depletion and global warming potentials of CF_3I. *J. Geophys. Res.,* **99**, 20929-20935.

Thompson, A.M., 1992: The oxidising capacity of the earth's atmosphere: probable past and future changes. *Science,* **256**, 1157-1165.

Wang, W-C., M.P. Dudek, X-Z. Liang and J.T. Kiehl, 1991: Inadequacy of effective CO_2 as a proxy in simulating the greenhouse effect of other radiatively active gases. *Nature,* **350**, 573-577.

Wang, W-C., Y-C. Zhuang and R.D. Bojkov, 1993: Climate implications of observed changes in ozone vertical distributions at middle and high latitudes of the northern hemisphere. *Geophys. Res. Lett.,* **20**, 1567-1570.

Wigley, T.M.L., 1994a: The effect of carbon cycle uncertainties on global warming potentials. *Geophys. Res. Lett.* (Submitted)

Wigley, T.M.L., 1994b: How important are carbon cycle model uncertainties? In:*Climate Change and the Agenda for Research*, T. Hanisch (Ed.), Westview Press, Boulder, Colorado, USA, pp169-191.

WMO 1990; Scientific Assessment of Stratospheric Ozone, 1989: *Global Ozone Research and Monitoring Project Report No. 20.* World Meteorological Organisation, Geneva.

WMO 1992; Scientific Assessment of Ozone Depletion, 1991: *Global Ozone Research and Monitoring Project Report No. 25.* World Meteorological Organisation, Geneva.

WMO 1995; Scientific Assessment of Ozone Depletion, 1994: *Global Ozone Research and Monitoring Project Report No. 37.* World Meteorological Organisation, Geneva.

Wuebbles, D.J., J.S. Tamaresis and K.O. Patten, 1994a: Quantified estimates of total GWPs for greenhouse gases taking into account tropospheric chemistry. UCRL – ID-115850.

Wuebbles, D.J., A.K. Jain, K.O. Patten and K.E. Grant, 1994b: Sensitivity of direct global warming potentials to key uncertainties. *Climatic Change.* (In press)

Zhang, Z., S. Padmaja, R.D. Saihi, R.E. Huie and M.J. Kurylo, 1994: Reactions of hydroxyl radicals with several hydrofluorocarbons : the temperature dependencies of the rate constants for HFC-245ca, HFC-236ea, HFC-227ea and HFC-356ffa. *J. Phys. Chem.,* **98**, 4312-4315.

Climate Change 1994

Part II

An Evaluation of the IPCC IS92 Emission Scenarios

Prepared by Working Group III

Preface to WGIII Report

This section of the Special Report of the IPCC was prepared by Working Group III. It consists of two parts: a Summary for Policymakers and a Technical Report, Chapter 6. The reader should be aware of the IPCC procedures which were followed in producing each part.

The Technical Report was prepared by a writing team of six experts from developed and developing countries in different regions of the world. It was subject to extensive peer review and government review by correspondence. The resulting draft was accepted by Working Group III as an underlying technical document. However, the content of the Technical Report remains the responsibility of the Writing Team.

The Summary for Policymakers (SPM) was prepared by the Working Group III Technical Support Unit in consultation with the Writing Team. In addition to the extensive review and modification referred to above, the SPM was subject to line by line approval at plenary meetings of Working Group III. These intergovernmental meetings were attended by representatives of approximately 70 countries, with widely varying views on matters concerning greenhouse gases and climate change. In addition, several NGOs participated.

The resulting Summary for Policymakers is thus an intergovernmentally negotiated text. In the course of these negotiations some of the draft recommendations by the Working Group III Bureau were deleted and in a few places, where agreement could not be reached, differing views of the findings are presented.

The reader wishing to have a short resumé of the findings of the Writing Team is referred to the Summary at the beginning of the Technical Report.

James P. Bruce Hoesung Lee
Richard Odingo Lorents Lorentsen

Working Group III Bureau

Summary for Policymakers:

An Evaluation of the IPCC IS92 Emission Scenarios

A Report of Working Group III of the Intergovernmental Panel on Climate Change

CONTENTS

What are the IPCC IS92 Emission Scenarios?	241
How were the IS92 Emission Scenarios Evaluated?	242
What was not Covered by the Evaluation?	242
What are the Findings about Scenarios in General?	242
What are the Uses and Limitations of the IPCC IS92 Scenarios?	245
Addressing the Gaps in Knowledge	246
Present Scenarios - Examples of Gaps and Needs	246
New Scenarios - Examples of Gaps and Needs	246

What are the IPCC IS92[1] Emission Scenarios?

In 1992 the IPCC sponsored development of emission scenarios as inputs to global climate models because of the scarcity of comprehensive emission scenarios available in the literature at the time. This report by Working Group III presents an evaluation of the emission scenarios and especially a discussion of their most important input assumptions (see Table 1 which summarises these assumptions of the IS92 Scenarios) and results. Other scenarios are included by way of comparison to the IPCC Scenarios. Also covered are the appropriate uses of these emission scenarios in scientific assessment and policy making.

Emission scenarios in this evaluation are projections of anthropogenic emissions of gases that have the potential to affect climate, based on assumptions about future trends of key determinants such as population, economic growth, technological change, land use and emission control policies. The six scenarios are among the most comprehensive available, providing estimates of direct and indirect greenhouse gases by source for four regions over the period 1990 to 2100, and covering a wide range of values for the key input assumptions. It is emphasised that this evaluation deals mostly with non-intervention scenarios, that is, scenarios that do not assume any climate policies to reduce greenhouse gas emissions (although they

Table 1: Summary of assumptions in the six IPCC 1992 alternative IS92 Scenarios.

Scenario	Population	Economic growth	Energy supplies	Other	CFCs
IS92a	World Bank (1991) 11.3 billion by 2100	1990-2025: 2.9% 1990-2100: 2.3%	12,000 EJ conventional oil 13,000 EJ natural gas. Solar costs fall to $0.075/kWh. 191 EJ/year of biofuels available at $70/barrel†	Legally enacted and internationally agreed controls on SO_x, NO_x and NMVOC emissions. Efforts to reduce emissions of SO_x, NO_x and CO in developing countries by middle of next century.	Partial compliance with Montreal Protocol. Technological transfer results in gradual phase out of CFCs in non-signatory countries by 2075.
IS92b	World Bank (1991) 11.3 billion by 2100	1990-2025: 2.9% 1990-2100: 2.3%	Same as "a"	Same as "a" plus commitments by many OECD countries to stabilise or reduce CO_2 emissions.	Global compliance with scheduled phase out of Montreal Protocol.
IS92c	UN Medium-Low Case 6.4 billion by 2100	1990-2025: 2.0% 1990-2100: 1.2%	8,000 EJ conventional oil 7,300 EJ natural gas Nuclear costs decline by 0.4% annually	Same as "a"	Same as "a"
IS92d	UN Medium-Low Case 6.4 billion by 2100	1990-2025: 2.7% 1990-2100: 2.0%	Oil and gas same as "c" Solar costs fall to $0.065/kWh 272 EJ/year of biofuels available at $50/barrel	Emission controls extended worldwide for CO, NO_x, NMVOC and SO_x. Halt deforestation. Capture and use of emissions from coal mining and gas production and use.	CFC production phase out by 1997 for industrialised countries. Phase out of HCFCs.
IS92e	World Bank (1991) 11.3 billion by 2100	1990-2025: 3.5% 1990-2100: 3.0%	18,400 EJ conventional oil Gas same as "a" Phase out nuclear by 2075	Emission controls which increase fossil energy costs by 30%.	Same as "d"
IS92f	UN Medium-High Case 17.6 billion by 2100	1990-2025: 2.9% 1990-2100: 2.3%	Oil and gas same as "e" Solar costs fall to $0.083/kWh Nuclear costs increase to $0.09/kWh	Same as "a"	Same as "a"

† Approximate conversion factor: 1 barrel = 6 GJ.
Source of Table: Leggett *et al.*, IPCC, 1992.

---

[1] The emissions scenarios developed by the IPCC in 1992 are commonly referred to as the IS92 Scenarios. The six scenarios are labelled IS92a through IS92f.

may assume emission controls for other environmental reasons). This applies to five out of the six IS92 Scenarios, and to all other scenarios that we call non-intervention scenarios.

Emission scenarios and the assumptions upon which they are based "are inherently controversial because they reflect different views of the future. The results of scenarios can vary considerably from actual outcomes even over short time horizons. Confidence in scenario outputs decreases substantially as the time horizon increases, because the basis for the underlying assumptions becomes increasingly speculative. Considerable uncertainties surround the evolution of the types and levels of human activities (including economic growth and structure), technological advances, and human responses to possible environmental, economic and institutional constraints. Consequently, emission scenarios must be constructed carefully and used with great caution."[1]

Therefore, an informed assessment of the possible consequences of uncertain future emissions paths necessitates the use of a range of scenarios.

How were the IS92 Emission Scenarios Evaluated?

The factors outlined in the Terms of Reference and the Work Plan for Working Group III were considered, but it must be recognised that there are no accepted criteria for evaluating scenarios in the scientific literature.[2] Since scenarios deal with the future, they cannot be compared to observations. As an alternative, the "reasonableness" of their input assumptions and methodology can be examined. This can be done in part by seeking to establish the internal consistency of various assumptions over time, including possible interactions between assumptions such as those that might relate the evolution of systems for energy use, economic growth, land use and population. "Reasonableness" can also be examined by comparing several models in an organised intercomparison exercise. Such an intercomparison was outside the scope of this evaluation.

Instead, the IS92 emission scenarios were evaluated by comparing the range of their emission projections and their major input assumptions with those of other published non-intervention scenarios. While this comparison does not answer all questions about the scenarios, it does provide information about the position of the scenarios relative to other work in the scientific community, and it was feasible in the short time available for this evaluation.

Since few other scenarios are as comprehensive as the IS92 Scenarios, the evaluation concentrates on a limited number of important greenhouse gases and source categories. Much of the evaluation centres on CO_2 emissions from human activities: the most important category of emissions. Also covered are emissions of methane (CH_4), nitrous oxide (N_2O) and sulphur (S) emissions from the energy sector, as well as CO_2 from deforestation, and CH_4 and N_2O from selected land use sources. Both global and regional emission scenarios are evaluated.

What was not Covered by the Evaluation?

Neither the types of models used for scenario development nor the sensitivity of input variables to government actions were evaluated. Both of these topics can best be addressed by comparing the different models used to compute emission scenarios. This intercomparison requires the full cooperation of different modelling groups and extensive analysis, and there was neither the time nor funds available in the evaluation for this task. The scenarios also were not compared to the emission inventories or emissions projections of countries because there were too few country estimates available to make substantial improvements in regional and global emission scenarios.

What are the Findings about Scenarios in General?

Almost all published reference scenarios (including the IS92 Scenarios) show an increase in annual greenhouse gas emissions over the next century. Only one of the six IS92 Scenarios, IS92c, shows a decline in annual CO_2 emissions between 1990 and 2100. Even it projects increases in emissions of almost all of the other direct and indirect greenhouse gases (e.g., methane, nitrous oxide) except the halocarbons controlled by the Montreal

---

1 J.T. Houghton, B.A. Callander and S.K. Varney, *Climate Change 1992: The Supplementary Report to the IPCC Scientific Assessment*, Cambridge University Press, Cambridge, 1992, page 9.

2 Working Group III's Terms of Reference referred to in the Work Plan require that its evaluation of the IS92 Scenarios should result in a report on the "validity, appropriateness and utility" of "a range of internally consistent scenarios for future emissions based on reasonable economic, demographic and technological projections, and taking account of gaps and uncertainties in available knowledge, especially concerning the evolution of socio-economic development and technology; where possible, policy assumptions should reflect their economic and social consequences."

This report evaluates the IS92 scenarios in the light of most of the specific factors outlined in Part III, Emissions Scenarios, Chapter 5, "Consideration of Consistent Scenarios," of the approved WG III Work Plan but does not evaluate the IS92 Scenarios according to all of the above factors because of the great difficulty in doing so in the available time.

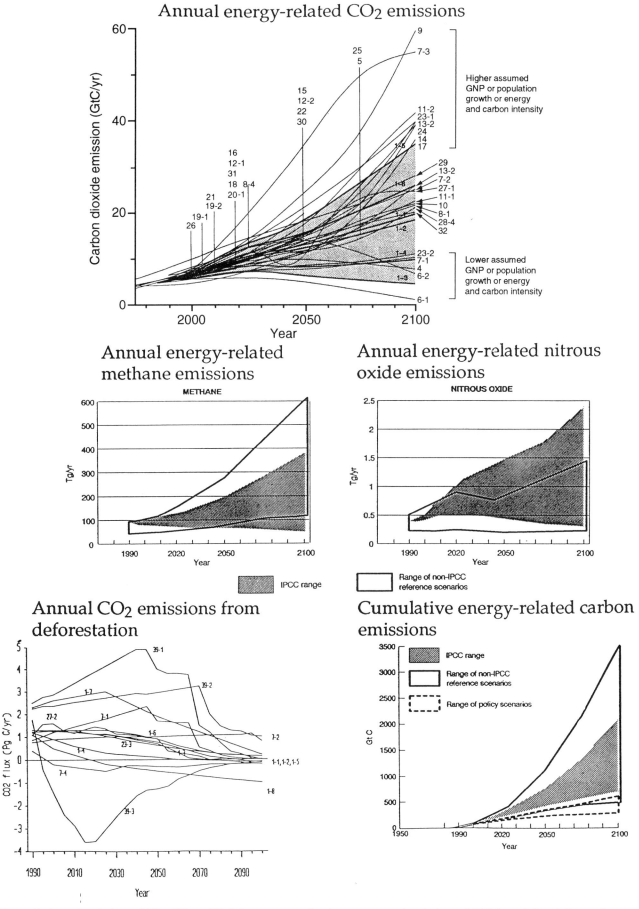

Figure 1: Annual emissions of CO_2, CH_4 and N_2O from energy-related sources, annual emissions of CO2 from deforestation, and cumulative CO_2 emissions from energy related sources.

Protocol. Global estimates of CH_4 and N_2O emissions from energy in 2100 vary by a factor of six or seven (Figure 1).

The set of published scenarios show a very wide range of projected annual CO_2 emissions by 2100. For energy-related CO_2 emissions, the extremes in the literature span a huge range, from 1.2 to 60 GtC per year[1] in 2100, and diverge greatly from 1990 to 2100. The IS92 Scenarios are representative of the range of views found in the published scenarios regarding energy-related CO_2 emissions. The IS92 Scenarios bracket a large part of this range, spanning from 4.6 to 34.9 GtC per year. It has been suggested by some, however, that comparing the results of the IS92 Scenarios is not informative, since this report has not analysed the reasonableness of the other scenarios or their assumptions. They further believe that the 110 year IS92 Scenarios and the other 100 year scenarios are equally speculative. Energy-related emissions currently dominate other types of emissions, and are expected to do so for the next century.

Figure 1 also shows emissions of CO_2 related to deforestation. Net deforestation contributes about 18% to current total CO_2 emissions, and most scenarios estimate that this percentage will decline in the future. Unlike energy emissions which diverge with time, land use emissions have their widest range (0.8 to 4.8 GtC per year) in the middle of the 21st century. Most scenarios of this type then converge on zero by the end of the century because they project that driving forces of deforestation equilibrate or forests are depleted. However, the amount of information available for non-energy-related CO_2 is more limited than information available for energy-related CO_2 emissions. Very few global scenarios of CH_4 and N_2O emissions from land use are available. This could be remedied by additional research.

The wide range of projected emissions of the IS92 and other scenarios stems in large part from differences in scenario input assumptions. For energy-related CO_2 emissions, population and economic growth assumptions usually play a key role as well as assumptions that influence the amount of carbon-based fuels used in the economy. For land use CO_2 emissions, the deforestation rate and the carbon content of vegetation are particularly important input assumptions. Future emission estimates can also vary because of differences in model structures and methods used to compute emissions. This analysis has not examined the relative importance of modelling factors compared to input assumptions in determining results and, therefore, it seems unwise to draw strong conclusions. However, EMF-12 intercomparisons[2] for energy CO_2 emission with harmonised input assumptions about population and economic activity provide information on how future emission estimates can vary because of differences in models as well as differences in input assumptions.

What input assumptions are important to the IS92 Scenarios in particular? In addition to the factors mentioned above, other assumptions are important in influencing the estimates of energy-related CO_2 emissions in the IS92 Scenarios. These include labour productivity, the growth of demand for energy in developing countries as incomes rise, and the rate of technological improvement of energy efficiency.

Cumulative emissions have a smaller range than annual emissions. Cumulative energy-related emissions are an important factor in the determination of future atmospheric concentrations of long-lived greenhouse gases. Accordingly, the annual emissions as given in the IS92 Scenarios have been summed over the periods 1990 to 2100 (long-term perspective) and 1990 to 2025 (short-term perspective).

- Long-term perspective (1990 to 2100): the cumulative emissions of the IS92 Scenarios range between 700 and 2080 GtC (the corresponding range for non-IPCC scenarios is 490-3450 GtC) (see Figure 1). The cumulative net emissions from deforestation and changing land use range between 30 and 320 GtC.

- Short-term perspective (1990 to 2025) the cumulative emissions as given by IS92 Scenarios range between 230 and 330 GtC.

It is of interest to note that emissions due to fossil fuel use between 1860 and 1990 amount to 215 GtC and due to deforestation and changing land use 80-120 GtC.

Reference is made to the Working Group I Summary for Policymakers for a discussion of the range of future changes of carbon dioxide concentrations in the atmosphere on the basis of the IS92 Scenarios.

Current commitments have a relatively small impact on the IS92 emission scenarios. A number of governments have emphasised the need to assess the impact on emissions scenarios of commitments to limitation of greenhouse gases referred to in the Framework Convention on Climate Change. In response, an analysis has been undertaken with the assumptions and results presented in the box on "Scenarios of Current Commitments".

---

1 The symbols "GtC" stand for gigatonnes of carbon dioxide emissions in units of carbon (C). a gigatonne is 109 tonnes and it is the same as a petagram (1015 grams) of emissions. National inventories prepared for the Conference of Parties measure emissions in thousands of tonnes (Gg) of carbon dioxide (CO2). 1Pg = 106Gg. To convert Gt of carbon (C) to Gt of carbon dioxide (CO_2) multiply by 3.67 (i.e. by 44/12).

2 The Energy Modelling Forum (EMF) at Stanford University in the USA organises model comparisons.

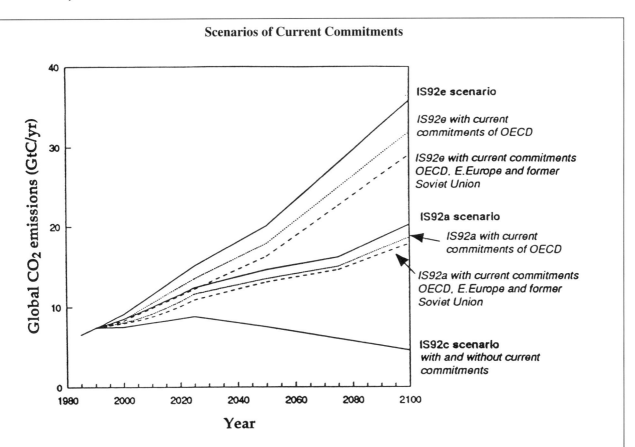

Scenarios of Current Commitments

This graph depicts a set of optimistic scenarios for global CO_2 emissions based on current commitments under Article 4 of the Framework Convention on Climate Change. These are optimistic scenarios because they assume that the countries making commitments under the Convention will freeze their emissions after 2005 at their committed levels.

The current commitment scenarios are compared to the lowest, highest and intermediate IPCC reference scenarios. For each reference scenario, two cases of current commitments are shown:

- **Case 1**: All OECD countries comply with commitments noted in Climate Change Policy Initiatives, Volume 1 OECD Countries, 1994 Update (OECD, 1994) and freeze emissions at these levels after year 2005;

- **Case 2**: All OECD countries comply with their commitments as in Case 1, plus countries in Eastern Europe and the former USSR freeze their emissions in year 2000 at their 1990 levels.

As a result of these assumptions, CO_2 emissions are reduced by about 8 to 12% (depending on the year considered) relative to the intermediate IPCC reference scenario (IS92a). Percentage reductions are slightly larger for the highest IPCC Scenario (IS92e). Current commitments have no effect on global emissions of the lowest IPCC Scenario (IS92c) because emissions in this reference case are lower than they would be for either Case 1 or 2.

What are the Uses and Limitations of the IPCC IS92 Scenarios?

Is one IS92 Scenario more likely than another? The likelihood that scenarios are close to real future developments depends mainly on the actual evolution of key input variables. The IS92a Scenario, for instance, provides a reasonable central case projection of global emissions. Its world population projections are close to the medium projections of three international organisations, and its global economic growth assumptions are close to the average of other published scenarios. However, it should be clear that being an intermediate scenario is not the same as being the most likely scenario.

The intermediate inputs of the IS92a Scenario yield, at the *global* level, annual CO_2 emission estimates that are near the centre of the range of published scenarios designed for similar purposes, at least up to 2050. IS92a is not, however, representative of the central case of other scenarios at the *regional* level.

The full range of scenarios should be used for climate analysis. Considering the degree of uncertainty, the wide range of views about future emissions, and the absence of a most likely scenario, it is unwise to use only one scenario for climate analysis. Rather, it is recommended to use the full range of IS92 Scenarios for this purpose. Users of the IPCC Scenarios are cautioned, however, that the lowest (IS92c) has emission levels and some input assumptions that are more characteristic of a policy, rather than a reference, scenario.

The IS92 emission scenarios are not suitable for other purposes. The IS92 Scenarios were not designed for other purposes, and indeed they are not suitable for assessing consequences of intervention to reduce greenhouse gas emissions, to examine the feasibility and costs of mitigating greenhouse gases, nor as input to negotiating possible emission reductions. However, they do help set the context for such negotiations by indicating the possible range of emissions if no mitigation takes place. New scenarios will be needed for these other purposes.

Addressing the Gaps in Knowledge

Present Scenarios - Examples of Gaps and Needs

The estimation of non-CO_2 emissions is weak in current scenarios and needs to be improved. There is a lack of knowledge or agreement on how to represent the processes of non-CO_2 emissions in emission models. Work is needed to improve the estimation of emissions, based on a comprehensive all-gas approach.

Emissions inventory data should be used to improve baseline data of scenarios. One of the largest sources of uncertainty in CO_2 and non-CO_2 emission scenarios is the uncertainty of base year emission estimates. Countries and organisations are beginning to compile inventories of their anthropogenic emissions of greenhouse gases under the Framework Convention on Climate Change, and these data should be used to improve base year estimates in new scenarios.

More detailed work has to be done to compare the driving forces, particularly the economic assumptions, behind the IS92 Scenarios, their sensitivities, and the validity of the assumptions used.

New Scenarios - Examples of Gaps and Needs

- *New reference scenarios are needed to explore a variety of economic development pathways.*
- *Policy scenarios are needed to explore a variety of climate policies, instruments and programmes, already developed or yet to be developed, and their results in terms of greenhouse gas emission reductions.* For emission scenarios to be used for estimation of feasibility and costs of mitigation measures, it is not only important to analyse the costs of emission reductions, but also to assess the costs and benefits of the impacts of emissions on the natural environment and society. Therefore scenarios developed for this purpose should be coupled with their impact on the environment and society in the form of consistent, comprehensive, sustainable development scenarios of global change.

 Policies should be tested for robustness against a wide range of reference emission estimates. Given the wide divergence of possible future emissions, it is wise to assess whether proposed policies will achieve their emissions objectives under a wide range of scenarios.

 Policy studies should focus on cumulative emissions over a period of time rather than annual emissions of a particular year, as climate impacts are related to cumulative emissions rather than annual emissions and cumulative emissions have a lower range of uncertainty than annual emissions.

- *Endogenous capacity-building is crucial.* Special effort is needed to improve the capabilities of researchers to analyse and develop scenarios, especially in developing countries and in countries with economies in transition. With this in mind, it is especially important to build their capabilities to (i) estimate base year emissions, (ii) prepare data on the driving forces of emissions, and (iii) run computer models of emissions.

6

An Evaluation of the IPCC IS92 Emission Scenarios

J. ALCAMO, A. BOUWMAN, J. EDMONDS, A. GRÜBLER,
T. MORITA, A. SUGANDHY

CONTENTS

Summary 251

6.1 Background and Scope 257
    6.1.1 Terms of Reference 257
    6.1.2 Scope of Evaluation 257
    6.1.3 Approach to the Evaluation 258

6.2 The IPCC IS92 Scenarios 259
    6.2.1 The ASF Model 260
    6.2.2 Overview of the IS92 Scenarios 260
    6.2.3 GDP Assumptions 262
    6.2.4 Population Assumptions 263
    6.2.5 The Relationship between GDP per capita and Population 264

6.3 Comparison of Energy-Related Scenarios 265
    6.3.1 Global Energy Scenarios 265
        6.3.1.1 Results from global CO_2 emission scenarios 265
        6.3.1.2 Components of global CO_2 scenarios 268
    6.3.2 Regional CO_2 Scenarios 270
        6.3.2.1 China and Centrally Planned Asia 277
        6.3.2.2 Eastern Europe and former USSR (EEFSU) 277
        6.3.2.3 Africa 280
        6.3.2.4 USA 280
    6.3.3 Scenarios for CH_4 and N_2O 281
    6.3.4 Scenarios for Sulphur Emissions 281
    6.3.5 Different Variables: their Relative Importance and the Sensitivity of Emissions to them 282
    6.3.6 Discussion of Scenarios for Energy-Related Emissions 283

6.4 Comparison of Land-Use-Related Scenarios 284
    6.4.1 Results from CO_2 Global Scenarios 286
    6.4.2 Results from CO_2 Regional Scenarios 286
    6.4.3 Factors Affecting CO_2 Emissions from Deforestation and Afforestation 289
        6.4.3.1 Rate of deforestation 289
        6.4.3.2 Carbon content of vegetation 290
    6.4.4 Results from Global Scenarios of CH_4 and N_2O 290
    6.4.5 Results from Regional Scenarios for CH_4 and N_2O 291
    6.4.6 Factors Affecting CH_4 and N_2O Emissions from Land-Use 292
        6.4.6.1 Factors affecting emission of CH_4 from rice production 292
        6.4.6.2 Factors affecting emission of CH_4 from enteric fermentation 293
        6.4.6.3 Factors affecting emission of N_2O from fertilised agricultural soils 293
    6.4.7 Discussion of Scenarios of Land-Use Emissions 293

6.5 Discussion and Recommendations 294
    6.5.1 What Are the Uses and Limitations of the IS92 Scenarios? 296
    6.5.2 The Question of New Scenarios 297
    6.5.3 The Process of Developing Scenarios 297
    6.5.4 Integrated Scenarios 298
    6.5.5 Dissemination of Scenarios 298

Acknowledgements 298

Supplementary Table: Scenarios Reviewed in this Report 299

References 301

SUMMARY[1]

Scenarios of greenhouse gas emissions play an important role in the analysis of potential climate change. From a scientific standpoint, they provide a departure point for analysing the potential occurrence and impacts of climate change; from a policy perspective they provide information about the consequences of action or inaction to reduce greenhouse gas emissions. Recognising the pivotal role of these scenarios, the IPCC undertook the development of a comprehensive and international set of emission scenarios as part of its First Assessment (1990). In 1992 the original reference scenario, SA90, was updated with new information and six new reference scenarios were developed: IS92 a-f (reported in IPCC (1992)).

This report presents an evaluation of the six IS92 Scenarios. Other published scenarios are also taken into account, but only by way of comparison with the IS92 Scenarios. It is emphasised that this evaluation deals mostly with non-intervention scenarios, that is, scenarios that do not assume any climate policies to reduce greenhouse gas emissions (although they may assume emission controls for other environmental reasons). This applies to five of the six IS92 Scenarios and to all other scenarios that we call non-intervention scenarios. Policy scenarios are only briefly discussed in the evaluation.

Because of the limited time available for this evaluation, it focuses exclusively on CO_2, CH_4, N_2O and S emission scenarios from energy and land-use activities. Other sources of greenhouse gases and sinks are not discussed.

Time constraints also made it impossible to evaluate scenario results for all regions. Therefore, the evaluation was limited to a few case study regions from the industrialised and the developing world.

It is also important to note that no other scenarios other than the IS92 Scenarios were evaluated. Therefore the reader should be aware that any shortcomings identified for the IS92 Scenarios in this evaluation may or may not apply to other scenarios.

Despite the somewhat limited focus of the evaluation, it is believed that a wider scope would not have led to significantly different conclusions.

Approach to Evaluation

Because there are no generally accepted procedures for evaluating these or other kinds of scenarios, a new approach is proposed. The approach consists of a set of purposes of scenarios, together with criteria to satisfy those purposes. The following purposes of scenarios are suggested:

- Purpose 1. As input to evaluating the environmental/climatic consequences of "non-intervention", i.e. no action to reduce greenhouse gas emissions.

- Purpose 2. As input to evaluating the environmental/climatic consequences of intervention to reduce greenhouse gas emissions.

- Purpose 3. As input to examining the feasibility and costs of mitigating greenhouse gases from different regions and economic sectors, and over time. This purpose can include setting emission reduction targets and developing scenarios to reach these targets. It can also include examining the driving forces of emissions and sinks to identify which of these forces can be influenced by policies.

- Purpose 4. As input for negotiating possible emission reductions for different countries and geographic regions.

Because the IS92 Scenarios were only designed for Purpose 1, this evaluation centres on the suitability of the scenarios for this purpose.

The mixed scientific/policy nature of the scenarios makes it difficult to separate scientific from policy objectives. Indeed, Purposes 1 and 2 have both scientific and policy goals. From the scientific side, they explore future states of the global climate system, and from the policy side, they assess the consequences on climate and society of different emission levels.

No accepted methods exist for evaluating whether the IS92 Scenarios are satisfactory for Purpose 1 above, or any

---

[1] This Summary was accepted as a Technical Summary by the IPCC Working Group III Plenary.

of the other purposes.[1] Scenarios deal with the future, so they cannot be compared with observations. The "reasonableness" of a scenario can be evaluated by its assumptions and methodology, but this is best done by comparing several models in an organised exercise. However, this approach was outside the scope of this evaluation. We have chosen instead to evaluate the IS92 emissions scenarios by comparing the range of their emission projections and their major input assumptions with those of other non-intervention scenarios. Although this comparison does not answer all questions about the scenarios, it provides information about their position relative to other work in the scientific community, and it was feasible in the short time available for this evaluation.

Energy-Related Scenarios — How Different Are Their Results?

Global Results

There are many more scenarios related to energy than to land-use because the models and methods used to estimate energy emissions are more detailed and advanced than those for land-use emissions. In addition, existing energy models could easily be adapted to compute CO_2 emissions. As a result, much of the scenario comparison focuses on energy-related emissions with the following results:

- Energy-related emissions currently dominate other types of CO_2 emissions and will continue to do so in the future under all scenarios considered.
- Almost all published non-intervention scenarios (including the IS92 Scenarios) show an increase in energy-related CO_2 emissions over the next century. Only one of the six IS92 Scenarios, IS92c, shows a decline in total CO_2 emissions between 1990 and 2100.
- For energy-related CO_2 emissions, the extremes in the literature span a very wide range, from 1.2 to 60 GtC/yr in 2100. (1 GtC stands for one gigatonne (10^9 tonnes) of carbon. A gigatonne is equal to a petagram.) The IS92 Scenarios bracket a large part of this range, spanning the range 4.6 to 34.9 GtC/yr.
- The estimates of emissions diverge greatly with time. The difference between the high and low IS92 Scenarios is less than a factor of 2 in 2025, but by 2100 it is nearly a factor of 8. The highest and lowest non-IPCC scenarios reviewed here vary by a factor of 3 in 2025 and a factor of 50 in 2100.
- This wide range can be explained mostly by differences in the input assumptions for the scenarios, particularly assumptions regarding economic and population growth, and the amount of carbon-based fuels in the economy. Another factor is the uncertainty of the models used to compute the scenarios.

- The cumulative energy-related emissions of the scenarios from 1990 to 2100 range from 492 to 3447 GtC (a factor of 7), which is considerably smaller than the range of annual emissions in 2100. The cumulative emissions of the IS92 Scenarios fall within this range (700 to 2078 GtC).
- The cumulative emissions of energy-related CO_2 from 1860 to 1990 were about 215 GtC. This amount is exceeded by even the IS92c Scenario (the lowest IPCC scenario) which would result in energy-related emissions of 700 GtC from 1990 to 2100.
- Cumulative emissions are important for climate analysis because CO_2 and other direct greenhouse gases can affect climate for decades to centuries after they are emitted because of the integrative nature of the atmosphere-ocean system.
- Whether or not CO_2 emissions will accumulate in the atmosphere depends on many factors that are partly taken into account by carbon cycle models. Model calculations of this sort from Working Group I indicate that the emissions of the IS92c Scenario could raise atmospheric concentrations of CO_2 to approximately 480 ppmv[2] by 2100. Hence, even the lowest IPCC Scenario (which would result in lower CO_2 emissions in 2100 than at present) could lead to a build-up of CO_2 in the atmosphere.

Compared with CO_2, far fewer scenarios are available for CH_4, N_2O and other greenhouse gas emissions. For these scenarios an additional source of uncertainty is important, namely, the uncertainty of *baseline* emission estimates. Thus, non-CO_2 scenarios have an overall lower reliability than CO_2 scenarios. IS92 estimates of energy-related CH_4 and N_2O emissions have a range of a factor of 6 to 7 in the year 2100. The range of the IS92 Scenarios for non-CO_2 emissions overlap with the few other published scenarios of this type.

---

1 The Terms of Reference referred to in the Work Plan require that the evaluation of the IS92 Scenarios by Working Group III should result in a report on the "validity, appropriateness and utility" of "a range of internally consistent scenarios for future emissions based on reasonable economic, demographic and technological prjections, and taking account of gaps and uncertainties in available knowledge, especially concerning the evolution of socio-economic development and technology; where possible, policy assumptions should reflect their economic and social consequences." This chapter evaluates the IS92 Scenarios in the light of most of the specific factors outlined in Part III, "Emissions Scenarios", Chapter 5, "Consideration of Consistent Scenarios," of the approved WG III Work Plan, but does not evaluate the IS92 Scenarios according to all the above factors because of the great difficulty in doing so in the time available .

2 1 ppmv = 1 part per million by volume.

Global scenarios of sulphur emissions are virtually unknown outside of the IS92 Scenarios. Baseline emission estimates are quite good for the OECD regions but less reliable elsewhere. In the IS92 Scenarios, estimates of energy-related sulphur emissions in the year 2100 range by somewhat less than a factor of 5.

Regional Results

There are many regional and country estimates of energy-related CO_2 emissions, but the estimates do not correspond to the IS92 regions, so comparison was not possible. The following comments can be made about the few scenarios available for comparison:

- The range of results is about the same as that of the global scenarios.
- For the region Eastern Europe and Former Soviet Union, the IS92 Scenarios have a higher trend than more recent scenarios that take into account the decline in economic activity in this region.
- The range of IS92 Scenarios is in the lower half of the range of non-intervention estimates for the USA.

Energy-Related Scenarios — How Do Their Components Compare?

Both the IS92 Scenarios and other scenarios examined in this report rely on large numbers of input assumptions. It was not possible in this evaluation to assess the differences between all these assumptions. Rather, in this evaluation the projections of energy-related CO_2 emissions were broken down into four components, and these components were compared across the scenarios. The four components were: population growth, economic growth, energy intensity (the amount of primary energy used per unit GDP), and carbon intensity (the amount of carbon emitted per unit primary energy consumed).

- Some of these components are specified as model inputs, and some are derived from model calculations.
- Nearly all scenarios show a downward trend in energy intensity. The energy intensities of the IS92 Scenarios have a narrower range than those of other scenarios.
- The carbon intensity of all IS92 Scenarios shows a downward trend, whereas the carbon intensity of some non-IPCC scenarios shows an upward trend.
- Assumed changes in energy and carbon intensities have an important influence on estimates of emissions, especially in the second half of the next century.
- The range of economic growth assumptions for the IS92 Scenarios is representative of the range in other global scenarios (this statement applies to scenarios of both energy-related and land-use emissions). The economic growth rates used for the IS92 Scenarios are generally lower than historical rates, but would nevertheless lead to large increases in *per capita* GDP in all regions.
- Recent population estimates support the plausibility of the medium population growth rate assumed in the IS92 Scenarios.
- The IS92 Scenarios have the lowest and highest global population growth rates of all the scenarios reviewed. The low projection (IS92c) results in declining global population in the 21st century. These figures are taken from UN projections and are not inconsistent with other recent high and low estimates.
- It is outside the scope of this report to evaluate the various critical issues related to population projections, such as population ageing, the Earth's carrying capacity, and the relationship between economic development and demographic trends. More research is needed to improve our understanding of these issues so that we can better judge the consistency and likelihood of population projections and in turn improve our estimates of future emissions.
- In general, the components of the IS92 Scenarios are fairly representative of other *global* scenarios but not of *regional* scenarios. For example, the IS92 Scenarios span a much smaller range of assumptions of future energy intensity for China and the USA than the range found in non-IPCC scenarios. Hence the IS92 Scenarios only represent part of the range of regional perspectives contained in the scenarios examined here.

Energy-Related Scenarios — To What Factors Are Emission Estimates Most Sensitive?

Some of the uncertainty regarding future emission projections can be attributed to the different methods and models used to compute emissions. However, the evidence from model comparisons indicates that model differences explain only a small part of the wide range of emission estimates found in the literature. Most of this range arises from differences in the input assumptions of the scenarios. The input assumptions that have the most effect on the range of emission estimates are those concerning population growth, economic growth and the amount of carbon-based fuel in use in the economy.

The IS92 Scenarios are also sensitive to assumptions about labour productivity, the growth of demand for energy in developing countries as incomes rise, and the rate of improvement in end-use energy efficiency not related to fuel prices.

Land-use-Related Scenarios — How Different Are Their Results?

Global Results

Far fewer scenarios are available for land-use emissions than for energy emissions, and global estimates are more common in the literature than regional estimates. The following comments can be made about global CO_2 scenarios related to land-use:

- Net deforestation accounts for about 18% of current total CO_2 emissions, and most scenarios assume that this percentage will decline in the future.
- Projected emissions of CO_2 resulting from deforestation differ widely. Unlike energy-related emissions which diverge with time, land-use emissions have their widest range (0.2 to 4.8 GtC/yr) before the middle of the 21st century. Most scenarios converge to zero by the end of the century because they project that either the forces causing deforestation equilibrate or forests are depleted.
- The IS92 Scenarios fall within the lower half of the range of other global land-use CO_2 scenarios.
- The range of cumulative emissions over the period 1990 to 2100 is 30 to 320 GtC. Cumulative emissions have an upper bound set by the maximum available amount of terrestrial biomass that could be oxidised (about 600 GtC neglecting afforestation). No similar upper limit is foreseen for energy-related emissions in the next century.

Regional Results

The few published regional land-use scenarios for CO_2 and other greenhouse gases exhibit considerable differences. The IS92 Scenarios sometimes fall inside and sometimes fall outside the range of other scenarios. The range of emission estimates is about the same for regional scenarios as it is for global scenarios, and the regional and global scenarios converge in a similar way by the year 2100.

Land-use-Related Scenarios — How Do Their Inputs Compare and To What Are Their Emission Estimates Sensitive?

- The assumed rate of deforestation is the most important factor influencing emissions of CO_2 related to deforestation. Most scenarios make simple assumptions about the causes of deforestation, although some use models to simulate land conversion. The different ways of representing deforestation lead to different estimates of deforestation rates: the scenarios examined agree that forest clearing will diminish by 2100 for the reasons cited above.
- Assumptions about the carbon content of vegetation also strongly influence estimates of CO_2 emissions.
- The scenarios examined take a wide variety of approaches to estimating future emissions. This variety reflects lack of knowledge or agreement on how best to represent emission-generating processes related to land-use.
- Not only are these processes poorly understood, but baseline estimates differ between scenarios. Differences in baseline estimates lead to even greater differences in future emission estimates. Results from studies of current emission inventories are needed to improve these baseline estimates.
- The IS92 Scenarios do not account for the feedback of climate to land-use emissions, and this is one reason why their results differ from those of other scenarios. However, climate feedbacks are quite uncertain; thus including them in estimates of land-use emission scenarios may or may not improve these estimates.
- In general, the IS92 Scenarios do not cover the range of input assumptions found in other scenarios related to land-use emissions.

What Are the Uses and Limitations of the IS92 Scenarios?

In evaluating the IS92 Scenarios it is important to recall that they were developed only to accomplish Purpose 1. The IS92 Scenarios fulfil the most important criteria for this purpose. Specifically, their global results span the range of results from other scenarios, and in this sense can be said to reflect the current range of views about future emission estimates. Moreover, the IS92 Scenarios are among the most comprehensive scenarios available in their coverage of all important greenhouse gases and precursors of ozone and sulphate aerosol. Also the scenarios are sufficiently documented in reports and in digital form to allow their comparison with other scenarios, and to be used as input files for climate models. As a consequence, they are useful for Purpose 1, namely as an input to atmosphere/climate models for examining the environmental/climatic consequences of not acting to reduce greenhouse emissions.

Although the IS92 Scenarios were designed with Purpose 1 in mind, it is interesting to ask whether they are useful for any of the other purposes. They should not be used for Purpose 2 (evaluating the consequences of intervention to reduce emissions), because five of the six IS92 Scenarios have emission estimates that are substantially higher than policy scenarios published in the literature.

The IS92 Scenarios are inadequate for Purpose 3 (estimation of feasibility and costs of policies) because five out of the six scenarios do not include mitigation measures specifically targeted towards climate change. The IS92 Scenarios have limited value as reference scenarios for mitigation studies because their input assumptions and results do not have sufficient detail for geographic regions and sectors. Another reason that the IS92 Scenarios are inadequate for purpose 3 is that regional results and regional input assumptions of the IS92 Scenarios do not cover the full range of views found in other scenarios.

The IS92 Scenarios are not appropriate for Purpose 4 (as input to negotiating possible emission reductions) because of the high uncertainty of current greenhouse gas estimates from many sectors and countries. However, the IS92 Scenarios do help set the context for such negotiations by indicating the possible range of emissions if no mitigation takes place. It is again noted that the IS92 scenarios were not intended for Purpose 4.

For the reasons noted above, it is recommended that researchers use the IS92 Scenarios solely as input to atmosphere/climate models for examining the consequences of not acting to reduce greenhouse gas emissions (Purpose 1).

Is One Scenario More Likely Than Another?

In some ways IS92a is an intermediate scenario. Its global CO_2 emissions fall near the middle of other scenarios at least up to the year 2050, and some of its input assumptions are intermediate. For example, its population assumptions are equal or close to the medium projections recently published by three different international organisations. In other ways, however, IS92a is not intermediate, particularly in some of its input assumptions.

In any event, being an intermediate scenario is not the same as being the most likely scenario. Indeed, at this time there is no objective basis for assigning a likelihood to any of the scenarios. Furthermore, given the degree of uncertainty about future emissions, we recommend that analysts use the full range of IS92 Scenarios rather than a single scenario as input to atmosphere/climate models.

Users of the IS92 Scenarios are cautioned, however, that the lowest scenario (IS92c) has emission levels and some input assumptions that are more characteristic of an intervention than a non-intervention scenario.

The Question of New Scenarios

Because the IS92 Scenarios are neither designed nor suitable for Purposes 2, 3, and 4, new scenarios may be needed to fulfil these purposes. It is worth repeating that this report did not judge the suitability of non-IS92 Scenarios to satisfy these purposes.

New scenarios, if developed, should especially include improvements in the estimation of baseline and future non-CO_2 emissions, particularly from the land-use sector.

Although we did not judge the adequacy of all current scenarios, it is likely that new reference scenarios will be needed for the following:

(i) to take into account the latest information on economic restructuring, especially in the CIS, Eastern Europe, Asia and Latin America; (We note that economic restructuring in Eastern Europe and the CIS was partly taken into account in the IS92 Scenarios, but it would be worthwhile to include more recent information.)
(ii) to evaluate the possible consequences of the Uruguay Round amendments to the General Agreement on Tariffs and Trade;
(iii) to explore a variety of economic development pathways (e.g., a closing of the income gap between industrialised and developing regions);
(iv) to examine different trends in technological change; and
(v) to take into account the latest information about current emission commitments in connection with the Framework Convention on Climate Change.

The current commitments to stabilise emissions under the Convention, if maintained over the course of the next century, could result in large reductions of emissions in certain regions, but on a global basis a reduction of 10% or less is expected in CO_2 emissions (relative to the IS92a Scenario). This percentage reduction is somewhat greater or less depending on the year and scenario. This is an optimistic scenario, in particular because it assumes that countries will freeze their emissions after the year 2005 according to their current commitments.

New scenarios are also needed that are consistent with limits on atmospheric concentrations of greenhouse gases by specified dates. These scenarios should be related to the carbon stabilisation modelling exercises being performed by IPCC Working Group I.

The Process of Developing Scenarios

It is recommended that the IPCC or another suitable organisation act as an "umbrella" under which different groups can develop comparable, comprehensive emission scenarios. The process for developing scenarios should draw on current experience in harmonising scenarios and model calculations and should emphasise:

- *Openness:* scenario development should be open to wide participation by the research community, particularly from developing countries. Openness requires extensive documentation of modelling assumptions and inputs.

- *Pluralism:* A diversity of groups, approaches and methodologies should be encouraged, although they should be harmonised as noted below.
- *Comparability:* Reporting conventions for input and output should be standardised.
- *Harmonisation:* Input assumptions and methodologies for the scenarios should be harmonised to provide a common benchmark for scenarios from different groups.

To accomplish Purpose 3 for emission scenarios (estimation of feasibility and costs of mitigation measures), it is important not only to analyse the costs of emission reductions but also to assess the costs and benefits of the impacts of emissions on the natural environment and society. Therefore, scenarios developed for this purpose should be coupled with their impact on the environment and society in the form of consistent, comprehensive and *integrated* scenarios of global change.

Current and new scenarios should be widely disseminated to countries, international organisations, non-governmental organisations and the scientific community. As part of this effort, a central archive should be established to make available the results of new scenarios to any group. The archive should also make available some aspects of the models and the input assumptions used to derive the scenarios. In addition, special effort is needed to improve the capabilities of researchers to analyse and develop scenarios, especially in developing regions. The IPCC can play a role in these activities.

6.1 Background and Scope

6.1.1 Terms of Reference

This chapter presents an evaluation of the IPCC IS92 emission scenarios. The evaluation is a response to a request from the Chairman of the Intergovernmental Negotiating Committee for the Framework Convention on Climate Change that IPCC prepare "an evaluation of current scenarios of greenhouse gas emissions " before the March 1995 meeting of the Conference of Parties (United Nations General Assembly document A/AC.237/30). According to the approved work plan, the key question for the evaluation is: "Which issues and uncertainties about key variables may critically affect future net greenhouse gas emissions and aerosol concentrations?" Among the issues to be explored are:

- the types of models available for scenario development;
- the assumptions made in developing the IS92 Scenarios;
- the relative importance and sensitivity of different variables in determining future emissions;
- the relationship of the scenarios to the emission inventories of countries;
- the appropriate uses, capabilities and limitations of emission scenarios;
- the need for future work on scenarios.

The Bureau of Working Group III further recommended that this evaluation "build heavily" on the scenarios published in the IPCC 1992 Supplement (First Session of Working Group III, Montreal, 3 May, 1993, WGIII/1st/Doc.1, Rev.2).

An appropriate question at the outset is: "Who is the audience for this evaluation?" The Work Plan of Working Group III implies both a scientific and a policy audience by asking for recommendations on new scenarios that would " serve the needs of climate modelling, planning of response measures, and provision of information to the Conference of Parties". Indeed, scenarios by their nature have both scientific and policy aspects: from a scientific standpoint, they provide a reference point for computing the potential occurrence and impacts of climate change; from a policy standpoint they provide information about the consequences of intervention or inaction to reduce greenhouse gas emissions. In practice, however, it is difficult to separate the scientific from the policy aspects of scenarios, and in this chapter we do not try.

In this chapter we present the scope and approach of the evaluation, give a brief overview of IS92 Scenario assumptions and results, compare the IS92 and other scenarios (at this stage we examine much more closely the assumptions and results of the IS92 Scenarios) and discuss the uses and limitations of scenarios, together with our recommendations.

6.1.2 Scope of Evaluation

Only six months were available to meet the INC deadline for this evaluation. This time constraint, together with the lack of a budget for further analysis, led us to narrow the focus of the evaluation:

- Only the most recent IPCC scenarios are evaluated. These scenarios are reported in the IPCC (1992) and in Pepper et al., (1992), and are referred to as the "IS92 Scenarios".
- The IS92 Scenarios are only compared with other published "non-intervention" scenarios because the IS92 Scenarios are intended to be "non-intervention" scenarios (that is, they do not make assumptions about climate policies that would reduce greenhouse gases). Some intervention (or "policy") scenarios are reviewed, but only to compare them with the lowest IS92 Scenarios.
- The evaluation focuses on emissions related to energy and land-use activities because they are the primary sources of greenhouse gases. Non-anthropogenic, "natural" emissions are not considered.
- Principal attention is given to CO_2 emissions because of the critical contribution of CO_2 to radiative forcing and less attention is given to CH_4 and N_2O emissions.
- Scenarios for sulphur emissions are investigated, but not scenarios for major precursors of ozone.
- Sinks of gases, such as carbon sequestration in forest plantations, are omitted. An exception is made when CO_2 from deforestation is discussed, because part of the discussion of deforestation concerns carbon uptake by vegetation.

Certain tasks are not part of the evaluation:

- This evaluation focuses almost exclusively on scenario results and input assumptions rather than on the methodology behind them. This approach was taken because much more time would be needed to assess properly the methodology of the scenarios and because the IS92 Scenarios appear to be more sensitive to the input assumptions than to the methodology.
- We did not perform a sensitivity analysis of the model used to generate the IS92 Scenarios because the funds needed for this work were not available. In any case, analysis would be best performed as part of a more extensive comparison involving several

models. This comparison would require the cooperation of various modelling groups and would take longer than the time available for the evaluation of the emission scenarios.

- We did not compare 1990 baseline emissions for the IS92 Scenarios with emission inventories of different countries because too few inventories were available to be compared. However, as more inventories become available, we recommend that they be used to examine and update the baseline emissions for the scenarios.

- We did not attempt to identify the likelihood of any of the scenarios. It is our view that there is no objective method of assessing the likelihood of all the key assumptions that will influence future emission estimates, especially over the 110-year time horizon of the IS92 Scenarios. The poor track record of economic and energy forecasts indicates our inability to predict even the near-term future. Thus, we do not attempt to identify the most likely emission scenario.

6.1.3 Approach to the Evaluation

A scenario, in the most general sense, is "a sequence of events", or "an account of a possible course of action or events" (Merriom-Webster's Collegiate Dictionary, 10th edition, 1993). An emission scenario is a projection of future emissions based on specific assumptions about key determinants such as population, economic growth, energy intensity, land-use or emission control policies. Sometimes scenarios are confused with emission profiles, targets, or other concepts (see Box 1). Emission scenarios are also sometimes confused with the methodology used to generate them. In most cases emission scenarios are generated with mathematical models or other computational tools, although in some cases they are developed in an *ad hoc* manner. As noted above, this chapter focuses almost exclusively on scenario results and input assumptions rather than on the methodology behind them.

Because emission scenarios fall in the grey area between science and policy, it is difficult to select an appropriate benchmark by which to judge them. Box 2 makes a first attempt at specifying the major purposes of emission scenarios. We recognise that there may be other valid purposes.

Criteria for achieving these purposes are ranked according to "very important", "important" and "desirable" in Table 6.14, presented later in the chapter.

The IS92 Scenarios were only developed with Purpose 1 in mind. Therefore, the scenarios are evaluated in this chapter mainly in the light of how they comply with this purpose. The other purposes are discussed in Section 6.5.2 which covers the need for new scenarios.

Purpose 1 is the most straightforward use of emission scenarios. In this case, a non-intervention emission scenario is developed, and then used as input for a climate model which computes the effect of the emissions pathway on the build-up of greenhouse gases, changes in temperature, precipitation and other climate variables. The results from the climate model are then sometimes used to evaluate the impacts of climate change on sea level rise, agricultural productivity and other indicators. This type of emission scenario also serves as a reference point for the intervention scenarios developed for Purpose 2 and for

---

BOX 1: TERMINOLOGY RELATED TO EMISSION SCENARIOS

Emission scenario — Projections of future emissions based on specific assumptions about key determinants, such as population, economic growth, technological change, land-use trends or emission control policies.

Non-intervention or reference scenario — An emissions scenario that does not make any assumptions about *climate* policies to reduce greenhouse gases. It is sometimes difficult to distinguish non-intervention and intervention scenarios. For example, assumed high prices for fossil fuels could stem from either resource scarcity or carbon taxes aimed at reducing greenhouse gas emissions.

Intervention or policy Scenario — An emission scenario that makes assumptions about *climate* policies to reduce greenhouse gases. Policies must be directed towards mitigating climate change.

Emission profile — Level of future emissions derived directly or indirectly from a specific objective.

Emission target — A target emission level for a particular date.

Emission trend — A projection of future emissions that is not based on specific determinants or objectives.

> **BOX 2. PROPOSED PURPOSES OF EMISSION SCENARIOS**
>
> **Purpose 1.** To evaluate the environmental/climatic consequences of "no intervention" to reduce greenhouse gas emissions. For this purpose a non-intervention scenario is devised, and then used as input for a climate or similar model to evaluate the scenario's environmental/climatic consequences.
>
> **Purpose 2.** To evaluate the environmental/climatic consequences of intervention to reduce greenhouse gas emissions. For this purpose an intervention scenario is devised, and then used as input for a climate or similar model to evaluate the scenario's environmental/climatic consequences.
>
> **Purpose 3.** To examine the feasibility and costs of mitigating greenhouse gases from different regions and economic sectors and over time. This purpose can include setting emission reduction targets and developing scenarios to reach these targets. It can also include examining the driving forces of emissions and sinks to identify which of these forces can be influenced by policies.
>
> **Purpose 4.** As input for negotiating possible emission reductions for different countries and geographic regions.

assessing the benefits of reducing greenhouse gas emissions.

In our view, a scenario or set of scenarios must comply with three important criteria to fulfil Purpose 1. The first two criteria relate to the use of the emission scenario as input for atmosphere/climate models while the last criterion concerns the credibility of the scenarios:

- *Comprehensiveness.* The scenarios must take into account all important greenhouse gases, including precursors of ozone and sulphate aerosol. Comprehensiveness is important so that radiative forcing is not incorrectly calculated by the climate model.

- *Documentation of output.* Sufficient detail should be provided for emissions by type, region (if available) and sector, to allow comparisons with other scenarios and to prepare inputs for atmosphere/climate models.

- *Emission estimates and input assumptions for the scenario should reflect a range of views.* A single reference scenario is inadequate for Purpose 1. Instead, a range of scenario results should be prepared which take into account the uncertainties in the basic driving forces of emissions. The emission estimates and input assumptions of the scenarios should reflect the current range of views.

We examine the comprehensiveness of the IS92 Scenarios in Section 6.2, and briefly discuss documentation of output in Sections 6.2 and 6.5. The third criterion, the question of whether the IS92 Scenarios represent the current range of views of experts, is a central topic of this chapter. The topic is covered in Sections 6.2, 6.3 and 6.4 by comparing the emission projections and major input assumptions of the IS92 Scenarios with those of other published non-intervention scenarios. We recognise that the published scenarios only give the views of a limited number of experts, principally from industrialised countries. Nevertheless, this comparison provides information about the position of the scenarios relative to other work in the scientific community; also, the comparison can be performed in the short time available for this evaluation.

All published scenarios used for our comparisons are treated equally. We recognise that some scenarios may have been prepared by a single researcher using a simple model, whereas others may have been developed using an elaborate methodology and been subjected to extensive review. However, an evaluation of methodology is outside the scope of this report.

Although we consider the three criteria discussed above to be the most important benchmarks for Purpose 1, three other criteria follow these in importance:

- Reproducibility
- Sensitivity analysis
- Internal consistency

These criteria are elaborated later in the chapter (see Table 6.14).

6.2 The IPCC IS92 Scenarios

This evaluation covers the six IS92 Scenarios. They are described in Leggett *et al.* (1992), and in more detail in Pepper *et al.* (1992). The reader is referred to these publications for details about the scenarios; only a brief overview is presented here.

The basic purpose of the scenarios is to explore possible

pathways of greenhouse gas emissions in the absence of new policies to reduce them. The scenarios are among the most comprehensive ever developed in their coverage of all important greenhouse gases – carbon dioxide (CO_2), methane (CH_4), nitrous oxide (N_2O), halocarbons, sulphur oxides (SOx), and precursors of tropospheric ozone including nitrogen dioxides (NO_x), volatile organic compounds (VOC) and carbon monoxide (CO). The scenarios take into account control of stratospheric ozone depletion and local air pollution problems, and this coverage leads to reductions in emissions of some greenhouse gases. Moreover, the IS92b Scenario includes the commitments of some OECD countries to stabilise or reduce CO_2 emissions, as the commitments were known up to 1992. Nevertheless, since these emission control commitments in IS92b do not have a large effect on global emissions, we will refer to the entire set of IS92 Scenarios as "non-intervention" scenarios.

6.2.1 The ASF Model

The USA Environmental Protection Agency model called the "Atmospheric Stabilisation Framework" (ASF) was used to develop the IS92 Scenarios (Lashof and Tirpak, 1990). The model performs calculations for nine regions. Results for the nine regions are available in electronic form, but in IPCC reports these results have been aggregated into four regions: OECD, Eastern Europe and former USSR, China and Centrally Planned Asia[1], and "Other". The ASF combines modules for computing emissions from energy, industry, agriculture, forests and land conversion using common assumptions for population, economic growth and structural change.

Thousands of input assumptions define a case in the ASF.[2] Fortunately, not all assumptions are equally important. The findings of the Energy Modelling Forum indicate that fossil fuel carbon emissions are highly sensitive to assumptions about the rate of improvement of end-use energy efficiency, but relatively insensitive to the form of the model used to generate the scenario. The uncertainty analysis conducted by Edmonds et al. (1986) on the energy component of the ASF ranked the three most important factors influencing the variation in reference case fossil fuel CO_2 emissions to be labour productivity (the prime determinant of GDP), the income elasticity of demand for energy in the developing world (reflecting structural change rates) and the rate of exogenous end-use energy intensity improvement. In addition, fossil fuel CO_2 emissions in the energy component of the ASF model have been shown to be highly sensitive to the assumed costs and supply of renewable energy sources (Edmonds et al., 1986) as well as the environmental costs of extracting fossil fuels.

The energy component of the ASF model is insensitive to population growth assumptions in the short-term, but sensitive to this assumption in the long-term.[3] In the long term, fossil fuel CO_2 emissions are roughly proportional to population.

The model is relatively insensitive to changes in assumed fossil fuel resources, the elasticity of interfuel substitution, the price elasticity of demand for energy and emission coefficients for fossil fuel carbon. The fossil fuel resource base provides no meaningful constraint on carbon emissions for the next century. Fuel price variation is constrained by energy transformation technologies and the variety of alternative energy forms available, so the model is insensitive to the elasticity of interfuel substitution. Although fossil fuel CO_2 emissions are insensitive to the assumed price elasticity of demand for energy, this parameter strongly influences the cost of achieving emissions reduction targets (Edmonds and Barns, 1992). Sensitivity analysis of emission estimates is taken up again in Section 6.3.

The land-use component of the ASF is also strongly influenced by assumptions about GDP and population.

6.2.2 Overview of the IS92 Scenarios

To explore a variety of possible pathways for future emissions, the IS92 Scenarios span a wide range of assumptions for population, GDP, agriculture, energy and other factors. Specific assumptions are made for each of nine regions. An overview of the key assumptions is given in Table 6.1.

Among other recent developments, the IS92 Scenarios take into account the London Amendments to the Montreal Protocol; the revised population forecasts of the World Bank and United Nations; the report of the Energy and Industry Subgroup of IPCC (IPCC-EIS, 1990); the political and economic changes in the former Soviet Union, Eastern Europe and the Middle East; and new data on tropical deforestation and sources and sinks of greenhouse gases.

---

1 In the ASF model this region includes Cambodia, Laos, Mongolia, North Korea and Vietnam.

2 A distinction is sometimes made between model parameters and input assumptions. Model parameters are assumptions that remain constant from case to case, whereas input assumptions are the set of assumptions whose variation defines cases. There are no formal rules for determining which variables are model parameters and which are input assumptions, and the boundary partitioning assumptions into these two sets can change.

3 Increases in population do not affect GDP for 15 years (the time required for an infant to reach labour force age). Therefore increases in energy demand due to increases in population are countered by decreases in energy demand due to lower per capita income levels.

An Evaluation of the IPCC IS92 Emission Scenarios

The IS92a Scenario adopts intermediate assumptions (as compared to the other IS92 Scenarios) about population and economic growth. The major difference between the IS92b and IS92a scenarios is, as noted above, that IS92b takes into account information available up to 1992 on the commitments of some OECD countries to stabilise their CO_2 emissions. The IS92a and b Scenarios give emission estimates that are intermediate compared with those of the other IS92 Scenarios (Table 6.2).

The IS92c Scenario assumes the lowest rates of population and economic growth (relative to the other IS92 Scenarios) and severe constraints on fossil fuel supplies. As a result, it is the lowest emission scenario and the only one showing a decreasing emission trend (Table 6.2). IS92d assumes the low population growth rate of IS92c but a higher economic growth rate. The net result is that it has the second lowest future emission estimates. At the other extreme, the IS92e Scenario assumes intermediate population growth and high economic growth rates with plentiful fossil fuels. Consequently, this scenario has the highest estimates of future emissions (Table 6.2). IS92f uses the highest population estimates of the IS92 Scenarios, but lower economic growth assumptions. It is the second highest emission scenario (Table 6.2).

We provide more detail about the assumptions and results of the IS92 Scenarios in the remainder of the report.

Table 6.1: Summary of assumptions in the six IPCC 1992 alternative IS92 Scenarios.

Scenario	Population	Economic growth	Energy supplies	Other	CFCs
IS92a	World Bank (1991) 11.3 billion by 2100	1990-2025: 2.9% 1990-2100: 2.3%	12,000 EJ conventional oil 13,000 EJ natural gas. Solar costs fall to $0.075/kWh. 191 EJ/year of biofuels available at $70/barrel†	Legally enacted and internationally agreed controls on SO_x, NO_x and NMVOC emissions. Efforts to reduce emissions of SO_x, NO_x and CO in developing countries by middle of next century.	Partial compliance with Montreal Protocol. Technological transfer results in gradual phase out of CFCs in non-signatory countries by 2075.
IS92b	World Bank (1991) 11.3 billion by 2100	1990-2025: 2.9% 1990-2100: 2.3%	Same as "a"	Same as "a" plus commitments by many OECD countries to stabilise or reduce CO_2 emissions.	Global compliance with scheduled phase out of Montreal Protocol.
IS92c	UN Medium-Low Case 6.4 billion by 2100	1990-2025: 2.0% 1990-2100: 1.2%	8,000 EJ conventional oil 7,300 EJ natural gas Nuclear costs decline by 0.4% annually	Same as "a"	Same as "a"
IS92d	UN Medium-Low Case 6.4 billion by 2100	1990-2025: 2.7% 1990-2100: 2.0%	Oil and gas same as "c" Solar costs fall to $0.065/kWh 272 EJ/year of biofuels available at $50/barrel	Emission controls extended worldwide for CO, NO_x, NMVOC and SO_x. Halt deforestation. Capture and use of emissions from coal mining and gas production and use.	CFC production phase out by 1997 for industrialised countries. Phase out of HCFCs.
IS92e	World Bank (1991) 11.3 billion by 2100	1990-2025: 3.5% 1990-2100: 3.0%	18,400 EJ conventional oil Gas same as "a" Phase out nuclear by 2075	Emission controls which increase fossil energy costs by 30%.	Same as "d"
IS92f	UN Medium-High Case 17.6 billion by 2100	1990-2025: 2.9% 1990-2100: 2.3%	Oil and gas same as "e" Solar costs fall to $0.083/kWh Nuclear costs increase to $0.09/kWh	Same as "a"	Same as "a"

† Approximate conversion factor: 1 barrel = 6 GJ.
Source of Table: Leggett *et al.*, IPCC, 1992.

Table 6.2: Selected results of the six IS92 Scenarios.

Scenario	Scenario year	Emissions per year[1]			
		CO_2 (GtC)	CH_4 (Tg)	N_2O (TgN)	S (TgS)
IS92a	1990	7.4	506	12.9	98
	2025	12.2	659	15.8	141
	2100	20.3	917	17.0	169
IS92b	2025	11.8	659	15.7	140
	2100	19.1	917	16.9	164
IS92c	2025	8.8	589	15.0	115
	2100	4.6	546	13.7	77
IS92d	2025	9.3	584	15.1	104
	2100	10.3	567	14.5	87
IS92e	2025	15.1	692	16.3	163
	2100	35.8	1072	19.1	254
IS92f	2025	14.4	697	16.2	151
	2100	26.6	1168	19.0	204

[1] The figures for CO_2 are anthropogenic emissions. The figures for CH_4, N_2O and S are combined natural and anthropogenic emissions. Natural emissions in 1990 are estimated as: CH_4 = 340 $TgCH_4$, N_2O = 4.7 TgN, and S = 74 TgS. It is generally assumed that natural emissions will remain constant; thus anthropogenic CH_4, N_2O and S emissions for any scenario and year can be estimated by subtracting the figures for natural emissions presented in the text from the numbers in the body of the table.

6.2.3 GDP Assumptions

Both the energy-related and land-use emission estimates are strongly influenced by the assumptions about GDP and population growth. Although a detailed analysis of these assumptions is outside the scope of this report, the following paragraphs highlight some tentative conclusions and important issues concerning these assumptions. We begin by briefly reviewing the total GDP growth assumptions of the IS92 Scenarios.

The assumptions of the *intermediate* IS92 Scenarios (IS92a and b) are partly based on the reference scenario of the Energy and Industry Subgroup of the First IPCC Assessment (IPCC-EIS, 1990) for the period up to 2025. Near-term growth rates in the IPCC-EIS scenario were adjusted downwards in Eastern Europe, CIS and Persian Gulf to reflect the consequences of events occurring at the beginning of the 1990s. As noted by the developers of the IS92 Scenarios (Leggett, *et al.*, 1992), the medium assumptions for near-term economic growth are generally below or at the low end of the range given by the World Bank for 1990–2000 (Table 6.3). In addition, the medium assumptions for economic growth in the period 1990–2025 are lower than most regions experienced from 1965 to 1989 with the exception of Africa, Eastern Europe and the CIS (Table 6.3). For these regions it was assumed that structural adjustments will lead to higher than historical growth in the *medium-term*. For the period 2050-2100, all the IS92 Scenarios assume that economic growth will slow because of saturation of consumption and expected slower population growth (Table 6.3).

Because of the uncertainty of future economic growth trends, the low (IS92c) and high (IS92e) estimates span a wide range of possibilities.

The following additional points can be made about the GDP growth assumptions of the IS92 Scenarios:

- As noted in Section 6.3, the high and low rates of growth of *global average* GDP in the IS92 Scenarios are quite representative of the wide range of views expressed in other published scenarios. However, somewhat higher and lower global growth rates can be found in the literature about energy scenarios.
- With regards to *regional* results, there are many cases in which the IS92 Scenarios are higher or lower than any other scenario. In the case of Eastern Europe and the CIS, the IS92 Scenarios take into account economic restructuring in these regions, but they assume higher economic growth rates in the near term than more recent scenarios.
- Although the medium assumptions of the IS92 Scenarios are lower than historical growth rates, they nevertheless imply a substantial increase in *per capita* GDP. For example, in the second half of the next century, *per capita* GDP in East Asia and Latin America will exceed current levels in OECD Europe. Nevertheless, a large gap will remain between

Table 6.3: Growth Assumptions of total GDP (average annual rate)

	World Bank†			IS92c		IS92a		IS92e	
	1965–1989	Low†† 1990–2000	High†† 1990–2000	1990–2025	1990–2100	1990–2025	1990–2100	1990–2025	1990–2100
OECD	3.2%	2.4%	3.1%	1.8%	0.6%	2.5%	1.7%	3.0%	2.2%
USSR/E.Europe†††	1.3%	3.2%	3.6%	1.5%	0.5%	2.4%	1.6%	3.2%	2.4%
China & CP‡ Asia	7.6%	5.6%	6.7%	4.2%	2.5%	5.3%	3.9%	6.1%	4.7%
Other	4.7%	3.8%	4.5%	3.0%	2.1%	4.1%	3.3%	4.8%	4.1%
Global	-	-	-	2.0%	1.2%	2.9%	2.3%	3.5%	3.0%

Source: Leggett, *et al*. (1992).

†   Source: World Bank (1991).
††   Estimated using projections of regional growth in GDP *per capita* and country estimates of GDP, population and population growth from 1990 to 2000.
†††   World bank data do not include all countries in Eastern Europe.
‡   CP = Centrally planned economies.

incomes of developed and developing regions, and there is the need to develop emission scenarios which assume a closing of this income gap.

- Emission estimates can be improved by GDP projections that take into account the economic restructuring under way in many regions.

Summary of the discussion of GDP assumptions: the range of total economic growth assumptions of the IS92 Scenarios is representative of the range of other global scenarios. However, these assumptions are not always representative on the regional level. The economic growth rate assumptions of the IS92 Scenarios are generally lower than historical rates, but they would still lead to large increases in per capita GDP in all regions.

6.2.4 Population Assumptions

The influence of population assumptions on future emission estimates is significant. The following can be said about the population assumptions of the intermediate IS92 Scenarios:

- The medium global assumption of the IS92 Scenarios (11.3 billion people in year 2100) is based on the latest projection of the World Bank (1991) (Table 6.4).
- For year 2025, the IS92 medium population projection is close to the latest medium projections of both the UN (1992) and the International Institute of Applied Systems Analysis (IIASA) (Lutz *et al.*, 1994) (Table 6.4).
- For year 2100 the three medium projections are still somewhat close (Table 6.4).

The projections of the World Bank, UN and IIASA were all made using the cohort-component approach to demographic projections, but they employed different assumptions about changes in demographic variables. Hence, it is significant that the medium projections are close. Examination of the low and the high population assumptions for the IS92 Scenarios indicates that:

- The low and high global assumptions of IS92 Scenarios (6.4 to 17.6 billion in year 2100) are the latest medium-low and medium-high projections of the UN (1992) (Table 6.4). (The World Bank did not provide high or low projections with its most recent medium projections.)
- The low and high projections of IS92 Scenarios are the highest and lowest projections of all published scenarios reviewed in this report. Hence, for emission scenarios, they cover, and expand, the full range of views about future population.
- Recent IIASA high projections are close to the high projections of the IS92 Scenarios (Table 6.4). The low IIASA projection, however, is somewhat above the IS92 low projections (Table 6.4).

Despite the importance of long-term population assumptions in estimating future emissions, there have been few reviews of these assumptions. In one such review, Morita *et al.* (1993) reviewed the assumptions of global models and concluded that a plausible range of global population projections for year 2100 was 5.66 to 19.16 billion people. This range is somewhat lower and higher than the range of the IS92 assumptions (Table 6.4). The lower bound is from Nordhaus and Yohe (1983), and the upper bound is the high projection of the UN (1992). Morita *et al.* (1993) point out that fertility rates corresponding to the low population projection actually occurred in Japan from 1975 to 1985, and rates corresponding to the high projection occured during the period from 1947 to 1950. From this perspective Morita *et al.* judged that these projections are not implausible.

Because population growth assumptions strongly influence future emission projections, it would be helpful to assess their relative likelihood. Unfortunately, this assessment is not possible at present because of many unresolved issues related to population projections. For example, the feasibility of low projections is related to the issue of a rapidly ageing population. Lutz *et al.* (1994) pointed out that the IIASA low population projection for 2100 (the IS92 low projection is even lower; see Table 6.4) is based on assumptions about low fertility and mortality rates, and these low rates would lead to an unprecedented ageing of the population. In its low population projection, the percentage of the population over 60 increases in industrialised countries from 17.4% to 44.9% between 1990 and 2100, and in developing countries it increases from 6.9% to 40.2%. To judge the likelihood of the low population projections of IS92 or any other scenario, it is important to assess the response of societies to the increasing burden of an ageing population on their social welfare systems (see, e.g. Lutz *et al.* 1994).

Another critical issue is the relationship between high population projections and the Earth's carrying capacity, i.e. the ability of societies to provide food and resources for their populations. The UN projections do not take this concern into account, since mortality rates are never assumed to increase even for the highest population increases. There is very wide disagreement over the size of the Earth's carrying capacity; estimates made since the 1980s range from 2 billion to over 33 billion people, depending on a large variety of assumptions (see review in Heilig, 1994). This range can be compared to the IS92 high population assumption of 17.6 billion people in 2100. If the low figure for carrying capacity is correct then even the current population cannot be sustained, whereas the high figure implies that IS92 projections are theoretically possible. To judge the feasibility of the high population assumptions of the IS92 Scenarios or any scenario, it would be beneficial to try and narrow the range of estimates about global carrying capacity.

These are only two of the many issues related to demographic studies that require further research and understanding before progress can be made in assessing the likelihood of different population assumptions. Another key issue, the relationship between population and GDP *per capita* is taken up in the next subsection.

Summary of the discussion of population assumptions: the plausibility of the medium population growth assumption of the IS92 Scenarios is supported by other recent population projections. The low and high projections of the IS92 Scenarios cover and expand the range of assumptions found in other published scenarios. The IS92 projections are taken from authoritative projections of world population growth and are not inconsistent with other recent population projections. However, to assess the likelihood of the low and high population assumptions, more research is needed to improve our understanding of critical issues related to population projections, such as carrying capacity and possible fertility responses to population ageing.

6.2.5 The Relationship between GDP per capita and Population

This section reviews the issue of the consistency between population and GDP *per capita* assumptions in the IS92 Scenarios. This topic was briefly raised by the developers of the scenarios themselves (Leggett *et al.*, 1992). The IS92 Scenarios combine high population with high GDP

Table 6.4: Population projections (in millions)

Source	1990	2025	2100
IS92a, b & e from World Bank(1991)	5,252	8,414	11,312
IS92c & d from UN Medium-Low	5,252	7,591	6,415
UN Medium	5,252	8,472	11,186
IS92f from UN Med-High	5,252	9,445	17,592
IIASA Low	5,252	8,093	9,126
IIASA Central	5,252	8,955	12,562
IIASA High	5,252	9,872	16,090

Source: Leggett *et al.* (1992), except IIASA projections from Lutz *et al.* (1994).

per capita assumptions and low population with low GDP *per capita* assumptions. Although these combinations are logical for performing a sensitivity analysis of key input assumptions, they also imply that high population growth can accompany high economic growth and conversely low population growth can accompany low economic growth. However, the developers of the scenarios offer the opinion that GDP *per capita* and population are most likely negatively correlated (Leggett *et al.*, 1992).

Actually, there are many opinions about the connection between population and GDP *per capita*. Three of these are:

(i) *High population growth is compatible with high economic growth (or low population growth with low economic growth)*. As noted, this opinion is implied, if not intended, by the IS92 Scenarios. In support of this view, Boserup (1981) argued that a larger population density allows for investments in infrastructure and stimulates technological progress and the growth of financial institutions.

(ii) *Population growth is negatively correlated to economic growth*. The reasoning behind this argument is that improved economic status leads to lower birth rates and eventually lower population growth. This belief is embedded in conventional thinking about the "demographic transition" of societies, which holds that fertility in a society goes down as its standard of living goes up. A related point of view is that population growth leads to resource and capital constraints that in turn reduce GDP *per capita* growth.

In support of a negative correlation between population and GDP *per capita*, Morita *et al.* (1994) derived a relationship between total fertility rate and *per capita* income from Japanese data from the period 1950 to 1985. Applying this relationship globally, they back-calculated the annual *per capita* GDP growth rate that corresponds to the population growth rates of the IS92a scenario. In this way they estimated that the IS92a Scenario should have an effective annual *per capita* GDP growth rate of 3% rather than the 1.7% actually used. Hence, Morita *et al.* (1994) suggest that the population and economic assumptions of this scenario may be inconsistent.

(iii) *Population growth and economic growth are independent*. Recent field surveys in developing countries have led demographers to begin to argue that the empirical evidence does not support either of the preceding points of view, and that the fertility rate in many countries is much more clearly related to the educational level of women and the existence of family planning programmes than any economic indicators (see, e.g., UN, 1987; Lutz 1994; Westhoff, 1994).

As a final comment on this issue, we note that it is possible that the relationship between population and income can vary between cultures and that the three different relationships may actually correspond to different stages in a society's development.

Summary of the discussion of the relationship between population and GDP per capita: there are at least three points of view about this relationship, and it is beyond the scope of this evaluation to pass judgement on their validity. It is important to note, however, that the issue of population and GDP per capita is relevant to judging the consistency and likelihood of population and economic assumptions, which in turn affect future emission projections. Hence, it would be beneficial to improve our understanding of the relationship between population and GDP per capita.

6.3 Comparison of Energy-Related Scenarios

In this section we compare the IS92 energy-related emission scenarios to other published non-intervention scenarios. The goal of this comparison is to determine the degree to which the IS92 Scenarios reflect the range of emissions and major input assumptions found in other scenarios at both the global and the regional level.

We analyse global and regional emissions. Scenarios of CO_2, CH_4 and N_2O are examined, although the IS92 Scenarios consider the full spectrum of greenhouse gases. Because of the limited scope of this evaluation, the regional analysis is limited to four representative regions: China and Centrally Planned Asia, Eastern Europe and the former Soviet Union, Africa, and the United States. These regions are selected because they are at different stages of economic development.

For global and regional emissions we first assess the accuracy of the base year (1990) emission data in comparison to the latest emission inventory as described in UN energy statistics (Marland *et al.*, 1989 and 1993). Next, we compare future emission paths for IS92 to the range of emissions of other published non-intervention scenarios with similar methodologies and the range spanned by the 16th and 84th percentile (median plus/minus one standard deviation) of the poll of energy scenarios assembled by the International Energy Workshop (Manne *et al.*, 1991; Manne and Schrattenholzer, 1993, 1994). Finally, we compare the main input assumptions and scenario components of the IS92 with similar aspects of other published scenarios.

6.3.1 Global Energy Scenarios

6.3.1.1 Results from global CO_2 emission scenarios

Of all the greenhouse gas emissions, CO_2 is the most

studied. Of all the CO_2 sources, fossil fuel combustion is the most studied. This interest arises because fossil fuel CO_2 emissions contribute more to current and potential future climate change than any other single gas released by any other single human activity. CO_2 emissions dominate the IS92 Scenarios (IPCC, 1992). There is also more literature on scenarios of global fossil fuel carbon emissions than on those of land-use emissions. This is because the models and methods used to estimate energy-related emissions are more detailed and advanced than those used to compute land-use related emissions, and because existing energy models can easily be adapted to generate CO_2 emissions.

It is interesting that the scenarios developed in the past decade (beginning in 1983) have a narrower range of future emission estimates than those developed in the previous decade. Scenarios no longer anticipate emissions of greater than 90 GtC/yr in the year 2050, as considered, for example, by Siegenthaler and Oeschger (1978), Jason (1979), and Bacastow and Keeling (1981). The highest estimate of the post 1983 period projects emissions of 35 GtC/yr in the year 2050, (Lashof and Tirpak, 1990).

The IS92 scenario estimate of base year (1990) emissions, 6 GtC/yr, is close to the estimate of 5.9 GtC/yr by Boden *et al.* (1992) and Marland *et al.* (1993).

Estimates for 1990 from other published scenarios range from 5.4 GtC/yr to 6.8 GtC/yr.

Future estimates of emissions diverge substantially with time (Figure 6.1) and only resemble each other in that they almost all increase with time. A notable exception is the IS92c Scenario, which declines from 1990 to 2100. For the year 2025, the spread of non-IS92 scenario emissions ranges from 6.7 GtC/yr in Ausubel *et al.* (1988), to 19.8 GtC/yr in Lashof and Tirpak (1990). The IS92 Scenarios range from 7.4 (IS92c) to 13.5 GtC/yr (IS92e). As another indication of the range of views of the energy community, the 16th and 84th percentile range from the International Energy Workshop (IEW) scenario poll is between 6.9 GtC/yr and 10.9 GtC/yr for the year 2020.

By the year 2100 the range of non-intervention scenarios expands significantly, with the IS92 Scenarios extending from 4.6 GtC/yr to 35 GtC/yr, a factor of more than seven. The lowest non-IS92 scenario examined is given by Ausubel *et al.* (1988), 1.2 GtC/yr, and the highest is given by Ogawa (1990) at 60 GtC/yr, a range of a factor of 50. This range also encompasses the range of 12 reference scenarios compared by the Energy Modelling Forum (the "EMF-12" study) (Weyant, 1993). These scenarios span from 20 GtC/yr (Global-Macro) to 43 GtC/yr (CETA). The EMF-12 results are particularly

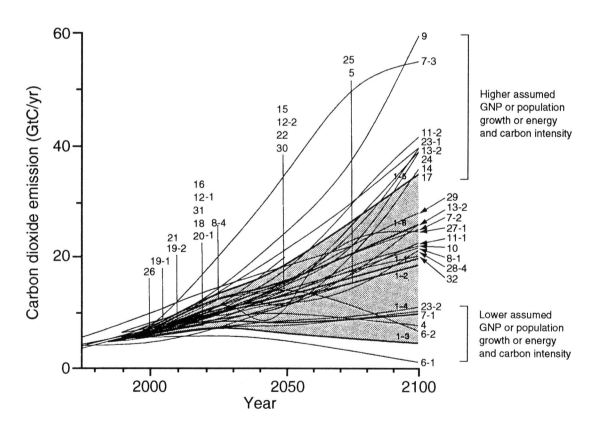

Figure 6.1: Energy-related global CO_2 emissions for various scenarios. Shaded area indicates coverage of IS92 Scenarios. Numbers correspond to list of scenarios in the Supplementary Table.

An Evaluation of the IPCC IS92 Emission Scenarios

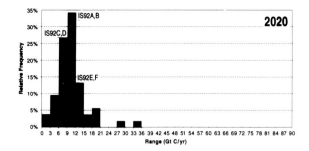

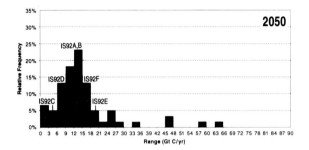

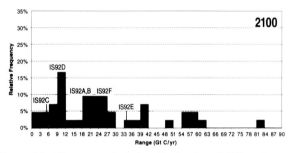

Figure 6.2: Histograms depicting distribution of different scenario estimates of energy-related global CO_2 emissions for three scenario years.

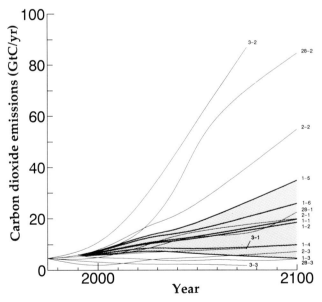

Figure 6.3: Results from uncertainty analyses showing 5th and 9th percentile of global CO_2 emissions. Also shown are IS92 estimates (code numbers beginning with "1"), with their range indicated by the shaded area (IS92 results are not from uncertainty analysis). Numbers correspond to list of scenarios in the Supplementary Table.

interesting because they were obtained from models with harmonised input assumptions about population and economic activity. Consequently, these results highlight how future emission estimates can vary because of differences in models and in input assumptions.

Histograms of projected emissions (Figure 6.2) show that emission estimates are relatively closely clustered in the year 2020, but also diverge subsequently. The frequency of the mode also becomes smaller and smaller with time, indicating that the scenarios agree less and less about the central estimate of emissions. The interval that contains each of the IS92 cases is also indicated, and shows that the IS92 Scenarios bracket a majority of other scenarios, but never capture their full range. This observation is particularly apt for the year 2100. Throughout the period, there are always scenarios with substantially more or less fossil fuel carbon emissions than the IS92 Scenarios.

Additional information about the range of future emissions can be obtained by examining results from detailed uncertainty analyses. These were conducted by Nordhaus and Yohe (1983), Edmonds *et al.* (1986), Manne and Richels (1993), and de Vries *et al.* (1994).

It is noteworthy that despite the enormous range of results produced by the uncertainty analyses (Figure 6.3), emissions almost always increase with time, with the exception of the extremes of results. The uncertainty analysis of Edmonds *et al.* (1986) shows a range of 2 GtC and 87 GtC/yr for the 5th and 95th percentile cases respectively in the year 2075, and a range of 4 GtC to 27 GtC/yr between the 25th and 75th percentiles. The uncertainty analysis of Nordhaus and Yohe (1983) spans the range 12 GtC to 27 GtC/yr for the year 2100 between its 25th and 75th percentiles, and 7 GtC to 55 GtC/yr between its 5th and 95th percentile cases.

It should be emphasised that these results stem from differences among models *and uncertainties about input assumptions, such as population and economic activity.* By comparison, when input assumptions are fixed between scenarios, a much lower range of emission estimates is found. For example, it was noted above that when key model inputs were harmonised in the EMF-12 scenarios, emission estimates in year 2100 varied by only a factor of two rather than over a factor of 40, as found in the uncertainty studies. This finding is consistent with the recent results of de Vries *et al.* (1994), who only assigned uncertainty to model input parameters, while fixing other input assumptions such as population and economic growth relative to a base case. As a consequence, they computed emission estimates for the year 2050 ranging by only a factor of about two, from 11.3 GtC to 24.4 GtC/yr (5th and 95th percentiles).

Although published energy-related emission estimates for 2100 vary by a factor of 50 between the highest and

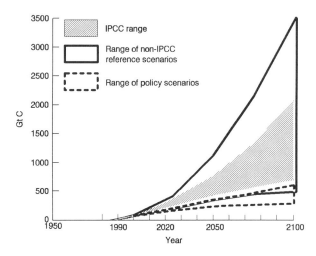

Figure 6.4: Range of cumulative energy-related CO_2 emissions from IS92 and other published scenarios.

lowest reference scenarios and by a factor more than 7 in the IS92 Scenarios, cumulative emissions have a much smaller variation. The cumulative emissions of the published scenarios over the period 1990 to 2100 (Figure 6.4) vary by a factor 7 (from 492 GtC in Ausubel *et al.*, 1988, to 3447 GtC in Lashof and Tirpak, 1990). The cumulative emissions of the IS92 Scenarios vary by a factor of 3 between IS92c (700 GtC) and IS92e (2078 GtC).[1]

The above cumulative emission figures can also be compared to the cumulative energy related carbon emissions between 1860 to 1990 of about 215 GtC (Marland *et al.*, 1989; IPCC, 1990; Nakicenovic *et al.*, 1993). Thus, even the lowest reference and policy scenarios indicate that more energy related carbon will be emitted from 1990 to 2100, than in the last 130 years.

It is important to note that cumulative emissions are not equivalent to the accumulation of carbon in the atmosphere. The relationship between emissions and atmospheric accumulation is described by carbon cycle models which are discussed in the IPCC Working Group I report. Carbon cycle models can be used to translate any time profile of carbon emissions into a corresponding time profile of atmospheric concentrations of CO_2. In a modelling exercise of this type, Working Group I of the IPCC estimated that the IS92c scenario could raise atmospheric CO_2 levels to 480 ppmv by 2100 from approximately 350 ppmv in 1990.

We also note that a few scenarios, for example Cline and Manne and Richels (1994), have explored emissions over even longer time periods. These cases extend beyond the time frame of the IS92 Scenarios, but we have not examined the implications of these scenarios beyond the year 2100. We note that cumulative emissions in these scenarios may be significantly greater than the cases examined in this report. These scenarios consider depletion of all available oil, gas and coal. The magnitude and distribution of these fuels is discussed in the chapter on energy supply mitigation in the IPCC Working Group II report.

Summary of the discussion of results of global energy-related CO_2 scenarios: the scenarios diverge with time and almost all have an upward trend. Their range is much narrower in 2020 and 2050 than in 2100. The IS92 Scenarios follow the same pattern, and fall within the range of the majority of published scenarios. The lowest and highest published scenarios reviewed in this report vary by a factor of 50 in 2100. Much of this large range, however, can be attributed to the wide range of input assumptions, such as economic activity, end-use energy intensity and the energy supply mix. The range of estimates for cumulative emissions between 1990 and 2100 is much narrower than for annual emissions (a factor of 7 versus a factor of 50).

6.3.1.2 Components of global CO_2 scenarios

The limited scope of this report makes it impossible to carefully examine and compare all the input assumptions of the published scenarios. Instead, selected energy-related scenarios are compared by breaking down the calculated emissions into different components:

CO_2 Emissions = Population*(GDP/Population) (Energy/GDP)*(CO_2/Energy)

In this equation, the ratio "Energy/GDP" represents *energy intensity* (primary energy consumed/unit of GDP). Energy intensity data are indexed to 1990 data because of the lack of comparability of GNP/GDP data in different studies. The ratio "CO_2/Energy" is the *carbon intensity* (carbon emissions/unit of primary energy consumed), which reflects the amounts of carbon-based fuels used in the energy economy. Carbon intensity is given as kg C/GJ.

Such an approach for analysing the components of future CO_2 emissions has been proposed by Kaya (1990) and applied by Edmonds and Barns (1992), Ogawa (1991) and Grübler *et al.* (1993) among others.

The decomposition of CO_2 emission growth into the four components provides an accounting of the sources of future emissions. The decomposition is useful because it indicates where to look for differences in scenario assumptions that may account for differences in projected

---

[1] The cumulative emissions of policy scenarios are obviously lower, ranging from 290 GtC (Lazarus et al., 1993) to 600 GtC (Lashof and Tirpak, 1990) in the scenarios reviewed.

emissions. Decomposition, however, neither identifies the key assumptions nor provides an explanation of the fundamental causes of differences in emission projections. Decomposition is also useful because it coincides with four broad areas of public policy concern that can potentially influence future emissions: demographics, economic growth and development, energy conservation, and energy supply.

Population and *per capita* income growth are an input assumptions in most models. Typically, energy intensity is partly specified and partly computed by the model. Carbon intensity is usually determined by the interaction of a variety of assumptions, including those concerning availability and cost of fossil energy resources, relative fuel prices, and the availability and cost of renewable energies.

Two additional caveats should be noted. First, for the purposes of this comparison, all components, not just population and income, are treated as independent of one another. This assumption may be inconsistent with some models which assume that components, such as *per capita* income and energy intensity are related to one another. Second, the influence of some components on emissions may depend on the aggregation level of the analysis (global versus regional or national). Hence, an analysis of the components of emissions on the global level may mask decisive differences between regions. For this reason, the component analysis should only be used to compare the input assumptions and the results of different scenarios, not, for example, to project emissions on the basis of population growth in different regions.

Table 6.5 shows the growth rates of components of selected scenarios, including the IS92 Scenarios to 2020[1].

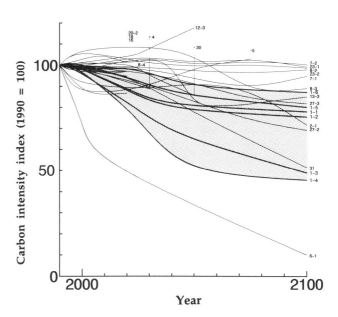

Figure 6.6: Trends of carbon intensity for various global scenarios. Shaded area indicates range of IS92 scenarios. Numbers correspond to list of scenarios in the Supplementary Table.

The IS92 Scenarios are in most instances within the range of the alternative scenarios, although they are not always representative of the range. They span a representative range of global economic growth rates, but the population growth assumptions of IS92c and d are lower and those of IS92f are higher than any of the published scenarios. Population and GDP were discussed in Section 6.2 and will not be discussed further here.

Figures 6.5 and 6.6 depict the ranges of energy intensity and carbon intensity for published global scenarios. It is noteworthy that all the scenarios reviewed show decreasing energy intensities, with a median decrease of about 1%/yr. The IS92 Scenarios fall within the range of a majority of published scenarios, although they cover a narrower range, especially up to 2020.[2]

Most scenarios assume that carbon intensity decreases at a lower rate than energy intensity (Figure 6.6); in some cases the carbon intensity increases, indicating increased use of carbon-rich fuels such as synthetic fossil fuels or coal (Table 6.5). However, on average, carbon intensity in the non-intervention scenarios decreases at a rate of about 0.2%/yr (1990 to 2020 average), a rate that is somewhat lower than the long-term historical average of 0.3%/yr (Nakicenovic *et al.*, 1993). Up to 2020, the rate of decline of carbon intensity assumed in the IS92 Scenarios agrees

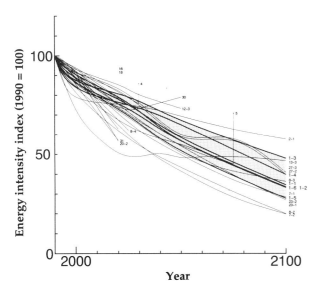

Figure 6.5: Trends of energy intensity for various global scenarios. Shaded area indicates range of IS92 scenarios. Numbers correspond to list of scenarios in the Supplementary Table.

---

1 For an analysis of the period beyond 2020, see Grubler (1994).

2 There is a significant overlap, particularly before 2020, between the ranges of reference and policy scenarios, although the policy scenarios are generally lower.

with the IEW poll and other reference scenarios. However, all the IS92 Scenarios have declining carbon intensities. Beyond the year 2020, the range of published reference scenarios shows a much wider variation than the IS92 cases. In some non-IS92 Scenarios, carbon intensities increase with time owing to a rapid expansion in coal production and use, a case not considered in the IS92 Scenarios.[1]

The decomposition of global emissions of the IS92 and the non-IS92 emissions tell a generally consistent story about the sources of emission growth and the uncertainty of future emissions. In all scenarios, population growth and *per capita* income growth contribute to rising global CO_2 emissions over time. Growing world population and incomes increase the consumption of energy, giving rise to higher emissions. The relative contribution of *per capita* income growth to global CO_2 emissions is slightly greater than the contribution of population growth in most but not all scenarios.

Decreases in energy intensity mitigate emission growth in all scenarios, which is consistent with historical observations. With a decrease in energy intensity, energy consumption for a given population and *per capita* income declines. Consequently, carbon emissions also decline. Declining energy intensity offsets roughly one-fifth to one-half of the effects of population and *per capita* income growth, depending on the scenario. The IS92 Scenarios reflect a much narrower range of changes in future energy intensity than other scenarios. This narrower range is the most important source of the narrower range of carbon emissions displayed in the IS92 Scenarios relative to the range of emissions in other published scenarios.

Carbon intensity is projected to decline in all IS92 Scenarios and the vast majority of other scenarios, further mitigating the effects of growing population and *per capita* incomes. This decline reflects a shift toward greater reliance on energy resources that emit less carbon per unit of energy. The offset from declining carbon intensity is typically much less than the offset from declining energy intensity. As in the case of energy intensity, the IS92 Scenarios display a narrower range of changes in carbon intensity relative to other scenarios. A small number of non-IS92 Scenarios allow for the possibility of increases in carbon intensity over time, adding another source of emission growth. This possibility is not allowed for in any of the IS92 Scenarios.

The decomposition of CO_2 emission growth into the components of population growth, *per capita* income growth, energy intensity and carbon intensity identifies the general sources of emission growth from energy use, but the decomposition does not explain why the contributions of the four components differ across scenarios. To understand the fundamental causes of differences in emissions from one scenario to the next, it is necessary to explore and evaluate differences in assumptions regarding the factors that determine growth in population, *per capita* income, energy intensity and carbon intensity.

In most emission scenarios, population and *per capita* income growth, or their composite effect on total income, are input assumptions. These input assumptions are important sources of differences in projected emissions. However, the determinants of population and *per capita* income growth, and feedbacks between them are not explicitly incorporated into emission scenarios (see section 6.2). Identifying the factors influencing population and *per capita* income growth, and exploring the uncertainties regarding their change over time, are therefore important research tasks.

The input assumptions that drive differences in energy and carbon intensity across scenarios include different estimates of the reserves and supply costs of carbon and non-carbon-emitting energy resources; the income and price elasticities of household demands for energy, the production of energy-intensive goods and services, and many others. A particularly important component is technological change at both levels of energy production and energy end use. Because of time limitations, differences in assumptions regarding factors that influence energy and carbon intensity have not been compared across scenarios. Consequently, another important task to explore the causes of differences in energy intensity and carbon intensity trends and their role in shaping future emissions.

Summary of the discussion of the components of global CO_2 emissions, especially for energy and carbon intensity: the energy intensity assumptions of the IS92 Scenarios have a narrower range than similar assumptions in other published scenarios. The carbon intensity assumptions of the IS92 Scenarios all show a downward trend, but some of the published scenarios assume an upward trend (although most show an eventual decrease over time). An important task for new scenario work is to explore future trends for energy and carbon intensity.

6.3.2 Regional CO_2 Scenarios

The background report for the IS92 Scenarios gives results for four regions: OECD, Eastern Europe and the Former Soviet Union, China and Centrally Planned Asia, and the Rest of the World (Pepper *et al.*, 1992). The original IS92

---

[1] Carbon intensities assumed in policy scenarios are generally lower than those assumed in the IS92 Scenarios and other non-intervention scenarios, although some reference scenarios have quite low carbon intensities as in the case of the "methane economy" of Ausubel *et al.* (1988).

Table 6.5: 1990 to 2020 – World average annual growth rates[1] (%/yr) for selected non-intervention (reference) scenarios and range for intervention (policy) scenarios.

Non-intervention scenarios[3]	POP	GDP capita	GDP	Energy GDP	Carbon energy	Carbon emissions
1-1 IS92a	1.40	1.53	2.92	-0.97	-0.24	1.68
1-2 IS92b	1.40	1.53	2.92	-1.08	-0.26	1.54
1-3 IS92c	1.10	0.92	2.00	-0.86	-0.55	0.56
1-4 IS92d	1.10	1.75	2.83	-1.10	-0.68	1.00
1-5 IS92e	1.40	2.29	3.71	-1.16	-0.11	2.39
1-6 IS92f	1.75	1.36	3.10	-0.85	-0.03	2.19
8-4 IPCC-EIS	1.24	1.74	3.00	-0.80	0.05	2.23
20-3 IEW-84%[2]			3.36	-0.75	0.05	1.95
20-2 IEW-50%[2]			2.78	-1.16	-0.18	1.44
20-4 IEW-16%[2]			2.46	-1.71	-0.67	0.63
18 ECS '92	1.42	0.78	2.21	-0.75	-0.16	1.28
6-1 CH_4-efficiency	1.29	0.83	2.13	-0.78	-0.85	0.47
16-1 CHALLENGE			2.55	-0.94	-0.18	1.41
31-2 WEC A	1.43	2.36	3.82	-1.50	-0.01	2.25
31-1 WEC B	1.43	1.85	3.30	-1.84	-0.22	1.17
3-1 E&R B	1.13	1.68	2.85	-1.03	-0.23	1.54
7-1 EPA-SCW	1.43	0.35	1.79	-0.72	-0.07	0.98
7-2 EPA-RCW	1.29	1.35	2.65	-0.70	0.04	1.98
12-2 GREEN			2.71	-0.53	0.14	2.31
30-1 12RT			2.50	-0.99	0.00	1.49
27-2 IMAGE 2.0-CW	1.40	1.61	3.00	-0.56	-0.25	2.20
Minimum[2]	**1.10**	0.35	1.79	-1.84	-0.85	0.47
Median[2]	1.40	1.53	2.83	-0.94	-0.18	1.54
Mean[2]	1.35	1.46	2.79	-0.99	-0.21	1.56
Maximum[2]	**1.75**	2.36	3.82	-0.53	0.14	**2.39**
Policy Scenarios						
Minimum[2]	*1.17*	*0.35*	*1.79*	*-2.40*	*-1.94*	*-1.31*
Median[2]	*1.29*	*1.35*	*2.65*	*-1.78*	*-1.19*	*-0.32*
Mean[2]	*1.32*	*1.37*	*2.66*	*-1.74*	*-1.16*	*-0.29*
Maximum[2]	*1.43*	*2.19*	*3.50*	*-1.08*	*-0.52*	*0.24*

[1] Component growth rates do not add exactly to (sub)totals due to independent rounding errors.
[2] Component growth rates do not add as calculated from original frequency distributions.
[3] Reference for scenarios given in the Supplementary Table.
BOLD denotes IPCC Scenarios

Table 6.6: 1990 to 2020 – China and centrally planned Asia average annual growth rates[1] (%/yr) for selected non-intervention (reference) scenarios and range from intervention (policy) scenarios

Non-intervention scenarios[3]	POP	GDP capita	GDP	Energy GDP	Carbon energy	Carbon emissions
1-1 IS92a	1.03	3.91	4.98	-1.73	-0.32	2.84
1-2 IS92b	1.03	3.91	4.98	-1.73	-0.32	2.84
1-3 IS92c	0.69	2.98	3.70	-1.61	-0.51	1.51
1-4 IS92d	0.69	4.09	4.81	-1.88	-0.57	2.25
1-5 IS92e	1.03	4.83	5.91	-2.00	-0.21	3.57
1-6 IS92f	1.31	3.75	5.12	-1.57	-0.22	3.24
8-4 IPCC-EIS	0.78	4.49	5.30	-2.44	0.30	3.04
20-3 IEW-84%[2]			5.74	-0.96	-0.29	3.15
20-2 IEW-50%[2]			4.79	-1.61	-0.38	2.77
20-4 IEW-16%[2]			3.88	-2.49	-0.38	2.33
18 ECS '92	0.97	2.01	3.00	-1.02	-0.18	1.76
33-2 ESCAPE S1			5.50	-1.42	-0.20	3.78
37 He *et al.* (c)	0.68	3.96	4.67	-1.60	-0.34	2.75
31-2 WEC A	0.94	5.07	6.06	-2.85	-0.24	2.79
31-1 WEC B	0.94	4.08	5.06	-2.40	-0.03	2.51
3-1 E&R B	0.91	2.34	3.28	-0.22	-0.52	2.52
7-1 EPA-SCW	0.90	2.25	3.17	-0.81	-0.58	1.74
7-2 EPA-RCW	1.02	4.11	5.17	-1.21	-0.47	3.41
12-2 GREEN			4.40	-0.27	0.25	4.38
30-1 12RT			4.03	-1.36	-0.38	2.22
27-2 IMAGE 2.0-CW	1.02	4.15	5.21	-0.82	-0.21	4.13
Minimum[2]	0.67	2.01	3.00	-2.85	-0.58	**1.51**
Median[2]	0.94	3.91	4.90	-1.59	-0.30	2.85
Mean[2]	0.93	3.73	4.66	-1.52	-0.28	2.83
Maximum[2]	**1.31**	5.07	6.06	-0.22	**0.30**	4.38
Policy Scenarios						
Minimum[2]	*0.81*	*2.25*	*2.34*	*-4.32*	*-1.72*	*-0.82*
Median[2]	*0.94*	*4.08*	*4.48*	*-1.89*	*-1.06*	*1.24*
Mean[2]	*0.94*	*3.70*	*4.24*	*-2.24*	*-1.03*	*0.86*
Maximum[2]	*1.02*	*4.11*	*5.17*	*-0.53*	*-0.52*	*2.29*

[1] Component growth rates do not add exactly to (sub)totals due to independent rounding errors.
[2] Component growth rates do not add as calculated from original frequency distributions.
[3] Reference for scenarios given in the Supplementary Table.
BOLD denotes IPCC scenarios

An Evaluation of the IPCC IS92 Emission Scenarios 273

Table 6.7: 1990 to 2020 – Eastern Europe and ex-USSR average annual growth rates[1] (%/yr) for selected non-intervention (reference) scenarios and range from intervention (policy) scenarios.

Non-intervention scenarios[3]	POP	GDP capita	GDP	Energy GDP	Carbon energy	Carbon emissions
1-1 IS92a	0.43	1.49	1.93	-0.66	-0.24	1.01
1-2 IS92b	0.43	1.49	1.93	-0.66	-0.24	1.01
1-3 IS92c	0.31	0.95	1.26	-0.69	-0.50	0.06
1-4 IS92d	0.31	1.88	2.21	-1.03	-0.67	0.48
1-5 IS92e	0.43	2.83	3.27	-1.37	-0.19	1.67
1-6 IS92f	0.67	1.73	2.31	-0.88	-0.13	1.28
8-4 IPCC-EIS	0.70	2.49	3.20	-1.14	-0.17	1.84
20-3 IEW-84%[2]			2.98	-1.06	-0.34	0.46
20-2 IEW-50%[2]			2.16	-1.42	-0.36	0.07
20-4 IEW-16%[2]			1.51	-2.01	-0.41	-0.25
18 ECS '92	0.49	2.18	2.68	-1.52	-0.24	0.88
34-1 Bashmakov Base	0.58	0.77	1.36	-0.93	-0.05	0.37
35-1 Sinyak *et al.* DAU	0.56	1.44	2.00	-1.09	-0.41	0.49
31-2 WEC A	0.52	1.85	2.38	-1.81	-0.62	-0.10
31-1 WEC B	0.52	1.85	2.38	-2.13	-0.38	-0.18
3-1 E&R B	0.48	1.62	2.11	-1.07	-0.40	0.61
7-1 EPA-SCW	0.52	1.73	2.27	-1.29	-0.17	0.78
7-2 EPA-SCW	0.45	3.80	4.27	-2.97	-0.16	1.01
12-2 GREEN			2.51	-0.63	0.32	2.19
30-1 12RT			2.29	-2.00	-0.29	-0.05
27-2 IMAGE 2.0-CW	0.43	1.48	1.91	0.72	-1.62	0.99
Minimum[2]	**0.31**	0.77	**1.26**	-2.97	-1.62	-0.25
Median[2]	0.48	1.73	2.29	-1.12	-0.23	0.61
Mean[2]	0.49	1.85	2.63	-1.17	-0.29	0.76
Maximum[2]	**0.70**	3.80	4.25	0.72	0.32	2.42
Policy Scenarios						
Minimum[2]	*0.37*	*0.77*	*1.36*	*-4.01*	*-1.96*	*-1.79*
Median[2]	*0.52*	*1.85*	*2.38*	*-2.41*	*-1.24*	*-1.34*
Mean[2]	*0.50*	*2.26*	*2.68*	*-2.59*	*-1.15*	*-1.13*
Maximum[2]	*0.58*	*3.80*	*4.27*	*-1.45*	*-0.21*	*-0.25*

[1] Component growth rates do not add exactly to (sub)totals due to independent rounding errors.
[2] Component growth rates do not add as calculated from original frequency distributions.
[3] References for scenarios given in the Supplementary Table.
BOLD denotes IPCC scenarios

Table 6.8: *1990-2020 – Africa average annual growth rates[1] (%/yr) for selected non-intervention (reference) scenarios and range from intervention (policy) scenarios*

Non-intervention scenarios[4]	POP	GDP capita	GDP	Energy GDP	Carbon energy	Carbon emissions
1-1 IS92a	2.63	1.25	3.92	0.26	-0.21	3.98
1-2 IS92b	2.63	1.25	3.92	0.26	-0.21	3.98
1-3 IS92c	2.20	0.40	2.60	0.40	-0.39	2.61
1-4 IS92d	2.20	1.46	3.69	0.09	-0.44	3.32
1-5 IS92e	2.63	2.11	4.80	0.17	-0.19	4.78
1-6 IS92f	3.03	1.06	4.12	0.51	-0.12	4.53
8-4 IPCC-EIS	2.42	1.55	4.00	-0.48	0.46	3.99
20-3 IEW-84%[2]			3.82	-0.52	0.10	3.33
20-2 IEW-50%[2]			3.64	-0.28	-0.17	3.28
20-4 IEW-16%[2]			3.45	-0.63	-0.09	3.06
38-1 UNEP base	1.86	2.53	4.44	-1.40	0.47	3.46
31-2 WEC A	2.94	2.98	6.00	-0.58	1.17	6.61
31-1 WEC B	2.94	2.00	5.00	-1.67	0.26	3.51
3-1 E&R B	1.56	2.29	3.89	-1.46	0.07	2.44
7-1 EPA-SCW	2.72	0.01	2.73	-0.17	0.86	3.45
7-2 EPA-RCW	2.42	1.71	4.17	-0.10	0.72	4.81
27-2 IMAGE 2.0-CW	2.63	1.25	3.92	0.53	-0.16	4.30
Minimum[3]	1.56	0.01	**2.60**	-1.67	**-0.44**	2.44
Median[3]	2.63	1.50	3.92	-0.17	-0.09	3.51
Average[3]	2.49	1.56	4.00	-0.30	0.13	3.85
Maximum[3]	**3.03**	2.98	6.00	0.53	1.17	6.61
Policy Scenarios						
Minimum[3]	*1.86*	*0.01*	*2.73*	*-2.19*	*-4.01*	*-0.31*
Median[3]	*2.57*	*1.71*	*4.15*	*-1.46*	*-1.13*	*1.25*
Mean[3]	*2.51*	*1.53*	*4.08*	*-1.31*	*-1.47*	*1.21*
Maximum[3]	*2.94*	*2.53*	*5.00*	*-0.30*	*-0.11*	*2.43*

[1] Component growth rates do not add exactly to (sub)totals due to independent rounding errors.
[2] Refers to non-OPEC developing countries. Component growth rates do not add as calculated from original frequency distributions.
[3] Component growth rates do not add as calculated from original frequency distributions.
[4] References for scenarios given in the Supplementary Table.
BOLD denotes IPCC scenarios

An Evaluation of the IPCC IS92 Emission Scenarios

Table 6.9: *1990-2020 – USA average annual growth rates[1] (%/yr) for selected non-intervention (reference) scenarios and range from intervention (policy) scenarios.*

Non-intervention scenarios[3]	POP	GDP capita	GDP	Energy GDP	Carbon energy	Carbon emissions
1-1 IS92a	0.57	2.33	2.91	-1.81	-0.26	0.78
1-2 IS92b	0.57	2.33	2.91	-1.81	-0.26	0.78
1-3 IS92c	0.22	1.72	1.94	-1.59	-0.63	-0.31
1-4 IS92d	0.22	2.45	2.67	-1.94	-0.88	-0.21
1-5 IS92e	0.57	2.91	3.49	-1.98	-0.11	1.34
1-6 IS92f	0.90	2.09	3.00	-1.62	0.00	1.34
8-4 IPCC-EIS	0.56	0.90	1.46	-0.20	0.16	1.42
20-3 IEW-84%[2]			2.36	-0.88	0.29	1.37
20-2 IEW-50%[2]			2.20	-0.96	0.24	1.24
20-4 IEW-16%[2]			2.03	-1.18	-0.22	0.80
18 ECS '92	0.54	1.09	1.63	-1.21	-0.25	0.15
16-1 CHALLENGE			1.28	-0.66	0.03	0.80
36-1 EMF 12 (lowest)			2.20	-1.38	-0.24	0.49
36-2 EMF 12 (highest)			2.20	-1.04	0.13	1.24
31-2 WEC A	0.56	1.83	2.40	-1.94	-0.44	-0.02
31-1 WEC B	0.56	1.83	2.40	-2.08	-0.40	-0.13
3-1 E&R B	0.51	1.63	2.15	-1.09	0.22	1.26
7-1 EPA-SCW	0.53	1.13	1.67	-1.29	0.00	0.35
7-2 EPA-RCW	0.43	2.23	2.64	-1.83	0.02	0.81
32 NES			2.08	-0.32	-0.12	1.63
12-2 GREEN			2.27	-0.97	-0.05	1.23
30-1 12RT			2.20	-1.06	0.16	1.28
27-2 IMAGE 2.0-CW	0.56	2.33	2.91	-2.17	-0.22	0.46
Minimum[2]	0.22	**0.90**	1.28	-2.17	**-0.88**	**-0.31**
Median[2]	0.56	1.96	2.20	-1.29	-0.11	0.80
Mean[2]	0.52	1.91	2.31	-1.35	-0.12	0.79
Maximum[2]	**0.90**	**2.91**	**3.49**	**-0.20**	0.29	1.63
Policy Scenarios						
Minimum[2]	*0.43*	*1.04*	*1.19*	*-2.88*	*-2.98*	*-2.30*
Median[2]	*0.53*	*1.83*	*2.14*	*-1.91*	*-1.13*	*-0.97*
Mean[2]	*0.51*	*1.69*	*2.08*	*-1.87*	*-1.18*	*-1.08*
Maximum[2]	*0.56*	*2.23*	*2.67*	*-0.66*	*-0.53*	*0.73*

[1] Component growth rates do not add exactly to (sub)totals due to independent rounding errors.
[2] Component growth rates do not add as calculated from original frequency distributions.
[3] Reference for scenarios given in the Supplementary Table.
BOLD denotes IPCC scenarios

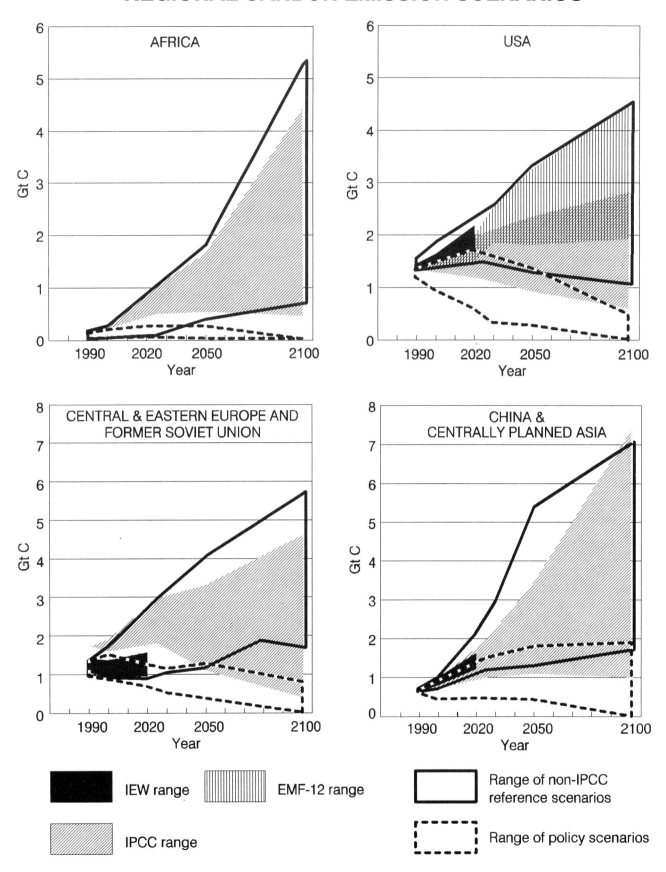

Figure 6.7: Range of regional CO_2 emissions from IS92 and other published scenarios.

analysis, however, was conducted for nine regions. For our analysis we use the following regions: China and Centrally Planned Asia, Central and Eastern Europe and the former Soviet Union (EEFSU), Africa, and the United States of America (USA). These regions were selected because they are at different stages of economic and demographic development. The USA was selected because there are a large number of published scenarios for this region.

Only CO_2 emissions are analysed because the literature contains few non-CO_2 regional scenarios. The main components of the different scenarios from 1990 to 2020 are given in Tables 6.5 to 6.9 and Figures 6.7 to 6.9.

6.3.2.1 China and Centrally Planned Asia

The carbon emissions of the IS92 Scenarios are within the range of the other reference scenarios. However, their underlying driving forces are not representative of other published scenarios. The IS92 Scenarios assume GDP increases that are on the high side of the other scenarios, and their range of assumptions about energy intensity and carbon intensity for China is both narrower and lower than the range in other scenarios.

Figure 6.7 shows the range of carbon emission scenarios up to the year 2100. The IS92a, IS92b, IS92d and IS92f scenarios are well within the 16th to 84th percentile range of the IEW poll for 2020. The IS92a is close to the IEW poll median and consequently can be considered as a "middle of the road" scenario. Conversely both IS92e and IS92c expand the range of all other long-term (2100) emission scenarios reviewed. IS92e is the highest, and IS92c is the lowest of all reference scenarios analysed.

Analysis of the components in the growth of carbon emissions (Table 6.6) indicates the importance of economic and population growth assumptions followed by decreases in energy intensity. By 2100 IS92e assumes the highest and IS92c the lowest GDP growth rates of all scenarios reviewed. In all scenarios the intensity of energy use decreases typically by about 1.5%/yr in reference cases and well above 2% in policy cases (Figure 6.8). The resulting improvements in energy efficiency across all scenarios are large, but within the range of both historical experience and calculations of the theoretical minimum requirements derived from energy analysis (cf. Nakicenovic et al., 1993). The IS92 Scenarios have a narrower range of assumed energy intensities than other scenarios. Most scenarios also anticipate further decarbonisation of the energy system, i.e. a decrease in carbon intensity of 0.2 to 0.5%/yr (Figure 6.9). The IS92 range is lower than other reference scenarios (cf. the IEW poll). Indeed, the low carbon intensities assumed in IS92c, IS92d and IS92e for China in the years after 2050 are more characteristic of policy scenarios than non-intervention scenarios.

6.3.2.2 Eastern Europe and former USSR (EEFSU)

The IS92 Scenarios do not fully capture current perceptions about trends in energy-related emissions in EEFSU in the medium term (up to 2020). First, the IS92 base year (1990) emission estimate (1.7 GtC/yr) exceeds by 30% the most recent estimate based on UN energy statistics (1.304 GtC/yr) (Marland et al., 1993). Second, the assumptions in the IS92 Scenarios about high growth rates (particularly for GDP) are probably unrealistic considering the current economic crisis in the region. These assumptions lead to medium term estimates of CO_2 emissions in the IS92 Scenarios that are higher and narrower in range than the estimates in other published scenarios (Figure 6.7). The range of IS92 Scenarios in 2020 span from 1.8 GtC/yr to 3.0 GtC/yr, whereas the more recent IEW poll spans a range of 1 to 1.6 GtC/yr by 2020. Because of uncertainty in economic trends, the range of medium-term emission estimates in the literature (Figure 6.7) is perhaps the largest of any region.

Most long-term scenarios (including IS92) represent more or less a pre-1990 "business as usual" point of view, and that does not fully take into account the economic crisis in EEFSU. By 2100 the published scenarios have a range that differs by a factor of eight. One reason for this wide spread is that both IS92c and IS92d are beyond the emission range of other published scenarios and are more characteristic of policy scenarios.

As expected, the rate of economic growth has the largest range of any input variable (Table 6.7). Recent scenarios (e.g., Bashmakov, 1992; Sinyak and Nagano, 1992) assume much lower values (lower by nearly a factor of two) than scenarios developed before 1991.

Almost all scenarios agree that both energy and carbon intensities will decline. The projected decline in energy intensity ranges from 1%/yr to 2%/yr (Figure 6.8), although there are significant differences in assumed short-term (up to 2000) and longer-term (2020 and beyond) trends. As indicated by results from the IEW poll, short-term energy intensities in the region could increase. In fact, between 1990 and 1992 the energy intensity of the former USSR increased by 23% (ECE, 1993) as economic output fell more rapidly than energy consumption. Over the longer term, significant improvements in energy intensities could occur following economic restructuring and the replacement of inefficient capital stock and the removal of price distortions. Current estimates of energy conservation in the former USSR, for instance, indicate a potential of more than 50% of actual consumption (Bashmakov, 1992 and Sinyak and Nagano, 1992). That economic restructuring results in significant uncertainty in future emissions is also confirmed by recent studies of other economies in transition, e.g., Hungary.

The extent and timing of efficiency improvements in the region are of course uncertain, and this uncertainty is

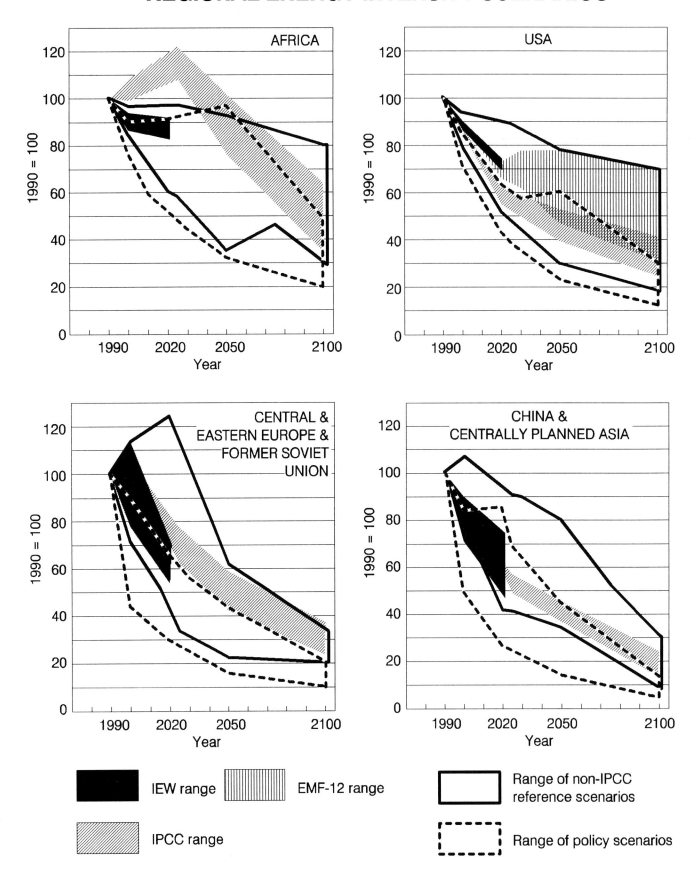

Figure 6.8: Range of regional energy intensities from IS92 and other published scenarios.

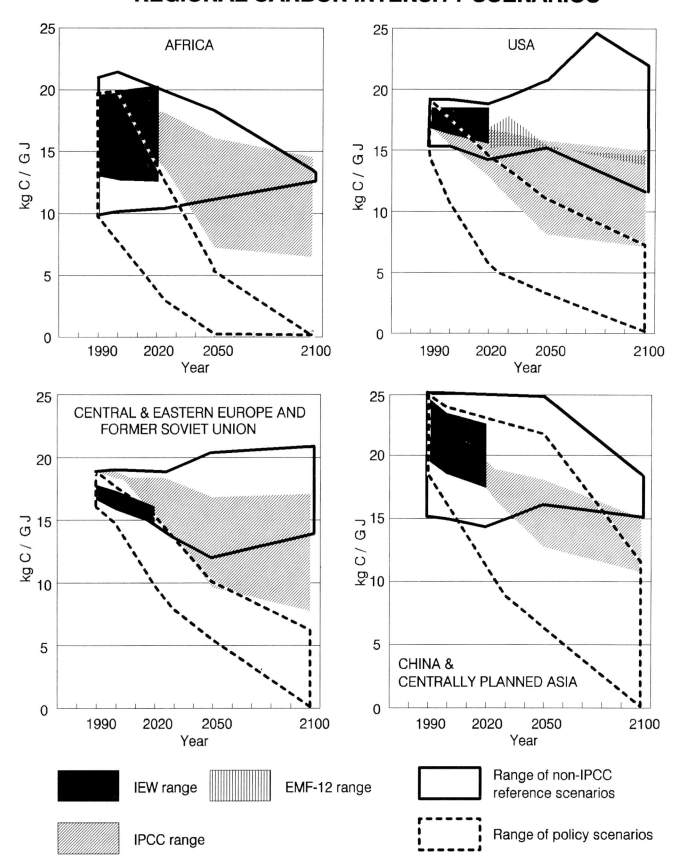

Figure 6.9: Range of regional carbon intensities from IS92 and other published scenarios.

reflected in the scenario ranges of the IEW poll. This range is wider than the range of assumptions used in the IS92 Scenarios (Figure 6.8). Most scenarios also project declining carbon intensities of around 0.3%/yr (Figure 6.9), although statistical data and base year calibration problems in some models/scenarios remain significant (illustrated by the wide range in 1990 carbon intensities). By 2020, the range spanned by the IS92 Scenarios is narrow, but continues to be representative of the "middle of the road" scenarios available in the literature. Over the long term (to 2100), the range of the IS92 Scenarios is much wider than, and at the low end of, published scenarios. The carbon intensities of the IS92c and d scenarios are close to the values of published policy scenarios (Figure 6.9).

6.3.2.3 Africa

The IS92 Scenarios account for half of all the available scenarios of long-term energy related emissions for Africa.

The range of emission scenarios is among the widest in both absolute and relative terms for all regions covered in this review (Figure 6.7). By 2020 regional emissions range from 0.3 GtC/yr to 0.8 GtC/yr, and the range increases to 0.5 GtC/yr to 5.2 GtC/yr (i.e., a factor of 10). However even for the highest scenario, *per capita* carbon emissions by 2100 would remain significantly below current OECD averages. Overall, the range of absolute emission projections trajectories spanned by the IS92 Scenarios is in agreement with the limited number of published scenarios.

Up to 2020, population growth rates dominate over *per capita* GDP growth (Table 6.8), but afterwards they become less important than other driving forces (Grübler, 1994). Up to 2020, assumed *per capita* economic growth rates show a large range (0 to 3%/yr), with GDP growth rates ranging from 2.6 to 6% annually. IS92c assumes the lowest GDP growth rate of all the reference scenarios reviewed.

Scenarios differ substantially in their assumptions about energy and carbon intensities up to 2020. Whereas the IS92 Scenarios all assume increasing energy intensities (Figure 6.8), all but one of the other scenarios project decreasing energy intensities. Conversely, the IS92 Scenarios all project declining carbon intensities (Figure 6.9), whereas all but one of the other scenarios anticipate increasing carbon intensities. The IS92 Scenarios also assume a much smaller range of energy intensities than is found in other scenarios (Figure 6.8). The overlap between reference and policy scenarios is another indication of disagreement over trends in Africa (Figures 6.7, 6.8 and 6.9).

6.3.2.4 USA

This region has the largest number of published scenarios available and this review covers over 50 of them. The IS92 Scenarios only partly reflect the range of views in the published scenarios.

The range spanned by scenarios of future emissions in the USA is large (Figure 6.7). For 2020, scenario results range from 1.2 GtC/yr (IS92c) to 2.3 GtC/yr (e.g., IPCC-EIS or NES, 1991). The IS92 Scenarios are skewed to the low side (Figure 6.7). By 2100, the range of the published scenarios widens from 0.6 GtC/yr (IS92c) to 4.5 GtC/yr (the highest scenario from the EMF-12 model runs). The IS92 Scenarios range by a factor of five by 2100 and continue to cover the low side of the range of published scenarios.

The upper range of the published scenarios implies a large increase in USA emissions by 2100 (up to 3.2 GtC/yr). However, an increase of this magnitude may be unlikely in a developed, service-oriented economy such as that of the USA, which already has one of the highest *per capita* emission levels in the world. The lower emission estimates of the majority of other published scenarios supports this view.

Assumptions about trends in energy intensity vary significantly between scenarios, although all scenarios assume a decline in intensity (Table 6.9 and Figure 6.8). The range of energy intensities spanned by the IS92 Scenarios (and the IEW poll) appears narrow, especially up to 2020. Over the long term (2050 and beyond), there is a significant overlap between reference and policy scenarios. One interpretation of this overlap is that the distinction between climate policies targeted towards improving energy efficiency and efficiency improvements due to other reasons becomes increasingly blurred in the long term.

Although all non-intervention scenarios assume that energy intensity declines, they do not make the same assumption for carbon intensity. Some scenarios show a decrease because of increased reliance on low- or no-carbon fuels, whereas other scenarios increase carbon intensity due to assumptions, for example, about greater use of coal. Both IS92c and IS92d assume much lower carbon intensities than other reference scenarios (Table 6.9).

Over longer time horizons, changes in energy intensity and carbon intensity become as important as determinants of uncertainty in future emissions as do changes in population and economic growth (Grübler, 1994). This situation implies that over the long run, differences in models and parameter assumptions may become more important than uncertainties about future population and economic growth. This conclusion is supported by the results of the EMF-12 exercise (Gaskins and Weyant, 1993; Weyant 1993). It also suggests that more scenarios are needed to represent a wider range of views on changes in the energy and carbon intensity of the future economy.

Summary of the discussion of regional energy scenarios: there are several instances where the emission estimates of

An Evaluation of the IPCC IS92 Emission Scenarios

the IS92 Scenarios are not representative of the range of other published scenarios. In the case of Eastern Europe and the former Soviet Union, IS92 estimates are considerably higher than other recently published scenarios that take into account the current economic situation in that region. For the USA, the IS92 Scenarios span a lower range of emission estimates than the range of other published scenarios. The input assumptions of the IS92 Scenarios (e.g. the economic growth and carbon intensity assumptions) are in many cases not representative of other published regional scenarios.

6.3.3 Scenarios for CH_4 and N_2O

Although fossil fuel CO_2 is not the only radiatively important gas associated with human activities, far less attention has been paid to non-CO_2 greenhouse gases. In this section we compare the IS92 Scenarios with the few others available for energy-related CH_4 and N_2O emissions (Lashof and Tirpak, 1990; Messner and Strubegger, 1991; Edmonds, 1993; Matsuoka *et al.*, 1993; Nakicenovic *et al.*, 1993 and Alcamo *et al.*, 1994). The IS92 Scenarios provide a broad range of possible future trajectories for both CH_4 and N_2O (Figure 6.10). By the year 2100 the difference between the high and low cases for CH_4 and N_2O is similar to that for CO_2 (that is, a factor of about seven or eight). Of particular importance to CH_4 and N_2O scenarios is the significant uncertainty in base year (1990) emissions, which reflects gaps in data collection and gaps in our knowledge about the processes which lead to these emissions.

6.3.4 Scenarios for Sulphur Emissions

Scientific studies reported by Working Group I indicate that the cooling effects of the atmospheric sulphate particles associated with sulphur emissions may be highly significant. Consequently, the IS92 Scenarios include sulphur emissions in addition to greenhouse-related gases. Global sulphur emissions were included as part of a comprehensive set of scenarios for the first time in the IS92 Scenarios.

The IS92 global scenarios of sulphur emissions range from 77 to 254 Tg/yr for the year 2100. These figures include natural sulphur emissions, which are estimated at 74 Tg/yr for 1990. Natural emissions are expected to remain constant. Thus the IS92 Scenarios imply anthropogenic sulphur emissions ranging from 3 Tg/yr to 180 Tg/yr. The main source of current and future anthropogenic emissions is estimated to be energy combustion, although there the estimation of current non-energy sources is highly uncertain.

Few global scenarios are available for comparison with the IS92 sulphur scenarios. The only global scenarios that

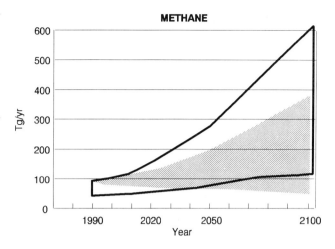

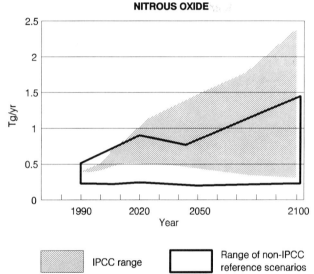

Figure 6.10: Range of CH_4 and NO_x energy-related global emissions from IS92 and other published scenarios.

we have been able to identify are those of Matsuoka (1992) developed for the AIM (Asian-Pacific Integrated Model) Programme (Morita *et al.*, 1993). Both the IS92 and the AIM scenarios assume that emissions in OECD regions will decline because of a shift away from coal use, and international agreements on air pollution control. These scenarios also indicate that emissions from developing regions will greatly increase due to economic development.

There has been considerable work on estimating current and near-term emissions of sulphur in some OECD regions. The near-term reference scenarios, under the assumption of no intervention, show a levelling of emissions in Europe (Alcamo *et al.*, 1990), and a decline in the USA (NAPAP, 1991) because of a shift away from coal use. These results are consistent with the IS92 assumptions. As an indication of the range of uncertainty, the 2030 reference estimate for the USA ranged from 10 Tg/yr to 25 Tg/yr (NAPAP, 1991). If recent legislation is taken into account (e.g., Amendments to the Clean Air Act in the USA and the 1994 Sulphur Protocol of the Geneva Convention on Transboundary Air Pollution) sulphur emissions in all OECD regions are expected to decline drastically in coming decades. The IS92 Scenarios only reflect some of this legislation.

The uncertainty surrounding baseline emissions in OECD regions is relatively small because of the development of detailed emission inventories in connection with air pollution control efforts. As an example, the uncertainty about current sulphur emissions in the USA is estimated to be only ±6% to 11% (NAPAP, 1991). The baseline estimates for OECD in the IS92 emissions are close to the latest official figures. Outside the OECD, baseline emissions are not well known because the first comprehensive emission inventories are only now becoming available. We note that for China the estimates from the AIM model (Matsuoka, 1992) and the IS92 Scenarios differ by a factor of two for 1985 emissions.

The baseline global sulphur emissions (1990) of the IS92 Scenarios are 73.5 Tg S/yr, which is within the wide range of 70 Tg S/yr to 80 Tg S/yr given by Leggett *et al.* (1992).

The main sources of uncertainty in estimating sulphur emissions from worldwide energy use is the sulphur content of coal (if it is used) and the amount of sulphur remaining in an energy-generating device after combustion. As already noted, estimates of emissions from biomass burning and natural sources are also highly uncertain.

6.3.5 Different Variables: their Relative Importance and the Sensitivity of Emissions to them

A sensitivity analysis investigated the relative influence of three variables, GDP, energy intensity, and carbon intensity on energy-related CO_2 emissions. Population was not included in the analysis because information about population assumptions is not available for all the scenarios reviewed. In addition, the sensitivity of emission estimates to varying assumptions about population growth is clearly illustrated by the IS92 Scenarios, which have higher and lower population assumptions than any other scenario reviewed. Sensitivity is expressed as the change in emissions relative to IS92a emissions, resulting from changes in GDP, energy intensity and carbon intensity derived from the extremes of the non-intervention scenarios in the sample presented in Tables 6.4 to 6.8

For the non-intervention scenarios over the medium term (up to 2020), assumptions about GDP growth are the most important, followed by energy intensity or carbon intensity (Figure 6.11). Over the longer term (to 2050 and 2100) energy and carbon intensity become increasingly important globally and in some regions, although GDP growth continues to be the most important factor. If policy

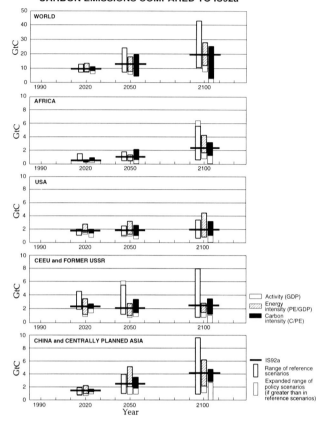

Figure 6.11: Uncertainty analysis of energy-related CO_2 emission estimates relative to the IS92a scenario.

scenarios are included in the analysis, assumptions about carbon intensity become as important as GDP growth assumptions (Figure 6.11).

A comparison of the reference and policy scenarios suggests that uncertainties in the driving forces of emissions within reference cases are at least as significant as the uncertainties between reference and policy cases. This comparison implies that over a time horizon of 50 to 100 years it is difficult to distinguish between the effect of climate policies and the effect of societal/technological changes irrespective of climate policies.

The sensitivity analysis suggests that varying the critical input assumptions and model parameters for future emission scenarios can have large impacts on emissions. These sensitivities are so large over the time horizon to 2100 that we recommend against the use of a single scenario as input to climate models. Rather a range of scenarios should be used as input. Moreover, the distribution of sensitivities for the different emission components of the non-IS92 Scenarios around IS92a is highly skewed and is particularly divergent at the regional level. Therefore, the IS92a Scenario may not be suitable as a "central" or "middle of the road" scenario, especially for regional emission estimates (see also Figure 6.2). Even if a scenario clearly represented a consensus view of the evolution of most critical input assumptions and driving forces affecting future emissions, the scenario would not necessarily have a higher likelihood than an alternative case. This situation is an additional reason to refrain from using a single emission scenario as input to assess the climate consequences of no action to reduce greenhouse gas emissions. Even if policies do not change, the future will be different from the present and is thus inherently uncertain.

Published uncertainty analyses also provide information about the relative importance of different input assumptions. We have already discussed the emission results of four published uncertainty analyses (see Section 6.3.1.2). In this section their findings regarding the importance of input assumptions are discussed. Nordhaus and Yohe (1983) identified ten key input parameters to which they assigned probability distributions. Their analysis singled out the elasticity of substitution between fossil and non-fossil energy use as the most important parameter in explaining the uncertainty in carbon emissions. They found that the rate of growth of total factor productivity, a measure which affects both labour and energy productivity, is the next most important parameter. Interestingly, population assumptions were low on the list of factors affecting baseline emissions. Edmonds *et al.* (1986) obtained a similar result. However, the range of population growth assumptions used in both of these uncertainty studies did not encompass the current range of population projections so their conclusions are limited.

Edmonds *et al.* (1986) performed an uncertainty analysis that was similar to Nordhaus and Yohe (1983), but the model used by Edmonds *et al.* contained much more detail about energy-producing and energy-consuming sectors. The carbon emission forecasts from this study formed a biased distribution with median values significantly below the mean values. Edmonds *et al.* identified the four most important factors influencing carbon emission uncertainty as labour productivity in developing countries, labour productivity in developed countries, end-use energy efficiency improvements and the income elasticity of energy demand for the world.

Lashof and Tirpak (1990) also performed a sensitivity analysis of the ASF model, which was used to generate the IS92 Scenarios. This study showed that the most important factors were higher prices for coal and cost reductions for non-fossil energy technologies, both of which lead to emission reductions. Conversely, the impact of increasing the resource base for oil or gas was found to have comparatively small impact.

Manne and Richels (1993) performed an uncertainty analysis which used a poll of experts to set the uncertainty distributions of key input parameters, such as GDP growth rates, elasticity of substitution between energy, labour and capital, rate of autonomous energy efficiency improvements (AEEI), commercial year for economically-competitive carbon-free electricity, and the cost of the non-electric backstop technology. They found that the most important parameter was the potential GDP growth rates, followed by the AEEI and (to a smaller extent) the two supply side technology parameters. At the same time, the elasticity of substitution between energy and capital and labour only had a minor influence in reference scenarios (assuming comparatively modest energy price increases), a situation that changes drastically in policy intervention cases.

As noted earlier, de Vries *et al.* (1994) did not evaluate the uncertainty of model driving forces, but concentrated instead on the influence of uncertainty in model parameters on scenario results. They found that the most important parameters were the elasticities of energy demand, and the parameters that describe consumer responses to fuel price increases.

Summary of the discussion of sensitivity analysis: variations in the rate of GDP growth across regions which reflect the combined population and per capita income growth are the most important sources of variation in CO_2 emissions in the near to medium term. Over time, however, variations in the rates of change in energy and carbon intensity are equally important to GDP growth globally and in some regions. The sensitivity analysis did not suggest that the IS92a or any of the IS92 Scenarios is a "central" or "medium" case or has higher likelihood than alternative scenarios. In addition, uncertainty analyses demonstrate that emission estimates are sensitive to the structure and processes used in models to represent the real world. Consequently, to obtain a full range of views about future emissions, it is also important to use a wide range of models and methodologies.

6.3.6 Discussion of Scenarios for Energy-Related Emissions

The global CO_2 estimates of the IS92 Scenarios bracket the majority of other published non-intervention scenarios, and therefore can be said to reflect a reasonable range of fossil fuel scenarios. Although some plausible scenarios fall above or below the range spanned by the IS92 Scenarios, the number of such scenarios is comparatively small.

Compared to the literature on CO_2 emissions, the literature on non-CO_2 greenhouse gases is limited. Few other studies consider the variety of gaseous emissions reported by the IS92 Scenarios. An important reason for the divergence of the emission estimates for non-CO_2 gases is the lack of agreement on base year emissions, which reflects a lack of knowledge or agreement on how to represent emission-causing processes.

Although the IS92 Scenarios were designed as non-intervention reference cases, the emission estimates of the lowest scenario (IS92c) fall within the range of published policy scenarios. Nevertheless, as a whole, the IS92 Scenarios do not reflect the range of policy scenarios and they should not be interpreted as policy scenarios. It is important to note that some sources of uncertainty are more subject to control by policy than others.

At the regional level there are several instances where the IS92 Scenarios do not reflect the range of views represented by other scenarios. Notably, future emissions in Central and Eastern Europe and the Former USSR (EEFSU) are higher in the IS92 Scenarios than in more recent scenario studies that take the effects of the economic crisis in the region into account. In addition, the IS92 base year (1990) estimate of CO_2 emissions is about 30% higher than the latest estimate. For the USA, the IS92 Scenarios span a lower range than the full spectrum of scenarios available in the literature.

The analysis of energy scenarios does not provide any evidence that IS92a or any other individual IS92 Scenario is a "medium" or "central" case. IS92a sometimes covers the "middle of the road" alternative regional scenarios, but is often above or below the median of other scenarios.

The IS92 Scenarios are within the range of most emission estimates and input assumptions for global energy scenarios. However, they are not always representative of the range of driving forces, such as population and economic development, energy intensity and carbon intensity, found in published regional scenarios. In some cases the IS92 Scenarios span a much narrower range than other scenarios. For example, the IS92 Scenarios assume a much narrower range of energy intensities in the USA and China plus Centrally Planned Asia than the range of published scenarios. In other cases, the IS92 Scenarios fall outside the range of other scenarios, for example the GDP growth rate assumed in IS92e, which is higher for the USA and China plus Centrally Planned Asia, than any other scenario. The trends for driving forces in the IS92 Scenarios also deviate in some circumstances from the other published scenarios, as in the case of Africa, where the IS92 Scenarios assume increasing energy intensity up to year 2020, even though all but one other scenario assume a downward trend. Thus the IS92 Scenarios only partly reflect the range of regional views expressed in the published scenarios. Regional perspectives can be included in global scenarios through a decentralised approach to scenario-building which includes the participation of scientists and policymakers familiar with national and regional circumstances. Examples of the inclusion of regional perspectives are the scenario-building exercises sponsored by the World Energy Council, the Energy Modelling Forum, the International Energy Workshop, UNEP and the CHALLENGE project of IIASA.

6.4 Comparison of Land-Use-Related Scenarios

Few scenarios of emissions related to land-use and agricultural activities are available for comparison to the IS92 Scenarios. However, in contrast to the energy-related scenarios, most land-use-related scenarios include a range of greenhouse gas emissions rather than just CO_2. The scenarios reviewed in this section include:

- The previously described IS92 Scenarios plus two sensitivity scenarios included in the IS92 document: S1 (high estimates of deforestation and the carbon content of vegetation) and S4 (low estimates of deforestation, high afforestation rate).
- The following scenarios of the US Environmental Protection Agency (Lashof and Tirpak, 1990): "Slowly Changing World" (SCW), "Rapidly Changing World" (RCW) and two "Stabilising Policies" scenarios (SCWP and RCWP, where P stands for policy). These scenarios were developed from the Atmospheric Stabilisation Framework (ASF) of the EPA, which includes calculations from the global food system model entitled the "Basic Linked System" (BLS), which was developed at the International Institute for Applied Systems Analysis (IIASA).
- The IMAGE 2.0 "Conventional Wisdom" (CW) scenario (Alcamo et al., 1994b).
- The Asian Pacific Integrated Model (AIM) (Matsuoka and Morita, 1994).
- The three scenarios (H1-H3) developed by Houghton (1991).

Scenarios S4, SCWP and H3 assume policy intervention to reduce emissions, whereas the other scenarios can be considered "non-intervention" reference scenarios.

Because of the time constraints under which this evaluation was performed, this section focuses on only a few categories of land-use emissions, namely, emissions of CO_2 from deforestation, CH_4 from wetland rice fields and enteric fermentation, and N_2O from soils underlying fertilised agricultural areas and soils underlying natural vegetation. These categories were selected because they illustrate the sensitivity of emission estimates to differences in the input assumptions for agricultural variables. The scenario comparisons focus on both global scenarios and scenarios from Latin America, Africa and Asia. These regions are of particular interest because they are experiencing significant changes in the driving forces of land-use emissions (e.g., agricultural expansion, agricultural productivity, deforestation and population growth). The scenarios concerning emissions of CO_2 from deforestation include the fluxes of carbon resulting from changes in the area of forests, but not climatic feedbacks to the terrestrial carbon cycle.

Table 6.10: Major driving forces for land cover/use change.

Process	IS92 (IPCC)	SCW/RCW	CW (IMAGE 2.0)
Deforestation	Proportional to population increase, with a lag time of 25 years.	External scenario based on Terrestrial carbon model (Houghton *et al.*, 1985) for tropics only. Deforestation is not a result of demand for agricultural products.	Forest clearing results from change in demand for crop and animal products and industrial wood and fuelwood. Changes in land productivity take into account soil and changing climate.
Demand for agricultural products	See SCW/RCW	Demand consists of human consumption, feed, seed, industrial usage and waste. Human consumption is based on past trends (reflecting income, habits, tastes). Income changes allow changes in consumption. The model allows changes in arable land and pastures. Technical progress is captured in yield functions for 9 commodity classes.	Demand is determined by population and income through elasticities. This affects the required areas of both arable land and grassland. Arable crops are divided into 9 classes of vegetable products including rice and 5 classes of animal products.
Crop production	See SCW/RCW	Crop yields are function of fertiliser application. BLS extrapolates the trend in land-use *per capita* from 2025-2050 to the period 2050-2100.	Crop yields change as a consequence of the additions of agricultural inputs and technological development.
Rice acreage	See SCW/RCW	Various methods, e.g. China: based on past trends; India, Indonesia, USA: based on relative profitability of rice compared to other crops.	Rice acreage is modelled from demand for rice, climate and soil. The increase in harvested area is the result of increases in irrigated rice only, with the 1990 dryland area remaining constant.
Animal production and numbers	See SCW/RCW	The production of animal products is function of feeding intensity. For most regions the number of animals is not calculated explicitly.	In period 1990 to 2025 the production per animal increases along with GNP *per capita* to 1990 OECD-Europe productivity as GNP reaches 1990 OECD-Europe level. For industrialised countries increase in productivity is 10%. In 2025-2050 the increase in productivity is 50% of that in 1990-2025, in 2050-2100 the increase is zero.
Fertiliser use	Pre-specified. Scenario from Pepper *et al.* (1992).	Fertiliser use is explicit variable in most parts, except in regional models in BLS. In regional models agricultural production determines fertiliser use.	Fertiliser use is pre-specified, not a model result. Scenario for fertiliser use is from IS92a.

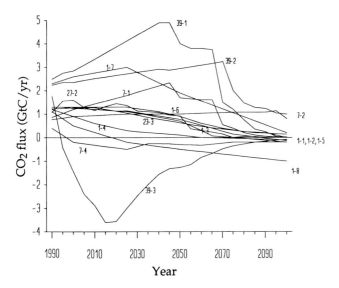

Figure 6.12: Comparison of global scenarios of CO_2 fluxes from deforestation. Numbers correspond to list of scenarios in the Supplementary Table.

For reference, Table 6.10 presents an overview of how each of the scenarios takes into account the driving forces of land-use emissions. Table 6.11 summarises other key assumptions of the scenarios. It should be noted that the IPCC methodology for national inventories may differ from the IPCC scenario assumptions for emission factors, the carbon contents of biomass, etc.

6.4.1 Results from CO_2 Global Scenarios

Before discussing scenarios for land-use-related CO_2 emissions, it should be noted that the scenarios substantially disagree on base year (1990) emissions;

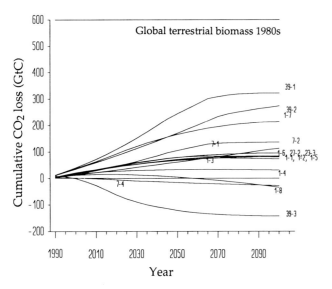

Figure 6.13: Comparison of cumulative global CO_2 fluxes from deforestation for different scenarios. Numbers correspond to list of scenarios in the Supplementary Table.

estimates range from 0.7 GtC to 2.5 GtC/yr. This disagreement reflects the lack of knowledge or agreement on how to represent the processes leading to CO_2 fluxes between the biosphere and atmosphere. It was noted in a previous section that a similar situation exists for regional-scale energy-related emissions.

Most "non-intervention" scenarios show an upward trend in emissions until about 2020, followed by a gradual decline (Figure 6.12). In comparison with energy-related scenarios which diverge up to 2100, land-use emissions show their greatest absolute differences just before the middle of the next century (with a maximum range of about 0.2 GtC to 4.8 GtC/yr), and then converge towards zero at the end of the century. Emissions decrease because either the driving forces of deforestation equilibrate or forests are depleted.

The lower scenarios are IS92 a,b,d and e, AIM, and CW. These scenarios all show decreasing global trends after 2010. Both H1 and SCW increase sharply until 2050, but their magnitude of emissions is different because they begin with different base year emissions. IS92 a,b and e show a net uptake in carbon from 2090 to 2100 because of their assumptions about forest plantations in Africa and Asia. In most of the scenarios, the release of carbon by forest biomass burning greatly exceeds the amount of carbon taken up in plantations.

The range of cumulative emissions for the non-intervention scenarios from 1990 to 2100 is 30 GtC to 320 GtC, indicating that the biosphere could continue to be a significant source of CO_2 in the atmosphere in the next century (Figure 6.13). The cumulative emissions of intervention scenarios assume considerable afforestation and result in a significant net uptake of atmospheric CO_2 (30 GtC to 150 GtC).

The upper limit in cumulative emissions of CO_2 from deforestation is formed by the amount of carbon stored in the global terrestrial biosphere, which is about 600 Gt (Atjay *et al.*, 1979; Olson *et al.* 1983) assuming no afforestation. According to the non-intervention scenarios reviewed in this chapter, deforestation in the next century may release between 5% and 53% of the carbon stored in the global terrestrial biosphere.

Summary of the discussion of global CO_2 scenarios for land-use-related emissions: emission estimates considered show their greatest absolute differences just before the middle of the next century (a factor of six). Emission estimates then converge to zero towards the end of the century because the driving forces of deforestation equilibrate or forests are depleted.

6.4.2 Results from CO_2 Regional Scenarios

The available regional scenarios for land-use-related emissions of CO_2 (IS92, SCW and CW) show considerable

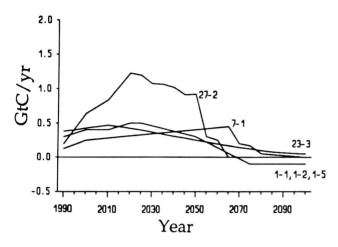

Figure 6.14a: CO_2 fluxes from deforestation for Africa for different scenarios. Numbers correspond to list of scenarios in the Supplementary Table.

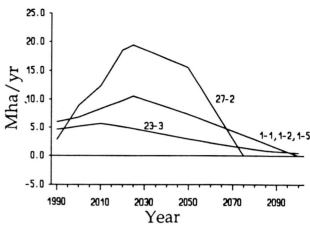

Figure 6.14b: Deforestation rates for Africa for different scenarios. Numbers correspond to list of scenarios in the Supplementary Table.

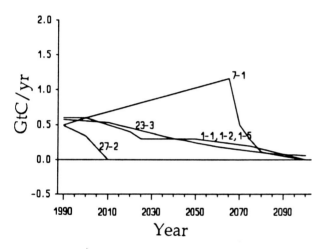

Figure 6.15a: CO_2 fluxes from deforestation for Latin America for different scenarios. Numbers correspond to list of scenarios in the Supplementary Table.

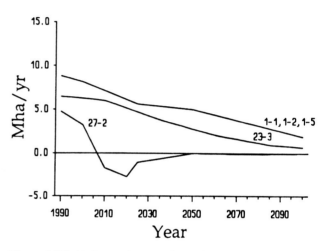

Figure 6.15b: Deforestation rates for Latin America for different scenarios. Numbers correspond to list of scenarios in the Supplementary Table.

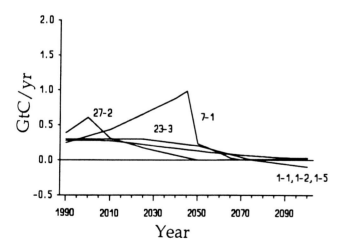

Figure 6.16a: CO_2 fluxes from deforestation for Asia for different scenarios. Numbers correspond to list of scenarios in the Supplementary Table.

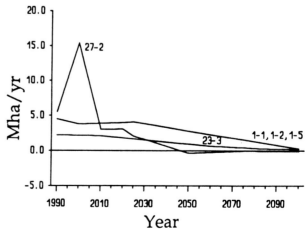

Figure 6.16b: Deforestation rates for Asia for different scenarios. Numbers correspond to list of scenarios in the Supplementary Table.

Table 6.11: Assumptions of land-use scenarios of emissions

Source	IS92a	SCW/RCW	CW
CO_2 from deforestation and afforestation	Low estimates of biomass based on OECD (1991)	Based on clearing rates and on low biomass estimates in vegetation and soils from Houghton (1991). High and low biomass values are specified for 6 types of forest (disturbed and undisturbed moist, seasonal and dry forests) and for 3 continents. The SCW scenario uses the low biomass estimates.	Carbon contents are result of soil and climate using BIOME model and Klein-Goldewijk *et al.* (1993)
CH_4 from wet rice cultivation	38 g2/yr	75 g2/yr	38 g2/yr
CH_4 from animals	Crutzen *et al.* (1986); Lerner *et al.* (1988). Emissions per unit of product decrease with productivity.	Crutzen *et al.* (1986); Lerner *et al.* (1988). Emissions per unit of product decrease with productivity.	Gibbs and Leng (1993); Crutzen *et al.* (1986). Emissions per unit of product decrease with productivity.
CH_4 from animal waste	1990 emission of 26 Tg/yr (source not specified)	Not included in calculations	Based on Gibbs and Woodbury (1993), giving total of ~13 Tg/yr for 1990
Biomass burning emissions (CH_4, CO, NO_x, N_2O, VOC)	Crutzen and Andreae (1990)	Various literature sources (Cicerone and Oremland, 1988; Seiler and Crutzen, 1980 and other references)	Crutzen and Andreae (1990)
N_2O from fertiliser and animal excreta	Fertiliser induced emission of 2.6% of applied mineral fertiliser-N. Background and animal excreta not considered.	Fertiliser induced emission of 2.5% of applied mineral fertiliser-N. Background and animal excreta not considered.	Fertiliser induced emission of 1% of applied mineral N and 1% of N in animal excreta, based on Bouwman *et al.* (1994.). Background emission from arable lands and pastures are based on climate and NPP.
N_2O from natural ecosystems	6.1 TgN/yr, constant in time.	6.0 TgN/yr, constant in time.	Depending on temperature, precipitation, NPP (Based on Bouwman *et al.* (1993).
N_2O from gain in agricultural land	Luizao *et al.* (1989) for current flux of 0.8 TgN/yr.	Bolle *et al.* (1986) for current flux of 0.4TgN/yr.	Modelled, based on Keller *et al.* (1993) and Bouwman *et al.* (1994).
Other sources	Calculated as complement to assumed global source	Calculated as complement to assumed global source	Calculated as complement to assumed global source

differences (Figures 6.14a to 6.16a). For Africa, CW greatly exceeds the other scenarios until about 2060, when African forests are depleted according to the scenario (Figure 6.14). After 2050, IS92 a,b and e show a net uptake in Africa as a result of forest plantations.

In Latin America, SCW is much higher than the others because it assumes a higher forest exploitation rate; CW is much lower than the others because it projects a stabilisation of the driving forces that have led to deforestation (Figure 6.15).

For Asia, SCW and CW differ the most from the IS92 Scenarios and AIM (Figure 6.16). Part of the difference may be caused by the definition of the region. For example, the deforestation estimates in IS92 and SCW do not include China, but CW does. For Asia, CW peaks around 2000, mainly because of high deforestation rates in India. Both the IS92 Scenarios and CW assume that wood from forest clearings is not burned in China and the Centrally-Planned Asian countries. Hence, deforestation in this part of Asia is assumed to result in a low, long-term flux of carbon from enhanced soil respiration rather than a higher short-term flux due to wood burning of cleared forests. Thus, even though the global results for CO_2 fluxes are consistent, the regional results show some inconsistency.

Summary of the discussion of regional CO_2 scenarios for deforestation: the differences in the various emission estimates of the regional scenarios are quite large. The most significant inconsistencies occur in Africa and Asia. However, some of these regional differences apparently offset each other so that differences in the global results of the scenarios are not as great as might be expected from the differences in the regional scenarios.

6.4.3 Factors Affecting CO_2 Emissions from Deforestation and Afforestation

The main factors used to estimate CO_2 emissions from deforestation are:
 (i) the rate of deforestation;
 (ii) the carbon content of vegetation;
 (iii) the allocation of the cleared biomass to different carbon pools on the deforested land.

The last factor will influence the timing of the release of CO_2 but will have only a small effect on the cumulative emissions. Hence the following sections will only discuss the first two factors.

6.4.3.1 Rate of deforestation
The IS92 Scenarios and CW use similar global average deforestation rates, but incorporate different rates within regions (Figures 6.14b, 6.15b, 6.16b). The IS92 Scenarios

for Africa increase up to 2020, and decline slowly afterwards (Figure 6.14b). CW increases deforestation rates more quickly than IS92, but also peaks around 2020. After 2020 the rate of deforestation for CW declines at a faster rate than in the IS92 Scenarios, reaching zero around 2070, when all African forests are depleted. AIM peaks somewhat earlier than the other two scenarios and maintains a lower rate of deforestation throughout much of the period.

For Latin America the deforestation rate of the IS92 Scenarios gradually decreases towards 2100 (Figure 6.15b). Meanwhile, the net rate of CW decreases much more rapidly, reaching zero in 2005. Afterwards, the scenario assumes the net afforestation of abandoned agricultural and rangeland until 2050, when net afforestation also reaches zero. AIM parallels the trend of the IS92 Scenarios, but at a lower rate.

Deforestation in Asia slowly decreases in the IS92 Scenarios, reaching near zero in 2100 (Figure 6.16b). Again, AIM follows the trend of the IS92 Scenarios, but with a lower deforestation rate. CW, however, differs from these scenarios, peaking around the year 2000 in China and India, and then rapidly decreasing by 2045 as forests are depleted.

Most scenarios base their deforestation rates on population estimates, under the assumption that population is related to demand for agricultural and forestry products which, in turn, leads to deforestation. The population scenarios used in the IS92 Scenarios and CW (World Bank, 1991), SCW and RCW (USBC, 1987) and AIM (UN, 1992) are only slightly different in their estimates of population trends to 2100. These small differences cannot explain the large differences in deforestation rates between some published scenarios. Houghton (1991) does not report the population estimates used for scenarios H1 and H2.

The IS92 Scenarios and AIM set deforestation rates proportional to population increase, with a 25 year time lag. Hence, as population growth declines (which it ultimately does according to the assumed population scenarios), deforestation rates also decline. SCW and H1 have similar emission trends, and it is presumed that their underlying deforestation assumptions are similar. Deforestation in SCW is a function of population growth, whereas in H1 it is extrapolated from historical trends of tropical deforestation in FAO/UNEP (1991).

Compared to SCW and H1, the deforestation rates of RCW and H2 are based on higher rates of agricultural productivity, which leads to lower rates of agricultural expansion and deforestation. The CO_2 emissions estimated by the RCW and H2 differ because of different assumptions about the carbon content of vegetation rather than because they assume different deforestation rates.

Deforestation rates are computed indirectly in CW:

> **BOX 3. UNCERTAINTIES IN ESTIMATING CO_2 FROM DEFORESTATION - AN EXAMPLE FROM INDONESIA**
>
> Country estimates help illuminate the uncertainties in estimating the driving forces behind the CO_2 emissions from deforestation. JEE(1993) estimates an Indonesian deforestation rate of 1.3 Mha/yr for 1982 to 1990. Deforestation was attributed to the expansion of agricultural land, infrastructure, and urban centres, as well as to shifting cultivation, forest fires and logging. The carbon content assumed for Indonesian forests was 45-90 t C ha-1 (JEE, 1993). The integrated flux is 60 to 124 TgC/yr (1989/90) for Indonesia. This is substantially different from Houghton's estimate of 192 Tg[t]C/yr (Houghton, 1991) which is based on assumptions about the carbon content of different types of vegetation.
>
> JEE (1993) estimate that the carbon uptake of vegetation was around 103 TgC/yr to 224 TgC/yr based on an annual carbon uptake of 3.0-6.6 t C ha-1/yr for 17 Mha of forest fallow and 4 Mha of forest plantations. This total area is equivalent to about 20% of the total forested area in Indonesia according to Myers (1991). These widely varying estimates illustrate the uncertainty involved with estimating CO_2 from deforestation.

specifically, population and GNP scenarios are used to compute regional agricultural demands, which are then used to estimate the expansion or abandonment of agricultural and rangeland on a geographically detailed basis for the whole world. The expansion of land for crops or livestock leads to a certain amount of forest clearing, from which a deforestation rate can be derived.

Some non-intervention scenarios assume afforestation. For example, IS92a, b and e assume a constant rate of afforestation of 1.3 Mha/yr, and CW converts abandoned agricultural land to its natural ecosystem.

6.4.3.2 Carbon content of vegetation

Another major factor influencing the range of CO_2 emission estimates is uncertainty about the carbon content of vegetation. Pepper *et al.* (1992) and Lashof and Tirpak, (1990) point out the sensitivity of the IS92 Scenarios, SCW and RCW to this uncertainty (see also Box 3).

Two different methods may be used to estimate the carbon content of vegetation. Higher estimates are obtained by destructive sampling of biomass, usually in small areas. Lower estimates of carbon content are obtained by estimating timber volumes and wood densities for forest stands; this information is usually derived from inventories covering large forest areas. The mean carbon content for all tropical forests is about 65% higher when using the destructive sampling method than when using the wood-volume-based method (Houghton, 1991). The scenarios reviewed above use different premises for their estimates of the carbon content of vegetation. The IS92 Scenarios use OECD (1991) estimates, which are mid-ranges of values based on the estimates of wood volumes used by Houghton (1991). H1 to H3 are based on the mean of the two methods: destructive sampling and wood-volume-based estimates. SCW, RCW, SCWP and RCWP are identical to the low estimates of Houghton (1991). The continental mean carbon content of vegetation in CW is similar to the low estimates in the IS92 Scenarios, RCW and SCW. However, unlike the other scenarios, the carbon content of vegetation in CW changes over time in accordance with changes in climate and atmospheric CO_2 concentration.

Summary of the discussion of factors influencing the estimation of CO_2 emissions from deforestation: there are relatively few studies to compare, and their assumptions about deforestation rates vary considerably. However, all scenarios agree that deforestation will diminish by the end of the 21st century because of the equilibration of the driving forces of deforestation or the depletion of the forests. The carbon content of vegetation is another important assumption in estimating CO_2 emissions, and the various scenarios use a wide range of estimates for this variable.

6.4.4 Results from Global Scenarios of CH_4 and N_2O

The IS92 Scenarios, SCW and CW include global and regional estimates of emissions other than CO_2. This section highlights global scenarios for some of the more important emission categories of these gases.

For CH_4 from rice fields, SCW assumes a base year (1985) emissions twice as large as those in the IS92 Scenarios and CW (Table 6.12). However, the trend of SCW is close to that of the IS92 Scenarios in that it increases until around the middle of next century, and declines slowly thereafter. CW, by comparison, declines until around 2025, and then increases slightly to the end of the century. These trends reflect the harvested area of wetland rice assumed in the scenarios. This issue will be discussed later in this chapter.

The scenarios for CH_4 from enteric fermentation agree on base year (1985) emissions and trends to 2000. Subsequently, the emission estimates of the IS92 Scenarios and SCW increase up to the end of the century, although the SCW does so at a slower rate. By contrast, CW emissions peak in about 2050 and decline thereafter. The lower emissions of CW result from its assumptions about increases in meat productivity which lead to estimates of lower numbers of animals than the other scenarios.

The various scenarios for N_2O emissions from soils of

Table 6.12: Emissions of CH_4 from wetland rice cultivation in Tg CH_4/yr

	1985	1990	2000	2010	2020	2025	2050	2075	2100
SCW									
Africa	2.6		3.9			5.5	6.6	7.6	7.8
Asia	96.0		110.0			131.6	148.5	155.3	148.1
LA	6.0		6.5			7.1	6.8	6.8	6.5
World	109		125			149	165	176	169
CW									
Africa	1.2	1.4	1.8	2.4	3.5	4.2	8.2	11.9	15.7
Asia	53.8	54.3	46.8	40.7	38.5	38.5	39.4	38.3	37.7
LA	2.5	2.4	2.6	2.6	2.5	2.4	2.5	2.7	3.0
World	58	59	53	48	46	47	52	55	54
IS92 a,b,e									
Africa	1.4	1.6	2.0	2.5	2.8	2.9	3.4	3.9	4.1
Asia	50.5	52.9	58.0	62.9	66.7	69.2	78.2	76.3	74.2
LA	3.1	3.1	3.4	3.7	3.7	3.7	3.4	3.3	3.2
World	58	60	66	69	76	78	87	86	84

natural vegetation (not shown) differ greatly. The IS92 Scenarios, RCW and SCW assume constant global emissions, whereas CW assumes slow increases. Part of the difference in the scenarios arises because CW takes into account climatic feedback to N_2O emissions. By comparison, scenarios of N_2O emissions that consider the soils of fertilised agricultural areas do not vary as much because they are all based on the same global scenario of fertiliser use. In this case the difference between scenarios stems from differences in the assumed emission factors.

6.4.5 Results from Regional Scenarios for CH_4 and N_2O

This discussion of CH_4 from rice production focuses on scenario results for Asia, which produces about 90% of the world's rice. The three available scenarios differ considerably. The base year (1985) emissions of SCW are nearly twice as high as base year emissions for CW and the IS92 Scenarios (Table 6.12). The IS92 Scenarios show a 24 TgCH$_4$/yr increase by 2100, SCW a 52 TgCH$_4$/yr increase, and CW a 4 TgCH$_4$/yr decrease. However, both SCW and the IS92 Scenarios project an increase of about 50% in CH_4 emissions, because they make identical assumptions about the increase in harvested rice area (Table 6.12). The IS92 Scenarios indicate a nearly constant emissions level for China plus Centrally Planned Asian countries (not shown), whereas the CW estimates a reduction by a factor of 2 caused by a major increase in rice yields per hectare. Emissions from India (not shown) also decrease by 20% in CW by 2100 because rice yield per unit area increases faster than rice demand.

There are large differences in the estimates of CH_4 emissions from enteric fermentation between the IS92 Scenarios, SCW and CW for Asia, but smaller differences for Africa (Table 6.13). For example, in China plus Centrally Planned Asia, the IS92 Scenarios have low and nearly constant emissions whereas CW has a rapid increase in emissions between 1985 and 2025, followed by a decline of 30% between 2025 and 2100.

The IS92 Scenarios for CH_4 from enteric fermentation for Latin America show a large increase from 16 TgCH$_4$/yr to 43 TgCH$_4$/yr between 1985 and 2100, whereas CW gives a doubling in emissions between 1990 and 2025 and a decline thereafter (Table 6.13).

Regional results for N_2O from the soils underlying natural vegetation are only provided by CW, which estimates that increases will be most important in temperate climates because of the importance of climate feedbacks in those areas.

Different regional assumptions about fertiliser use lead to different estimates of N_2O emissions from fertilised soils. For example, the growth in fertiliser use and associated N_2O loss for Africa is higher in IS92a,b and e than in CW. The increase in fertiliser-related N_2O is similar for Asia as a whole, but different for specific regions. For Latin America the growth of N_2O emission from N-fertiliser is faster in CW than in IS92a,b or e.

Table 6.13: *Emissions of CH_4 from enteric fermentation in $TgCH_4/yr$*

	1985	1990	2000	2010	2020	2025	2050	2075	2100
SCW									
World	75		94			125	151	172	179
CW									
Africa	8.1	8.5	11.2	14.9	21.5	25.4	40.5	43.5	47.5
Asia	22.8	24.1	38.9	53.6	68.0	75.1	82.3	69.9	63.1
CP Asia	6.7	7.2	16.1	24.2	31.4	34.8	31.1	27.5	24.8
LA	8.1	17.7	19.5	19.0	18.5	18.2	13.8	9.7	7.5
USA	5.8	5.6	4.3	3.2	3.0	2.9	2.8	2.8	2.2
World	75	79	96	112	131	142	161	147	142
IS92 a,b,e									
Africa	8.6	9.8	13.2	16.5	20.2	22.6	33.1	42.4	48.1
Asia	22.0	23.8	28.6	34.1	40.3	43.7	61.1	66.5	69.9
CP Asia	4.9	5.2	5.7	6.2	6.8	7.0	8.6	8.5	8.7
LA	16.0	18.0	22.0	26.0	30.0	32.0	40.0	42.0	43.0
USA	7.1	7.8	8.8	8.9	9.3	9.2	6.6	6.0	6.0
World	77	84	99	114	129	138	173	188	198

6.4.6 Factors Affecting CH_4 and N_2O Emissions from Land-Use

For reference, the main driving forces for the non-CO_2 emissions resulting from land-use are reviewed in Table 6.10 and key assumptions of the scenarios are reviewed in Table 6.11.

6.4.6.1 Factors affecting emission of CH_4 from rice production

This section discusses the assumptions of the different scenarios and how they lead to estimates of CH_4 emissions from rice fields. The main factors influencing the estimation of emissions are the assumed rate of CH_4 emissions per unit area of rice field and the total area of rice fields. The area of rice fields is in turn influenced by the assumed rice yield per hectare. The scenarios reviewed in this report all assume that the emission rate of CH_4 does not change over time.

RCW and SCW give emission rates for wetland rice that are twice as high as the value used in the IS92 Scenarios. CW uses the same wetlands emission rate as in the IS92 Scenarios, but unlike the IS92 Scenarios, distinguishes between wetland and dryland rice areas. This distinction contributes to the difference between the IS92 Scenarios and CW. In addition, the IS92 Scenarios estimate the area of rice fields using a priori assumptions about *per capita* rice consumption and rice yield per hectare, whereas the estimate of rice area in CW is based on economic and technological development. The rice yields of the IS92 Scenarios are similar to those of CW and SCW for South and Southeast Asia (Figure 6.17b), but are lower than those of CW for China and Centrally Planned Asia (Figure 6.17a). The higher yields in CW lead to smaller areas of rice fields compared to the IS92 Scenarios, and hence to lower CH_4 emission estimates. In India, CW shows that rice yield improvements exceed the increase in rice demand up to 2025, leading to a decrease in rice area and CH_4 emissions. After 2025 demand exceeds yield improvement, and both rice area and CH_4 emissions increase. Emissions in the IS92 Scenarios are relatively constant up to 2100.

The global increase in rice yield is assumed to be about 0.9%/yr in the IS92 Scenarios, SCW and RCW for the period 1990 to 2010. FAO (1993) assumes a global yield increase of 1.5%/yr and increase in harvested area of 0.5%/yr for the period 1990 to 2010, leading to a rise in production of 2%/yr. The total yield increase over the period is about 35%. This increase is even more optimistic than the yield increase of 25% for Asia assumed by CW, and it is substantially different from the IS92 Scenarios, SCW and RCW. For the period beyond 2010, it is not certain whether high growth rates in rice yields of the order of 1.5% can be maintained, and it has been suggested that for continued growth a genetic breakthrough is needed.

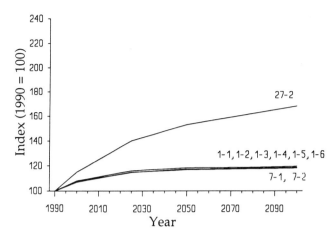

Figure 6.17a: Comparison of development of rice yields for China and Centrally Planned Asian countries for different scenarios. Numbers correspond to list of scenarios in the Supplementary Table.

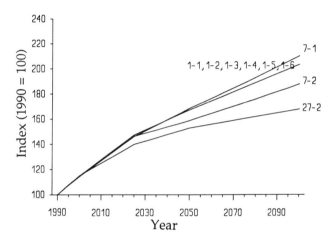

Figure 6.17b: Comparison of development of rice yields for South and Southeast Asia for different scenarios. Numbers correspond to list of scenarios in the Supplementary Table.

National projections can be used to examine the assumptions for different regions. For example, in Indonesia a number of different scenarios were developed to analyse different combinations of development in rice yields and harvested and planted rice area (DOAI, 1993). The highest yield increase was 1.5%/yr, with an increase in harvested area of 0.6%/yr which resulted in a 2.2%/yr growth in rice production. The same 2.2%/yr rise in production could also be achieved with a yield increase of 0.9%/yr combined with an increase of 1.2%/yr in harvested area. The latter scenario was based on historical increases in rice yield and is much less optimistic about rice yields than FAO (1993), the IS92 Scenarios, CW and RCW and SCW. This Indonesian example shows the important effect of assumptions about rice yields on the development of harvested area.

6.4.6.2 Factors affecting emission of CH_4 from enteric fermentation

Estimates of CH_4 emissions from enteric fermentation depend on assumptions about the number of animals, and these assumptions in turn depend on the size of the demand for meat and dairy products and the amount of meat and dairy products produced by each animal (animal productivity).

The major difference between the IS92 Scenarios and CW is the estimation of demand for meat and dairy products. This estimate is based on demographic, economic and technological factors in CW and on a priori assumptions in the IS92 Scenarios.

Of the scenarios reviewed, only CW makes explicit assumptions about meat productivity. Over the period 1990 to 2010, CW assumes a 0.2%/yr to 1.5%/yr increase in meat productivity in developing regions, whereas FAO (1993) assumes a much higher 0.9%/yr to 1.9%/yr for the same period. Scenarios for animal populations are linked to these productivity assumptions. For the period 1990 to 2010 CW shows an increase in the number of cattle by a factor of 1.9 for Africa, 2.2 for the Middle East, 5.6 for China, 1.2 for India and 1.3 for Latin America. The corresponding estimates from FAO (1993) are much lower, namely, 1.3 for Africa, 1.4 for the Middle East, 2.3 for East Asia (including China), 1.2 for South Asia (including India) and 1.1 for Latin America.

6.4.6.3 Factors affecting emission of N_2O from fertilised agricultural soils

The amount of N_2O from fertilised soils depends on the amount of fertiliser used and the N_2O loss rate from soil. It was already pointed out that the scenarios reviewed in this report assume the same amount of future fertiliser use. Hence most differences between scenarios arise from different assumptions about N_2O loss rates from soils.

The assumed growth of fertiliser up to 2010 is substantially lower than that given in FAO (1993). For the period 1990 to 2010, FAO assumes a 3.3%/yr to 4.8%/yr growth in fertiliser use in Africa, Latin America and Asia.

Summary of the discussion of factors influencing the land-use-related emissions of non-CO_2 gases: the scenarios reviewed take a wide variety of different approaches in estimating future emissions. The variety of approaches reflects the lack of knowledge or agreement on how to represent emission-generating processes. In general, the IS92 Scenarios only partly cover the range of input assumptions found in other published scenarios.

6.4.7 Discussion of Scenarios of Land-Use Emissions

Different clusters of scenarios for CO_2 emissions from deforestation are distinguishable. Starting with global scenarios, the IS92 Scenarios and AIM make similar assumptions about the relationship between population growth and emissions. Consequently, as population growth slows, they estimate similar downward trends in emissions. CW considers growth of population and GNP, technological development and other factors, and calculates a global emission trend close to the IS92 Scenarios and AIM. These scenarios are also in the lower range of the non-intervention scenarios. The remaining non-intervention scenarios, SCW, RCW, H1, and H2, extrapolate past trends in deforestation and estimate much higher emission rates than the other scenarios.

For regional scenarios, CW differed greatly from the IS92 Scenarios and AIM because of its different methods for estimating future trends in deforestation. Although nearly all scenarios relate the deforestation rate only to population growth, it is known that deforestation is more precisely related to pressure on agricultural land, consumption of wood products, settlement policy and other socio-economic factors (Turner *et al.* 1993). Accounting for these factors would permit the construction of more diverse scenarios of CO_2 emissions. Moreover, including these socio-economic factors would tend to dampen the rapid rates of deforestation found in nearly all scenarios.

Much of the variation between scenarios on both the global and the regional level can be attributed to different estimates of base year emissions. This difference indicates the lack of scientific agreement on the processes leading to CO_2 fluxes from deforestation. Another important source of variation is differing estimates of the carbon content of vegetation and soil. For example, SCW and H1 differ only because of their different assumptions about the carbon content of forests.

The scenarios for CH_4 emissions from rice fields and enteric fermentation are sensitive to assumptions about future technological improvement in rice and animal productivity and herd composition. Estimates derived from the different scenarios disagree with FAO (1993) and perhaps should be re-evaluated. Another source of uncertainty for CH_4 estimates from rice fields is the rate of emissions per unit area of rice field. Although the scenarios surveyed in this report assume a global, constant emission factor, the rate of emissions is known to depend on site conditions, rice variety and crop management. Some of this knowledge should be built into new scenarios.

A drawback of the IS92 Scenarios and other scenarios is that they do not take into account important climatic and atmospheric feedbacks in land-use emissions (e.g., N_2O from soils and CO_2 from soil respiration). The only scenario to consider this feedback is CW (of course there is great uncertainty in its treatment of these feedbacks).

Analysis shows that the IS92 Scenarios are on the low side of the wide range of global scenarios for land-use-related CO_2 emissions from deforestation and overlap with the few regional scenarios available. The IS92 Scenarios also partly overlap with the few scenarios available for N_2O and CH_4 emissions. In addition, the IS92 Scenarios cover most of the greenhouse gases originating from land-use and agriculture as well as precursors of ozone and sulphate aerosol. Emissions are reported for four world regions and by source and sector.

The documentation of certain land-use assumptions is not quite adequate to reproduce the land-use-related scenarios for IS92 Scenarios, SCW and RCW. For example, it would be helpful to have a more detailed description of the assumptions about afforestation, rice production and animal production. There are few published sensitivity analyses of land-use emissions. In one of the few studies available, Pepper *et al.* (1992) analysed the effect of the assumed carbon content of cleared vegetation and the rate of deforestation on CO_2 emissions. Although this report is not strictly a sensitivity analysis, it illustrates the simplicity of the assumptions. The calculated CO_2 emissions are directly related to the carbon content of cleared vegetation, population growth and deforestation rates in previous periods. The carbon content is specified for three forest types and for three continents. The deforestation rates in the IS92 Scenarios are calculated from human population growth and the deforestation rate, with a lag time of 25 years.

No analysis of sensitivity to input assumptions is made for the deforestation rates presented in SCW and RCW in Lashof and Tirpak (1990) and AIM (Matsuoka and Morita, 1994). Zuidema *et al.* (1994) examined the sensitivity of changes in land cover in CW to changes in crop and animal productivity. The authors showed that deforestation rates are more sensitive to crop yields than to animal productivity. Table 6.10 indicates that deforestation in CW is based on the assumption that it is driven by demand for food and wood products. Hence, in that scenario, deforestation depends on human population, GNP (through income elasticities of food products) and the land's productivity. Land productivity is based on grid-based maps of land cover, climate, soils and region-specific management factors. Further sensitivity analysis should examine all these factors.

In conclusion, it should be noted that one reason for the large differences in land-use scenarios for CO_2, CH_4 and N_2O emissions is the lack of dependable geo-referenced information on land cover and land-use. Systematic land cover and land-use mapping is needed to decrease this uncertainty, to improve emission scenarios, and to develop cost-effective mitigation measures.

6.5 Discussion and Recommendations

In this section, the IS92 Scenarios are discussed in the light of the purposes for scenarios laid out in Section 6.1. It should again be emphasised that time constraints led to a narrowing of the focus for the evaluation. Also, in fairness, it should be noted that other scenarios were not judged by the criteria that were used to judge the IS92 Scenarios. Therefore, the reader should be aware that any shortcomings of the IS92 Scenarios identified in this evaluation may or may not apply to other scenarios.

As a method for judging whether the IS92 Scenarios comply with criteria in Section 6.1, they have been compared to other scenarios. This comparison does not answer all questions about the requirements of the scenarios, but it provides important information about the position of the IS92 Scenarios relative to other work in the scientific community. This section first examines the uses and limitations of the IS92 Scenarios, and then discusses the need for new scenarios and the procedure for developing them.

Table 6.14: Suggested criteria for using emission scenarios for different purposes.

Criteria	Purpose of Scenario (See Box 2, Section 6.1)			
	1	2	3	4
Comprehensive. The scenarios take into account all important greenhouse gases, including precursors of ozone and sulphate aerosol.	xxx	xxx	xxx	xxx
Documentation of output. Sufficient detail is provided for emissions by type, region (if available) and sector, to allow comparisons with other scenarios.	xxx	xxx	xxx	xxx
Emission estimates and input assumptions of the scenario reflect a range of views. A single reference scenario is inadequate for this purpose. Instead, a range of scenario results are presented that take into account the uncertainties of the basic driving forces of the scenarios. Data on driving forces, as well as parameters in the case that a model is used, reflect the current range of views. In addition, derived variables, for example, carbon intensity of the economy, also reflect current understanding.	xxx	xxx	xxx	xxx
Reproducible. Input data and methodology are documented adequately enough to allow other researchers to reproduce the scenarios.	xx	xx	xx	xxx
Sensitivity analysis. Information is provided on the sensitivity of scenario output to variation and uncertainty in input data and model parameters (if a model is used).	xx	xx	xx	xxx
Internal consistency. The various input assumptions and data of the scenarios are internally consistent. For example, assumptions about biomass in the energy sector are consistent with assumptions in the land-use sector.	xx	xx	xxx	xxx
Documentation of assumptions about policy intervention. All assumptions about policy actions are documented adequately enough to allow policy interpretation of the scenario results.	NA	xxx	xxx	xxx
Disaggregation of data. Input and results of scenarios are sufficiently disaggregated into sectors and geographic regions to allow analysis of policies. The geographic breakdown is the country level, but the regional level may be adequate as a first step.	x	x	xxx	xxx
Documentation of input data and methodology. Input data and methodology are sufficiently documented to allow analysis of feasibility and costs of reduction measures.	NA	NA	xxx	xxx
Reference scenario. The set of emission scenarios includes a "reference scenario" for comparison with the policy scenarios in order to ensure consistency.	NA	xxx	xxx	xxx
Temporal calculations. The scenarios describe the time aspect of implementing mitigation measures. This is critical information for assessing the costs of control.	NA	x	xxx	xxx
Range of policy actions. The set of scenarios include a wide range of policy alternatives.	NA	x	x	x

Key to codes:
xxx   Very Important
xx   Important
x   Desirable
NA   Not Applicable

6.5.1 What are the Uses and Limitations of the IS92 Scenarios?

In our judgement, the IS92 Scenarios clearly fulfil the criteria for Purpose 1 listed as "very important" in Table 6.14. Specifically, Sections 6.2 to 6.4 showed that the global results and many of the input assumptions of the IS92 Scenarios span the range of the results and assumptions of other scenarios; in this sense it can be said that the scenarios reflect the current range of views about future emissions, at least on the global level. (For particular regions, however, the IS92 Scenarios do not represent the range of views.) Moreover, the IS92 Scenarios are among the most comprehensive scenarios available in their coverage of all important greenhouse gases and precursors of ozone and sulphate aerosol. Also, the IS92 Scenarios are sufficiently documented in reports and in digital form to allow their comparison with other scenarios and their use as input files for climate models. As a consequence, they are useful for Purpose 1, namely as input to atmosphere/climate models for examining the environmental/climatic consequences of not acting to reduce greenhouse emissions.

It is noted that the IS92 Scenarios do not fulfil all the criteria listed as "important" and "desirable" requirements for Purpose 1 in Table 6.14. The documentation of land-use assumptions concerning afforestation, rice production and animal production may not be adequate to allow other researchers to reproduce the IS92 calculations. Also, the ASF model used to generate the scenarios is not readily available, which affects reproducibility. Only a limited amount of sensitivity analysis work is reported in Pepper *et al.* (1992) for the IS92 Scenarios. The criterion of the internal consistency of the IS92 Scenarios could not be evaluated because there was no access to the ASF model.

Although these shortcomings should be addressed in new scenarios, it is not considered that they detract from the current usefulness of the IS92 Scenarios as input to atmospheric/climate models.

Even though the IS92 Scenarios were designed with Purpose 1 in mind, it is interesting to examine whether they are useful for the other three purposes for scenarios. Table 6.14 ranks the importance of different criteria to fulfil these purposes. There follows a brief review of the suitability of the IS92 Scenarios to fulfil them.

Purpose 2. To evaluate the environmental/climatic consequences of intervention to reduce greenhouse gas emissions. For this purpose, an intervention scenario is devised, then used as input for a climate or similar model to evaluate the scenario's environmental/climatic consequences.

This purpose is basically the same as Purpose 1 except that the intervention scenario takes into account climate policies for reducing greenhouse gas emissions. For example, the intervention scenario might assume that a carbon tax is imposed to reduce CO_2 emissions, or that a programme to reduce CH_4 leakage in natural gas pipelines is carried out.

The IS92 Scenarios make assumptions about controlling local air pollution problems and these measures also have an effect on greenhouse gas emissions. Moreover, IS92b includes information available in 1992 about the emission commitments of OECD countries. This information, however, must be updated with the latest information on commitments in connection with the Climate Convention. In general, it was found that the emission estimates of the IS92 Scenarios are substantially higher than policy scenarios published in the literature. Hence, they should not be used as intervention scenarios.

Purpose 3. To examine the feasibility and costs of mitigating greenhouse gases from different regions and economic sectors, and over time. This purpose can include setting emission reduction targets and developing scenarios to reach these targets. It can also include examining the driving forces of emissions and sinks to identify which of these forces can be influenced by policies.

This purpose requires more information about the scenarios than the first two purposes because its goal is a realistic assessment (technical and financial) of mitigation measures. This purpose also requires estimation of all important greenhouse gases from specific economic sectors and geographic regions. In addition, input assumptions must also be described with a similar level of detail. Under ideal circumstances, these data would be available by country, although for preliminary calculations data at the level of world regions may be adequate.

Analysing the feasibility and costs of mitigation also requires that the scenarios have a high level of internal consistency, so that costs can be accurately assessed. In addition, since the evaluation of mitigation measures must take into account the time period over which measures are put into place, the scenarios must provide information about changing emissions over time rather than for a single point in the future (e.g. emissions in 2025). To enhance the usefulness of scenarios for Purpose 3, it is also recommended that a range of policy alternatives be explored.

Scenarios that fulfil Purpose 3 also fulfil Purpose 2, but scenarios that fulfil Purpose 2 do not necessarily have enough detail to comply with Purpose 3.

The IS92 Scenarios are inadequate for Purpose 3 primarily because five out of six scenarios do not include climate policies to reduce greenhouse gas emissions (they were not intended for that purpose). In addition, the IS92 Scenarios have somewhat limited use as reference scenarios for mitigation studies because their regional emission estimates and regional input assumptions are in many cases outside the range of views found in published scenarios.

Purpose 4. As input for negotiating possible emission reductions for different countries and geographic regions. To achieve this purpose a scenario must gain broad international acceptance by being based on input assumptions and methodology that are widely acknowledged and accepted. Consequently, this scenario has the most stringent requirements.

Purpose 4 is difficult to achieve at this time because of high uncertainty about current greenhouse emissions from many emission categories and in many countries (especially CO_2 in non-energy sectors and other greenhouse gases from virtually all sectors). This uncertainty is one reason why the IS92 Scenarios are not appropriate for this purpose, although they can help set the context for negotiations by indicating the possible range of emissions if no mitigation takes place.

It should be noted that scenarios for Purpose 4 will soon be needed for discussions related to the Climate Convention.

6.5.2 The Question of New Scenarios

At this time, it is recommended that researchers use the IS92 Scenarios as input to atmosphere/climate models for examining the consequences of not intervening to reduce greenhouse gas emissions (Purpose 1).

Should one IS92 Scenario be singled out for use over the others? In some ways IS92a is intermediate among the scenarios. Its global CO_2 emissions fall near the middle of other scenarios at least to 2050, and some of its input assumptions are intermediate. For example, its population assumptions are equal or close to the medium projections recently published by three different international organisations. The sensitivity analysis showed, however, that there are no strong reasons for selecting it as the "medium case" scenario.

An intermediate scenario is not the same as the most likely scenario. Indeed, at this time there is no objective basis for assessing that one scenario is more likely to occur than another. Considering the degree of uncertainty about future emissions, the use of the full range of IS92 Scenarios as input to atmosphere/climate models is recommended, rather than the use of any single scenario.

Users of the IS92 Scenarios are cautioned, however, that the lowest scenario (IS92c) has the emission levels and some input assumptions that are more characteristic of an intervention scenario than a reference scenario.

Because the IS92 Scenarios were not designed and are not suitable for Purposes 2, 3 and 4, new scenarios may be needed to fulfil these purposes. It is worth repeating that this evaluation did not judge the suitability of non-IS92 Scenarios to satisfy these purposes.

New scenarios, if developed, should improve estimates of base year and future non-CO_2 emissions, particularly from the land-use sector.

Although the adequacy of all currently available scenarios was not judged in this study, it is likely that new reference scenarios will be needed for the following:

- to take into account the latest information on economic restructuring, especially in the CIS, Eastern Europe, Asia and Latin America (restructuring in Eastern Europe and the CIS was partly taken into account in the IS92 Scenarios, but it would be worthwhile to include more recent information about this restructuring);
- to evaluate the possible consequences of the Uruguay Round amendments to the General Agreement on Tariffs and Trade;
- to explore a variety of economic development pathways, for example, a closing of the income gap between industrialised and developing regions;
- to examine different trends in technological change;
- to take into account the latest information about current emission commitments in connection with the Climate Convention.

The current commitments to stabilise emissions under the Climate Convention (see Box 4) could result in large reductions of emissions in certain regions, but on a global basis a reduction of 10% or less (in annual emissions relative to the IS92a Scenario) is expected in CO_2 emissions. Current commitments will have no effect on the lowest IPCC Scenario (IS92c) because emissions in this scenario are lower than would be expected by current commitments.

New scenarios are also needed that are consistent with limits on atmospheric concentrations of greenhouse gases by a specified date. These scenarios should be related to the carbon stabilisation modelling exercises being performed by IPCC Working Group I.

6.5.3 The Process of Developing Scenarios

If new or updated scenarios are developed, it is recommended that efforts be guided by the following approach. The IPCC or other suitable organisation should act as an "umbrella" under which different groups can develop comparable, comprehensive emission scenarios. The process for developing scenarios should draw on current experience in harmonising scenarios and model calculations by emphasising the following:

- *Openness.* Scenario development should be open to wide participation by the research community, particularly from developing countries. Openness

> **BOX 4: SCENARIOS OF CURRENT COMMITMENTS**
>
>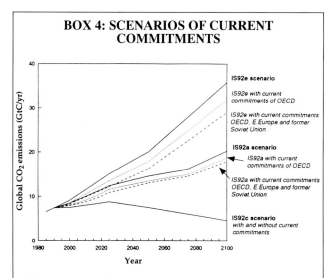
>
> This graph depicts a set of optimistic scenarios for global CO_2 emissions based on current commitments under Article 4 of the Framework Convention on Climate Change. These are optimistic scenarios because they assume that the countries making commitments under the Convention will freeze their emissions after 2005 at their committed levels.
>
> The current commitment scenarios are compared to the lowest, highest and intermediate IPCC reference scenarios. For each reference scenario, two cases of current commitments are shown:
>
> *Case 1:* All OECD countries comply with commitments noted in *Climate Change Policy Initiatives, Volume 1. OECD Countries, 1994 Update* (OECD, 1994) and freeze emissions at these levels after year 2005.
>
> *Case 2:* All OECD countries comply with their commitments as in Case 1, plus countries in Eastern Europe and the former USSR freeze their emissions in 2000 at their 1990 levels.
>
> As a result of these assumptions, CO_2 emissions are reduced by about 8 to 12% (depending on the year considered) relative to the intermediate reference scenario of IPCC (IS92a). Percentage reductions are slightly larger for the highest IPCC scenario (IS92e). Current commitments have no effect on global emissions of the lowest IPCC scenario (IS92c) because emissions in this reference case are lower than they would be for either Case 1 or 2.

requires extensive documentation of modelling assumptions and inputs.
- *Pluralism.* A diversity of groups, approaches and methodologies should be encouraged, although they should be harmonised as noted below.
- *Comparability.* Reporting conventions for input and output should be standardised.
- *Harmonisation.* Input assumptions and methodologies for the scenarios should be harmonised to provide a common benchmark for scenarios from different groups.

6.5.4 Integrated Scenarios

To accomplish Purpose 3 for emission scenarios (estimation of feasibility and costs of mitigation measures), it is important not only to analyse the costs of emission reductions but also to assess the costs and benefits of the impacts of emissions on the natural environment and society, Therefore, the next generation of emission scenarios should be coupled with their impact on the environment and society in the form of consistent, comprehensive and integrated scenarios for global change. These scenarios would cover not only the trend of emissions but also changes in the atmosphere and ocean and the impact of these changes on society, the terrestrial environment and the marine environment. It is recommended that the IPCC encourages the development of integrated scenarios.

6.5.5 Dissemination of Scenarios

Current and new scenarios should be widely disseminated to countries, international organisations, non-governmental organisations and the scientific community. As part of this effort, a central archive should be established to make available the results of new scenarios to any group. The archive should also make available some aspects of the models and the input assumptions used to derive the scenarios. In addition, special effort is needed to improve the capabilities of researchers to analyze and develop scenarios, especially in developing regions. The IPCC can play a role in these activities.

Acknowledgements

For assistance with this chapter, the authors are indebted to Y. Matsuoka, C. MacCracken, N. Nakicenovic, L. Schrattenholzer, S. Toet and M. Wise. We are particularly grateful to Mrs Judy Lakeman of the WGI Technical Support Unit for her careful and thorough work in preparing the text for final publication.

Supplementary Table: Scenarios Reviewed in this Report

Code	Scenario Identification	Type (*)	Reference
1-1	IS92a IPCC 1992	G R	Leggett *et al.* (1992)
1-2	IS92b IPCC 1992	G R	Leggett *et al.* (1992)
1-3	IS92c IPCC 1992	G R	Leggett *et al.* (1992)
1-4	IS92d IPCC 1992	G R	Leggett *et al.* (1992)
1-5	IS92e IPCC 1992	G R	Leggett *et al.* (1992)
1-6	IS92f IPCC 1992	G R	Leggett *et al.* (1992)
1-7	IS92 S1: IPCC 1992 Sensitivity 1 (High Deforestation, High Biomass)	G R	Leggett *et al.* (1992)
1-8	IS92 S4: IPCC 1992 Sensitivity 4 (Halt Deforestation, High Plantation)	G R	Leggett *et al.* (1992)
2-1	Nordhaus 50th% (50th%ile)	G	Nordhaus *et al.* (1983)
2-2	Nordhaus 95th% (95th%ile)	G	Nordhaus *et al.* (1983)
2-3	Nordhaus 5th% (5th%ile)	G	Nordhaus *et al.* (1983)
3-1	Edmonds 50th% (Scenario B, 50th%ile)	G R	Edmonds *et al.* (1986)
3-2	Edmonds 95th% (Scenario B, 95th%ile)	G R	Edmonds *et al.* (1986)
3-3	Edmonds 5th% (Scenario B, 5th%ile)	G R	Edmonds *et al.* (1986)
4	Rogner	G R	Rogner (1986)
5	Mintzer	G R	Mintzer (1988)
6-1	Methane Economy Efficiency Scenario	G	Ausubel *et al.* (1988)
6-2	Methane Economy Long-wave Scenario	G	Ausubel *et al.* (1988)
7-1	EPA Slowly Changing World (SCW)	G R	Lashof and Tirpak (1990)
7-2	EPA Rapidly Changing World (RCW)	G R	Lashof and Tirpak (1990)
7-3	EPA RCW with Accelerated Emissions (RCWA)	G R	Lashof and Tirpak (1990)
7-4	EPA High Reforestation (Halt Deforestation, High Plantation)	G R	Lashof and Tirpak (1990)
7-5	EPA SCW with Stabilisation Policies	G R P	Lashof and Tirpak (1990)
7-6	EPA RCW with Stabilisation Policies	G R P	Lashof and Tirpak (1990)
8-1	IPCC 1990 (SA:Scenario A)	G R	IPCC (1990)
8-2	IPCC 1990 (Scenario A, High)	G R	IPCC (1990)
8-3	IPCC 1990 (Scenario A, Low)	G R	IPCC (1990)
8-4	IPCC-EIS Energy Industry Sub-group 1990 (reference)	G R	IPCC (1990)
9	Ogawa (Institute of Energy Economics, Japan)	G R	Ogawa (1990)
10	Bach	G	Bach and Jain (1991)
11-1	Edmonds-Reilly (Scenario 1)	G R	Barns *et al.* (1991)
11-2	Edmonds-Reilly (Scenario 2)	G R	Barns *et al.* (1991)
12-1	GREEN (1991)	G R	Burineaux *et al.* (1992)
12-2	GREEN (1992)	G R	Burineaux *et al.* (1992)
12-3	GREEN (1993) reference	G R	Oliveira Martins *et al.* (1992)
12-4	GREEN (1993) 200 $/tc carbon tax	G R P	Oliveira Martins *et al.* (1992)
13-1	Global 2100 (Scenario 1)	G R	Manne and Richels (1991)
13-2	Global 2100 (Scenario 2)	G R	Manne and Richels (1991)
13-3	Global 2100 (1992 Scenario)	G R	Manne (1992)
14	TEC	G R	
15	Anderson (World Bank)	G R	Anderson and Bird (1992)
16-1	CHALLENGE reference	G R	Schrattenholzer (1992)
16-2	CHALLENGE 200 $/tc carbon tax	G R P	Schrattenholzer (1992)
17	CRTM	G R	Rutherford (1992)
18	ECS '92 (Dynamics as usual)	G R	
19-1	IEA 1992	G R	Vouyoukas (1992)

19-2	IEA 1993	G R	Vouyoukas (1993)
20-1	IEW 1991	G R	Manne et al. (1991)
20-2	IEW Poll 50th%ile	G R	Manne et al. (1994)
20-3	IEW Poll 84th%ile	G R	Manne et al. (1994)
20-4	IEW Poll 16th%ile	G R	Manne et al. (1994)
21	ITF-D4	G R	Shishido (1991)
22	AGE	G R	Manne et al. (1993)
23-1	AIM High Scenario	G R	Morita et al. (1993)
23-2	AIM Low Scenario	G R	Morita et al. (1993)
23-3	AIM land-use emission scenarios	G R	Matsuoka et al. (1994)
23-4	AIM sulphur emission scenarios	G R	Matsuoka (1992)
24	CETA	G R	Peck et al. (1993)
25	DICE	G	Nordhaus (1993)
26	FUGI 7.0	G R	Onishi (1993)
27-1	IMAGE 2.0 preliminary	G R	Alcamo et al. (1994b)
27-2	IMAGE 2.0 CW (Conventional Wisdom Scenario)	G R	Alcamo et al. (1994b)
27-3	IMAGE 2.0 No Biofuels (No Biofuels Scenario)	G R	Alcamo et al. (1994b)
28-1	Manne-Richels-50th%-(50th%ile)	G R	Manne and Richels (1994)
28-2	Manne-Richels-95th%-(95th%ile)	G R	Manne and Richels (1994)
28-3	Manne-Richels-5th%-(5th%ile)	G R	Manne and Richels (1994)
28-4	Global 2100, 1994	G R	Manne and Schhratenholzer (1994)
29	MERGE	G R	Manne et al. (1993)
30-1	12RT reference	G R	Manne (1993)
30-2	12RT 200 $/tc carbon tax	G R P	Manne (1993)
31-1	WEC reference (Scenario B)	G R	WEC (1993)
31-2	WEC A High Growth	G R	WEC (1993)
31-3	WEC C Ecologically Driven	G R P	WEC (1993)
32	NES current policies	R	NES (1991)
33-1	ESCAPE	G R	Rotmans et al. (1994)
33-2	ESCAPE S1 (Business as Usual)	R	Rotmans et al. (1994)
33-3	ESCAPE S3 (Emission Control)	R P	Rotmans et al. (1994)
34-1	Bashmakov-Base (Russia)	R	Bashmakov (1993)
34-2	Bashmakov-Efficiency	R P	Bashmakov (1993)
35-1	Sinyak DAU (Dynamics as Usual Scenario)	R	Sinyak and Nagano (1992)
35-2	Sinyak Efficiency (Energy Efficiency Scenario)	R P	Sinyak and Nagano (1992)
36-1	EMF-12	R	Gaskins and Weyant (1993)
36-2	EMF-12 lowest/highest reference scenarios	R	Weyant (1993)
36-3	EMF-12 20% 20% emission reduction case	R P	Gaskins and Weyant (1993)
36-4	EMF-12 20% -20% emission reduction case	R P	Weyant (1993)
37	He et al. (Gradual Changing Scenario)	R	He et al. (1993)
38-1	UNEP-base (Baseline Scenario)	R	Christensen et al. (1994)
38-2	USEP-abate (Abatement Scenario)	R P	Christensen et al. (1994)
39-1	Houghton-Population	G	Houghton (1991)
39-2	Houghton-Exponential Extrapolation	G	Houghton (1991)
39-3	Houghton-Reforestation	G	Houghton (1991)
40	RIGES (Renewable Intensive Global Energy System)	G R P	Johansson et al. (1993)
41	FFES (Fossil Free Energy System)	G R P	Lazarus et al. (1993)

* G = Global, R = Regional (National), P = Policy Scenarios

References

Alcamo, J., R. Shaw, and L. Hordijk (eds), 1990: *The RAINS Model of Acidification: Science and Strategies in Europe*. Kluwer Academic Publishers, Dordrecht/Boston/London.

Alcamo, J., G.J.J. Kreileman, M.S. Krol and G. Zuidema, 1994a: Modeling the global society-biosphere-climate system: Part 1: Model description and testing. *Water, Air, Soil Pollution*, **76**, 1-35.

Alcamo, J., G.J. van den Born, A.F. Bouwman, B.J. de Haan, K. Klein Goldewijk, O. Klepper, J. Krabrc, R. Leemans, J.G.J. Olivier, A.M.C. Toet, H.J.M. de Vries and H.J. van der Woerd, 1994b: Modeling the global society-biosphere-climate system: Part 2: Computed scenarios. *Water, Air, Soil Pollution*, **76**, 37-78.

Anderson, D. and C.D. Bird, 1992: Carbon accumulations and technical progress - A Simulation study of costs. *Oxford Bulletin of Economics and Statistics*, **54**(1).

Ajtay, G.L., P. Ketner and P. Duvigneaud, 1979: Terrestrial primary production and phytomass. In: *The Global Carbon Cycle*, B. Bolin, S. Kempe and P. Ketner (eds.), SCOPE Vol. 13, Wiley and Sons, New York, pp129-181.

Ausubel, J.H. and W.D. Nordhaus. 1983: Review of estimates of future carbon dioxide emissions. In: *Changing Climate*, Report of the Carbon Dioxide Assessment Committee, 153-185, National Academy Press, Washington DC.

Ausubel, J.H., A. Grübler and N. Nakicenovic, 1988: Carbon dioxide emissions in a methane economy. *Climatic Change*, **12**, 245-263.

Bacastow, R. and C. Keeling, 1981: Hemispheric airborne fractions difference and the hemispheric exchange times. In: *Carbon Cycle Modeling*, B. Bolin (ed.), SCOPE 16, Wiley, New York.

Bach, W. and A.K. Jain, 1991: The global warming challenge. *CMDC*, 65-124.

Barns, D.W., J.E. Edmonds and J.M. Reilly, 1991: Use of the Edmonds-Reilly Model to model energy related greenhouse gas emissions for inclusion in an OECD survey volume. OECD Workshop on Cost of Reducing CO_2 Emissions: Evidence from Global Models, 10-11 September, 1991, Paris.

Bashmakov, I., 1993: *The System of Statistical Indexes for World Energy*. Moscow Center for Energy Efficiency, Moscow (in Russian).

Boden, T.A., R.J. Sepanski and F.W. Stoss, 1992: *Trends '91: A Compendium of Data on Global Change*. ORNL CDIAC-46, Carbon Dioxide Information Center, Oak Ridge National Laboratory, Oak Ridge, TN.

Bolle, H.J., W. Seiler and B. Bolin, 1986: Other greenhouse gases and aerosols: assessing their role in atmospheric radiative transfer. In *The Greenhouse Effect, Climatic Change And Ecosystems*, B. Bolin, B.R. Döös, J. Jäger and R.A. Warrick (eds.), pp157-203, SCOPE Vol.29, Wiley and Sons, New York.

Bouwman, A.F., I. Fung, E. Matthews and J. John. 1993: Global analysis of the potential for N_2O production in natural soils. *Global Biogeochem. Cycles*, **7**, 557-597.

Bouwman, A.F., K.W. van der Hoek and J.G.J. Olivier, 1994: Uncertainty in the global source distribution of nitrous oxide. *J. Geophys. Res.* (in press)

Burniaux, J.M., G. Nicoletti and J.O. Martins, 1992: *GREEN: A Global Model for Quantifying the Costs of Policies to Curb CO_2 Emissions*. OECD Economic Studies No. 19, OECD, Paris.

Christensen, J.M., K. Halsnaes, G.A. Mackenzie, J. Swisher and A. Villavicencio, 1994: *UNEP Greenhouse Gas Abatement Costing Studies*. UNEP Collaborating Centre on Energy and Environment, Riso National Laboratory, Roskilde, Denmark.

Cicerone, R.J. and R.S. Oremland, 1988: Biogeochemical aspects of atmospheric methane. *Global Biogeochem. Cycles*, **2**, 299-327.

Cleland, J., 1994: Fertility trends in developing countries: 1960 to 1990. In: *Alternative Paths of Future World Population Growth*, W.Lutz (ed.), International Institute for Applied Systems Analysis, Laxenburg, Austria.

Crutzen, P.J. and M.O. Andreae, 1990: Biomass burning in the tropics: impact on atmospheric chemistry and biogeochemical cycles. *Science*, **250**, 1669-1678.

Crutzen, P.J., I. Aselmann and W. Seiler, 1986: Methane production by domestic animals, wild ruminants, other herbivorous fauna and humans. *Tellus*, **38B**, 271-284.

De Vries, H.J.M., J.G.J. Olivier, R.A. van den Wijngaart, G.J.J. Kreileman and A.M.C. Toet, 1994: Model for calculating regional energy use, industrial production and greenhouse gas emissions for evaluating global climate scenarios. *Water, Air, Soil Pollution*, **76**, 79-131.

DOAI (Department of Agriculture, Indonesia), 1993: *The Projection of the Target Rice Productivity and Planted Area of Paddy Field in the Second Long Term Development Plan of Indonesia*. Directorate of Food Crop Programme Development, Department of Agriculture, Indonesia.

ECE (United Nations Economic Commission for Europe), 1993: *Enhancing Energy Efficiency in the ECE Region: Recent Developments, Policies, International Trade and Co-operation*. ENERGY/R.88 12 November 1993, ECE, Geneva.

Edmonds, J.A., 1993: Carbon coalitions: the cost and effectiveness of energy agreements to alter trajectories of atmospheric carbon dioxide emissions, Climate Treaties and Models: Office of Technology Assessment (OTA) Workshop April 23, 1993, OTA, Washington DC.

Edmonds, J.A. and D.W. Barns, 1992: Factors affecting the long-term cost of fossil fuel CO_2 emissions reductions. *Int. Journal of Global Energy Issues*, **4**, No. 3, 140-166.

Edmonds, J., J. Reilly, J.R. Trabalka, D.E. Reichle, D. Rind, S. Lebedeff, J.P. Palutikof, T.M.L. Wigley, J.M. Lough, T.J. Blasing, A.M. Salomon, S. Seidel, D. Keyes and M. Steinberg, 1986: *Future Atmospheric Carbon Dioxide Scenarios and Limitation Strategies*. Noyes Publications, Park Ridge, NJ.

ESCAP (United Nations Economic and Social Commission for Aisia and the Pacific), 1991: *Climate Effects of Fossil Fuel Use in the Asia Pacific Region*. ST/ESCAP/1007, ESCAP, Bangkok.

FAO, 1993: *Agriculture Towards 2010*. Report C 93/24, November 1993, FAO, Rome, Italy.

Gaskins, D.W. and J.P. Weyant, 1993: Model comparison of the costs of reducing CO_2 emissions. *American Economic Review*, **83**(2), 318-323.

Gibbs, M.J. and R.A. Leng, 1993: Methane emissions from livestock. In: *Proceedings of the International Workshop on*

Methane and Nitrous Oxide. Methods for National Inventories and Options for Control, A. Van Amstel, (ed.), RIVM report 481507003, National Institute of Public Health and Environmental Protection, Bilthoven, The Netherlands, pp73-79.

Gibbs, M.J. and J.W. Woodbury, 1993: Methane emissions from livestock manure. In: *Proceedings of the International Workshop on Methane and Nitrous Oxide. Methods for National Inventories and Options for Control*, A. Van Amstel, (ed.), RIVM report 481507003, National Institute of Public Health and Environmental Protection, Bilthoven, The Netherlands, pp81-91.

Grübler, A. 1994: *A Comparison of Global and Regional Emission Scenarios*. Working Paper, IIASA, Laxenburg, Austria. (In press.)

Grübler, A., S. Messner, L. Schrattenholzer and A. Schäfer, 1993: Emission reduction at the global level. *Energy,* **18**(5), 539-581.

He, J., Z. Wei and Z. Wu, 1993: Study of China's energy system for reducing CO_2 emission. In: *Costs, Impacts, and Benefits of CO_2 Mitigation*, Y. Kaya, N. Nakicenovic, W.D. Nordhaus and F.L. Toth (eds.), CP-93-2, IIASA, Laxenburg, Austria, pp485-506.

Heilig, G. 1994: How Many People Can Be Fed On Earth? In: *Alternative Paths of Future World Population Growth*, W. Lutz (ed.), International Institute for Applied Systems Analysis, Laxenburg, Austria.

Houghton, R.A., 1991: Tropical deforestation and atmospheric carbon dioxide. *Climatic Change,* **19**, 99-118.

Houghton, R.A., R.D. Boone, J.M. Melillo, C.A. Palm, G.M. Woodwell, N. Myers, B. Moore III and D.L. Skole, 1985: Net flux of carbon dioxide from tropical forests in 1980. *Nature,* **316**, 617-620.

IPCC, 1990: *Climate Change: The Scientific Assessment*. J.T. Houghton, G.J. Jenkins and J.J. Ephraums (eds.), Cambridge University Press, Cambridge, UK.

IPCC, 1992: *Climate Change 1992: The Supplementary Report to the IPCC Scientific Assessment*, J.T. Houghton, B.A. Callander and S.K. Varney (eds). Cambridge University Press, Cambridge, UK.

IPCC-EIS (Intergovernmental Panel on Climate Change, Energy and Industry Subgroup), 1990: *Energy and Industry Subgroup Report May 31, 1990*. IPCC, Geneva (21P-2001 US EPA, Washington DC.).

Jason, A. 1979: *Long-term Impact of Atmospheric Carbon Dioxide on Climate*, JSR-78-07, SRI International, Arlington, VA.

JEE (Japan Environment Agency), 1993: *The Study on the Response Actions against the Increasing Emission of Carbon Dioxide in Indonesia*. Final report. Japan Environment Agency, Overseas Environment Cooperation Centre and Ministry of State for Population and Environment.

Johansson, T.B., H. Kelly, A.K.N. Reddy and R.H. Williams, 1993: A renewables-intensive global energy scenario. In: *Renewable Energy: Sources for Fuels and Electricity*, T.B. Johansson, H. Kelly, A.K.N. Reddy and R.H. Williams (eds.), Island Press, Washington DC, pp1071-1142.

Kaya, Y., 1990: Impact of Carbon Dioxide Emission Control on GNP Growth: Interpretation of Proposed Scenarios. Paper presented to the IPCC Energy and Industry Subgroup, Response Strategies Working Group, Paris, France (mimeo).

Keller, M., E. Veldkamp, A.M. Weitz and W.A. Reiners, 1993: Pasture age effects on soil-atmosphere trace gas exchange in a deforested area of Costa Rica. *Nature*, **365**, 244-246.

Klein-Goldewijk, K., J.G. Van Minnen, G.J.J. Kreileman, M. Vloedbeld and R. Leemans, 1994: Simulating the carbon flux between the terrestrial environment and the atmosphere, *Water, Air and Soil Pollution* **76**, 199-230.

Lashof, D.A., 1991: EPA's scenarios for future greenhouse gas emissions and global warming. *Energy Journal,* **12**(1), 125-146.

Lashof, D.A. and D.A. Tirpak, 1990: *Policy Options for Stabilizing Global Climate*. 21P-2003, US Environmental Protection Agency, Washington DC.

Lazarus, M., L. Greber, J. Hall., C. Bartels, S. Bernow, E. Hansen, P. Raskin and D. von Hippel, 1993: Towards a fossil free energy future in the next energy transition: A technical analysis for Greenpeace International. Stockholm Environment Institute - Boston Center, Boston.

Leggett, J., W.J. Pepper and R.J. Swart, 1992: Emission Scenarios for the IPCC: an Update. In: *Climate Change 1992: The Supplementary Report to the IPCC Scientific Assessment*, J.T. Houghton, B.A. Callander and S.K. Varney (eds.), Cambridge University Press, Cambridge, UK.

Lerner, E., E. Matthews and I. Fung, 1988: Methane emission from animals: a global high resolution database. *Global Biogeochem. Cycles*, **2**, 139-156.

Luizao, F., P. Matson, G. Livingston, R. Luizao and P. Vitousek, 1989: Nitrous oxide flux following tropical land clearing. *Global Biogeochem.Cycles*, **3**, 281-285.

Lutz, W., 1994: Future world population growth. *Population Bulletin*. (In press)

Lutz, W., C. Prinz, J. Langgassner, 1994: The IIASA world population scenarios. In: *Alternative Paths of Future World Population Growth*, W. Lutz, (ed.), International Institute for Applied Systems Analysis, Laxenburg, Austria.

Manne, A., 1992: Global 2100: Alternative scenarios for reducing carbon emissions. *Economics Department Working Papers*, (111), OECD, Paris.

Manne, A. 1993: International trade - the impact of unilateral carbon emission limits. *The Economics of Climate Change*. OECD/IEA, Paris.

Manne, A. and R.G. Richels, 1991: Global CO_2 emission reductions - the impacts of rising energy costs. *Energy Journal,* **12**, 87-101.

Manne, A. and T.F. Rutherford, 1993: International trade in oil, gas and carbon emission rights: An intertemporal general equilibrium model. In: *Cost, impacts, and benefits of CO_2 mitigation*, Y. Kaya *et al.* (eds.) Proceedings of a Workshop held on 28-30 September 1992 at IIASA, Laxenburg, Austria, pp315-340.

Manne, A. and L. Schrattenholzer. 1993: Global scenarios for carbon disoxide emissions. *Energy,* **18**(12), 1207-1222.

Manne, A. and R. Richels, 1994: *The Costs of Stabilizing Global CO_2 Emissions - a Probabilistic Analysis Based on Expert Judgements*. Electric Power Research Institute, Palo Alto, CA.

Manne, A. and L. Schrattenholzer. 1994: *International Energy Workshop, Part 1: Overview of Poll Responses, Part 2: Frequency Distributions, Part 3: Individual Poll Responses*. IIASA, Laxenburg, Austria.

Manne, A., L. Schrattenholzer and K. Marchant, 1991: The 1991

International Energy Workshop: the poll results and a review of papers. *OPEC Review*, **XV**(4), 389-411.

Manne, A., R. Mendelsohn and R. Richels, 1993: MERGE - A model for evaluating regional and global effects of GHG reduction policies. *Papers of International Workshop on Integrative Assessment of Mitigation, Impacts and Adaptation to Climate Change,* 13-15 October, 1993, IIASA, Laxenburg, Austria.

Marland, G., 1993: *National, Regional and Global CO_2 Emission Estimates 1950-1991. NDP-030/R5,* Carbon Dioxide Information Center, Oak Ridge National Laboratory, Oak Ridge, TN.

Marland, G., T.A. Boden, R..C. Griffin, S.F. Huang, P. Kanciruk and T.R. Nelson, 1989: Estimates of CO_2 emissions from fossil fuel burning and cement manufacturing. Based on the United Nations Energy Statistics and the U.S. Bureau of Mines Cement Manufacturing Data. *ORNL/CDIAC-25 NDP-030*, Oak Ridge National Laboratory, Oak Ridge, TN.

Matsuoka, Y., 1992: Future projection of global anthropogenic sulphur emission and its environmental effect. *Environmental Systems Research*, **20**, 142-151.

Matsuoka, Y. and T. Morita, 1994: *Estimation Of Carbon Dioxide Flux From Tropical Deforestation.* Center for Global Environmental Research, National Institute for Environmental Studies, CGER-I013-'94.

Matsuoka, Y., M. Kainuma and T. Morita, 1993: On the uncertainty of estimating global climate change. In: *Costs, Impacts, and Benefits of CO_2 Mitigation,.* Y. Kaya, N. Nakicenovic, W.D. Nordhaus and F.L. Toth (eds.), 371-384, CP-93-2, IIASA, Laxenburg, Austria.

Messner, S. and M. Strubegger, 1991: Potential effects of emission taxes on CO_2 emissions in the OECD and LDCs. *Energy*, **16**(11/12), 1379-1395.

Mintzer, I., 1988: *Projecting Future Energy Demand In Industrialized Countries: An End-Use Oriented Approach.* World Resource Institute, Washington, DC.

Morita, T., Y. Matsuoka, M. Kanuma, H. Harasawa and K. Kai, 1993: AIM - Asian-Pacific integrated model for evaluating policy options to reduce GHG emissions and global warming impacts. Paper for Workshop on Global Warming Issues in Asia, 8-10 September 1993, Bangkok, Thailand. In: . *1994: Global Warming Issue In Asia,* S.C. Bhattacharya et al. (eds.), Asian Institute of Technology, Bangkok, Thailand, pp254-273.

Morita, T., Y. Matsuoka, I. Penna, M. Kainuma. 1994: *Global Carbon Dioxide Emission Scenarios and Their Basic Assumptions: 1994 Survey.* CGER-1011-94. Center For Global Environmental Research, National Institute for Environmental Studies, Tsukuba, Japan.

Myers, N. 1991: Tropical forests: present status and future outlook. *Climatic Change* **19**, 3-32.

Nakicenovic, N., A. Grübler, A. Inaba, S. Messner, S. Nilsson, Y. Nishimura, H.-H. Rogner, A. Schäfer, L. Schrattenholzer, M. Strubegger, J. Swisher, D. Victor and D. Wilson, 1993: Long-term strategies for mitigating global warming. *Energy*, **18**(5), 401-609.

NAPAP (The U.S. National Acid Precipitation Assessment Program), 1991. *Acid Deposition: State of Science and Technology,* The NAPAP Office of the Director, 722 Jackson Place, NW, Washington, DC.

NES (National Energy Strategy), 1991: *National Energy Strategy: Powerful Ideas for America.* US Department of Energy, Washington, DC.

Nordhaus, W.D., G.W. Yohe, 1983: Future paths of energy and carbon dioxide emissions. In: *Changing Climate: Report of the Carbon Dioxide Assessment Committee,* National Academy Press, Washington, DC.

Nordhaus, W.D., 1994: Rolling the DICE: An optimal transition path for controlling greenhouse gases. *Quarterly Journal of Economics.*

OECD., 1991: Estimation of greenhouse gas emissions and sinks. OECD experts meeting, 18-21 February 1991, revised in August 1991. Prepared for the Intergovernmental Panel on Climate Change.

Ogawa, Y., 1990: Global warming issues and future response. *Proceedings of symposium on Energy and Economics 6-7 December 1990.* Institute for Energy Economics, Tokyo (in Japanese).

Ogawa, Y., 1991: Economic activity and the greenhouse effect. *Energy Journal,* **12**(1), 23-34.

Okada, K. and K. Yamaji, 1993: Simulation study on tradable CO_2 emission permits. In: *Cost, impacts, and benefits of CO_2 mitigation,* Y. Kaya et al. (eds.), Proceedings of a Workshop held on 28-30 September 1992 at IIASA, Laxenburg, Austria, pp341-352.

Oliveira Martins, J., J.M. Burniaux, J.P. Martin and G. Nicoletti, 1992: The cost of reducing CO_2 emissions: a comparison of carbon tax curves with GREEN. *Economics Department Working Papers*, (118), OECD, Paris.

Olson, J.S., J.A. Watts and L.J. Allison, 1983: Carbon in live vegetation of major world ecosystems. ORNL 5862, Environmental Sciences Division, Publication No. 1997, Oak Ridge National Laboratory, Oak Ridge, Tennessee. National Technical Information Service, U.S. Department of Commerce.

Onishi, A., 1993: FUGI Global Model 7.0 - Economic financial computing. *A Journal of the European Economic and Financial Centre*, **3**(1), Spring, 1-67.

Parikh, J.K., 1992: IPCC strategies unfair to the South. *Nature*, **360**, 507-508.

Peck, S.C. and T.J. Tiesberg. 1993: Global warming uncertainties and the value of information: An analysis using CETA. *Resource and Energy Economics*, **15**(1), 71-97.

Pepper, W., J. Leggett, R. Swart, J. Wasson, J. Edmonds and I. Mintzer. 1992: *Emission Scenarios for the IPCC, An Update, Assumptions, Methodology, and Results.* Prepared for the Intergovernmental Panel on Climate Change, Working Group I.

Rogner, H.H., 1986: Long-term energy protections and novel energy systems. In: *The Changing Global Carbon Cycle: A Global Analysis,* J.R. Trabalka and D.E. Reichle (eds.), Springer-Verlag, NY.

Rotmans, J., M. Hulme and T.E. Downey, 1994: Climate change implications for Europe: an application of the ESCAPE model. *Global Environmental Change.*(In press)

Rutherford, T.F., 1992: The welfare effects of fossil carbon reductions: Results from a recursively dynamic trade model. *Economics Department Working Papers*, (112), OECD/GD(92)89, Paris.

Schrattenholzer, L., 1994: Guest editorial: Global carbon emissions and energy scenarios based on the results of country studies. *Int. Journal of Global Energy Issues*, **6**(1/2), 1-8.

Shishido, S., 1991: Global impacts of carbon tax: A simulation analysis with a global econometric model. mimeo.

Seiler, W. and P.J. Crutzen, 1980: Estimates of gross and net fluxes of carbon between the biosphere and the atmosphere from biomass burning. *Climatic Change* **2**: 207-247.

Siegenthaler, U. and H. Oeschger, 1978: Predicting future atmospheric carbon dioxide levels. *Science*, **199**, 388-395.

Sinyak, Y. and K. Nagano, 1992: *Global Energy Strategies to Control Future Carbon Dioxide Emissions.* SR-92-04, IIASA, Laxenburg, Austria.

Turner, B.L., R.H. Moss and D.L. Skole, 1993: *Relating Land Use and Global Land Cover Change: A Proposal for an IGBP-HDP Core Project.* IGBP Report No. 24. Stockholm.

UN 1987: *Fertility Behavior in the Context of Development.* Population Studies 100.ST/ESA/SER.A/100. Dept of International Economic and Social Affairs: New York.

UN1992: *Long Range World Population Projections.* United Nations Population Division, New York.

USBC(United States Bureau of Census),1987: World Population Profile: 1987. U.S. Department of Commerce, Washington, DC.

Vouyoukas, L., 1992: Carbon taxes and CO_2 emission targets: result from the IEA model. *OECD Economics Department Working Paper*, (114).

WEC (World Energy Council) (1993). *Energy for Tomorrow's World.* Kogan Page Ltd., London.

Westoff, C.F., Reproductive preferences and future fertility in developing countries. In:*Alternative Paths of Future World Population Growth,* W. Lutz,(ed.), International Institute for Applied Systems Analysis, Laxenburg, Austria.

Weyant, J.P. 1993: Costs of reducing global carbon emissions. *Journal of Economic Perspectives*, **7**(4), 7-46.

World Bank, 1991: *World Development Report 1991.* Oxford University Press, New York.

Zuidema, G., G.J. Van Den Born, J. Alcamo, and G.J.J. Kreileman, 1994: Simulation land cover changes as affected by economic factors and climate, *Water, Air and Soil Pollution*, **76**, 163-198.

Appendix 1

ORGANISATION OF IPCC

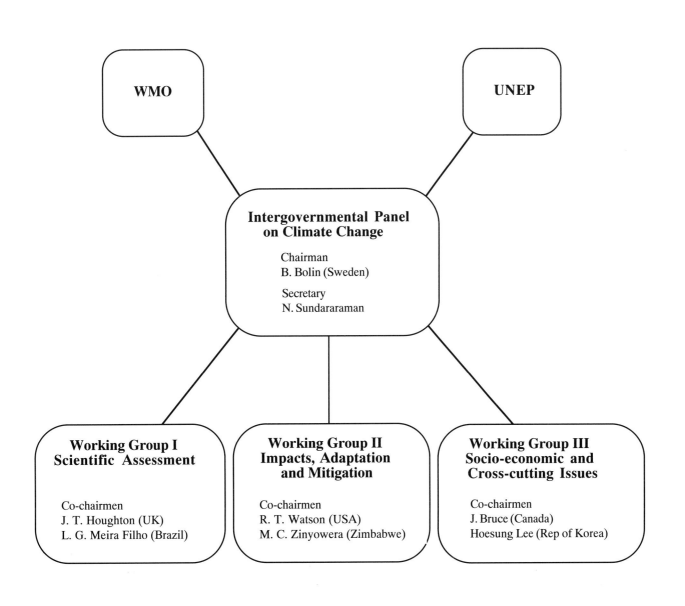

Appendix 2

LIST OF MAJOR IPCC REPORTS (in English unless otherwise stated)

Climate Change - The IPCC Scientific Assessment. The 1990 report of the IPCC Scientific Assessment Working Group (*also in Chinese, French, Russian and Spanish*).

Climate Change - The IPCC Impacts Assessment. The 1990 report of the IPCC Impacts Assessment Working Group (*also in Chinese, French, Russian and Spanish*)

Climate Change - The IPCC Response Strategies. The 1990 report of the IPCC Response Strategies Working Group (*also in Chinese, French, Russian and Spanish*)

Emissions Scenarios. Prepared for the IPCC Response Strategies Working Group, 1990.

Assessment of the Vulnerability of Coastal Areas to Sea Level Rise - a Common Methodology, 1991 (*also in Arabic and French*).

Climate Change 1992 - The Supplementary Report to the IPCC Scientific Assessment. The 1992 report of the IPCC Scientific Assessment Working Group.

Climate Change 1992 - The Supplementary Report to the IPCC Impacts Assessment. The 1992 report of the IPCC Impacts Assessment Working Group.

Climate Change: The IPCC 1990 and 1992 Assessments - IPCC First Assessment Report Overview and Policymaker Summaries, and 1992 IPCC Supplement.

Global Climate Change and the Rising Challenge of the Sea. Coastal Zone Management Subgroup of the IPCC Response Strategies Working Group, 1992.

Report of the IPCC Country Studies Workshop, 1992.

Preliminary Guidelines for Assessing Impacts of Climate Change, 1992.

IPCC Guidelines for National Greenhouse Gas Inventories (3 volumes), 1994 (*also in French, Russian and Spanish*).

IPCC Technical Guidelines for Assessing Climate Change Impacts and Adaptations, 1995 (*also in Arabic, Chinese, French, Russian and Spanish*).

Enquiries: IPCC Secretariat, c/o World Meteorological Organisation, P O Box 2300, CH1211, Geneva 2, Switzerland.

Appendix 3

CONTRIBUTORS TO IPCC WG1 REPORT, 1994, ON THE RADIATIVE FORCING OF CLIMATE CHANGE

CHAPTER 1: CO_2 AND THE CARBON CYCLE

Convening Lead Author
David Schimel — National Center for Atmospheric Research, USA

Lead Authors
D. Alves — Instituto Nacional de Pesquisas Espaciais (INPE), Brazil
I.G. Enting — CSIRO Division of Atmospheric Research, Australia
M. Heimann — Max-Planck Institut für Meteorologie, Germany
D. Raynaud — CNRS - Laboratoire de Glaciologie, France
U. Siegenthaler — University of Bern, Switzerland
T. Wigley — Office for Interdisciplinary Earth Studies @ UCAR, USA

Contributors
S. Brown — University of Illinois, USA
W. Emanuel — Oak Ridge National Laboratory, USA
M. Fasham — James Rennell Centre (NERC), UK
C. Field — Carnegie Institute of Washington, USA
P. Friedlingstein — BISA, Belgium
R. Gifford — CSIRO Division of Plant Industry, Australia
R. Houghton — Woods Hole Research Center, USA
A. Janetos — NASA Headquarters, USA
S. Kempe — University of Hamburg, Germany
R. Leemans — RIVM (Netherlands Institute for Public Health and Environment), Netherlands
E. Maier-Reimer — Max-Planck Institut für Meteorologie, Germany
G. Marland — Oak Ridge National Laboratory, USA
R. McMurtrie — University of New South Wales, Australia
J. Melillo — Woods Hole Oceanographic Institution, USA
J. F. Minster — GRGS, France
P. Monfray — Centre des Faibles Radioactivités, France
M. Mousseau — Université de Paris Sud, France
D. Ojima — Colorado State University, USA
D. Peel — British Antarctic Survey, UK
D. Skole — University of New Hampshire, USA
E. Sulzman — National Center for Atmospheric Research, USA
P. Tans — NOAA, Climate Monitoring & Diagnostics Laboratory, USA

I. Totterdell	Chilworth Research Centre, UK
P. Vitousek	Stanford University, USA

Modellers

J. Alcamo	RIVM (Netherlands Institute for Public Health and Environment), Netherlands
B.H. Braswell	National Center for Atmospheric Research, USA
B.C. Cohen	UNECE Committee on Energy, Switzerland
W.R. Emanuel	Oak Ridge National Laboratory, USA
I.G. Enting	National Center for Atmospheric Research, USA
G.D. Farquhar	Australian National University, Australia
R.A. Goldstein	Electric Power Research Inst, USA
L.D.D. Harvey	University of Toronto, Canada
M. Heimann	Max-Planck Institut für Meteorologie, Germany
A. Jain	Lawrence Livermore University, USA
F. Joos	University of Bern, Switzerland
J. Kaduk	Max-Planck Institut für Meteorologie, Germany
A.A. Keller	Electric Power Research Inst, USA
M. Krol	RIVM (Netherlands Institute for Public Health and Environment), Netherlands
K. Kurz	Max Planck Institut für Meteorologie, Germany
K.R. Lassey	National Inst for Water and Atmospheric Research, New Zealand
C. Le Quere	Princeton University, USA
J. Lloyd	Australian National University, Australia
E. Meier-Reimer	Max-Planck Institut für Meteorologie, Germany
B. Moore III	University of New Hampshire, USA
J. Orr	Laboratoire de Modelisation du Climate et de l'Environment, France
T.H. Peng	Oak Ridge National Laboratory, USA
J. Sarmiento	Geophysical Fluid Dynamics Laboratory, USA
U. Siegenthaler	University of Bern, Switzerland
J. A. Taylor	Australian National University, Australia
J. Viecelli	Lawrence Livermore National Laboratory, USA
T.M.L. Wigley	Office for Interdisciplinary Earth Studies @ UCAR, USA
D. Wuebbles	Lawrence Livermore National Laboratory, USA

CHAPTER 2: OTHER TRACE GASES AND ATMOSPHERIC CHEMISTRY

Convening Lead Author

Michael Prather	University of California @ Irvine, USA

Lead Authors

R. Derwent	Meteorological Office, UK
D. Ehhalt	Institut für Chemie der KFA Jülich GmbH, Germany
P. Fraser	CSIRO Division of Atmospheric Research, Australia
E. Sanhueza	Instituto Venezolano de Investigaciones Cientificas, Venezuela
X. Zhou	Academy of Meteorological Sciences, China

Contributors

F. Alyea	Georgia Institute of Technology, USA
J. Bradshaw	Georgia Institute of Technology, USA
J. Butler	NOAA, Climate Monitoring & Diagnostics Laboratory, USA
M.A. Carroll	University of Michigan, USA
D. Cunnold	Georgia Institute of Technology, USA
E. Dlugokencky	NOAA ERL @ Boulder, USA
J. Elkins	NOAA ERL @ Boulder, USA

Appendix 3

D. Etheridge	CSIRO Division of Atmospheric Research, Australia
D. Fisher	DuPont, Wilmington, USA
P. Guthrie	Systems Applications International, USA
N. Harris	European Ozone Research Coordination Unit, UK
I. Isaksen	University of Oslo (Geophysics), Norway
D.J. Jacob	Harvard University, USA
C.E. Johnson	Harwell Laboratory, UK
J. Kaye	NASA Headquarters, USA
S. Liu	NOAA Aeronomy Laboratory, USA
C.T. McElroy	Atmospheric Environment Service (ARQX), Canada
P. Novelli	NOAA, Climate Monitoring & Diagnostics Laboratory, USA
J. Penner	Lawrence Livermore National Laboratory, USA
R. Prinn	Massachusetts Institute of Technology, USA
W. Reeburgh	University of California @ Irvine, USA
B. Ridley	National Center for Atmospheric Research, USA
J. Rudolph	Institut für Atmosphärische Chemie, Germany
P. Simmonds	Bristol University, UK
L.P. Steele	CSIRO Division of Atmospheric Research, Australia
F. Stordal	Norwegian Institute for Air Research, Norway
R. Weiss	Scripps Institute of Oceanography, USA
A. Volz-Thomas	Institut für Chemie der Belasteten, Germany
A. Wahner	Institut für Chemie der KFA Jülich GmbH, Germany
D. Wuebbles	Lawrence Livermore National Laboratory, USA

CHAPTER 3: AEROSOLS

Convening Lead Author
Peter R. Jonas   UMIST, UK

Lead Authors
R.J. Charlson   University of Washington, USA
H. Rodhe   University of Stockholm, Sweden

Contributors
T.L. Anderson   University of Washington, USA
M.O. Andreae   Max-Planck Institut für Chemie, Germany
E. Dutton   NOAA Climate Monitoring & Diagnostics Laboratory, USA
H. Graf   Max Planck Institut für Meteorologie, Germany
Y. Fouquart   LOA/Université des Science & Technologie de Lille, France
H. Grassl   Max-Planck Institut für Meteorologie, Germany
J. Heintzenberg   University of Stockholm, Sweden
P.V. Hobbs   University of Washington, USA
D. Hofmann   NOAA Climate Monitoring & Diagnostics Laboratory, USA
B. Huebert   University of Hawaii, USA
R. Jaenicke   Johannes Gutenberg-Universität, Mainz, Germany
M. Jietai   Peking University, China
J. Lelieveld   University of Wageningen, Netherlands
M. Mazurek   Brookhaven National Laboratory, USA
M.P. McCormick   Langley Research Center, USA
J. Ogren   NOAA Climate Monitoring & Diagnostics Laboratory, USA
J. Penner   Lawrence Livermore National Laboratory, USA
F. Raes   CEC Joint Research Centre, Italy
L. Schütz   Johannes Gutenberg Universität, Mainz, Germany

S. Schwartz	Brookhaven National Laboratory, USA
G. Slinn	Pacific Northwest Laboratories, USA
H. ten Brink	Netherland Energy Research Foundation, Netherlands

CHAPTER 4: RADIATIVE FORCING

Convening Lead Author

Keith P. Shine	University of Reading, United Kingdom

Lead Authors

Y. Fouquart	LOA/Université des Science & Technologie de Lille, France
V. Ramaswamy	Geophysical Fluid Dynamics Laboratory, USA
S. Solomon	NOAA Aeronomy Laboratory, USA
J. Srinivasan	Langley Research Center, USA

Contributors

M.O. Andreae	Max-Planck Institut für Chemie, Germany
J. Angell	NOAA ERL @ Silver Springs, USA
G. Brasseur	National Center for Atmospheric Research, USA
C. Brühl	Max-Planck Institut für Chemie, Germany
R.J. Charlson	University of Washington, USA
M.D. Chou	NASA, Goddard Space Flight Centre, USA
J.R. Christy	University of Alabama @ Huntsville, USA
T. Dunkerton	Northwest Research Associates, USA
E. Dutton	NOAA Climate Monitoring & Diagnostics Laboratory, USA
B.A. Fomin	Institute of Molecular Physics, USA
C. Granier	NCAR, USA
H. Grassl	Max-Planck Institut für Meteorologie, Germany
J. Hansen	Goddard Institute for Space Studies, USA
Harshvardhan	Purdue University, USA
D. Hauglustaine	Service d'Aeronomie du CNRS, France
P. Hobbs	University of Washington, USA
D.J. Hofman	NOAA Climate Monitoring & Diagnostics Laboratory, USA
L. Hood	University of Arizona, USA
N. Husson	CNRS, Laboratoire de Météorologie Dynamique, France
I. Karol	Main Geophysical Observatory, Russia
Y.J. Kaufman	NASA, Goddard Space Flight Centre, USA
J. Kiehl	National Center for Atmospheric Research, USA
S. Kinne	NASA Ames Research Center, USA
M.K.W. Ko	Atmosphere & Environment Research Inc, USA
K. Labitzke	Free University of Berlin, Germany
H. Le Treut	CNRS, Laboratoire de Météorologie Dynamique, France
A. McCulloch	AFEAS / ICI C&P, UK
A.J. Miller	NOAA National Meteorological Center, USA
M. Molina	Massachusetts Institute of Technology, USA
E. Nesme-Ribes	Observatoire de Paris, France
A.H. Oort	Geophysical Fluid Dynamics Laboratory, USA
J.E. Penner	Lawrence Livermore National Laboratory, USA
S. Pinnock	University of Reading, UK
V. Ramanathan	Scripps Institute of Oceanography, USA
A. Robock	University of Maryland, USA
E. Roeckner	Max-Planck Institut für Meteorologie, Germany
M.E. Schlesinger	University of Illinois @ Urbana-Champaign, USA

Appendix 3

K. Sassen	University of Utah, USA
G. Y. Shi	Institute of Atmospheric Physics, China
A.N. Trotsenko	Institute of Molecular Physics, Russia
W.C. Wang	State University of New York @ Albany, YSA

CHAPTER 5: TRACE GAS RADIATIVE FORCING INDICES

Convening Lead Author
Daniel L. Albritton   NOAA Aeronomy Laboratory, USA

Lead Authors

R.G. Derwent	Meteorological Office, UK
I.S.A. Isaksen	University of Oslo (Geophysics), Norway
M. Lal	Max-Planck Institut für Meteorologie, Germany
D.J. Wuebbles	Lawrence Livermore National Laboratory, USA

Contributors

C. Brühl	Max-Planck Institut für Chemie, Germany
J.S. Daniel	NOAA Aeronomy Laboratory, USA
D. Fisher	DuPont, Wilmington, USA
C. Granier	NCAR, USA
S.C. Liu	NOAA Aeronomy Laboratory, USA
K. Patten	Lawrence Livermore National Laboratory, USA
V. Ramaswamy	Geophysical Fluid Dynamics Laboratory, USA
T.M.L. Wigley	Office for Interdisciplinary Earth Studies @ UCAR, USA

Appendix 4

REVIEWERS OF THE IPCC WGI REPORT

The IPCC Working Group I Report went through a peer review process, the purpose of which was to involve the widest possible scrutiny of IPCC work and to ensure that the whole range of scientific opinion was taken into account. Conduct of the review followed guidelines in *IPCC Procedures for Preparation, Review, Acceptance, Approval and Publication of its Reports*. The persons named below all contributed to the peer review of the Report and its supporting scientific evidence. The list may not include some experts who contributed to the peer review only through their governments.

AUSTRALIA
G. Ayres	CSIRO (Division of Atmospheric Research)
R. Boers	CSIRO (Division of Atmospheric Research)
W. Bouma	CSIRO (Division of Atmospheric Research)
P. Cheng	Department of the Environment, Sport & Territories
K. Colls	Bureau of Meteorology
M. Dix	CSIRO (Division of Atmospheric Research)
B. Dixon	Bureau of Meteorology
I. G. Enting	CSIRO (Division of Atmospheric Research)
D. Etheridge	CSIRO (Division of Atmospheric Research)
G. Farquhar	Australian National University
B. Forgan	Bureau of Meteorology
R. Francy	CSIRO (Division of Atmospheric Research)
P.J. Fraser	CSIRO (Division of Atmospheric Research)
I. E. Galbally	CSIRO (Division of Atmospheric Research)
R. M. Gifford	CSIRO (Division of Plant Industry)
S. Godfrey	CSIRO (Division of Oceanography)
J. Gras	CSIRO (Division of Atmospheric Research)
I. Noble	Australian National University
B. Pittock	CSIRO (Division of Atmospheric Research)
T. Weir	Department of Primary Industries & Energy
J. Zillman	Bureau of Meteorology

BARBADOS
C. A. Depradine	Caribbean Meteorological Institute

BELGIUM
A. Berger	Université Catholique de Louvain
D. Crommelynck	Royal Belgium Meteorological Institute
D. De Muer	Royal Belgium Meteorological Institute

BENIN
E. D. Ahlonsou   Service Météorologique

BRAZIL
P. Fearnside   National Institute of Research on the Amazon

CANADA
H. Barker   Canadian Climate Centre
L. A. Barrie   Canadian Climate Centre
J. P. Blanchet   Canadian Climate Centre
I. J. Fung   University of Victoria
E. F. Haites   IPCC WGIII Technical Support Unit
L. D. Harvey   University of Toronto
H. Hengeveld   Canadian Climate Centre
M. Hewson   Environment Canada
K. Higuchi   Atmospheric Environment Service
W.R. Leaitch   Canadian Climate Centre
G.A. McBean   Atmospheric Environment Service
J. McConnell   York University
J.M.R. Stone   Canadian Climate Centre
D. Tarasick   Atmospheric Environment Service
C.S. Wong   Institute of Ocean Sciences

CHINA
G.-Y. Shi   Institute of Atmospheric Physics
M.-X. Wang   Institute of Atmospheric Physics
D. Yihui   Academy of Meteorological Sciences
X. Zhou   Academy of Meteorological Sciences

DENMARK
J. Fenger   Ministry of the Environment (Research Inst)
E. Friis-Christensen   Danish Meteorological Institute
J. Gundermann   Danish Energy Agency
C. Hansen   Danish Meteorological Institute
A.-M.K. Jørgensen   Danish Meteorological Institute
P. Laut   Engineering Academy of Denmark
S. Struwe   University of Copenhagen

EL SALVADOR
L. Merlos   Servicio de Meteorología e Hidrología

FINLAND
E. Holopainen   University of Helsinki
E. Jatila   Finnish Meteorological Institute
I. Savolainen   Technical Research Centre of Finland

FRANCE
G. Bergametti   LPCA-Université Paris VII
B. Bonsang   CFR
H. Cachier   CNRS/CEA
M.-L. Chanin   CNRS, Service d'Aeronomie
G. Dedieu   LERTS/CNES/CNRS
R. Delmas   LGGE
F. Dulac   Centre des Faibles Radioactivités

Appendix 4

J. C. Duplessy	Centre des Faibles Radioactivités
D. Hauglustaine	CNRS, Service d'Aeronomie
J. Jouzel	CEA/DSM
M. Kanakidou	Centre des Faibles Radioactivités
G. Lambert	CNRS/CFR
H. Le Treut	CNRS, Laboratoire de Météorologie Dynamique
P. Monfray	Centre des Faibles Radioactivités
E. Nesme-Ribes	Observatoire de Paris
D. Raynaud	CNRS-Laboratoire de Glaciologie
R. Sadourny	CNRS, Laboratoire de Météorologie Dynamique
B. Saugier	Université Paris Sud

GERMANY

C.H. Brühl	Max-Planck Institut für Chemie
D.H. Ehhalt	Institut für Chemie der KFA Jülich GmbH
G. Esser	Justus-Liebig University
H. Grassl	Max-Planck Institut für Meteorologie
V. Hesshaimer	Institute of Heidleberg
K. Labitzke	Free University of Berlin
I. Levin	Institute of Heidleberg

GHANA

P.C. Acquah	Environmental Protection Council

HUNGARY

T. Faragó	Ministry for Environment & Regional Policy
T. Pálvölgyi	Ministry for Environment & Regional Policy

INDIA

A.P. Mitra	National Physical Laboratory
P.C. Pandey	ISRO
S. Sadasivan	Bhabha Atomic Research Centre

ITALY

F. Raes	CEC Joint Research Centre
G. Vialetto	ENEA-Environmental Evaluations Project
G. Visconti	Università Degli Studi dell'Aquila

JAPAN

H. Akimoto	University of Tokyo
K. Fushimi	Japan Meteorological Agency
M. Hirota	Japan Meteorological Agency
H.Y. Inoue	Japan Meteorological Agency
Y. Iwasaka	Nagoya University
K. Minami	National Institute of Agro-Environmental Sciences
Y. Nozaki	University of Tokyo
T. Okita	Obirin University
S. Tsunogai	University of Hokkaido
M. Yoshino	Aichi University

KENYA

J.K. Njihia	Kenya Meteorological Department

KIRIBATI
N. Teuatabo Ministry of Environment & Natural Resource Development

NETHERLANDS
A.P.M. Baede Royal Netherlands Meteorological Institute(KNMI)
P. Builtjes University of Utrecht
W. Fransen Royal Netherlands Meteorological Institute(KNMI)
R. Guicherit TNO/IMW
M. Krol RIVM/MTV
L.A. Meyer Ministry of the Environment
M.G.M. Roemer TNO/IMW
S. Slanina Netherlands Energy Research Foundation
H.M. ten Brink Netherlands Energy Research Foundation
R. van Dorland KNMI
B. Weenink Ministry of the Environment

NEW ZEALAND
M. Manning National Inst of Water & Atmospheric Research
R. McKenzie National Inst of Water & Atmospheric Research

NORWAY
J. S. Fuglestvedt University of Oslo (CICERO)
F. Stordal Norwegian Institute for Air Research

RUSSIA
B.A. Fomin Institute of Molecular Physics
I.L. Karol Main Geophysical Observatory
V. Meleshko Main Geophysical Observatory
A.N. Trotsenko Institute of Molecular Physics

SAUDI ARABIA
A. Al-Gain Meteorological & Environ Protection Admin

SPAIN
J.M. Cisneros National Meteorological Institute
J.L.-G. Merayo National Meteorological Institute
M.A. Pastor National Meteorological Institute

SWEDEN
B. Bolin University of Stockholm, IPCC Chairman
S. Craig University of Stockhom
T. Hedlund Swedish National Committee on Climate Change
J. Heintzenberg University of Stockholm
W. Josefsson Swedish Meteorological & Hydrological Institute
C. Prentice Lund University
H. Rodhe University of Stockholm
L. Zetterberg Swedish Environmental Research Institute(IVL)

SWITZERLAND
A. Ohmura Dept of Geography
B. Sevruk ETH Institute of Geography
J. Staehelin ETH Atmospheric Physics Laboratory
T. Stocker University of Bern

Appendix 4

TUNISIA
Y. Labane						Institut National de la Météorologie

UNITED KINGDOM
D. Bennetts					Hadley Centre for Climate Prediction and Research, Meteorological Office
M. Beran						Institute of Hydrology
D. Burdekin					Forestry Commission
B.A. Callander					IPCC WGI Technical Support Unit
I. Colbeck					University of Essex
C.K. Folland					Hadley Centre for Climate Prediction and Research, Meteorological Office
J. Haigh						Imperial College
J.T. Houghton					IPCC WGI
P.G. Jarvis					University of Edinburgh
C. Johnson					Harwell Laboratory
S.P. Long					University of Essex
P. Mallaburn					Dept of the Environment
K. Maskell					IPCC WGI Technical Support Unit
D.E. Parker					Hadley Centre for Climate Prediction and Research, Meteorological Office
D. Peel						British Antarctic Survey
C. Reeves					University of East Anglia
D.L. Roberts					Hadley Centre for Climate Prediction and Research, Meteorological Office
P. Rowntree					Hadley Centre for Climate Prediction and Research, Meteorological Office.
N. Shackleton					University of Cambridge
K. Shine						University of Reading
R. Toumi						University of Cambridge
A. Watson					Plymouth Marine Laboratory
D. Webb						Institute of Oceanographic Sciences
I. Woodward					University of Sheffield

UNITED STATES OF AMERICA

J.K. Angell					NOAA ERL @ Silver Springs
A. Arking					Johns Hopkins University, USA (WCRP)
L. Bishop					Center for Applied Mathematics
J. Butler						NOAA, Climate Monitoring & Diagnostics Laboratory
M.A. Carroll					University of Michigan
T. Charlock					NASA
J.R. Christy					University of Alabama @ Huntsville
E. Dutton					NOAA Climate Monitoring & Diagnostics Laboratory
C. Field						Carnegie Institute of Washington
J.E. Hansen					Goddard Institute for Space Studies
P.V. Hobbs					University of Washington
L. Hood						University of Arizona
Y. Kaufman					NASA/GSFG
J. Kaye						NASA Headquarters
R. Keeling					Scripps Institute of Oceanography
M. Ko						Atmospheric & Environment Research Inc
J. Lean						Naval Research Laboratory
S. Liu						NOAA Aeronomy Laboratory
J. Logan						Harvard University
J. Mahlman					Geophysical Fluid Dynamics Laboratory
M. Mazurek					Brookhaven National Laboratory
G. Meehl						National Center for Atmospheric Research
P.J. Michaels					University of Virginia

A.J. Miller	NOAA National Meteorological Center
P. Novelli	NOAA Climate Monitoring & Diagnostics Laboratory
J. Ogren	NOAA Climate Monitoring & Diagnostics Laboratory
J. Penner	Lawrence Livermore National Laboratory
M. Prather	University of California @ Irvine
R.G. Prinn	Massachusetts Institute of Technology
V. Ramaswamy	Geophysical Fluid Dynamics Laboratory
D. Rind	Goddard Institute of Space Studies
A. Robock	University of Maryland
J.L. Sarmiento	Geophysical Fluid Dynamics Laboratory
J. Seinfeld	California Institute of Technology
J. Shlaes	Global Climate Coalition
K.R. Smith	Environmental Risk
J. Srinivasan	Langley Research Center
W.-C. Wang	State University of New York @ Albany
R. Weiss	Scripps Institute of Oceanography
L.V. Zawyalova	Main Adminsistration of Hydrometeorology

NON-GOVERNMENTAL ORGANIZATIONS

J.G. Owens	3M Company
E.A. Reiner	3M Environmental Engineering & Pollution Control
A. McCulloch	AFEAS/ICI Chemicals & Polymers Limited, UK
G. Fynes	British Coal (World Coal Institute)
R.S. Whitney	Coal Research Association, New Zealand
V.R. Gray	Coal Research Association, New Zealand
W. Hennessey	Coal Research Association, New Zealand
E. P. Olaguer	Dow Chemical Co
M. Stroben	Duke Power Company
R. A. Beck	Edison Electric Institute, USA
J. Kinsman	Edison Electric Institute, USA
J. Shiller	Ford Motor Company
W. Hare	Greenpeace International, Amsterdam
P. Womeldorf	Illinois Power
B. Flannery	IPIECA & Exxon, USA
K. Tate	Landcare Research NZ Ltd
L. Coleman	Mobil Oil
J. Karaganis	National Coal Association, USA
C. Holmes	National Coal Association, USA
D. Lashof	Natural Resources Defense Council, USA
D.H. Pearlman	The Climate Council, USA
R. Gehri	The Southern Company
K. Gregory	World Coal Inst & Coal Research Est, UK
I.S.C. Hughes	World Coal Inst & Coal Research Est, UK
P. Sage	World Coal Inst & Coal Research Est, UK
M. Jefferson	World Energy Council

Appendix 5

REVIEWERS OF THE IPCC WGIII REPORT

The persons named below all contributed to the peer review of the IPCC Working Group III Report and its supporting scientific evidence. Whilst every attempt was made by the Lead Authors to incorporate their comments, in some cases these formed a minority opinion which could not be reconciled with the larger consensus. Therefore, there may be persons below who still have points of disagreement with areas of the Report.

ALBANIA
E. Demiraj   Hydrometeorological Institute, Academy of Sciences

AUSTRALIA
J. Daley   Business Council of Australia
T. Weir   Australian Government (Department of Primary Industries and Energy)

AUSTRIA
T. Balabanov   Institute for Advanced Studies
P. Gilli   Institut fur Warmetechnik
I. Ismail   OPEC
B. Okogu   OPEC
Y. Sinyak   IIASA
J. van de Vate   IAEA (Division of Nuclear Power)
D. Victor   IIASA

BANGLADESH
M. Asaduzzaman   Bangladesh Institute of Developmental Studies

BELGIUM
J. Delbeke   Commission of the European Communities
J. Dreze   Center for Operations Research & Econometrics
G. Koopman   European Commission
M. Mors   European Commission
C. Pimenta   Globe E.C.

BRAZIL
J. Moreira   Biomass Users Network
L. Rosa   COPPE/UFRJ
R. Schaeffer   COPPE/UFRJ

BULGARIA
A. Yotova   Bulgarian Academy of Sciences

CANADA
N. Beaudoin   Environment Canada (Economic Analysis Branch)
J. Bruce   IPCC, WG III
K. Hare
M. Jaccard   BC Utilities Commission
J. Last   University of Ottawa, Faculty of Medicine
R. Poddington   Bedford Institute of Oceanography(Marine Chemistry Division)
J. Robinson   University of British Columbia (Sustainable Development Research Institute)
G. Wall   University of Waterloo, Faculty of Environmental Studies

CHILE
E. Fogueroa   Univeristy of Chile

CHINA
D. Yihui   Chinese Academy of Meteorological Sciences

DENMARK
K. Halsnaes   UNEP Center (Riso National Laboratory)
B. Sorensen   Roskilde University (Physics Department)

FRANCE
G. Bernier   OECD
C. Blondin   METEO FRANCE
J. Caneill   EDF/DER
S. Faucheux   C3E - University of Paris
J.C. Hourcade   CIRED/CNRS
P. Sturm   OECD

GERMANY
J. Blank   Universität Oldenburg
H. Gottinger   IIEEM
P. Henncike   Wuppertal Institut
E. Jochem   Fraunhoffer-Institute fur Systemtechnik und Innovationsforschung
M. Leimbach   Potsdam-Institut for Climate Impact Research
R. Pethig   Universität - Gesamthochschule Siegen
W. Strobele   Universität Oldenburg

GREECE
D. Katochianou   Center of Planning and Economic Research

HUNGARY
T. Palvolgyi   Ministry for Environment and Regional Policy
T. Farago   Ministry for Environment and Regional Policy

INDONESIA
K. Abdullah   Institut Pertanian Bogor

ITALY
R. Brinkman   Food and Agriculture Organization of the UN
M. Contaldi   ENEA
M. Farrell   Food and Agriculture Organization of the UN

Appendix 5

D. Siniscalco   Fondazione ENI Enrico Mattei
A. Tudini   Fondazione ENI Enrico Mattei
G. Tosato   ENEA

KENYA
Y. Adebayo   United Nations Environment Programme

JAPAN
N. Goto   Kanazawa University
Y. Matshoka   Faculty of Engineering, Kyoto University
S. Mori   Science University of Tokyo
Y. Tanaka   Environment Agency
I. Tsuzaka   GISPRI/TSU, WGIII

LATVIA
I. Shteinbuka   Ministry of Finance

MALAYSIA
H. Chan   Malaysian Institute of Economic Research

MALDIVES
A. Majeed   Department. of Meteorology

MALI
M. Diallo Iam   Centre National de la Recherche Scientifique et Technologique

NETHERLANDS
K. Blok   Utrecht University
G. Gelauff   Central Planning Bureau
E. van Imhoff   NIDI
W. Lenstra   Ministry of Housing, Spatial Planning and Environment
C.W. Lee   University of Groningen, Faculty of Economics
E. Tellegen   University of Amsterdam
R. Tol   Vrje Universiteit
H. Vollebergh   Erasmus University
D. Wolfson   Scientific Council for Government Policy
E. Worst   Utrecht University

NEW ZEALAND
P. Maclaren   New Zealand Forest Research Institute
P. Read   Massey University

NORWAY
A. Aaheim   CICERO
O. Benestad   University of Oslo
L. Lorentsen   Royal Ministry of Finance

POLAND
B. Greinert   Technical University of Gdansk
W. Kamrat   Technical University of Gdansk
M. Lissowska   Central School of Economics

PORTUGAL
T. Moreira   University of Evora

RUSSIA
K. Kondratyev   Russian Academy of Sciences

SPAIN
V. Alcantara   Autonomous University of Barcelona
S. Lopez   Autonomous University of Barcelona
J. Pacqual   Autonomous University of Barcelona

SWEDEN
B. Bolin   IPCC
P. Soderbaum   Swedish University of Economics

SWITZERLAND
M. Beniston   IPCC WGII,
S. Kypreos   Paul Scherrer Institute
G. Fritz   Paul Scherrer Institute
N. Sundararaman   IPCC
W. Seifritz   University of Stuttgart
R. Slooff   World Health Organization

UGANDA
J. Mubazi   Makerere University

UNITED KINGDOM
N. Adger   CSERGE
A. Atkinson   University of Cambridge
R. Booth   Shell International Petroleum Company Limited
S. Boehmer-Christiansen   University of Sussex
N. Collins   World Conservation Monitoring Center
T. Cooper   Global Commons Institute
G. Dupont   Shell International Petroleum Company
P. Ekins   Birkbeck College, University of London
K. Gregory   CGS Centre for Business and the Environment
M. Grubb   RIIA
J.T. Houghton   IPCC WGI
M. Jefferson   World Energy Council
I. Hughes   Coal Research Establishment (World Coal Institute)
H. Merkus   Ministry of Housing, Physical Planning and Environment
S. Mansoob Murshed   Northern Ireland Economic Research Centre
A. McCulloch   ICI Chemicals & Polymers Limited
J. Penman   UK Department. of Environment (Global Atmosphere Division)
F. Yamin   University of London

USA
J. Ausubel   Rockefeller University
A. Baker   ARCO
W. Barron   University of Hong Kong
R. Beck   Edison Electric Institute
R. Borgwardt   Environmental Protection Agency
J. Byrne   University of Delaware
R. Engelman   Population Action International
B. Flannery   Exxon Research and Engineering Company
R. Ford   Westminster College
D. Hall   California State University

Appendix 5

W. Harrison	
D. Hill	ETSAP, International Energy Agency
C. Holmes	National Coal Association
R. Jones	American Petroleum Institute
W. Kempton	University of Delaware
L. Kozak	Southern Company Services Inc.
N. Leary	Environmental Protection Agency
G. Marland	Oak Ridge National Laboratory
R. Mendelsohn	Yale University
A. Miller	University of Maryland
A. Olende	United Nations (DPCSD)
D. Pearlman	The Climate Council
R. Promboin	Amoco Corporation
R. Schmalensee	MIT (Center for Energy and Environmental Policy Research)
D. Shelor	Department of Energy
J. Shlaes	Global Climate Coalition
T. Siddiqi	East-West Centre
D. Spencer	Simteche
J. de Steiguer	Southern Global Change Progam, US Forest Service
M. Steinberg	Brookhaven National Laboratory
T. Teisberg	Tesiberg Associates
S. Vavrik	Yale Law School
G. Yohe	Wesleyan University

VENEZUELA

L. Perez	Ministerio de Energia y Minas

VIETNAM

N. Van Hai	HMS Vietnam
N. Thi Loc	Hydrometeorological Service

YUGOSLAVIA

R. Pesic	University of Belgrade

Appendix 6

ACRONYMS

AASE	Airborne Arctic Stratospheric Expedition
ABLE	Atmospheric Boundary Layer Experiment
ACHEX	Aerosol Characterisation Experiment
ACRIM	Active Cavity Radiometer Irradiance Monitor
AEEI	Autonomous Energy Efficiency Improvements
AER	Atmospheric and Environmental Research, Inc
AFEAS	Alternative Fluorocarbons Environmental Acceptability Study
AGU	American Geophysical Union
AGWP	Absolute Global Warming Potential
AIM	Asian-Pacific Integrated Model
ALE/GAGE	Atmospheric Lifetime Experiment/Global Atmospheric Gases Experiment
ASF	Atmospheric Stabilisation Framework
ASI	Agenzia Spaziale Italiana
BLS	"Basic Linked System" Scenario
CETA	Carbon Emissions Trajectory Assessment
CCC	Canadian Climate Centre
CCN	Cloud Condensation Nuclei
CDIAC	Carbon Dioxide Information Analysis Center
CFC	Chlorofluorocarbon
CIS	Commonwealth of Independent States
CITE	Chemical Instrumentation Test and Evaluation
CLIMAP	Climatic Applications Project (WMO)
CMDL	Climate Monitoring and Diagnostics Laboratory (NOAA)
CN	Condensation Nuclei
CNRS	Centre National de la Recherche Scientifique, France
CRC	Cyclic Redundancy Check
CSIRO	Commonwealth Scientific & Industrial Research Organisation, Australia
CTM	Chemistry /Transport Models
CW	"Conventional Wisdom" scenario, World Bank, 1990
DICE	Dynamic Integrated Climate Economy
DOAI	Department of Agriculture, Indonesia
ECE	United Nations Economic Commission for Europe

ECHAM	European Centre/Hamburg Model (ECMWF/MPI)
ECMWF	European Centre for Medium-Range Weather Forecasts
ECS	EOSDIS Core System
EEFSU	Eastern Europe and Former Soviet Union
EMEP-CCC	European Monitoring and Evaluation Programme - Chemical Coordination Centre
EMF	Energy Modelling Forum
EOSDIS	Earth Observing System Data and Information System
EPA	Environmental Protection Agency, USA
ERBE	Earth Radiation Budget Experiment
ESCAP	United Nations Economic and Social Commission for Asia and the Pacific
EUV	Extreme UltraViolet
FAO	Food and Agriculture Organization (of the UN)
FDH	Fixed Dynamical Heating
FFES	Fossil Free Energy System
FIRE IFO	First ISCCP Regional Experiment
GAGE	Global Atmospheric Gases Experiment
GCM	General Circulation Model
GDP	Gross Domestic Product
GEIA	Global Emissions Inventory Activity
GEISA	Gestion et Étude des Information Spectroscopiques Atmosphériques
GFDL	Geophysical Fluid Dynamics Laboratory, (NOAA)
GISS	Goddard Institute of Space Sciences (NASA)
GNP	Gross National Product
GRIP	Greenland Icecore Project
GWP	Global Warming Potential
HAMOCC	Hamburg Ocean Carbon Cycle Model
HILDA	High Latitude Diffusive/Advective
HITRAN	High Resolution Transmission Molecular Absorption Database
IAMAP	International Association for Meteorology and Atmospheric Physics
IEA	International Energy Agency
IEW	International Energy Workshop
IGAC	International Geosphere-Biosphere Programme
IAHS	International Association of Hydrological Science
IIASA	Institute for Applied Systems Analysis
IMAGE	Integrated Model to Assess the Greenhouse Effect
INPE	Instituto Nacional de Pesquisas Espaciais, Brazil
IPCC	Intergovernmental Panel on Climate Change
IPCC-EIS	Intergovernmental Panel on Climate Change, Energy and Industry Subgroup
IS92	IPCC Scenarios, 1992
IUPAC	International Union of Pure and Applied Chemistry
JEE	Japan Environment Agency
JPL	Jet Propulsion Laboratory (NASA)
KNMI	Royal Netherlands Meteorological Institute
LaRC	Langley Research Center (NASA)
LGM	Last Glacial Maximum
LIMS	Limb Infrared Monitor of the Stratosphere

Appendix 6 329

LLNL	Lawrence Livermore National Laboratory, USA
LMD	Laboratoire de Météorologie Dynamique du CNRS
LODYC	Laboratoire d'Oceanographie Dynamique et de Climatologie
LWC	Liquid Water Content
MERGE	Model for Evaluating Regional and Global Effects of GHG reduction policies
MIT	Massachusetts Institute of Technology, USA
MLOPEX	Mauna Loa Observatory Photochemistry Experiment
MOGUNTIA	Model of the General Universal Tracer transport In the Atmosphere
MPI	Max-Planck Institut für Meteorologie
MSU	Microwave Sounder Unit
NACNEMS	North American Cooperative Network of Enhanced Measurement Sites
NAPAP	National Acid Precipitation Assessment Program, (USA)
NASA	National Aeronautics and Space Administration, USA
NATO	North Atlantic Treaty Organisation
NCAR	National Center for Atmospheric Research, USA
NCSU	North Carolina State University, USA
NES	National Energy Strategy
NILU	Norwegian Institute For Air Research
NMHC	Non-Methane Hydrocarbons
NOAA	National Oceanic and Atmospheric Administration, USA
NPP	Net Primary Production
ODP	Ozone Depleting Potential
OECD	Organisation for Economic Cooperation and Development
OGCMs	Oceanic General Circulation Models
ORNL/CDIAC	Oak Ridge National Laboratory
PEM	Particle Modulator Radiometer
POC	Particulate Organic Carbon
PSC	Polar Stratospheric Clouds
QBO	Quasi-Biennial Oscillation
RCW	"Rapidly Changing World" Scenario (EPA)
RCWP	"Rapidly Changing World" Policy Scenario (EPA)
RH	Relative Humidity
RIGES	Renewable Intensive Global Energy System
RIVM	Institute of Public Health and Environmental Protection, Netherlands
SA90	IPCC Scenario A, 1990
SAGA	The Soviet-American Gas and Aerosol Experiment
SAGE	Stratospheric Aerosol and Gas Experiment
SAMS	Stratospheric and Mesospheric Sounder
SBUV	Solar Backscattered UltraViolet Instrument
SCOPE	Scientific Committee On Problems of the Environment
SCW	"Slowly Changing World" Scenario (EPA)
SCWP	"Slowly Changing World" Policy Scenario (EPA)
SMIC	Study of Man's Impact on Climate
SOS/SONIA	Southern Oxidants Study/Southeast Oxidant and Nitrogen Intensive Analysis
STRATOZ	Stratospheric Ozone (Campaign)
TOMCAT	Toulouse Off-Line Model of Chemistry and Transport

TOR	Tropospheric Ozone Research Project
TRACE A	Transport and Atmospheric Chemistry near the Equator - Atlantic
TROPOZ	Tropospheric Ozone (Campaign)
UN	United Nations
UCI	University of California at Irvine, USA
UCRL	University of California Radiation Laboratory
UGAMP	United Kingdom Global Atmospheric Modelling Project
UKMO	United Kingdom Meteorological Office
UEA	University of East Anglia, UK
UNEP	United Nations Environment Programme
UNFCC	United Nations Framework Convention on Climate Change
UV	Ultraviolet
VOC	Volatile Organic Compounds
WCRP	World Climate Research Programme
WEC	World Energy Council
WMO	World Meteorological Organisation
WOCE	World Ocean Circulation Experiment

Appendix 7

UNITS

SI (Systeme Internationale) Units:

Physical Quantity	Name of Unit	Symbol
length	meter	m
mass	kilogram	kg
time	second	s
thermodynamic temperature	kelvin	K
amount of substance	mole	mol

Fraction	Prefix	Symbol	Multiple	Prefix	Symbol
10^{-1}	deci	d	10	deka	da
10^{-2}	centi	c	10^2	hecto	h
10^{-3}	milli	m	10^3	kilo	k
10^{-6}	micro	μ	10^6	mega	M
10^{-9}	nano	n	10^9	giga	G
10^{-12}	pico	p	10^{12}	tera	T
10^{-15}	femto	f	10^{15}	peta	P
10^{-18}	atto	a			

Special Names and Symbols for Certain SI-Derived Units:

Physical Quantity	Name of SI Unit	Symbol for SI Unit	Definition of Unit
force	newton	N	$kg\ m\ s^{-2}$
pressure	pascal	Pa	$kg\ m^{-1}s^{-2} (=N\ m^{-2})$
energy	joule	J	$kg\ m^2 s^{-2}$
power	watt	W	$kg\ m^2 s^{-3} (=J\ s^{-1})$
frequency	hertz	Hz	s^{-1} (cycle per second)

Decimal Fractions and Multiples of SI Units Having Special Names:

Physical Quantity	Name of Unit	Symbol for Unit	Definition of Unit
length	ångstrom	Å	$10^{-10}\ m = 10^{-8}\ cm$
length	micrometre	μm	$10^{-6}\ m$
area	hectare	ha	$10^4\ m^2$
force	dyne	dyn	$10^{-5}\ N$
pressure	bar	bar	$10^5\ N\ m^{-2} = 10^5\ Pa$
pressure	millibar	mb	$10^2\ N\ m^{-2} = 1\ hPa$
weight	tonne	t	$10^3\ kg$

Non-SI Units:

°C	degrees Celsius (0 °C = 273 K approximately)
	Temperature differences are also given in °C (=K) rather than the more correct form of "Celsius degrees".
ppmv	parts per million (10^6) by volume
ppbv	parts per billion (10^9) by volume
pptv	parts per trillion (10^{12}) by volume
bp	(years) before present
kpb	thousands of years before present
mbp	millions of years before present

The units of mass adopted in this report are generally those which have come into common usage, and have deliberately not been harmonised, e.g.,

kt	kilotonnes
GtC	gigatonnes of carbon (1 GtC = 3.7 Gt carbon dioxide)
PgC	petagrams of carbon (1PgC = 1 GtC)
MtN	megatonnes of nitrogen
TgC	teragrams of carbon
TgN	teragrams of nitrogen
TgS	teragrams of sulphur

Appendix 8

SOME CHEMICAL SYMBOLS USED IN THIS REPORT

O	atomic oxygen
$O(^1D)$	an energetic form of atomic oxygen
O_2	molecular oxygen
O_3	ozone
N_2	molecular nitrogen
N_2O	nitrous oxide
N_2O_5	dinitrogen pentoxide
NO	nitric oxide
NO_2	nitrogen dioxide
NO_3	nitrate radical
NO_x	the sum of NO and NO_2
HNO_3	nitric acid
PAN: $CH_3CO_3NO_2$	peroxyacetylnitrate
HONO	nitrous acid
HO_2NO_2	pernitric acid
NO_y	total reactive nitrogen
NH_3	ammonia
NH_4^+	ammonium ion
H	atomic hydrogen
H_2	molecular hydrogen
H_2O	water
H_2O_2	hydrogen peroxide
OH	hydroxyl
HO_2	hydroperoxyl
HO_x	the sum of OH and HO_2
C	carbon: there are 3 isotopes: ^{12}C, ^{13}C, ^{14}C
CO	carbon monoxide
CO_2	carbon dioxide
CFC	chlorofluorocarbon
CFC-11	$CFCl_3$, or equivalently CCl_3F (trichlorofluoromethane)
CFC-12	CF_2Cl_2, or equivalently CCl_2F_2 (dichlorodifluoromethane)
CFC-13	CF_3Cl, or equivalently $CClF_3$ (chlorotrifluoromethane)
CFC-113	$C_2F_3Cl_3$, CCl_2FCClF_2

	(trichlorotrifluoroethane)
CFC-113a	CCl_3CF_3, isomer of CCl_2FCClF_2
CFC-114	$C_2F_4Cl_2$, $CClF_2CClF_2$
	(dichlorotetrafluoroethane)
CFC-115	C_2F_5Cl
	(chloropentafluoroethane)
HFC	hydrofluorocarbon
HFC-23	CHF_3
HFC-32	CH_2F_2
HFC-43-100 mee	$C_5H_2F_{10}$
HFC-125	C_2HF_5
HFC-134	CHF_2CHF_2
HFC-134a	CH_2FCF_3
HFC-143	CHF_2CH_2F
HFC-143a	CF_3CH_3
HFC-152a	$C_2H_4F_2$
HFC 227ea	C_3HF_7
HFC-236fa	$C_3H_2F_6$
HFC-245ca	$C_3H_3F_5$
HCFC	hydrochlorofluorocarbon
HCFC-22	CF_2HCl
	(chlorodifluoromethane)
HCFC-123	$C_2F_3HCl_2$
HCFC-124	C_2F_4HCl
HCFC-141b	$C_2FH_3Cl_2$
HCFC-142b	$C_2F_2H_3Cl$
HCFC 225ca	$C_3F_5HCl_2$
HCFC 225cb	$C_3F_5HCl_2$
HALON 1211	CF_2BrCl ($CBrClF_2$)
	(bromodichloromethane)
HALON 1301	CF_3Br ($CBrF_3$)
	(bromotrifluoromethane)
CH_4	methane
C_2H_6	ethane
C_3H_8	propane
C_2H_4	ethylene (ethene)
C_2H_2	acetylene (ethyne)
CH_3O_2	methyl peroxy radical
CH_3OOH	methyl peroxide
NMHC	non-methane hydrocarbon
VOC	volatile organic compound
S	atomic sulphur
SO_2	sulphur dioxide
SO_4^{2-}	sulphate ion
SF_6	sulphur hexafluoride
H_2SO_4	sulphuric acid
^{222}Rn	radon (1 isotope of)
Cl	atomic chlorine
$CHClCCl_2$	trichloroethylene
CCl_4	carbon tetrachloride
CCl_2CCl_2	perchloroethylene (tetrachloroethene)
CH_3Cl	methylchloride
CH_3CCl_3	methylchloroform

Appendix 8

CH_2Cl_2	dichloromethane/ methylene chloride
$CHCl_3$	chloroform, trichloromethane
CH_2Br_2	dibromomethane
$CHBr_2Cl$	dibromochloromethane
$CHBr_3$	tribromomethane
CH_3Br	methylbromide
CF_3I	trifluoroiodomethane
CF_4	perfluoromethane
C_2F_6	hexafluoroethane/perfluoroethane
C_3F_8	perfluoropropane
C_4F_{10}	perfluorobutane
$c\text{-}C_4F_8$	perfluorocyclobutane
C_5F_{12}	perfluoropentane
C_6F_{14}	perfluorohexane

Index

Absolute Global Warming Potential (AGWP), definition, 216
absorption coefficient (of aerosols), 145
accumulation mode (of aerosols), 139
adjustment lifetime, 82–3, 115
aerosols, 7
   budgets, 140–1
   carbonaceous, 137, 181
   see also soot
   direct effect on radiative forcing, 30, 181
   effect on clouds, 150, 152, 153–4
   formation, 133
   from biomass burning, 137, 182–3
   from combustion, 136
   from oxidation of precursor gases, 136–7
   indirect effect on radiative forcing, 30, 183–4
   lifetimes, 133, 140, 143
   measurements, 140, 142–3, 156
   modelling, 148, 157
   optical depth, 146, 148–9
   organic, 136, 152
   properties, 138, 139, 140, 143–6
   radiative forcing by, 12, 31, 133–4, 144, 147
   sea-salt, 135–6, 152
   scenarios, 154
   sinks, 139, 143
   size distribution, 133, 138–9, 139, 143
   soil dust, 134–5, 152
   sources, 134, 136
   stratospheric, 98, 142
   transformation of, 137
   trends, 137, 141–2
   volcanic, 136, 140, 142–3, 186
AGWP, *see* Absolute Global Warming Potential
aircraft
   as sources of aerosols, 143, 144, 153
   as sources of gaseous pollutants, 99, 103, 112
Atmospheric Stabilization Framework, ASF, 260

biomass burning, 137
biosphere
   marine, 49, 58
   terrestrial, 51, 52, 56
brominated species
   measurements of, 93
   ozone depletion by, 96

   sinks, 95
   sources, 94, 95
   stabilisation, 117

carbon
   budget, 17–19, 46, 55
   cycle, 20, 41
   isotopes, 42, 46–7
carbon dioxide, CO_2
   concentrations, 11–12, 13, 16–17, 19–20, 25, 42, 43
   emission scenarios related to energy, 265–84
   emission scenarios related to land-use, 286–9
   emissions, 48–9, 52–3, 60
   fertilisation, 18, 52, 54, 56, 57, 59, 65
   oceanic uptake, 47–8, 51
   radiative forcing by, 12, 16, 172, 194
   recent anomalies, 42
   spatial distributions, inferred from concentration measurements, 48
   stabilisation of concentration, 11, 13, 21, 22, 60–1
   stabilisation of emissions, 13, 21
   terrestrial uptake, 53–5
carbon intensity (in emission scenarios), 268
carbon monoxide, CO
   used to infer OH concentrations, 83, 103
   measurements of, 103, 104
   sources and sinks, 103
carbon tetrachloride, CCl_4, 92
CCN, *see* Cloud Condensation Nuclei
CFCs
   Global Warming Potentials, 221–2
   industrial production, 94
   measurements, 25, 92
   ozone depletion, 96
   radiative forcing, 194
   sinks, 95
   stabilisation, 117
   substitutes, 12, 29
   see also HFCs, HCFCs
chemical transport models, 105–16
climate feedbacks, 24–5, 27, 55–8
climate sensitivity, 169–71, 211
Cloud Condensation Nuclei, 138, 139, 142, 150, 151, 153
CO_2 fertilisation, *see* carbon dioxide fertilisation

deforestation, 51, 52–3, 56–7

DMS, dimethyl sulphide, 136–7
dust, *see* aerosols

El Chichon, 136
emission scenarios, 60, 65
   criteria for evaluation, 258
   developing new scenarios, 297
   IS92 scenarios, overview, 260
   IS92 scenarios, uses and limitations, 254–5, 296–7
   terminology, 258
emissions profile(s), 62
energy intensity (in emission scenarios), 270
energy-related emission scenarios
   comparisons, 252–3, 265–84
   components and sensitivities, 253, 268–70, 282–3
   regional, 270–81
enteric fermentation, 87, 293

FCCC,
feedbacks
   chemical, 83, 85
   marine, 57–8
   terrestrial, 56–7
forcing/response relationships, 197
forest regrowth, 51–2, 52, 53, 59

GDP, *see* Gross Domestic Product
Global Warming Potential
   and natural sources of greenhouse gas, 229
   definition, 32, 215
   indirect effects in, 82, 223
   product with emissions, 226
   reference molecule, 217, 221
   sensitivity to background atmosphere, 214, 218
   sensitivity to clouds and water vapour, 219
   sensitivity to time horizon, 229
   uncertainties, 221, 226, 229
   uses and limitations, 211, 227
   values, 223
Gross Domestic Product, GDP, 262–3
   relationship to population, 264–5
GWP, *see* Global Warming Potential, *also* radiative forcing index

halocarbons, 92
   measurements of, 25, 28
   ozone depletion, 96–8
   radiative forcing by, 12, 28
   stabilisation (of concentrations), 117
halons, 93–4, 117
HCFCs
   stabilisation of, 117
   Global Warming Potentials, 221, 222
   industrial production, 94–5
   measurements, 93
   radiative forcing by, 172–3, 194
   sinks, 95
HFCs, 117, 172–3, 194
HITRAN database, 171
hydroxyl (OH) in the troposphere, 79, 81, 82, 83, 103

index of radiative forcing, *see* Global Warming Potential

Krakatoa, 143

land-use changes, 51, 52–3, 53, 56–7
land-use emission scenarios
   base year estimates, 286
   comparisons, 254, 284–94
   components and sensitivities, 254, 289–90, 292–3
   regional, 291
lifetime, atmospheric
   definition, 82–3
   of trace gases, 82–3, 84, 84–5, 220

Mauna Loa, 43
methane, CH_4
   adjustment time, 109, 113–6
   atmospheric chemistry, 82, 85
   emission scenarios related to energy, 281
   emission scenarios related to land-use, 290–3
   from enteric fermentation, 87, 293
   from rice production, 87, 292–3
   global budget, 25, 86
   Global Warming Potential, 11, 34, 222, 225
   indirect effect on radiative forcing, 25–6
   measurements, 24, 25, 87–9
   pre-industrial level, 88–9
   radiative forcing, 12, 25, 172, 194
   seasonal cycle, 87
   sinks, 25, 87
   sources, 25, 85
   stabilisation (of concentrations), 14, 116–7
methyl chloroform,
   used to infer OH concentrations, 83–4
   measurements of, 12, 13, 93
   sinks, 95
   sources, 94
   stabilisation (of concentrations), 117
Montreal Protocol, 98, 117
Mount Pinatubo, 12, 13, 32, 98, 136, 143, 186

nitrogen fertilisation, 19, 52, 54–5
nitrogen oxides, NO_x
   measurements, 30, 100–3
   ozone formation, 80–2, 109–10, 112–13
   sinks, 100
   sources, 99, 112–13

Index

nitrous oxide, N_2O, 281
   emission scenarios related to energy, 290–3
   emission scenarios related to land-use, 13, 25, 27–8, 91–2
   measurements, 13, 25, 27–8, 91–2
   radiative forcing by, 12, 28, 194
   sinks, 27–8, 91–2
   sources, 27–8, 89–90, 92, 293
   stabilisation (of concentrations), 14, 28, 117
non-methane hydrocarbons, NMHCs, 104–5, 131

ocean, circulation, 57–8
Ozone Depleting Potential (ODP), 227
ozone, O_3, stratospheric
   ozone depletion, 97–8, 117
   radiative forcing by, 29–30, 175–7, 193, 194
   WMO/UNEP assessments, 211
ozone, O_3, tropospheric
   chemical processes, 80–2
   formation, 109–10, 111–16
   measurements, 98–9, 114
   modelling, 109–10, 111–16
   radiative forcing by, 30, 193
   stabilisation (of concentrations), 117

perflourocarbons, 25
   measurements of, 94, 171
   radiative forcing by, 173
   sinks, 95
   sources, 95
   stabilisation (of concentrations), 117
photochemistry, processes, 80–2
POC, particulate organic carbon, 136
population, 263–5, 268–70
precursors of aerosols, 136–7

radiative forcing
   adjusted forcing, 169–70
   as a measure of induced climate change, 211
   definition, 15, 169–70
   instantaneous forcing, 169–70
radiative forcing index(see also Global Warming Potential)
   definition, 212

   indirect effects, 214
   limitations, 216
   various formulations, 215
reference molecule, *see* Global Warming Potential
respiration (in the soil), 57
rice production (factors affecting methane emissions), 292–3

scattering coefficient of aerosols, 145
ship tracks, 150
soils, factors affecting nitrous oxide emissions, 293
solar output, 13
solar variability, 32, 189, 194–5
soot, 136, 142, 152, 183
spectroscopy, 171
stabilisation of concentrations, see entries under individual gases and aerosols
stratosphere, *see also* Ozone
   temperatures, 81–2, 98, 177–8
   water vapour, 173, 180, 194
sulphur, emission scenarios, 281–2
sulphur dioxide, SO_2, 136–7, 155
sulphur fluxes, 140
sulphur hexafluoride, SF_6
   measurements of, 94
   sinks, 95
   source, 95
   stabilisation (of concentrations), 117

temperature, glacial-interglacial link with carbon dioxide, 45

ultraviolet light, UV, 80, 100
unidentified sink, 55

volatile organic compounds, VOCs
   definition, 104
   sinks, 105
   sources, 104–5
volcanic aerosols, *see* aerosols
volcanoes, 13, 32, 143, 186

WEC scenarios, 60